ANNUAL REVIEW OF NUCLEAR AND PARTICLE SCIENCE

EDITORIAL COMMITTEE (1991)

ANNUAL REVIEW OF NUCLEAR AND PARTICLE SCIENCE

VOLUME 41, 1991

J. D. JACKSON, *Editor*
University of California, Berkeley

HARRY E. GOVE, *Associate Editor*
University of Rochester

VERA LÜTH, *Associate Editor*
Stanford Linear Accelerator Center

ANNUAL REVIEWS INC   4139 EL CAMINO WAY   P.O. BOX 10139   PALO ALTO, CALIFORNIA 94303-0897

ANNUAL REVIEWS INC.
Palo Alto, California, USA

International Standard Serial Number: 0163-8998
International Standard Book Number: 0-8243-1541-3
Library of Congress Catalog Card Number: 53-995

Annual Review and publication titles are registered trademarks of Annual Reviews Inc.

∞ The paper used in this publication meets the minimum requirements of American National Standard for Information Sciences—Permanence of Paper for Printed Library Materials, ANSI Z39.48-1984.

Annual Reviews Inc. and the Editors of its publications assume no responsibility for the statements expressed by the contributors to this *Review*.

TYPESET BY BPCC-AUP GLASGOW LTD., SCOTLAND
PRINTED AND BOUND IN THE UNITED STATES OF AMERICA

Annual Review of Nuclear and Particle Science
Volume 41, 1991

CONTENTS

SOME RELATED ARTICLES IN OTHER *ANNUAL REVIEWS*

From the *Annual Review of Astronomy and Astrophysics*, Volume 29 (1991):

The Interplanetary Plasma, Bruno Rossi

Collective Plasma Radiation Processes, D. B. Melrose

Inflation for Astronomers, J. V. Narlikar and T. Padmanabhan

Radioactive Dating of the Elements, John J. Cowan, Friedrich-Karl Thielemann, and James W. Truran

From the *Annual Review of Earth and Planetary Sciences*, Volume 19 (1991):

Nuclear Winter: Physics and Physical Mechanisms, R. P. Turco, O. B. Toon, T. P. Ackerman, J. B. Pollack, and C. Sagan

From the *Annual Review of Energy and the Environment*, Volume 16 (1991):

Energy Applications of Superconductivity, Thomas R. Schneider

Safety and Environmental Aspects of Fusion Energy, John P. Holdren

From the *Annual Review of Fluid Mechanics*, Volume 23 (1991):

The Theory of Hurricanes, Kerry A. Emanuel

Symmetry and Symmetry-Breaking Bifurcations in Fluid Dynamics, John David Crawford and Edgar Knobloch

Fractals and Multifractals in Fluid Turbulence, K. R. Sreenivasan

From the *Annual Review of Physical Chemistry*, Volume 42 (1991):

Electron Correlation Techniques in Quantum Chemistry: Recent Advances, Krishnan Raghavachari

Annu. Rev. Nucl. Part. Sci. 1991. 41: 1–28

DECAYS OF B MESONS

Karl Berkelman

Laboratory of Nuclear Studies, Cornell University, Ithaca,
New York 14853

Sheldon L. Stone

Department of Physics, Syracuse University, Syracuse, New York 13244

KEY WORDS: B decays, weak interactions of b quarks

CONTENTS

0163–8998/91/1201–0001$02.00

1. INTRODUCTION

1.1   *The Standard Model and the b Quark*

According to the Standard Model of six quarks and six leptons, the charged-current decays of quarks can couple any of the "up" quarks, u, c, or t, to any of the "down" quarks, d, s, or b. Several of these couplings can be measured only in B-meson decays. It is important to measure them accurately in order to expand our knowledge of the model, to test its validity, and more importantly, to take us to a formulation of what is beyond. In the following, we review the measurements that have been made on the weak decays of B mesons and discuss their implications for the matrix of quark couplings.

1.2   *Electron-Positron Production of B$\bar{B}$ Near Threshold*

Just above the threshold for $e^+e^- \to B\bar{B}$, there is a resonance in the annihilation cross section, called the $\Upsilon(4S)$. The B's are produced in a two-body B^+B^- or $B^0\bar{B}^0$ final state of $J^P = 1^-$. Because the B's are produced near threshold in a two-body final state with an energy that is accurately determined by the energy of the beam, the mass uncertainty in kinematic reconstructions is only 2 to 3 MeV out of 5.3 GeV. Tagging one of the B's always uniquely identifies the other that was produced along with it. These advantages explain why most of what we know now about b quarks has come from experiments at the $\Upsilon(4S)$ resonance.

There is no direct measurement of the branching fractions, f_+ and f_0, for $\Upsilon(4S) \to B^+B^-$ and $\Upsilon(4S) \to B^0\bar{B}^0$. The naive expectation of equal charged and neutral rates could be modified by phase space if the masses were different, or by electromagnetic effects. A detailed model calculation (1) suggests that phase space is not relevant, and in any case, the latest measurements indicate equal charged and neutral B masses (see Section 2.4). Coulomb effects could enhance pointlike B^+B^- relative to $B^0\bar{B}^0$ by 18% near threshold (2), but form factor effects reduce the enhancement to between -3 and $+6\%$ at the $\Upsilon(4S)$ (3). Even if the calculations were completely reliable, it would be impossible now to predict f_+/f_0, since the expected ratio can vary by 7% with only a 1-MeV shift in the B^+, B^0, or $\Upsilon(4S)$ mass, and at present, masses are not that accurately known from experiment.

Recently CLEO (4) and ARGUS (5) detected inclusive ψ's from the $\Upsilon(4S)$ resonance with momenta above the kinematic limit for a ψ from the decay of one of the B's produced in the reaction $\Upsilon(4S) \to B\bar{B}$. The measured branching ratios for $\Upsilon(4S) \to \psi X$ with $p_\psi > 2$ GeV/c are $0.22 \pm 0.06 \pm 0.04\%$ (CLEO) and $0.24 \pm 0.06\%$ (ARGUS). Because of the u, d, s, c background under the $\Upsilon(4S)$ resonance, it is difficult to see the effects of non-$B\bar{B}$ decays into other particle species. By comparing lepton and dilepton rates at the $\Upsilon(4S)$, CLEO (6) has set a 95% confidence upper limit at $f_{\text{nonB}\bar{B}} = 1 - f_\pm - f_0 < 14\%$.

2. NONLEPTONIC DECAYS

2.1 *Inclusive Measurements*

The measured inclusive branching ratio B for B mesons to decay to some particular particle species z plus any other particles X is actually a sum:

$$B = \frac{f_\pm}{f_\pm + f_0} \frac{1}{2}[B(B^+ \to zX) + B(B^- \to zX) + B(B^+ \to \bar{z}X) + B(B^- \to \bar{z}X)]$$

$$+ \frac{f_0}{f_\pm + f_0} \frac{1}{2}[B(B^0 \to zX) + B(\bar{B}^0 \to zX) + B(B^0 \to \bar{z}X) + B(\bar{B}^0 \to \bar{z}X)].$$

B is obtained from the difference between the measured rate on the $\Upsilon(4S)$ resonance and rate below $B\bar{B}$ threshold, corrected for the ratio of accumulated luminosities and for the $1/s$ dependence of the nonresonant cross section. The unknown contamination from non-$B\bar{B}$ decays of the $\Upsilon(4S)$ can contribute both to the observed particle rates and to the normalization (the number of decaying B's). Continuing the past practice, we will ignore both effects; that is, we assume $f_\pm = f_0 = 1/2$.

Table 1 shows the CLEO (7–11) and ARGUS (12, 13) data on inclusive B decays to several particle species that can be readily identified experimentally. Where the relevant secondary particle is identified by its subsequent decay, we have assumed the decay branching ratios given in the 1990 Particle Data Group tabulation (14), except we use Mark III (15, 16) data for D and D* and CLEO and ARGUS data for D_S (17) and Λ_c (9, 13). Note that the charged kaon yields from B and $\bar{B}$ decays have been separated by lepton tagging (7), but not corrected for B^0-$\bar{B}^0$ mixing.

It is clear from the inclusive data that the b quark decays mainly to the c quark, through the $b \to cW^-$ vertex, and only rarely to the u quark. The dominance of $b \to cW^-$ would imply that the inclusive charm rate, obtained by summing the inclusive rates for D^+, D^0, D_S, Λ_c, and bound $c\bar{c}$ (doubled for the two charmed quarks and doubled again to account for the undetected χ states) should be about 115%. The extra 15% come from the fact that the W^- is expected to materialize as $s\bar{c}$ about 15% of the time. The CLEO branching ratios sum to $97 \pm 15\%$; the ARGUS sum is $89 \pm 13\%$.

The background subtraction is made easier by the fact that the spectra

Table 1   Inclusive nonleptonic B branching ratios (in %)

Particle	Signature	CLEO	ARGUS
K^- (from $\overline{B}$)		$66 \pm 5 \pm 7$	
K^+ (from $\overline{B}$)		$19 \pm 5 \pm 2$	
$K^0, \overline{K}^0$	$\pi^+\pi^-$	$63 \pm 6 \pm 6$	-
ϕ	K^+K^-	$2.3 \pm 0.6 \pm 0.5$	-
p (not from Λ)		$5.7 \pm 0.8 \pm 1.0$	$5.3 \pm 1.0 \pm 1.5$
Λ	$p\pi^-$	$3.9 \pm 0.4 \pm 0.6$	$4.2 \pm 0.5 \pm 0.5$
Ξ^-	$\Lambda\pi^-$	$0.28 \pm 0.05 \pm 0.04$	-
D^+	$K^-\pi^+\pi^+$	$22 \pm 5 \pm 3$	$23 \pm 5 \pm 3$
D^0	$K^-\pi^+$	$53 \pm 7 \pm 7$	$46 \pm 7 \pm 6$
D^{*+}	$D^0\pi^+$	$32 \pm 6^{+7}_{-4}$	$30 \pm 5 \pm 5$
D_S^+	$\phi\pi^+$, etc.	9.9 ± 1.5	$8 \pm 2 \pm 2$
Λ_c^+	$pK^-\pi^+$	$6.1 \pm 0.8 \pm 1.0$	7.6 ± 1.4
ψ	$e^+e^-, \mu^+\mu^-$	$1.12 \pm 0.10 \pm 0.23$	$1.07 \pm 0.16 \pm 0.19$
ψ'	$\psi\pi^+\pi^-, \ell^+\ell^-$	$0.33 \pm 0.08 \pm 0.12$	$0.46 \pm 0.17 \pm 0.11$

for charmed particles produced in non-$B\overline{B}$ events are peaked above 2.5 GeV/c, i.e. beyond the kinematic limit for particles from $B\overline{B}$ events. The rate of charmed particles from $B\overline{B}$ near 2.5 GeV/c suggests an appreciable contribution from two-body final states. The charmed meson yields and momentum spectra are compatible with predictions based on the factorization approximation (18).

B mesons can decay into baryons. Studies of the inclusive yields of $\bar{p}$, Λ, Ξ^-, and Λ_c^+ and of the $p\bar{p}$, $p\Lambda$, and $\Lambda\overline{\Lambda}$ correlations by CLEO (9) and ARGUS (13) suggest that the Λ yield comes mainly from B decays to Λ_c, and the measured Λ_c rate in Table 1 is based on the assumption that the baryon-antibaryon pair always includes a charmed baryon. Adding the proton and lambda yields and allowing for a comparable inclusive neutron yield, one can conclude that about 7% of B decays contain a baryon-antibaryon pair.

2.2   Exclusive Decay Modes

The study of B-meson hadronic decays allows accurate measurements of the charged and neutral B masses and tests models for B decay. Reconstructed B decays serve as an unbiased tag for the study of the other B produced in the same event, and two-body modes like $B \to \psi K$ serve as signatures for B production in experiments at hadron machines and higher energy e^+e^- colliders. The study of *CP* violation in B decays will require a large sample of reconstructed B decays.

Although the exclusive decays $B^- \to \psi K^-$ and $\bar{B}^0 \to \psi \bar{K}^{*0}$ have been reconstructed by the CDF collaboration (19) in $\bar{p}p$ collisions, the only measurements of branching ratios come from the CLEO and ARGUS e^+e^- experiments at the $\Upsilon(4S)$ resonance. To observe B decays into some exclusive final state, one selects events containing the desired particles and satisfying the constraints that the total measured energy match the beam energy and the invariant mass match that of the B. Depending on the mode and the detector performance, the resolution in energy is 20 to 30 MeV and the mass resolution is about 2.5 MeV. The backgrounds come from spurious combinations of particles, either from non-$B\bar{B}$ events or incorrectly reconstructed $B\bar{B}$ events. The sidebands in the energy spectrum, between the peak and the threshold for an additional pion, provide the best estimate of the mass dependence of the combinatoric background (20).

Tables 2 and 3 show the ARGUS (22) and CLEO (20) measurements of exclusive nonleptonic branching ratios of B^- and $\bar{B}^0$ mesons. Since the $\Upsilon(4S)$ decay fractions are unknown, the branching ratio data listed in the tables are actually the products $B \times 2f_\pm$ and $B \times 2f_0$. For decays of secondaries, we used the same branching ratios assumed for Table 1.

2.3 Comparison with Theoretical Predictions for Branching Ratios

At the quark level there are three tree diagrams for B-meson decay. Because of its rapid m_b^5 dependence on quark mass, the spectator mechanism (Figure 1a) is expected to dominate in B decays. The exchange (Figure 1c) and annihilation (Figure 1d) diagrams leading to a pair of light quarks may be suppressed by the same helicity argument that makes the $\pi \to e\nu$ branching ratio small. We also expect the annihilation amplitude to be suppressed by the V_{ub} factor. The details of the long-range processes taking the fragmentation of quarks into hadrons, including possible final-state rescattering among the hadrons, are important in determining the branching ratios of the various exclusive decay channels. We briefly discuss here the calculations of Bauer, Stech & Wirbel (21), based on a "factorization" hypothesis, that the short-range effects of perturbative gluon exchange and long-range quark binding can be dealt with separately. All amplitudes for two-body B decays are expressed as linear combinations of two amplitudes on the basis of "flavor flow," one amplitude a_1 corresponding to the case

Table 2 Exclusive nonleptonic B^- branching ratios (in %)[a]

Final state	CLEO '86	CLEO '88	ARGUS	Average	Model
$D^0\pi^-$	$0.54^{+0.18+0.12}_{-0.15-0.09}$	$0.50 \pm 0.07 \pm 0.09$	$0.20 \pm 0.08 \pm 0.06$	0.39 ± 0.06	$0.33(a_1 + 0.75a_2)^2$
$D^0\rho^-$	-	-	$1.3 \pm 0.4 \pm 0.4$	1.3 ± 0.6	$0.87(a_1 + 0.34a_2)^2$
$D^0 D_S^-$	-	$1.8 \pm 0.8 \pm 0.8$	-	1.8 ± 1.1	$0.51a_1^2\delta$
$D^{*0}\pi^-$	-	$0.72 \pm 0.18 \pm 0.25$	$0.40 \pm 0.14 \pm 0.12$	0.48 ± 0.16	$0.26(a_1 + 1.04a_2)^2$
$D^{*0}\rho^-$	-	-	$1.3 \pm 0.7 \pm 0.5$	1.3 ± 0.9	$0.82(a_1 + 0.79a_2)^2$
$D^{*+}\pi^-\pi^-$	$0.24^{+0.17+0.10}_{-0.16-0.06}$	< 0.4	$0.26 \pm 0.14 \pm 0.07$	0.25 ± 0.12	-
$D^{*+}\pi^-\pi^-\pi^0$	-	-	$1.8 \pm 0.7 \pm 0.5$	1.8 ± 0.9	-
ψK^-	$0.10 \pm 0.07 \pm 0.02$	$0.08 \pm 0.02 \pm 0.02$	$0.07 \pm 0.03 \pm 0.01$	0.08 ± 0.02	$0.70a_2^2$
ψK^{*-}	-	$0.13 \pm 0.09 \pm 0.03$	$0.16 \pm 0.11 \pm 0.03$	0.14 ± 0.07	$3.01a_2^2$
$\psi K^-\pi^+\pi^-$	-	$0.12 \pm 0.06 \pm 0.03$	< 0.16	0.12 ± 0.05	-
$\psi' K^-$	-	< 0.05	$0.18 \pm 0.08 \pm 0.04$	-	-

[a] The two CLEO data sets are independent and were obtained with different tracking chambers. The results are preliminary.

Table 3 Exclusive nonleptonic $\bar{B}^0$ branching ratios (in %)

Final state	CLEO '86	CLEO '88	ARGUS	Average	Model
$D^+\pi^-$	$0.51^{+0.28+0.13}_{-0.25-0.12}$	$0.26 \pm 0.06 \pm 0.05$	$0.48 \pm 0.11 \pm 0.11$	0.32 ± 0.07	$0.33a_1^2$
$D^+\rho^-$	-	-	$0.9 \pm 0.5 \pm 0.3$	0.9 ± 0.6	$0.87a_1^2$
$D^0\rho^0$	-	< 0.06	-	-	$0.05a_2^2$
$D^+D_S^-$	-	$0.75 \pm 0.21 \pm 0.32$	-	0.75 ± 0.38	$0.47a_1^2\delta$
$D^{*+}\pi^-$	$0.28^{+0.15+0.10}_{-0.12-0.06}$	$0.40 \pm 0.09 \pm 0.09$	$0.28 \pm 0.09 \pm 0.06$	0.32 ± 0.07	$0.26a_1^2$
$D^{*+}\rho^-$	-	$1.9 \pm 0.9 \pm 1.3$	$0.7 \pm 0.3 \pm 0.3$	0.8 ± 0.4	$0.82a_1^2$
$D^{*+}a_1^-$	-	$2.6 \pm 0.5 \pm 0.6$	-	2.6 ± 0.8	$1.13a_1^2$
$D^{*+}D_S^-$	-	$1.5 \pm 0.9 \pm 0.7$	-	1.5 ± 1.1	$0.21a_1^2\delta$
$D^{*+}\pi^-\pi^0$ [a]	-	-	$1.8 \pm 0.4 \pm 0.5$	1.8 ± 0.6	-
$D^{*+}\pi^-\pi^+\pi^-$ [b]	< 4.2	$1.5 \pm 0.4 \pm 1.0$	$1.2 \pm 0.3 \pm 0.4$	1.3 ± 0.5	-
$D^{*+}\pi^-\pi^-\pi^+\pi^0$	-	-	$4.1 \pm 1.5 \pm 1.6$	4.1 ± 2.2	-
$\psi\bar{K}^0$	-	$0.06 \pm 0.03 \pm 0.02$	$0.08 \pm 0.06 \pm 0.02$	0.06 ± 0.03	$0.71a_2^2$
$\psi\bar{K}^{*0}$	$0.35 \pm 0.16 \pm 0.03$	$0.11 \pm 0.05 \pm 0.03$	$0.11 \pm 0.05 \pm 0.02$	0.12 ± 0.04	$3.03a_2^2$
$\psi K^-\pi^+$ [c]	-	$0.10 \pm 0.04 \pm 0.03$	-	0.10 ± 0.05	-
$\psi'\bar{K}^{*0}$	-	$0.14 \pm 0.08 \pm 0.04$	< 0.23	0.14 ± 0.09	-

[a] Includes $D^{*+}\rho^-$. [b] Includes $D^{*+}a_1^-$, $a_1^- \to \rho^0\pi^-$. [c] Does not include $\psi\bar{K}^{*0}$.

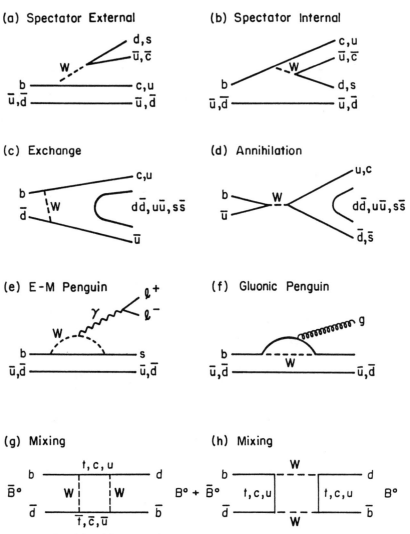

Figure 1 Quark level diagrams for B-meson decay: (a) spectator, (b) spectator with color mixing, (c) exchange, (d) annihilation, (e) electromagnetic penguin, (f) gluonic penguin, (g) and (h) box diagrams for B^0-$\bar{B}^0$ mixing.

in which the initial partner $\bar{u}$ or $\bar{d}$ is paired in a final-state hadron with the daughter quark of the initial heavy quark (the c or u from the b), and the other amplitude a_2 in which the initial $\bar{u}$ or $\bar{d}$ and the heavy quark daughter do not end up in the same final-state hadron.

The predicted exclusive branching ratios (21) in the last column of

Tables 2 and 3 assume the KM matrix element $V_{cb} = 0.041$ as determined below in Section 6 and a simple model for the effect of quark binding. The rates for the modes involving D_S contain the factor $\delta = (f_{D_S}/162 \text{ MeV})^2$ from $W^+ \rightarrow c\bar{s} \rightarrow D_S^+$. We have performed a least-squares fit of the measured branching ratios to the predictions of Bauer et al (21) to determine a_1, a_2, and f_{D_S}. We obtain two solutions distinguished by the relative signs of a_1 and a_2 (Table 4).

From the factorization hypothesis one expects $a_1 = C_1 + \xi C_2$ and $a_2 = \xi C_1 + C_2$, where the coefficients $C_1 = 1.1$ and $C_2 = -0.24$ are calculated from perturbative QCD at the b mass scale. Note that in the limit $C_2 \ll C_1$ the ψ modes would have amplitudes proportional to the QCD color-matching factor ξ, naively $1/3$. Taking C_1 and C_2 as given, one finds that the a_1 and a_2 measurements overdetermine ξ. We extract ξ from the ratio a_2/a_1, which is insensitive to the poorly predicted total B decay rate (see Table 4). The leading order in the $1/N_c$ expansion (23) (N_c is the number of colors) predicts $\xi = 0$, as does an analysis of D decays using QCD sum rules (24). The fit results for f_{D_S} are to be compared with other experimental determinations, ranging from 259 ± 74 MeV (25) to $276 \pm 45 \pm 44$ MeV (26).

We can try a variation of the same factorization approach for inclusive cross sections (5, 18). If we neglect the $b \rightarrow uW^-$ coupling and other Cabibbo-suppressed processes, we get the predictions of Table 5, normalized to $B_e = B(B \rightarrow Xe^-\bar{\nu}_e)$. For decays of $b\bar{u}$ states in category II we allow interference between the amplitudes corresponding to the two

Table 4 Best-fit values for the parameters a_1, a_2, and ξ of Bauer et al (21)

Parameter	Solution 1	Solution 2
a_1	1.01 ± 0.08	1.09 ± 0.08
a_2	0.22 ± 0.04	-0.20 ± 0.05
f_{D_S} (MeV)	206 ± 65	191 ± 61
$\chi^2/\text{d.f.}$	$10.3/10$	$17.2/10$
ξ	0.42 ± 0.04	0.03 ± 0.04
$B(B^- \rightarrow Xe^-\bar{\nu}_e)$	0.122 ± 0.005	0.161 ± 0.006
$B(\bar{B}^0 \rightarrow Xe^-\bar{\nu}_e)$	0.146 ± 0.006	0.133 ± 0.005
$B(B \rightarrow \langle \psi, \chi, \ldots \rangle X)$	0.008 ± 0.003	0.007 ± 0.003

Table 5 Factorization predictions for inclusive branching ratios

Mode category	$\mathcal{B}_\pm/\mathcal{B}_e$	$\mathcal{B}_0/\mathcal{B}_e$
I. $b \to c(\ell^- \bar{\nu})$	$1 + 1 + 0.23$	$1 + 1 + 0.23$
II. $b \to c(\bar{u}d)$	$3(a_1 + a_2)^2$	$3a_1^2 + 3a_2^2$
III. $b \to c(\bar{c}s) \to no\ c\bar{c}$	$1.3a_1^2$	$1.3a_1^2$
IV. $b \to c(\bar{c}s) \to c\bar{c}$	$1.3a_2^2$	$1.3a_2^2$

choices of pairing the two $\bar{u}$ quarks with the c and d before the hadronization. Categories III and IV differ depending on whether the c and $\bar{c}$ end up in the same hadron.

Table 4 shows some inclusive branching ratios predicted by this model using the a_1 and a_2 solutions derived from exclusive branching ratios. Unless a_1 is increased to a value near 1.3, the predicted semileptonic branching ratios are too high relative to the measured values (see Section 3.1). Also, the predicted inclusive charmonium branching ratios are lower than the $B \to \psi X$ measurement, which suggests that the value for $|a_2|$ obtained from the exclusive decays may be too low.

2.4 *Masses*

The peaks in the spectra of effective mass of exclusive decays measure the masses of the charged and neutral B mesons. The width of the peak (typically about 2.5 MeV) is dominated by the natural spread in beam energy. The systematic uncertainty in the storage ring energy calibration dominates the error in mass but cancels in the mass difference.

ARGUS and CLEO data on the masses of the neutral and charged B's used to indicate a difference of about 2 MeV (27); with the addition of more recent data (20, 22) the mass difference is consistent with zero (see Table 6). Although the mass of the b quark, and therefore of the B meson, is an arbitrary parameter in the Standard Model, the mass difference between neutral and charged mesons should be calculable. There are predictions ranging from 1.2 to 4.4 MeV (28).

3. SEMILEPTONIC DECAYS

3.1 *Inclusive Measurements*

Semileptonic decays proceed via the spectator diagram (Figure 1a), and models must account for strong interaction effects only at one vertex.

Table 6 Measured B masses (in MeV)

	$M(\overline{B}^0)$	$M(B^-)$	$M(\overline{B}^0) - M(B^-)$
CLEO '86	$5280.6 \pm 0.8 \pm 1.0$	$5278.6 \pm 0.8 \pm 2.0$	$2.0 \pm 1.1 \pm 0.3$
CLEO '88	$5278.0 \pm 0.4 \pm 2.0$	$5278.3 \pm 0.4 \pm 2.0$	$-0.4 \pm 0.6 \pm 0.5$
ARGUS	$5280.5 \pm 1.0 \pm 2.0$	$5279.6 \pm 0.7 \pm 2.0$	$-0.9 \pm 1.2 \pm 0.5$
Average	5279 ± 2	5279 ± 2	0.0 ± 0.5

Semileptonic decays have been used to find $B\overline{B}$ mixing and evidence for the b → u transition, and to measure V_{cb}. The charge of the lepton indicates the flavor, B versus $\overline{B}$, at the time of decay. The spectator model prediction for the semileptonic branching ratio, corrected for strong interaction effects (29, 30), is

$$B_e = \frac{\Gamma(\overline{B} \to Xe^-\bar{\nu}_e)}{\Gamma(\overline{B} \to All)} = \frac{\Gamma(\overline{B} \to X\mu^-\bar{\nu}_\mu)}{\Gamma(\overline{B} \to All)} = 14\%.$$

There are two categories of models that predict the lepton spectrum. The first, which treats the b and c quarks as quasi-free partons (31), predicts an effective mass distribution of c-quark fragmentation products peaking close to the D* mass (32). That is, the final-state hadrons should be predominantly a single charmed meson, D, D*, or D**. In the charmless case, the u-quark fragmentation spectrum peaks at 1.3 GeV, well above the low-lying π or ρ states. This suggests that the second category of models (33–37), the exclusive models that treat the decay of a B meson into single meson plus lepton and antineutrino, will work well for b → c transitions but not for b → u (38, 39).

The inclusive electron and muon spectra from CLEO (40, 41) are shown in Figure 2. The peak at lower momentum is due to the decay chain B → DX, D → Y$\ell\bar{\nu}$. The data are fit to the sum of two spectra, one from models of B decay (31, 34), the other derived from the measured lepton spectrum from D mesons produced at the ψ''. The electron and muon results agree and we have averaged them. The resulting semileptonic branching ratios from CLEO (40), ARGUS (41), and Crystal Ball (42) are given in Table 7. CLEO uses the radiative corrections of Atwood & Marciano (2), which raise B_e by about 2%. The best-fit branching ratio depends on the model that is used. The results are consistent with $B_e = 10.3 \pm 0.4\%$. Since the $\Upsilon(4S)$ decay fractions are unknown, the quoted result is actually $B_e = B_\pm f_\pm + B_0 f_0$.

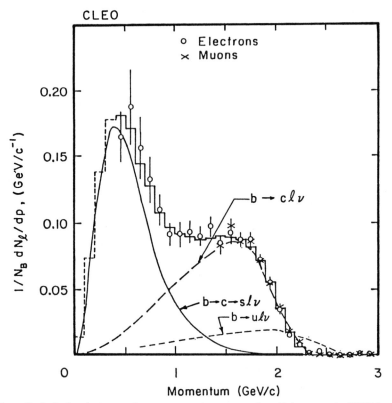

Figure 2 Inclusive electron and muon momentum spectra from B decays at the Υ(4S) from CLEO (40).

Table 7 B inclusive semileptonic branching ratio (%) from the Υ(4S), stating the model dependence of the interpretation of the data

Model[a]	ARGUS	CLEO	Crystal Ball	Average
ACM	10.2±0.5±0.2	10.5±0.3±0.4	12.0±0.5±0.7	10.7±0.4
ISGW	9.8±0.5±0.2	10.0±0.3±0.4	11.9±0.4±0.7	10.3±0.4
WSB	8.7±0.5±0.2	-	10.8±0.4±0.7	9.3±0.4

[a] References 31, 34, and 35.

If the process $B \to \Lambda_c^+ \bar{N} \ell^- \bar{v}$ had a significant rate, it would contribute mainly to the low end of the lepton momentum spectrum and would tend not to be included in the fits to the data. The inclusive Λ_c branching ratio (Table 1) indicates that the correction might be as high as 1.5% (43). However, ARGUS has found that $B(B \to X \bar{p} e^+ v_e) < 0.16\%$ at 90% confidence limit (41), which implies a smaller rate.

Table 8 shows the semileptonic branching ratio measurements at higher energies. The PEP and PETRA measurements (14) assume the fraction of events containing b quarks to be 1/11; measurements at the Z^0 peak (44–46) assume the Standard Model prediction, 0.217. The average is $11.3 \pm 0.4\%$; the error may be underestimated as we have assumed that the systematic errors are uncorrelated.

The branching ratio measurements are low compared to the quark model predictions. The theory should be accurate in the limit in which the final states include many exclusive channels; however, a few channels, e.g. $D\ell\bar{v}$ and $D^*\ell\bar{v}$, actually make up most of the semileptonic rate. Since the models that can predict the exclusive semileptonic rates cannot predict the inclusive hadronic rates and vice versa, it is difficult to formulate a

Table 8 B inclusive semileptonic branching ratio (%) from high energy measurements

Expt.[a]	Electron	Muon
Mark II	$11.6 \pm 2.1 \pm 1.7$	$12.0 \pm 0.5 \pm 0.7$
MAC	-	$15.5^{+5.4}_{-2.0}$
Mark J	-	$10.5 \pm 1.5 \pm 1.3$
DELCO	$14.9^{+2.2}_{-1.9}$	$11.9 \pm 0.4 \pm 0.7$
TASSO	$11.1 \pm 3.4 \pm 4.0$	$11.7 \pm 2.8 \pm 1.0$
JADE	-	$11.7 \pm 1.6 \pm 1.5$
TPC	$11.0 \pm 1.8 \pm 1.0$	$10.8 \pm 0.4 \pm 0.7$
ALEPH	$10.0 \pm 0.9 \pm 0.5$	$11.0 \pm 1.3 \pm 0.5$
L3	-	$11.0 \pm 0.8 \pm 0.8$
OPAL	-	$10.8 \pm 0.4 \pm 0.6$
Average	11.1 ± 0.8	11.4 ± 0.4

[a] References 14, 44, 45, and 46; A. Jawahery, private communication.

quantitative prediction. Nevertheless Bigi (private communication) has pointed out that B_e could be as low as 12% in a phenomenological approach that uses $\xi = 0$ rather than 1/3.

3.2 Inclusive B^0 Semileptonic Decays

CLEO (6) has constrained the ratio of neutral and charged semileptonic decay rates, $0.44 < B(\bar{B}^0 \to X^+\ell^-\bar{\nu})/B(B^- \to X^+\ell^-\bar{\nu}) < 2.05$ (90% confidence level), from the ratio of dilepton and single-lepton event rates. Alternatively, tagged samples of B^0 mesons found by partially reconstructing the decays $\bar{B}^0 \to D^{*+}\pi^-$ and $\bar{B}^0 \to D^{*+}\ell^-\bar{\nu}$ give raw values of $9.1 \pm 3.1^{+1.0}_{-1.5}\%$ and $8.6 \pm 4.5^{+0.60}_{-0.5}\%$ for the two samples (40). After correcting for mixing, we get $B(\bar{B}^0 \to X^+\ell^-\bar{\nu}) = 10.4 \pm 2.2^{+1.0}_{-1.1}\%$. This result is independent of the fraction of non-$B\bar{B}$ $\Upsilon(4S)$ decays.

3.3 Exclusive Semileptonic Decays

The decay of a pseudoscalar meson to a pseudoscalar meson can be described with a single form factor, while three are needed to describe the decay to a vector meson. Different authors choose different parametrizations for the form factors and calculate different values for their overall scale (33, 35). A new approach, derived from the limit where the initial meson is infinitely heavy, allows all of these form factors to be expressed in terms of one universal function (48). It is based on very general principles of QCD and its validity is generally accepted, although corrections for the finite masses of the quarks have not yet been done and may turn out to be important.

The exclusive semileptonic final states $D\ell\bar{\nu}$ and $D^*\ell\bar{\nu}$ can be identified at the $\Upsilon(4S)$ by selecting decays with low missing mass recoiling against the lepton and charmed meson, and by assuming that the B is at rest. The ARGUS (49) and CLEO (50) branching ratios (actually $2B_\pm f_\pm$ and $2B_0 f_0$) are given in Table 9.

Table 9 Exclusive B semileptonic branching ratios (in %), where ℓ^- stands for either an e^- or μ^-

Mode	CLEO	ARGUS	Average
$\bar{B}^0 \to D^{*+}\ell^-\bar{\nu}$	$4.6 \pm 0.5 \pm 0.7$	$5.4 \pm 0.9 \pm 1.3$	$4.8 \pm 0.4 \pm 0.7$
$\bar{B}^0 \to D^+\ell^-\bar{\nu}$	$1.8 \pm 0.6 \pm 0.3$	$1.7 \pm 0.6 \pm 0.4$	$1.75 \pm 0.42 \pm 0.35$
$B^- \to D^{*0}\ell^-\bar{\nu}$	$4.1 \pm 0.8 \pm 0.8$	-	$4.1 \pm 0.8 \pm 0.8$
$B^- \to D^0\ell^-\bar{\nu}$	$1.6 \pm 0.6 \pm 0.2$	-	$1.6 \pm 0.6 \pm 0.2$

Several interesting conclusions can be drawn from these data. The vector-to-pseudoscalar decay ratio is

$$\frac{\Gamma(B \to D^* \ell^- \bar{\nu})}{\Gamma(B \to D\ell^- \bar{\nu})} = 2.6^{+1.1+1.0}_{-0.6-0.8} \text{(CLEO)}, \quad 3.3^{+3.7}_{-1.1} \text{(ARGUS)}.$$

We can also determine the fraction f_{excl} of the semileptonic rate into the sum of $D\ell\bar{\nu}$ and $D^*\ell\bar{\nu}$ final states. Voloshin & Shifman (51) predicted that these final states, which occur without any spin flip between the outgoing c quark and spectator antiquark, comprise the entire semileptonic rate. CLEO (50) finds that the fraction of semileptonic decays that contain D^0's and D^+'s is $0.68 \pm 0.09 \pm 0.10$, and $0.26 \pm 0.07 \pm 0.04$, respectively. The sum, $0.94 \pm 0.11 \pm 0.11$, is consistent with $b \to c$ dominance. Then,

$$f_{\text{excl}} = \frac{f_{\pm}[B(B^- \to D^0\ell^- \bar{\nu} + D^{*0}\ell^- \bar{\nu})] + f_0[B(\bar{B}^0 \to D^+\ell^- \bar{\nu} + D^{*+}\ell^- \bar{\nu})]}{B(B \to D^0 X\ell^- \bar{\nu}) + B(B \to D^+ X\ell^- \bar{\nu})}$$

$$= 0.64 \pm 0.10 \pm 0.06.$$

Thus, although the D and D* final states comprise most of the rate, there is considerable room left for either D**, D*π, or Dπ final states. The prediction of Isgur et al (34) for f_{excl} is 85%, consistent with the data.

Of the several angular distributions that could be examined—the direction of the π^+ in the D*$^+$ rest frame, the lepton direction in the $\ell\bar{\nu}$ rest frame, and the correlation between these two decay planes—only the π^+ of the decay of the D*$^+$ has been investigated. The decay angular distribution of the π^+ can be parameterized as

$$W(\cos\theta) = \frac{3}{6 + 2\alpha}(1 + \alpha\cos^2\theta),$$

where $\alpha = 2\Gamma_L/\Gamma_T - 1$. CLEO (50) finds $\alpha = 0.65 \pm 0.66 \pm 0.25$ and ARGUS (49) finds $\alpha = 0.7 \pm 0.9$. The predicted value of α depends on the range of lepton momentum included. The CLEO result is for $p_\ell > 1.4$ GeV/c while ARGUS has a lower cut of 1.0 GeV/c. The model predictions are 0.28 (34), 0.46 (35), and 0.32 (36) for $p_\ell > 1.4$ GeV/c. The data are in agreement with all of the models.

3.4 Test of Vertex Factorization

By vertex factorization we mean the hypothesis that the amplitude for the semileptonic decay $B \to D\ell\bar{\nu}$ is the product of a hadronic current and a leptonic current, $\langle D|J_\mu|B \rangle \times \langle \nu|\gamma_\mu(1 - \gamma_5)|\ell \rangle$. In the two-body hadronic decay the lepton factor is replaced by another hadron factor. For example, the amplitude for $B \to D\pi^-$ is $\langle D|J_\mu|B \rangle \times \langle 0|A_\mu|\pi^- \rangle$. The test of the factorization concept consists of checking the equality

$$\frac{\Gamma(\bar{B}^0 \to D^{*+}\pi^-)}{\dfrac{d\Gamma}{dq^2}(\bar{B}^0 \to D^{*+}\ell^-\bar{\nu})\bigg|_{q^2=m_\pi^2}} = 6\pi^2 f_\pi^2 |V_{ud}|^2.$$

Recently, Bortoletto & Stone (26) used a combination of ARGUS and CLEO data on exclusive semileptonic and exclusive hadronic decays to make a test of vertex factorization proposed by Bjorken (52). They averaged the results of the q^2 distributions from both groups, shown in Figure 3. Combining the values of the $\bar{B}^0$ branching ratio to $D^{*+}\pi^-$ with the value of q^2 distribution at $q^2 = m_\pi^2$, they determine values of f_π^2 from the above equation that are consistent with the value measured in $\pi^\pm$ decay at the 25% level. Rosner (53) has fit the q^2 distribution to derive a value for the universal form-factor parameter (48), $w_0 = 1.1^{+0.3}_{-0.2}$.

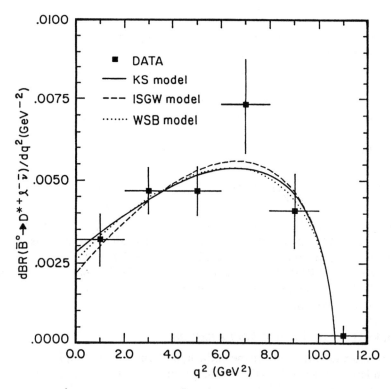

Figure 3   The q^2 distribution for the decay $\bar{B}^0 \to D^{*+}\ell^-\bar{\nu}$ in units of branching ratio $\times$ GeV^{-2}. This is a weighted average of CLEO and ARGUS data (see text). The curves are fits using various models of semileptonic decay.

4. LIFETIMES

4.1 *Lifetime for Average b Hadron*

Measurements of the average lifetime for hadrons containing the b quark are made by observing the decay length of b hadrons in jets produced in e^+e^- collisions at energies far above threshold. To enrich the event sample of $b\bar{b}$ jets the various experiments have used either a cut on the p_T of a lepton with respect to the jet axis or a cut on the product of the sphericities of the two jets evaluated in their respective rest frames; other experiments used no enrichment. The B lifetime has been estimated from measurements of the impact parameter of the lepton, the impact parameters of all particles, the dipole moment of vertex positions, or the reconstructed decay length. The sample purity and the dependence of the measured quantities on the mean lifetime are modeled by Monte Carlo calculations in order to extract the result.

There is good agreement among the different experiments and measurement techniques. The best estimate for the average B lifetime from the PEP and PETRA measurements is $\tau_b = 1.18 \pm 0.11$ ps (14). The ALEPH collaboration at LEP (54) obtained a measurement of the b lifetime, $\tau_b = 1.29 \pm 0.06 \pm 0.10$ ps, using the impact parameter of high-momentum, high-p_T electrons and muons with respect to the beam center. The world average is then $\tau_b = 1.24 \pm 0.09$ ps. It is important to note that this value is a weighted average over the mix of b hadrons produced in e^+e^- collisions at high energies, that is, B^-, $\bar{B}^0$, B_s, Λ_b, and so on. If the light quark in the b hadron influences the decay rates, as in the decay of the charmed hadrons, then the various lifetimes will not be equal.

4.2 *Separate Lifetimes for Charged and Neutral B Mesons*

If it is reasonable to assume that semileptonic B-to-charm decays occur through $b \to c\ell\bar{\nu}$ independent of the flavor of the partner antiquark, then the ratio of charged and neutral B lifetimes should be the same as the ratio of branching ratios for a pair of corresponding exclusive charged and neutral modes. CLEO (55) measures

$$\frac{\tau_+}{\tau_0} = \frac{\Gamma(\bar{B}^0 \to D^{*+}\ell^-\bar{\nu})B(B^- \to D^{*0}\ell^-\bar{\nu})}{\Gamma(B^- \to D^{*0}\ell^-\bar{\nu})B(\bar{B}^0 \to D^{*+}\ell^-\bar{\nu})} = \frac{B(B^- \to D^{*0}\ell^-\bar{\nu})}{B(\bar{B}^0 \to D^{*+}\ell^-\bar{\nu})}$$

$$= (0.89 \pm 0.19 \pm 0.13)f_0/f_\pm.$$

ARGUS uses the inclusive D^0, D^+, and D^{*+} measurements in semileptonic decays to derive the lifetime ratio (56) $\tau_\pm/\tau_0 = (1.00 \pm 0.23 \pm 0.14)f_0/f_\pm$. The idea here is that most of the semileptonic decays involve a change of charge between the B and D, that is, $\bar{B}^0 \to D^{*+}X$ or $B^- \to D^{*0}X$. The D^{**} component just increases the error. Averaging the two results, we conclude that the assumptions that $\tau^\pm = \tau_0$ and $f_\pm = f_0$ are consistent with the data within the present experimental accuracy of 17%.

The factorization model for inclusive decays (5, 18) predicts a difference in semileptonic branching ratios for charged and neutral B's (see Table 5), as a result of the interference between the a_1 and a_2 amplitudes. Solutions 1 and 2 for a_1 and a_2 (Table 4) predict $\tau_\pm/\tau_0 = 0.85$ and 1.18, respectively.

5. $B\bar{B}$ OSCILLATIONS

The B^0 and $\bar{B}^0$ (or the B_S^0 and $\bar{B}_S^0$) differ only in the sign of the beauty quantum number carried by the b quark. Since this is not a conserved quantity in weak decays, the eigenstates for which the lifetimes and masses are defined are not the b eigenstates created in the production reaction. A pure b eigenstate, which is a superposition of the two mass eigenstates propagating with different frequencies given by m_L and m_H, will then develop an oscillating admixture of the anti-b eigenstate with a beat frequency $\Delta m = m_L - m_H$.

The short lifetime of B mesons requires one to measure decay lengths in the tens of microns in order to observe the oscillation directly at the $\Upsilon(4S)$. It is therefore natural to integrate over time. Starting from $B^0\bar{B}^0$ pairs, as in $\Upsilon(4S)$ decay, one can observe B^0B^0 and $\bar{B}^0\bar{B}^0$ decays, tagged by like-sign semileptonic final states, for instance. In $\Upsilon(4S)$ decays the two B's are in a p-wave state, which is odd under exchange, so that Bose statistics implies the mixing ratio

$$r = \frac{\text{Prob}(B\bar{B} \to BB) + \text{Prob}(B\bar{B} \to \bar{B}\bar{B})}{\text{Prob}(B\bar{B} \to B\bar{B}) + \text{Prob}(B\bar{B} \to \bar{B}B)} = \frac{x^2}{2+x^2}.$$

Observations of $B\bar{B}$ oscillations at the $\Upsilon(4S)$ have been made by ARGUS (57) and CLEO (58), looking for both mesons to decay into final states that indicate whether the decaying meson was B or $\bar{B}$. Using the semi-leptonic decays, $B^0 \to \ell^+X$ and $\bar{B}^0 \to \ell^-X$, as an indicator, one obtains the like-sign/unlike-sign ratio, r_{obs}, corrected for leptons from non-$\Upsilon(4S)$ events, misidentified leptons, leptons from $b \to c \to \ell$ cascades, and leptons from $B \to \psi X \to \ell^+\ell^- X$:

$$r_{obs} = \frac{N(\ell^+\ell^+) + N(\ell^-\ell^-)}{N(\ell^+\ell^-) + N(\ell^-\ell^+)}.$$

Dilepton events can also come from $\Upsilon(4S) \to B^+B^-$. If we assume that the non-$B\bar{B}$ decays of the $\Upsilon(4S)$ do not contribute dileptons, the mixing ratio

r is given by

$$r = r_{\text{obs}} \frac{1+\lambda}{1-\lambda r_{\text{obs}}},$$

where $\lambda = f_{\pm} B_{\pm}^2 / (f_0 B_0^2)$. We will assume $\lambda = 1$.

Table 10 shows the dilepton results from ARGUS (57) and CLEO (58). Each of the two groups has confirmed its measurement of mixing by a second technique. ARGUS gets a result independent of λ by comparing rates for $D^{*+}\ell^-\ell^-$ and $D^{*+}\ell^-\ell^+$ events (and their charge conjugates), and CLEO derives a value of r from the rates for $K^-\ell^-\ell^-$ and $K^-\ell^-\ell^+$. The average of the ARGUS and CLEO results for the mixing ratio is $r = 0.18 \pm 0.05$. This implies $x = \Delta m/\Gamma = 0.66 \pm 0.11$ for the B^0.

The first evidence for $B\bar{B}$ mixing was actually reported by the UA1 collaboration (59). They saw an excess of like-sign muon pairs above the Monte Carlo calculation for the rate from mundane sources, such as pion, kaon, and charm decays and misidentification of other particles. The probability that a $b(\bar{b})$ hadron that decays semileptonically will decay through mixing to a $\ell^+(\ell^-)$ is measured from the observed ratio of rates for $\mu^+\mu^+$ and $\mu^-\mu^-$ relative to the total rate of dimuons from B^0 and B_S. It has contributions from B^0-$\bar{B}^0$ mixing and B_S-$\bar{B}_S$ mixing:

$$\chi = f_d \chi_d + f_s \chi_s, \qquad \chi_d = \frac{x^2}{2(1+x^2)};$$

χ_s is similarly related to x_s. The data from UA1 and later PETRA (60), PEP (61), and LEP (62) experiments are shown in Table 11. High energy jet fragmentation data are consistent with $f_d = 0.375, f_s = 0.15$. The measured average χ together with the $\chi_d = 0.15 \pm 0.04$ from the ARGUS and CLEO data imply $\chi_s > 0.36$, and therefore $x_s > 1.6$ at 90% confidence level.

Table 10   Data on mixing of $B\bar{B}$ from $\Upsilon(4S)$

	ARGUS	CLEO '86	CLEO '88
$N(\ell^\pm\ell^\pm)$, background corrected	$35 \pm \pm 12$	5 ± 5	32 ± 12
$N(\ell^+\ell^-)$, background corrected	381 ± 23	117 ± 12	426 ± 24
r from dileptons	0.20 ± 0.08	0.11 ± 0.11	0.17 ± 0.08
r from other techniques	0.25 ± 0.12	-	0.23 ± 0.12
r, average	0.22 ± 0.07	0.11 ± 0.11	0.17 ± 0.08

Table 11 Data on like-sign dileptons in high energy experiments

Experiment	Beam	c.m. GeV	χ
UA1*	$\bar{p}p$	540	0.121 ± 0.047
JADE	e^+e^-	34	$< 0.13(90\%c.l.)$
Mark II	e^+e^-	29	$< 0.12(90\%c.l.)$
MAC	e^+e^-	29	$0.21^{+0.29}_{-0.15}$
ALEPH*	e^+e^-	91	$0.132^{+0.027}_{-0.026}$
L3*	e^+e^-	91	$0.178^{+0.049}_{-0.040}$
*average			0.142 ± 0.020

6. THE CKM MATRIX

6.1 *Parametrizing the Matrix*

Flavor-changing weak decays in the six-quark Standard Model proceed through the xyW vertex, where x denotes a charge 2/3 quark u, c, or t and y is a charge $-1/3$ quark d, s, or b (see Figure 1). Besides the weak coupling g the vertex amplitude contains a factor V_{xy}, where V is the 3×3 Cabibbo-Kobayashi-Maskawa (CKM) unitary matrix (63, 64). The nine matrix elements V_{xy} can be expressed in terms of four independent quantities. A convenient form was suggested by Wolfenstein (65):

$$V = \begin{pmatrix} V_{ud} & V_{us} & V_{ub} \\ V_{cd} & V_{cs} & V_{cb} \\ V_{td} & V_{ts} & V_{tb} \end{pmatrix} \approx \begin{pmatrix} 1-\lambda^2/2 & \lambda & A\lambda^3(\rho-i\eta) \\ -\lambda & 1-\lambda^2/2 & A\lambda^2 \\ A\lambda^3(1-\rho-i\eta) & -A\lambda^2 & 1 \end{pmatrix},$$

where we have assumed that the off-diagonal elements are small. The Standard Model makes no prediction for the values of the parameters; until we have a more general theoretical understanding, they must be obtained from experiment. Data on strangeness-changing decays have fixed the λ parameter, essentially the Cabibbo angle (14), at the value $\lambda = 0.226 \pm 0.002$. The other three parameters can be determined from B decays.

6.2 V_{cb}: *Charm Decays*

To find V_{cb} we use the equation $|V_{cb}|^2 = B(B \to [D \text{ or } D^*]\ell\bar{\nu})/(\tau_b K)$, where K comes from a model calculation of the appropriate exclusive semileptonic decay process (34–37) and is defined by $\Gamma = K|V_{cb}|^2$. We take $\tau_b = 1.24 \pm 0.09$ ps. To minimize any effects of the difference in charged and neutral lifetimes we also average the charged and neutral branching ratios in the last column of Table 12.

The extracted values are the same for both the D and D* final states. Averaging over the models, a somewhat dubious procedure, gives a value for $|V_{cb}| = 0.041 \pm 0.003$. The quoted error here is statistical. There are three sources of systematic errors: (a) the model dependence, $\approx \pm 0.002$; (b) the non-B$\bar{B}$ Υ(4S) decays, which could raise the branching ratios by 17% and V_{cb} by 8%, an error of $+0.003$; and (c) the systematic error on τ_b that arises from the B_S and Λ_b mixture in the sample used to measure τ_b. We quote the final result as $|V_{cb}| = 0.041 \pm 0.003 \pm 0.004$. This gives $A = 0.8 \pm 0.1$.

6.3 V_{ub}: *Charmless Decays*

The quark-mixing explanation of *CP* violation (64) requires that every element in the KM matrix be nonzero. Otherwise, it is possible to express the independent variables (A, λ^2, ρ, and η) in such a way that the complex phase is not required and the matrix is real. It is therefore important to demonstrate a nonzero value for V_{ub}.

CLEO (66) and ARGUS (67) have observed evidence for the $b \to u$ transition by finding leptons from B decay with momentum greater than that allowed by $B \to D\ell\bar{\nu}$ decay (see Figure 2). Although these leptons could in principle arise from non-B$\bar{B}$ Υ(4S) decays, this source has been ruled out as a major contributor (6). The measured rates in several momentum ranges between 2.2 and 2.6 GeV/c are proportional to $|V_{ub}/V_{cb}|^2$ with

Table 12   Measurements of $|V_{cb}|$ according to different models

Model[a]	$D\ell\bar{\nu}$	$D^*\ell\bar{\nu}$	Average
ISGW	0.036±0.005	0.039±0.004	0.038±0.003
KS	0.042±0.005	0.039±0.004	0.040±0.003
WSB	0.042±0.005	0.042±0.004	0.042±0.003
Jaus 1	0.041±0.005	0.042±0.004	0.042±0.003
Jaus 2	0.042±0.005	0.042±0.004	0.042±0.003
Average	0.041±0.005	0.041±0.004	0.041±0.003

[a] References 34, 36, 35, and 37, respectively.

coefficients obtained from models (31, 34–36, 39). The model dependence of $|V_{ub}/V_{cb}|^2$ is more than a factor of three. The results are consistent with $|V_{ub}/V_{cb}| = 0.12 \pm 0.03$.

In another technique one tries to reconstruct exclusive B decays to modes that have no charmed (or strange) particle in the final state, modes simple enough to be easily identified experimentally and detected with high efficiency and low background. The lowest branching ratio upper limits reported by CLEO (68) and ARGUS (69) are at the level of 10^{-4} (for $p\bar{p}$, $\pi^+\pi^-$, $\rho^0\pi^-$, $\pi^-\pi^-\pi^+$ for instance). The theoretical predictions (21), with $|V_{ub}/V_{cb}| = 0.1$, are still at least a factor of four lower. The original ARGUS report of the observation of charmless decays $B^- \to p\bar{p}\pi^-$ and $\bar{B}^0 \to p\bar{p}\pi^+\pi^-$ (70) has not been confirmed by later CLEO (71) and ARGUS (72) data. Table 13 shows the measured upper limits for modes that are less than 10^{-3}.

6.4 V_{td}: Mixing

Until we are able to observe the decays of hadrons containing the t quark and measure the rate for the suppressed $t \to d$ transition directly, we will have to use higher order weak processes containing the t quark in intermediate states. In the Standard Model the B^0-$\bar{B}^0$ transition takes place through the box diagram (Figures 1g and 1h). Although the intermediate state can involve the u, c, or t quark, the dominant contribution comes from the heaviest quark. With $m_t \gg m_c$, m_u the predicted ratio x of B^0-$\bar{B}^0$ oscillation rate to decay rate is given by

$$x = \frac{\Delta m}{\Gamma} = \frac{G_F^2}{6\pi^2} m_t^2 g\left(\frac{m_t^2}{M_W^2}\right) \tau_B f_B^2 B_B \eta_{qcd} |V_{tb} V_{td}^*|^2,$$

where

$$g(y) = \frac{1}{4}\left[1 + \frac{3-9y}{(y-1)^2} + \frac{6y^2 \ln y}{(y-1)^3} \right],$$

which is one for $m_t \ll M_W$ and decreases to about 1/2 at $m_t = 2M_W$ (73). The QCD correction factor η_{qcd} is about 0.85 (74). V_{tb} should be close to one if there are only three families.

At present there are obstacles to extracting V_{td} from a measurement of x. The mass of the t quark is not known; it is probably between 90 and 200 GeV (75). The bound-state dependence may eventually be calculated reliably by lattice gauge theory, but for now $f_B^2 B_B = (129 \pm 23 \text{ MeV})^2$ is our best guess (76). The measured x from ARGUS and CLEO (see Section 5) and the box diagram formula imply

Table 13 Branching ratio upper limits (90%) for charmless B decays (in 10^{-4})[c]

$\bar{B}^0 \rightarrow$	Data	Model	$B^- \rightarrow$	Data	Model
$\pi^+\pi^-$	0.9^a	0.20	$\pi^-\pi^0$	2.4^b	0.06
$\rho^{\pm}\pi^{\mp}$	5.2^b	0.6	$\rho^0\pi^-$	1.5^a	0.002
-	-	-	$f_0\pi^-$	1.2^a	0.04
-	-	-	$f_2\pi^-$	2.1^a	-
$a_1^{\pm}\pi^{\mp}$	5.7^a	0.6	$a_1^0\pi^-$	9.0^b	-
-	-	-	$a_1^-\rho^0$	5.4^a	0.3
$a_2^{\pm}\pi^{\mp}$	3.5^a	-	$a_2^-\rho^0$	6.3^b	-
$\rho^0\pi^0$	4^b	0.02	$\rho^-\pi^0$	5.5^b	0.2
$\omega\pi^0$	4.6^b	0.001	$\omega\pi^-$	4.0^b	0.02
$\rho^0\rho^0$	2.8^b	0.01	$\eta\pi^-$	7.0^b	0.03
$\pi^+\pi^-\pi^0$	7.2^b	0.6	$\pi^+\pi^-\pi^-$	4.5^b	0.6
$\pi^+\pi^+\pi^-\pi^-$	6.7^b	1	$\pi^-\pi^0\pi^0$	8.9^b	0.6
-	-	-	$2(\pi^+\pi^-)\pi^-$	8.6^b	2
$p\bar{p}$	0.4^a	-	$\Delta^0\bar{p}$	3.3^a	-
$\Delta^0\bar{\Delta}^0$	17.6^a	-	$\bar{\Delta}^{--}p$	1.3^a	-
$\Delta^{++}\bar{\Delta}^{--}$	1.3^a	-	-	-	-
$p\bar{p}\pi^+\pi^-$	2.9^a	-	$p\bar{p}\pi^-$	1.4^a	-

[a] CLEO data (68).
[b] ARGUS data (69).
[c] Model is that of Bauer et al (21).

$$|V_{td}| = (0.013 \pm 0.003) \times \left(\frac{m_t}{140\,\text{GeV}}\right)^{-0.75}.$$

6.5 V_{ts}: B_S Mixing and Loop Decays

Until there are data on the decay of hadrons containing the t quark from higher energy colliders, we will have to wait for information on direct t → s transitions. Meanwhile, there are higher order s-quark processes involving t quarks in intermediate states. In principle, we can extract $|V_{ts}|$ from a measurement of x_s (or from x_s/x_d), but for now we have only the dilepton data (see Section 5), which imply $x_s > 1.6$ and the limit

$$|V_{ts}| > 0.020 \times \left(\frac{m_t}{140\,\text{GeV}}\right)^{-0.75},$$

at 90% confidence level. CESR has taken data at the $\Upsilon(5S)$ resonance, which should be above $B_S\bar{B}_S$ threshold, but the statistical evidence for B_S is not yet compelling.

One can construct loop diagrams (see Figures 1e and 1f) for processes $b \to s\gamma$ or sg, called "penguin" modes. Although the theoretical predictions (77, 78) for the inclusive radiative decays,

$$B(b \to s\gamma) = 3.5 \times 10^{-4} \times |V_{ts}/0.04|^2 (m_t/140\,\text{GeV})^{0.5},$$

$$B(b \to s\ell^+\ell^-) = 6 \times 10^{-6} \times |V_{ts}/0.04|^2 (m_t/140\,\text{GeV})^{1.2},$$

are subject to large QCD corrections, they are considered reliable, within $\pm 10\%$ uncertainties coming from Λ_{qcd}. The corresponding hadronic rate, $B(b \to sg, sgg, sq\bar{q}) = 1.7 \times 10^{-2} \times |V_{ts}/0.04|^2$, is probably less reliable, although it is not expected to vary significantly with m_t. Inclusive processes are difficult to identify experimentally; the best limit we have is on flavor-changing neutral current decays (79), $B(B \to \ell^+\ell^- X) < 0.12\%$, still far above the level expected.

Predictions are much less reliable for exclusive modes, especially for the nonradiative ones (78, 80, 81). Their branching ratios should be of the order of 10^{-4}, somewhat lower than the sensitivity level of the present experiments (69, 82, 83) (see Tables 14 and 15). If we take seriously the prediction $B(B \to K^*\phi)/B(b \to sg) \approx 0.3\%$ (78), the measured branching ratio limit (Table 15) implies a model-dependent upper limit, $|V_{ts}| < 0.11$, which is much larger than the lower limit implied by the dilepton data. Of

Table 14 Branching ratio upper limits (90%) of electromagnetic $b \to s$ decays of B mesons (in 10^{-4})

$\bar{B}^0 \to$	Data	Model[a]	$B^- \to$	Data	Model[a]
$\bar{K}^{*0}\gamma$	2.8^b	0.2	$K^{*-}\gamma$	5.2^c	0.2
$\bar{K}_1^0(1270)\gamma$	7.8^c	-	-	-	-
$\bar{K}_2^{*0}(1430)\gamma$	4.4^c	-	-	-	-
$\bar{K}^{*0}e^+e^-$	0.8^b	0.02	$K^-e^+e^-$	0.5^b	0.006
$\bar{K}^{*0}\mu^+\mu^-$	1.9^b	0.02	$K^-\mu^+\mu^-$	1.5^b	0.006

[a] Reference 78.
[b] CLEO data (82).
[c] ARGUS data (69).

Table 15 Branching ratio upper limits (90%)
of hadronic b → s decays of B mesons (in 10^{-4})

$\overline{B}^0 \rightarrow$	Data	$B^- \rightarrow$	Data
$K^-\pi^+$	0.9^b	$\overline{K}^0\pi^-$	0.9^b
$\overline{K}^0\rho^0$	3.2^c	$K^-\rho^0$	0.7^b
$\overline{K}^0\phi$	4.9^b	$K^-\phi$	0.8^b
$\overline{K}^0 f_0$	4.2^b	$K^- f_0$	0.7^b
$K^{*-}\pi^+$	4.4^b	$\overline{K}^{*0}\pi^-$	1.3^b
$\overline{K}^{*0}\rho^0$	4.6^c	-	-
$\overline{K}^{*0}\phi$	3.2^c	-	-
$\overline{K}^{*0} f_0$	2.0^b	-	-
$K_1^-(1400)\pi^+$	6.6^c	-	-
-	-	$K^-\pi^+\pi^-$ ^{a}	1.7^b
-	-	$\Lambda\overline{p}$	0.5^b

^{a} Not including charm final states.
^{b} CLEO data (82).
^{c} ARGUS data (83).

course, if the CKM matrix is indeed a 3×3 unitary matrix, $V_{ts} \approx V_{cb} \approx 0.04$.

6.6 Conclusions

One consequence of the unitarity of the KM matrix is that the sum of squares of the elements in each column or row must equal one. In the first row we have $|V_{ud}|^2 + |V_{us}|^2 = (0.9744 \pm 0.0010)^2 + (0.2205 \pm 0.0018)^2 = 0.9976 \pm 0.0021$. This is consistent with unitarity without the contribution from $|V_{ub}|^2$, which is about 2×10^{-5}. Applying unitarity to the last column of the matrix, one obtains an estimate

$$|V_{tb}| = \sqrt{1 - |V_{cb}|^2 - |V_{ub}|^2} = 0.9991 \pm 0.0002$$

showing a dominant coupling of the t quark to the b quark.

Other unitarity relationships can be formed by taking the product of one column with the complex conjugate of another column. For example, $V_{ud}^* V_{ub} + V_{cd}^* V_{cb} + V_{td}^* V_{tb} = 0$, which can be represented as the triangle shown in Figure 4a. The triangle can be plotted in the plane defined by the Wolfenstein parameters ρ and η by normalizing the dimensions (dividing by $A\lambda^3$) so that the base is defined by $(\rho, \eta) = (0,0)$ and $(1,0)$ (Figure 4b). The upper two sides are then $V_{ub}^*/(A\lambda^3) = \rho + i\eta$, measured in charmless B decays, and $V_{td}/(A\lambda^3)$, which can be determined from B^0-$\bar{B}^0$ mixing once the top mass is known. The angles of the triangle can be measured by observing CP violation in B decay.

Figure 4c shows the constraints from the measurements of the matrix elements. The V_{ub} measurement gives an annulus centered at $(0,0)$. The B^0-$\bar{B}^0$ mixing constraints are arcs centered at $(1,0)$, while the constraints based on the measurement of the CP violation parameter ε in K^0 decay are almost horizontal. Thus the B mixing and ε constraints, which depend on m_t, are nearly orthogonal. We have drawn a line as a function of m_t that gives allowed values of ρ and η. Allowing for theoretical and experimental uncertainties expands this line into a larger region. A fit including all K and B decay information has been performed by Kim et al (84). If we assume that $m_t < 200$ GeV, the fit implies $\rho = -0.40^{+0.28}_{-0.24}$, $\eta = 0.23^{+0.11}_{-0.08}$, and $m_t = 120^{+45}_{-30}$ GeV.

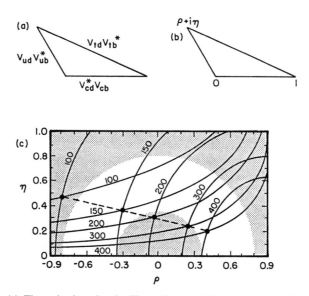

Figure 4 (a) The unitarity triangle, illustrating a relation among quark-mixing matrix elements implied by unitarity. (b) The same triangle rescaled (divided by λV_{cb}) and plotted in the (ρ, η) plane. (c) Constraints in ρ and η from B^0-$\bar{B}^0$ mixing (arcs centered at 1), and ε. The numbers are values of m_t in GeV. The dashed line connects (ρ, η) values implied by the combination of mixing and ε for various m_t. The shaded regions are excluded by the measurement of $|V_{ub}/V_{cb}|$ in charmless semileptonic B decay.

The data on the decays of B mesons are therefore consistent with the Standard Model of six quarks and six leptons (85), and the range of acceptable values of the top-quark mass derived from the decays of B's and lighter mesons is consistent with the range derived from intermediate vector boson measurements at LEP and the Fermilab $\bar{p}p$ collider (86).

ACKNOWLEDGMENTS

This work was supported by the National Science Foundation. We are grateful to our colleagues in the CLEO collaboration for conversations and for their contribution to the work summarized here. We thank E. Blucher and V. Sharma for informing us of the most recent LEP results.

Literature Cited

1. Ono, S., *Acta Phys. Pol.* B15: 201–11 (1984)
2. Atwood, D., Marciano, W. J., *Phys. Rev.* D41: 1736–46 (1990)
3. Lepage, G. P., *Phys. Rev.* D42: 3251–61 (1990); Byers, N., Eichten, E., UCLA preprint UCLA/90/TEP/46 (1990)
4. Alexander, J., et al. (CLEO), *Phys. Rev. Lett.* 64: 2226–29 (1990)
5. Schröder, H. (ARGUS), *Int. Conf. on High Energy Physics*, Singapore, Aug. 1990. Singapore: World Sci. (1990)
6. Poling, R. (CLEO), See Ref. 5; CLEO preprint, Jan. 1991
7. Alam, M. S., et al. (CLEO), *Phys. Rev. Lett.* 58: 1814–17 (1987)
8. Bortoletto, D., et al. (CLEO), *Phys. Rev. Lett.* 56: 800–3 (1986)
9. Alam, M. S., et al. (CLEO), *Phys. Rev. Lett.* 59: 22–25 (1987)
10. Chen, W.-Y., et al. CLEO preprint CBX 89-48; submitted to the *Int. Symp. on Lepton and Photon Interactions at High Energies*, Stanford, July 1989
11. Bortoletto, D., et al. (CLEO), *Phys. Rev. Lett.* 64: 2117–20 (1990)
12. Hoffman, W. (ARGUS), presented at *UCSB Conf. on Weak Decays of Heavy Quarks*, Santa Barbara, May 21, 1990
13. Albrecht, H., et al. (ARGUS), *Phys. Lett.* B210: 263–66 (1988)
14. Particle Data Group, *Phys. Lett.* B239: 1–516 (1988)
15. Adler, J., et al. (Mark III), *Phys. Lett.* B208: 153–56 (1988)
16. Adler, J., et al. (Mark III), *Phys. Rev. Lett.* 60: 89–93 (1988)
17. Alexander, J., et al. (CLEO), *Phys. Rev. Lett.* 65: 1531–34 (1990); Ball, S. (ARGUS), See Ref. 5
18. Wirbel, M., Wu, Y.-L., *Phys. Lett.* B228: 430–33 (1989)
19. Rohaly, T. F. (CDF), See Ref. 5
20. Lewis, J., CLEO internal report CBX 90-11A (1990)
21. Bauer, M., Stech, B., Wirbel, M., *Z. Phys.* C34: 103–23 (1987)
22. Albrecht, H., et al. (ARGUS), In *10th Int. Conf. on Physics in Collision*, Duke Univ., June 1990; See Ref. 5
23. Buras, A. J., Gerard, J.-M., Rückl, R., *Nucl. Phys.* B268: 16–26 (1986)
24. Blok, B., Shifman, M. A., *Yad. Fiz.* 45: 211, 478, 841 (1987); ITEP Reports 9, 17, 37 (1986)
25. Rosner, J. L., *Phys. Rev.* D42: 3732–42 (1990)
26. Bortoletto, D., Stone, S., *Phys. Rev. Lett.* 65: 2951–54 (1990)
27. Bebek, C., et al. (CLEO), *Phys. Rev.* D36: 1289–1301 (1987)
28. Singh, C. P., Sharma, A., Khanna, M. P., *Phys. Rev.* D24: 788–98 (1981); Chan, L.-H., *Phys. Rev. Lett.* 51: 253–56 (1983); Tiwari, K. P., Singh, C. P., Khanna, M. P., *Phys. Rev.* D31: 642–52 (1985); Kim D. Y., Sinha, S. N., *Ann. Phys. NY* 42: 47–67 (1985)
29. Leveille, J. P., Univ. Mich. preprint UMHE81-18 (1981)
30. Stone, S., In *Proc. 1983 Int. Symp. on Lepton and Photon Interactions at High Energies*, ed. D. G. Cassel, D. L. Kreinick. Ithaca: Cornell Univ. Press (1983), pp. 203–43
31. Altarelli, G., Cabibbo, N., Corbo, G., Maiani, L. *Nucl. Phys.* B207: 365–85 (1982)
32. Bareiss, A., Paschos, E. A., *Nucl. Phys.* B327: 353–63 (1989)
33. Grinstein, B., Isgur, N., Wise, M. B., *Phys. Rev. Lett.* 56: 258–61 (1986)
34. Isgur, N., Scora, D., Grinstein, B., Wise, M. B., *Phys. Rev.* D39: 799–809 (1989)

35. Wirbel, M., Stech, B., Bauer, M., Z. Phys. C29: 637–57 (1985)
36. Körner, J. G., Schüler, G. A., Z. Phys. C38: 511–21 (1988)
37. Jaus, W., Phys. Rev. D41: 3394–3404 (1990)
38. Nussinov, S., Wetzel, W., Phys. Rev. D36: 130–40 (1987)
39. Ramirez, C., Donoghue, J. F., Burdman, G., Phys. Rev. D41: 1496–1506 (1990)
40. Alexander, J., et al. (CLEO), Submitted to SLAC Lepton-Photon Conference (1989)
41. Albrecht, H., et al. (ARGUS), Phys. Lett. B249: 359–65 (1990)
42. Wachs, K., et al. (Crystal Ball), Z. Phys. C42: 33–43 (1989)
43. Stone, S., Nucl. Phys. B(Proc. Suppl.)13: 261–69 (1989)
44. Dydak, F., See Ref. 5
45. Decamp, D., et al. (ALEPH), Phys. Lett. B244: 551–65 (1990); Adeva, B., et al., L3 preprint #6, Feb. 20, 1990
46. Jawahery, A. (OPAL), See Ref. 5
47. Deleted in proof
48. Isgur, N., Wise, M. B. Phys. Lett. B237: 527–30 (1990)
49. Albrecht, H., et al. (ARGUS), Phys. Lett. B219: 121–24 (1989); B197: 452–55 (1987); B229: 175–78 (1989)
50. Bortoletto, B., et al. (CLEO), Phys. Rev. Lett. 63: 1667–70 (1989)
51. Voloshin, M. B., Shifman, M. A., Sov. J. Nucl. Phys. 47: 511–21 (1988)
52. Bjorken, J. D., Nucl. Phys. B(Proc. Suppl.)11: 325–45 (1989)
53. Rosner, J. L., Univ. Chicago preprint EFI 90-80 (1990)
54. Decamp, D., et al. (ALEPH), CERN preprint PPE/90-116 (1990)
55. Fulton, R., et al. (CLEO), Phys. Rev. D43: 651–63 (1991)
56. Albrecht, H., et al. (ARGUS), Phys. Lett. B232: 554–57 (1989)
57. Albrecht, H., et al. (ARGUS), Phys. Lett. B192: 245–48 (1987); Albrecht, H. (ARGUS), See Ref. 5
58. Artuso, M., et al. (CLEO), Phys. Rev. Lett. 62: 2233–36 (1989); Bean, A., et al. (CLEO), Phys. Rev. Lett. 58: 183–86 (1987)
59. Albajar, C., et al. (UA1), Phys. Lett. B186: 247–50 (1987)
60. Bartl, W., et al. (JADE), Phys. Lett. B146: 437–40 (1984)
61. Schaad, T., et al. (Mark II), Phys. Lett. B160: 188–91 (1985); Band, H. R., et al. (MAC), Phys. Lett. B200: 221–24 (1988)
62. Decamp, D., et al. (ALEPH), CERN preprint PPE/90-194; Adeva, B., et al. (L3), L3 preprint 20, Nov. 2, 1990
63. Cabibbo, N., Phys. Rev. Lett. 10: 531–34 (1963)
64. Kobayashi, M., Maskawa, T., Prog. Theor. Phys. 35: 252–82 (1977)
65. Wolfenstein, L., Phys. Rev. Lett. 51: 1945–48 (1984)
66. Fulton, R., et al. (CLEO), Phys. Rev. Lett. 64: 16 (1990); Procario, M., In Heavy Quark Physics, ed. P. S. Drell, D. Rubin, New York: Am. Inst. Phys. (1989), pp. 122–29
67. Albrecht, H., et al. (ARGUS), Phys. Lett. B234: 409–12 (1990)
68. Bortoletto, D., et al. (CLEO), Phys. Rev. Lett. 62: 2436–39 (1989)
69. Albrecht, H. (ARGUS), See Ref. 5
70. Albrecht, H., et al. (ARGUS), Phys. Lett. B209: 119–22 (1988)
71. Bebek, C., et al. (CLEO), Phys. Rev. Lett. 62: 8–11 (1989)
72. Schubert, K., See Ref. 66
73. Inami, T., Lim, C. S., Prog. Theor. Phys. 65: 297–307 (1981)
74. Buras, A. J., Slominski, W., Steger, H., Nucl. Phys. B238: 529–39 (1984); B245: 369–79 (1984)
75. Abe, F., et al. (CDF), Phys. Rev. Lett. 64: 142–45 (1990); Langacker, P., Phys. Rev. Lett. 63: 1920 (1989)
76. Altarelli, G., Franzini, P. J., Z. Phys. C37: 271–81 (1988); Schubert, K. R., Karlsruhe preprint IEKP-KA/88-4, and in Proc. Conf. on Phenomenology in High Energy Physics, Trieste, 1988
77. Grigjanis, R., Navelet, H., Sutherland, M., O'Donnell, P., Phys. Lett. B213: 355–58 (1988); O'Donnell, P., In Proc. Workshop towards Establishing a b Factory, Syracuse, Sept. 1989, p. 1.66; Ali, A., Greub, C., preprint DESY 90-102 (1990)
78. Deshpande, N. G., Trampetic, J., Phys. Rev. Lett. 60: 2583–86 (1988); Deshpande, N. G., In Proc. Workshop towards Establishing a b Factory, Syracuse, Sept. 1989, p. 1.35
79. Bean, A., et al. (CLEO), Phys. Rev. D35: 3533–36 (1987)
80. Gavela, M. B., et al. Phys. Lett. B154: 425–28 (1985)
81. Chau, L.-L., Cheng, H. Y., Phys. Rev. Lett. 59: 958–61 (1987)
82. Avery, P., et al. (CLEO), Phys. Lett. B223: 470–73 (1989)
83. Reidenbach, M. (ARGUS), In EPS Conf. on High Energy Physics, Madrid, Sept. 6, 1989; MacFarlane, D. B. (ARGUS), In SLAC Summer Inst. Particle Physics, SLAC, July 27, 1988
84. Kim, C. S., Rosner, J. L., Yuan, C. P., Phys. Rev. D42: 96–116 (1990)
85. Gilman, F. J., Nir, Y., Annu. Rev. Nucl. Part. Sci. 40: 213–38 (1990)
86. Ellis, J., Fogli, G. L., Phys. Lett. B249: 543–50 (1990)

Annu. Rev. Nucl. Part. Sci. 1991. 41: 29–54

SUBTHRESHOLD PARTICLE PRODUCTION IN HEAVY-ION COLLISIONS

U. Mosel

Institut für Theoretische Physik, Universität Giessen, D-6300 Giessen, Germany

KEY WORDS: transport theory, antiprotons, dileptons, etas, kaons, photons, pions

CONTENTS

1. INTRODUCTION

The properties of nuclear matter are still not very well known (1). Only its saturation point at equilibrium (density $\rho_0 = 0.17$ fm^{-3}, binding energy per particle $E/A = -16$ MeV) is well determined; even the compressibility at equilibrium is only known to lie between 210 (2) and 300 MeV (3). For densities much larger (2–$4\rho_0$) no reliable information exists at all. This is exactly the range of densities that might be reached in the gravitational

29

0163–8998/91/1201–0029$02.00

collapse of massive stars (4). The corresponding stiffness of the nuclear equation of state (EOS) then determines whether a prompt explosion of the supernova can possibly take place. It would thus be highly desirable to know the nuclear equation of state at higher densities.

One of the exciting prospects of heavy-ion research in the energy regime of a few hundred MeV/u up to a few GeV/u is the prospect of obtaining just this information about the properties of dense and hot nuclear matter. Among these properties are the equation of state of nuclear matter and the behavior of nuclear interactions in dense matter that may lead to new phases of nuclear matter, e.g. pion condensates and the excitation of pisobars, that is, strongly coupled πN-Δh states (1). Besides being necessary for an understanding of astrophysical processes (supernova explosions, for example), knowledge of these features is also mandatory for any reliable analysis of the ongoing experimental searches for a quark-gluon plasma, a new state of matter in which the building blocks of nucleons and mesons become deconfined and move in an extended volume (5, 6). Such experiments will always, in the border zones of the reaction volume, lead to "colder," highly compressed nucleonic matter. Its properties must be well understood so that particles emitted from it will not be confused with those coming from the quark-gluon plasma.

The only conceivable laboratory experiments to compress nuclear matter to densities of up to 2–3 times ρ_0 and to temperatures of $T \approx 60$ MeV are heavy-ion experiments in the energy regime indicated above. This energy regime offers a unique window for a study of the phenomena cited in two respects. First, the mean-field effects in this regime have not yet completely disappeared so that experiments here are still sensitive to the density dependence of the mean field and thus to the nucleonic EOS. Second, with these energies one can expect to reach densities that are high enough for an investigation of the questions raised above.

There is, however, a major complication connected with such experiments. Whereas the term "hot and dense nuclear matter," about which we would like to gain information, refers to nuclear matter in equilibrium, the initial configuration in a heavy-ion collision is far from equilibrium. In momentum space it consists of two spheres, well separated by the relative momentum, whereas an equilibrated configuration is described by just one sphere in momentum space. Only the onset of nucleon-nucleon collisions during the course of a heavy-ion collision will (perhaps) create such an equilibrated configuration (7). One thus has to develop probes that are sensitive to specific periods during the time that the reaction develops and to specific phase-space regions. This is the primary motivation for investigating particle production in heavy-ion collisions. Particle-production measurements can yield valuable information complementary to that obtained from other hadronic variables such as sideward flow (8, 9).

Particles produced at bombarding energies (per projectile nucleon) less than their threshold energies in a collision of free nucleons are called subthreshold particles. They are particularly useful for all the purposes described above since their observation indicates that a considerable sharing of energy has taken place.

In the first part of this review, I discuss the theoretical framework for the description of subthreshold production processes. As a test I then briefly discuss the available hard-photon production data in the energy regime from roughly 20 to 100 MeV/u because it is here that many data already exist and that model assumptions can be checked; a more detailed account of these experiments can be found in another recent review (12). I then describe, more speculatively, the reactions at higher energy and discuss the predictions for particle production cross sections there. I pay particular attention to properties of hadrons, nucleons and mesons, in the nuclear medium. Several recent reviews on related topics contain more detailed discussions of specific problems (8, 10–14).

2. THEORY OF PARTICLE PRODUCTION

Two crucial ingredients are needed for a theoretical description of particle production in heavy-ion collisions. The first is a model for the production process itself, the second is a framework for the description of the many-body dynamics. Both are discussed in this section.

2.1 *Particle Production Mechanism*

Some of the first theoretical explanations of pion and gamma production data assumed collective bremsstrahlung as the main mechanism (15). However, it soon became clear in quantitative studies (16–18) that the radiation from a collective deceleration of the projectile in the collision can account for only a small part of the total photon cross section.

Bauer et al (19) were the first to show that the observed photon yield can be understood if one assumes that the radiation seen experimentally is comprised of an incoherent superposition of bremsstrahlung from individual nucleon-nucleon collisions. This picture has now been experimentally justified in the measured source velocities (20).

Crucial for this result was the development of methods to describe this deceleration microscopically without any free parameters; explanations of particle production in terms of collective radiation have the deceleration time as a free parameter, which allows one to adjust the total yield (15, 21). Collective deceleration always proceeds on a longer time-scale than individual nucleon-nucleon collisions; the energy spectra of emitted particles obtained from collective deceleration are thus necessarily more steeply falling than those caused by nucleon-nucleon collisions (12). This does not preclude that in the low energy parts of the spectrum the collective radiation dominates over the incoherent. This is particularly true when the "charge" that represents the source of the particles is large, because collective radiation increases with the square of this charge (16). Indeed, Koch et al (22) showed, for example, that for the emission of energetic photons the collective, coherent bremsstrahlung dominates for photon energies up to about 50–100 MeV when heavy nuclei collide at bombarding energies above about 1 GeV/u. On the other hand, the hard part of the spectrum is dominated by the incoherent radiation from nucleon-nucleon collisions.

It is thus natural to assume that this same mechanism also prevails in the production of other particles, such as mesons, which (because of their finite rest mass) are sensitive to the short-range, high-frequency (hard) part of the source spectrum, just as hard photons are. Indeed, as I discuss in this article, this mechanism works extremely well in comparison with experimental data. I therefore assume from now on that the particle yield seen in experiment is made up of an incoherent superposition of the yields from individual nucleon-nucleon (NN) collisions.

In a model that ascribes the observed yield of particles produced in a heavy-ion collision to individual NN collisions, we have to know the collision rates and the total available energy in each NN collision as well as the positions and momenta of all other nucleons so that the Pauli blocking of the final states of the two colliding nucleons, after they have produced a particle, can be evaluated. Both types of information are contained in the one-body phase-space distribution.

If this phase-space distribution, $f(\mathbf{x}, \mathbf{p}, t)$, is known then the particle yield from a collision of two nucleons with initial momenta $\mathbf{p}_1$ and $\mathbf{p}_2$ and final momenta $\mathbf{p}_3$ and $\mathbf{p}_4$ is given by the production probability corresponding to the initially available energy $\sqrt{s}$, summed over all NN collisions and integrated over the final momentum distribution. In addition, there is an integration over the heavy-ion impact parameter b. One thus obtains the following expression (12) for the production cross section for a particle y:

$$\frac{d^2\sigma^y}{dE_y \, d\Omega_y} = 2\pi \int b \, db \sum_{\text{NN-coll}} \int \frac{d\Omega_{34}}{4\pi} \frac{k_y}{k_y'} \frac{d^2 P^y(\sqrt{s})}{dE_y' \, d\Omega_y'}$$

$$\times \left[1 - f(\mathbf{x}, \mathbf{p}_3, t)\right]\left[1 - f(\mathbf{x}, \mathbf{p}_4, t)\right]. \quad 1.$$

Although not shown here explicitly, f is also dependent on the impact parameter b, its arguments $\mathbf{x}$ and t are the space-time coordinates of each

NN collision. P is the particle production probability, obtained by dividing the particle production cross section by the total nucleon-nucleon cross section; P is given in the NN center-of-mass system and is then boosted to some common frame, for example the nucleus-nucleus center-of-mass system. The last factors $(1-f)$, finally, contain the final-state Pauli blocking since the NN collisions take place in the nuclear medium. The blocking of the intermediate scattering states is contained in the in-medium production rate P, which may thus differ from that in free space.

According to Equation 1, there is no feedback from the particle production to the phase-space distribution function f. This perturbative treatment is justified only if the particle production probability is so small that the average time development of the nucleus-nucleus collision is not affected by particle emission. This is the case at bombarding energies in the subthreshold domain, well below the free particle production threshold.

2.2   Nucleus-Nucleus Dynamics

A crucial input in Equation 1 is the phase-space distribution $f(\mathbf{x}, \mathbf{p}, t)$. For a stationary nucleus in its ground state, it is given in a semiclassical description by the spatial density, multiplied by a space-dependent momentum distribution that reaches out to the local Fermi momentum. What is needed now is a theory to propagate f in time.

Such a theory can be derived from the equation of motion for the single-particle density matrix by cutting the density matrix hierarchy at the two-body level, that is by writing the two-particle density matrix as an antisymmetrized product of one-particle density matrices plus a two-body correlation function. Whereas the former leads to a theory describing particles that move independently in a time-dependent mean field (time-dependent Hartree-Fock field), the latter term (the correlation function) gives rise to explicit two-body correlations. If one performs a semiclassical approximation in which all wave functions are locally replaced by plane waves, one arrives at a Boltzmann-like equation. Such a formal derivation has, for example, been given in (23, 24). In particular, Cassing et al (25) showed how on this basis a time-dependent G-matrix theory can be derived.

Since this formal derivation is well documented in the literature (12) and was described extensively in a recent review (14), in the following few paragraphs I give only a heuristic derivation of the transport equation used to calculate the time dependence of the phase-space distribution function $f(\mathbf{x}, \mathbf{p}, t)$ (see also 11).

I thus assume that the nucleons can be described as moving in a mean-field potential and experiencing direct nucleon-nucleon collisions in addition. The time development of f is then determined by Liouville's theorem stating that the total time derivative of f is equal to a loss and gain term arising from nucleon-nucleon collisions, $C[f]$. One thus has

$$\frac{df(\mathbf{x}, \mathbf{p}, t)}{dt} = C[f], \qquad\qquad 2.$$

where the collision term $C[f]$ is obviously a functional of f. The total time derivative is to be taken for a phase-space cell that moves in the mean-field potential $U(\mathbf{x}, \mathbf{p}, t) = U[f(\mathbf{x}, \mathbf{p}, t)]$. This gives

$$\frac{df(\mathbf{x}, \mathbf{p}, t)}{dt} = \frac{\partial f(\mathbf{x}, \mathbf{p}, t)}{\partial t} + \frac{\partial f(\mathbf{x}, \mathbf{p}, t)}{\partial x}\left(\frac{p}{m} + \frac{\partial U}{\partial p}\right) - \frac{\partial f(\mathbf{x}, \mathbf{p}, t)}{\partial p}\frac{\partial U}{\partial x}. \qquad 3.$$

The collision term finally is given by

$$C[f] = -\frac{4}{(2\pi)^3}\int d^3 p_2 \iint d^3 p_3\, d\Omega_{34}\left\{ f(\mathbf{x}, \mathbf{p}, t)f(\mathbf{x}, \mathbf{p}_2, t)|\mathbf{v}_{12}|\frac{d\sigma}{d\Omega}\right.$$

$$\left. \times\, \delta^3(\mathbf{p}+\mathbf{p}_2-\mathbf{p}_3-\mathbf{p}_4)[1 - f(\mathbf{x}, \mathbf{p}_3, t)][1 - f(\mathbf{x}, \mathbf{p}_4, t)] - \text{gain term}\right\}. \qquad 4.$$

The structure of this collision term is quite similar to that of Equation 1. Notice again the in-medium correction in the form of the Pauli-blocking factors $(1-f_3)(1-f_4)$ for the final states.

Equations 2–4 together constitute the so-called Boltzmann-Uehling-Uhlenbeck (BUU) equation. In it the phase-space distribution function f is a semiclassical quantity averaged over quantal oscillations. In the so-called Quantum Molecular Dynamics model (26, 27), one approximates the phase-space distribution function $f(\mathbf{x}, \mathbf{p}, t)$ by Gaussians centered at the actual phase-space locations of the nucleons.

It is intuitively clear and the formal derivations also show that both the mean-field term and the collision cross section come from the same in-medium nucleon-nucleon interaction (23, 25). Since f and thus the degree of Pauli blocking change with time, a consistent calculation would require the simultaneous solution of the transport equation just given and a coupled G-matrix calculation, that is, a time-dependent G-matrix calculation (28).

Equations 3 and 4 are solved by a test particle method. In this method, many simulations of the nucleus-nucleus reaction with randomly chosen initial conditions are run in parallel and the phase-space density is obtained by averaging f in each time step over the full ensemble of parallel runs (for the details, see 11, 12). While the test particles move according to relativistic kinematics, the whole formalism as outlined so far is clearly not covariant. This deficiency is overcome by ad hoc techniques such as evaluating potential energies only in local rest frames, but the error inherent in such prescriptions is not well understood.

It is thus tempting to develop a manifestly covariant transport theory for nucleus-nucleus collisions in the GeV/u energy regime. The practical implementation of such a theory has indeed been achieved by Blättel et al (29, 30); see also the work by Ko and collaborators (31). Blättel et al based their work on earlier theoretical work by Gyulassy and collaborators (32). In this method, one starts from an effective, covariant Lagrangian (the Walecka model) and derives a transport equation for the Lorentz scalar part of the relativistic Wigner matrix; the Lorentz vector part can then be obtained by the usual local density approximation. For details of the formalism as well as of the results, see the review by Cassing & Mosel (13) as well as (29–31, 33).

For higher bombarding energies, where pions and delta isobars can be copiously produced, the perturbative method of calculating particle production is no longer appropriate because, for example, the emission of pions constitutes a major source of cooling for the nucleus-nucleus system. In this regime one thus has to generalize the transport equation just described to a coupled system of transport equations that include nucleons, pions, deltas, and possibly other resonances. Only the heavier mesons and baryons (that are clearly still subthreshold at the given bombarding energy) can be treated perturbatively. Such a coupled transport equation method was recently developed by Wolf et al (34) and Ko and coworkers (35); a covariant version was proposed by Wang et al (36).

When evaluating the cross sections according to Equation 1 or the collision term in Equation 4, one assumes that energy and momentum of the colliding particles are connected by the classical dispersion relation for free particles in a mean-field potential U. This allows the nucleons to pick up energy from the Fermi motion in a nucleon-nucleon collision. For proton-induced particle production reactions, this mechanism was discussed by McMillan & Teller (37) as early as 1947; for heavy-ion collisions it was used again by Bertsch (38) in 1977.

For heavy-ion collisions, the on-shell assumption finds its justification in detailed studies of the energy-momentum relation performed some years ago by Cassing (39). There it was found that for bombarding energies above about 30 MeV/u even particles in the momentum tails of one nucleus are put on the energy shell when, during the nucleus-nucleus collision, the potential wall between the two nuclei collapses and thereby allows these particles to move freely in the potential of the other nucleus.

Figure 1 shows the production thresholds for various particles as a function of the Fermi momentum p_F both for nucleus-nucleus and nucleon-nucleus reactions. It is seen that, in heavy-ion collisions in particular, pions can be produced quite easily far below their free nucleon-nucleon threshold of 290 MeV in the laboratory frame of reference.

thresholds

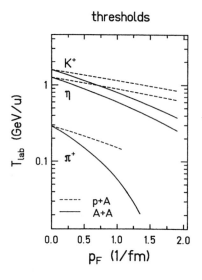

Figure 1 Thresholds for production of π^+, η, and K^+ mesons in primary nucleon-nucleon collisions for proton-nucleus (*dashed lines*) and nucleus-nucleus (*solid lines*) collisions as a function of Fermi momentum p_F.

2.3   *Elementary Cross Sections*

To evaluate the heavy-ion-induced particle production cross sections, one must know the elementary elastic and inelastic NN cross sections. As discussed in the introduction, all these quantities have to be known in the nuclear medium, the properties of which change during the course of the collision.

Only a few studies of the influence of density and temperature on nucleon-nucleon scattering cross sections exist so far (40–42). The usual assumption in these studies is that the NN interaction takes place in homogeneous nuclear matter at rest, in which the momentum-space configuration is that of a single Fermi sphere. The main difficulty here is that the self-energies of the mesons that mediate the nuclear force may themselves change in the nuclear medium (1). There are speculations that the pion mass should go to zero with increasing density; this has profound implications for the nucleon-nucleon cross sections (45, 46).

A one-sphere momentum-space configuration is appropriate only for the final, possibly equilibrated, state of a heavy-ion collision. Only recently were studies begun that work with a momentum-space configuration of two overlapping, shifted Fermi spheres. Faessler and collaborators (42–44) work with a frozen-configuration assumption in which the NN interaction is evaluated for various fixed densities and relative momenta of the Fermi spheres. The cross section to be used during the collision then depends on the instantaneous density and momentum-space configuration.

In view of the very large uncertainties on the role of medium corrections for the NN cross sections, in most analyses so far free cross sections have been used. Such a simplifying assumption is not unreasonable during the early stages of a collision at high bombarding energy, where the initial phase-space configuration consists of two well-separated Fermi spheres and the Pauli-blocked part of the total available phase-space is relatively small. Unfortunately, even these free cross sections are generally not measured in the full kinematical regime in which they are needed for the calculations. An additional complication arises because the subthreshold production of heavy particles proceeds largely through collisions of nucleon resonances, the cross sections of which are mostly totally unknown (discussed below).

One is thus forced to model all of these cross sections. In the following short sections I briefly discuss the models. A more detailed description can be found in (12).

2.3.1 PION PRODUCTION   For pion production in nucleon-nucleon collisions quite reasonable parametrizations guided by boson-exchange models exist. VerWest & Arndt (47) parametrized the NN → NNπ and the NN → NΔ cross sections; from the latter the cross section for ΔN → NN, a typical in-medium process, can be obtained by detailed balance. These parametrizations are used, for example, in the work of Wolf et al (34).

Bauer (48) demonstrated that in subthreshold production of pions the Δ resonance plays an essential role and hardens the spectrum considerably. This, as a side effect, opens up the possibility of studying the properties of the Δ resonance in dense nuclear matter.

2.3.2 ETA PRODUCTION   Eta (η) mesons have a large decay width of 39% for decay into two photons, so that high resolution photon spectrometers such as the new TAPS apparatus at the heavy-ion synchrotron (Schwer-Ionen Synchrotron, SIS) can be used for their detection. They are predominantly produced through the N*(1535) resonance, because their direct coupling to the nucleon is close to zero (49).

Exploiting this resonance production mechanism and using information on the η coupling constants obtained from earlier coupled channel analyses of the $\pi N \to \eta N$ reaction (49, 50), calculations of the η production cross section have recently been performed (51, 52). In the results of these calculations, both based on an effective one-boson-exchange (OBE) model, the influence of the resonance is clearly visible.

Since the N*(1535) resonance also acts as an absorber for η's initially produced, the η spectra may yield information on the properties of this resonance in dense nuclear matter. The absorption cross section can be calculated on the basis of a Breit-Wigner fit to the amplitude (12). Because the N*(1535) is an S_{11} resonance, there is no centrifugal barrier so at low η energies the absorption cross section diverges. On resonance ($T_\eta = 76$ MeV), it is still around 40 mb, in agreement with the coupled channel analysis of (49); the corresponding mean free path is only 1–2 fm at ρ_0. At higher energies the cross section seems to level off (53).

2.3.3   KAONS AND ANTIPROTONS   For kaons the total production cross sections are not well known close to threshold (54), for antiprotons there are no data below about 11 GeV bombarding energy, and for both particles no spectral information is available. To obtain the spectra of these particles, one can follow the phase-space arguments of Randrup & Ko for meson production (55) and assume that the spectra are determined by the final-state phase space. For antiproton production this requires the additional assumption that the production proceeds primarily to a four-particle final state ($p + p \to p\bar{p}pp$). As I discuss below, the lack of total cross-section data for $\bar{p}$ production in NN collisions close to threshold leads to order-of-magnitude uncertainties in the total subthreshold production cross section in heavy-ion reactions.

2.3.4   PHOTONS AND DILEPTONS   As discussed above, hard photons can also be viewed as subthreshold particles if their energy exceeds about one half of the beam energy per nucleon, the maximum energy possible in a collision of two free nucleons with this energy.

First studies of high energy photons used a semiclassical approximation for the nucleon-nucleon photon cross section obtained by considering the hard-sphere scattering of two classical charges (56, 57). In this model the radiation from pp collisions is suppressed relative to that from pn collisions (57); the suppression factor is about 20 at 200 MeV, and it still amounts to about 5 at 1 GeV (58).

Until very recently, only two sets of data were available for the pnγ process. Thus the needed cross sections had to be calculated. Schäfer et al (58), using an effective one-boson-exchange model, performed microscopic calculations by evaluating the fully relativistic Feynman diagrams for this process; the calculation is thus gauge invariant and current conserving (see also 59). The results of such a calculation agree very well with a very recent measurement of the pnγ cross section at 170 MeV by Pinston & Nifenecker (58, 60).

As discussed in detail by Cassing et al (12), real photons are difficult to observe in heavy-ion collisions at energies above about 300–400 MeV/u. A natural extension of electromagnetic probes to this energy regime is then to investigate the emission of correlated e^+e^- pairs, dileptons, effectively the emission of massive, timelike photons.

In the last few years data have become available on dilepton production in heavy-ion and proton-induced reactions (61). Experiments are also under way to measure the elementary pn-dilepton cross section (G. Roche, private communication, 1990) in the energy regime up to 5 GeV, for which at present no data exist.

Initial calculations of this dilepton yield were based on the so-called soft-photon approximation, which assumes that the dilepton production cross section can be factorized into a product of the nucleon-nucleon cross section and a term that describes the emission of the photon and the following pair creation (63). Schäfer et al (64) performed a calculation based on an effective OBE model that was also used for the calculation of production cross sections for real photons (58). They found (12) that the calculated spectra could indeed be quite well represented by the long-wavelength approximation if a correction for the final-state phase space of the nucleons (65) was applied. Similar results have also been obtained by Haglin et al (65).

In none of these calculations was a form factor for the electromagnetic nucleon-photon vertex used. The electromagnetic form factor of the nucleon is very well known for spacelike photons from elastic electron scattering experiments; in the timelike sector, however, there are only data for $q^2 > 4$ GeV^2 from p̄p annihilation. The whole momentum range between 0 and about 2 GeV is still terra incognita!

The Breit form factor of the proton for spacelike photons ($q^2 < 0$) follows very well the so-called dipole fit (66), which exhibits a pole close to the ρ-meson mass. This—and other evidence (67)—has been used to suggest that the coupling of the nucleon to the photon proceeds through a virtual ρ meson (vector-meson dominance).

Based on the chiral quark-bag model, Brown et al (68) give an expression for the electric form factor of the proton:

$$G_E^p(t) = \frac{1}{2}[F_{\text{core}}(t) + F_{\text{cloud}}(t)] + \frac{t}{4M_p^2} 3.63 \frac{1}{2} F_{\text{cloud}}(t). \qquad 5.$$

Here F_{core} is the form factor of the quark core of the nucleon, whereas F_{cloud} represents that of the vector-meson cloud to which the photon may couple. The model of Brown et al assumes that half of the total charge resides in the quark core, the other half in the meson cloud; thus the factor 1/2 in front of both the form factors. The factor 3.63 on the right-hand side of this equation is the sum of the isoscalar and two times the isovector anomalous magnetic moments of the nucleon. The two form factors $F_{\text{core}}(t)$ and $F_{\text{cloud}}(t)$, finally, are given by

$$F_{\text{core}}(t) = \frac{\lambda^2}{\lambda^2 - t} \qquad F_{\text{cloud}}(t) = \frac{m_v^2}{m_v^2 - t} \frac{2m_v^2}{2m_v^2 - t}. \qquad 6.$$

Here $\lambda \simeq 0.9$ GeV and $m_v \simeq 0.65$ GeV (68).

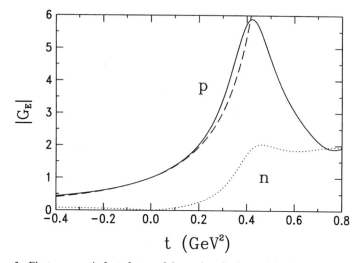

Figure 2   Electromagnetic form factor of the nucleon in the model of Brown et al (68). The dipole fit is shown by the dotted curve. The electric form factor of the proton is represented by the solid line, that of the neutron by the dashed one. In the calculation of these curves, reasonable widths have been added to the masses in the denominators of the form factors.

This form factor is shown in Figure 2 by the solid line. In the spacelike sector it agrees very well with the dipole fit; in the timelike sector it shows a clear resonance enhancement that reaches a maximum value of 6, very similar to the experimentally known electromagnetic form factor of the pion (66).

In this model the neutron's electric form factor is given by

$$G_E^m(t) = -\frac{t}{4M_n^2} 3.82 \frac{1}{2} F_{cloud}(t). \qquad 7.$$

In the spacelike sector, the neutron's electric form factor is about one order of magnitude smaller than that of the proton, but in the timelike sector it reaches a value of about 2 (dashed line in Figure 2). Its amplitude is thus $\sim 1/3$ that of the proton and causes a nonnegligible interference of the radiation from the proton and the neutron. This effect has not been taken into account in any of the calculations so far.

The dependence of the electromagnetic form factor on the vector-meson properties could conceivably permit the study of the properties of the vector mesons in compressed, hot nuclear matter. It seems quite plausible that the meson cloud contribution to the form factor, $F_{cloud}(t)$, is affected by the surrounding nuclear medium and that it is quenched at higher densities. A similar quenching of the axial form factor, even at normal nuclear densities, is experimentally well established (1). A very similar argument applies to $\pi^+\pi^-$ annihilation into dileptons. This process is experimentally known to proceed through a ρ meson (66, 67).

For a conclusive interpretation of heavy-ion data on dilepton production, it is therefore essential to determine the electromagnetic form factor of the nucleon in the timelike region, first in p + n and then in p + A reactions at energies of up to 2 GeV, so that the light vector-meson mass region is covered.

3. HARD-PARTICLE PRODUCTION IN HEAVY-ION REACTIONS

3.1 *Hard Photons*

The methods described above were first applied to the calculation of photon spectra produced in relatively low energy heavy-ion collisions (19, see also 69). At bombarding energies of a few tens of MeV/u photons with energies up to 150 MeV were observed, i.e. with energies clearly above those that could possibly be created in free NN collisions at the same bombarding energy (in this sense, these photons are also "subthreshold" particles). Data and theoretical results are reviewed elsewhere (12). The spectra are generally quite well reproduced; the calculated results show some sensitivity to the elementary cross section used (70).

In Figures 3 and 4 I therefore show only a summary of comparisons of the calculations with the data. Figure 3 compares the empirical probability per nucleon-nucleon collision to emit a γ ray, and Figure 4 shows the slope parameter of the spectra versus a Coulomb-barrier-corrected bombarding energy. In extracting the γ probabilities, a model expression given by Nifenecker & Pinston (71) for the number of neutron-proton collisions, averaged over impact parameter, has been used. There is an indication of a somewhat too weak dependence of the slope parameter on bombarding energy, but the overall systematic agreement between theory and experiment is quite good.

This systematic agreement, together with the measured source velocities (20), allows us to conclude that the mechanism for γ emission in heavy-ion collisions is indeed understood: the observed photons are due to an incoherent superposition of nucleon-nucleon bremsstrahlung.

Now that the mechanism has been clarified, one can use these particles as probes for phase-space distributions (72). Photons with the highest energies observed are produced in nucleon-nucleon collisions from the two opposing pole caps of the momentum spheres of the two nuclei.

Photons have a great advantage over hadronic probes in that they suffer essentially no final-state interaction. They would thus be ideal probes at

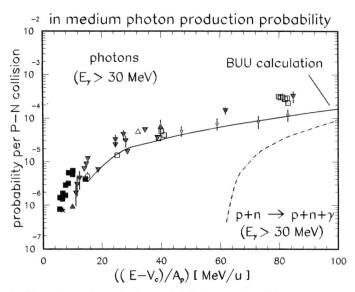

Figure 3 Comparison of γ production probabilities per NN collision extracted from data in comparison with the results of the BUU calculation (from 12).

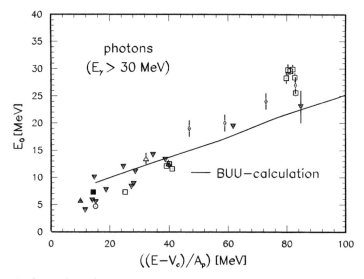

Figure 4 Comparison of empirical slope parameters of photon spectra with results of the BUU calculation (from 12).

higher bombarding energies also, where they could give an undistorted picture of the high density, high excitation phase of the collision. Unfortunately, existing γ spectrometers cannot distinguish these direct photons from those originating in the π^0 decay if the π^0 yield becomes too large. This occurs at bombarding energies above about 300 MeV/u (12).

3.2   Pion Production

Pions have a finite rest mass and thus can be produced only if the available energy per nucleon-nucleon collision exceeds roughly 140 MeV in the reaction volume. Subthreshold pions are produced at bombarding energies below 290 MeV/u in the laboratory frame of reference. It is amusing to recall that the first laboratory experiment ever to produce a pion was run at a subthreshold energy; the reaction used was ^{4}He$+$^{12}C at 95 MeV/u at the 184-inch cyclotron at Berkeley. More recently, subthreshold pion production in heavy-ion collisions was started again in the pioneering work of Benenson and collaborators (73) and, at lower energies (60–84 MeV/u), by Grosse and coworkers (74). These data were well summarized by Braun-Munzinger & Stachel (10).

The first calculations for these energies used the phase-space prescription mentioned in Section 2.3.1 to generate the final-state angular and momentum distributions. The total cross sections obtained were nearly correct (12), but the spectra were considerably too soft. Bauer (48) showed that the spectra could be considerably hardened by recognizing that pions are produced primarily through the Δ resonance; nevertheless, at the lowest bombarding energies the spectra were still significantly too steep.

A solution to this problem was recently found by Niita (75). He showed that the pion spectra in this energy regime are extremely sensitive to the tails of the momentum distribution. By modeling the momentum distributions of the two interacting nuclei very carefully down to their outmost tails, and by assuming that these high momentum tails are lifted on-shell during the collision, Niita was able to reproduce the measured pion spectra very well. This is illustrated in Figure 5, which shows both a calculation without (dashed line) and one with (full line) high-momentum tails. The agreement is obviously perfect.

While the slope parameters obtained in the calculation without the momentum tails fall off continuously at the lower bombarding energies, those that do include the tails stay constant for energies below about 100 MeV. The same effect is evident in the data (76, 77). While this behavior has been attributed before to some collective effect (76), it finds a natural explanation here. At very low bombarding energies, only the few-tail nucleons have enough energy to produce a pion so that they determine the

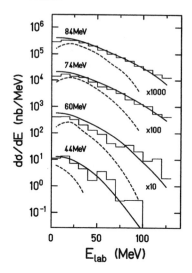

Figure 5   Energy spectra for π^0's produced in ^{12}C + ^{12}C collisions at 60, 74, and 84 MeV/u and in a ^{40}Ca + ^{40}Ca collision at 44 MeV/u. Solid lines include the high-momentum tails, dashed lines do not. Data are indicated by the histogram (from 75).

cross section. With increasing bombarding energy, the many nucleons in the bulk of the momentum distribution acquire enough energy so that the bombarding energy becomes the determining factor. There are indications for exactly the same process in hard-photon production, as recently pointed out by Metag (78).

Unfortunately the mean free path of slow pions in nuclei is not very well known (79); theoretical considerations indicate a very strong dependence of the pion mean free path on the pion energy [see Figure 4.7 in (12)]. In addition, its effects on heavy-ion-induced pion production data have to be calculated in a nuclear environment varying rapidly with time. This introduces a rather large uncertainty into quantitative predictions of pion yields in heavy-ion collisions, which thus makes pions a much less desirable probe than strange mesons or photons.

3.3   Eta Production

At higher bombarding energies in the GeV/u regime, heavier mesons can be produced subthreshold. For example, the η meson has a free threshold of about 1.2 GeV, whereas kaons are produced in NN collisions only at about 1.5 GeV bombarding energy. The kaon threshold is higher because of strangeness conservation, which requires the simultaneous production of a hyperon together with the kaon.

Eta production cross sections for heavy-ion collisions have been calculated by Schürmann & Zwermann (80) and DePaoli et al (81). Total cross sections of 1–10 mb at 1 GeV/u have been predicted; a summarizing discussion can be found in (12). Experiments to measure these cross sections at SIS were recently begun by the TAPS collaboration.

As discussed in Section 2.3.2, η's couple strongly to the N*(1535) resonance; η's with a kinetic energy of only 76 MeV in the nucleon rest frame can form this resonance when they collide with a nucleon. Since the mean full width of the resonance is about 150 MeV, all η's with kinetic energies up to about 150 MeV can easily be absorbed. Since the branching ratio of the N*(1535) into the η channel is about 50%, roughly one half of all η's absorbed will reemerge on the average.

As also discussed in Section 2.3.2 the mean free path of the η below the resonance region is very small [of the order of 0.5–2 fm at equilibrium density (12)]; η's are thus strongly reabsorbed and rescattered so that most of those seen experimentally will have been produced in the nuclear surface, i.e. at rather low densities. At bombarding energies below about 0.7 GeV, the η's can—because of energy limitations—only be produced in collisions between nucleons that have already undergone a number of collisions and thus acquired enough energy (12). At 500 MeV/u most of the nucleons producing an η have on average undergone 3–4 previous collisions; some nucleons have even had 8 prior collisions. Since about 3–5 collisions are needed to thermalize a system, at this low bombarding energy the η's are primarily produced in a hot, nearly equilibrated collision zone. Since the η production rate closely tracks the density compression ($\approx 2\rho_0$ at this energy), these particles are good probes for compressed, nearly thermalized nuclear matter, in contrast to pions, most of which are created in the first collisions when target and projectile nucleons are still well separated in momentum space (12).

The measurement of η spectra at energies below about 800 MeV/u could thus yield information on the properties of the N*(1535) resonance in hot nuclear matter. On the experimental side this would require high current heavy-ion accelerators; on the theoretical side methods are urgently needed for a quantitative treatment of the reabsorption and rescattering effects on the η's produced.

3.4 *Kaon Production*

As mentioned above, kaons show features qualitatively similar to those of η's because of their rather high mass; at sufficiently low energies they are produced predominantly by nucleons that have already undergone many collisions.

Positively charged kaons, however, have a great advantage over η's: they have a rather large mean free path of about 6–8 fm at ρ_0 so that— especially in collisions of light nuclei—K^+ reabsorption plays only a minor role. K^+ mesons then have the same advantage as electromagnetic probes;

they give information about the compressed state largely free of final-state interactions.

The main interest in a study of the K^+ production stems from this feature. Many authors have discussed the production processes; for references to theory and data, see (12). In all of these studies it has proved essential to include not only the production of K^+ in NN collisions, but also in $N\Delta$, $\Delta\Delta$, πN, and $\pi\Delta$ secondary reactions. At an energy of about 800 MeV/u to 1 GeV/u the elementary cross section increases rapidly with energy and therefore small changes in the available kinetic energy will cause large changes in the kaon production in heavy-ion collisions. Based on this argument, Aichelin & Ko (82) predicted that the K^+ yield should be a particularly good tool for studying the nuclear equation of state; the calculated yields change roughly by a factor of 2 when the equation of state is varied from a hard to a soft one (12). Quantitatively reliable results, however, still suffer from the fact that momentum-dependent effects in the mean-field potential have so far not been included (27). Since these will also tend to take energy out of the system and since the effects of a momentum dependence on the interaction vertices are totally unknown, their effect on the K^+ production cross sections can easily be of the same order of magnitude as that of the EOS.

3.5   *Antiproton Production*

Antiproton production at energies of a few GeV/u is clearly the most extreme subthreshold process of all particle production processes. Nevertheless, it was first observed in proton-nucleus collisions more than 25 years ago (83). Recently, experiments performed at JINR (84) and at the Bevalac (85–87) have produced the first measurements of subthreshold antiproton production in nucleus-nucleus collisions. Various descriptions for these data have been proposed. Based on thermal models it was suggested that the antiproton yield contains large contributions from $\Delta N \rightarrow \bar{p}+X$, $\Delta\Delta \rightarrow \bar{p}+X$, and $\rho\rho \rightarrow \bar{p}N$ production mechanisms (88–90). Other models have attempted to explain these data in terms of multiparticle interactions (91).

The BUU method is able to describe the full transition from the early, preequilibrium stages of the collision to a stopped, possibly equilibrated configuration. Batko et al (92) therefore recently studied the problem of antiproton production in this framework by evaluating the cross section as

$$E_{\bar{p}}\frac{d^3N(b)}{d^3p_{\bar{p}}} = \sum_{\text{BBcoll}} \int d^3p_3' \, d^3p_4' \, d^3p_5' \frac{1}{\sigma_{\text{BB}}(\sqrt{s})} E_{\bar{p}}' \frac{d^{12}\sigma_{\text{BB}\rightarrow\bar{p}}(\sqrt{s})}{d^3p_{\bar{p}}' \, d^3p_3' \, d^3p_4' \, d^3p_5'}$$
$$\times [1-f(\mathbf{x},\mathbf{p}_3';t)][1-f\mathbf{x},\mathbf{p}_4';t)][1-f(\mathbf{x},\mathbf{p}_5';t)]. \quad 8.$$

Note here the triple Pauli blocking of the final state. The antiproton cross section is obtained by integrating the differential multiplicity over the impact parameter b.

The differential elementary antiproton cross section is taken to be proportional to the final-state phase space. Unfortunately, not even the total $BB \to \bar{p}$ cross section is known experimentally at low energies ($\sqrt{s} < 4M_p + 1$ GeV). This uncertainty in the elementary cross section causes a rather large uncertainty in the heavy-ion-induced cross section. As an example I show in Figure 6 the spectrum of antiprotons for the reaction $^{28}Si + ^{28}Si$ at 2.1 GeV/u, calculated by Batko et al, neglecting reabsorption (92). The two dotted lines give the predictions obtained with two different extrapolations of the elementary $\bar{p}$ production cross section

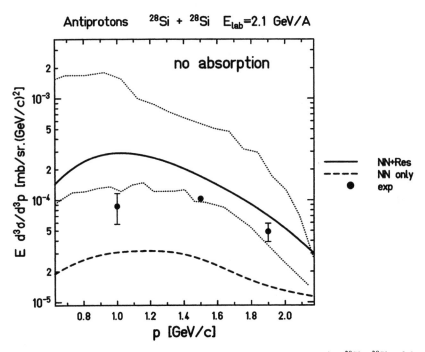

Figure 6 Invariant cross section for antiproton production in the reaction $^{28}Si + ^{28}Si$ at 2.1 GeV/u for $\theta = 0$ as a function of the laboratory momentum **p**. Data are taken from (84). The dashed line contains results obtained by including only collisions between nucleons, whereas the solid line also contains the contribution from collisions involving resonances. The two dotted lines correspond to two different extrapolations of the elementary $\bar{p}$ cross section down to low energies (from 85).

down toward the threshold; the uncertainty amounts to roughly one order of magnitude and is thus comparable to that arising from the neglect of final-state reabsorption in this calculation.

Nevertheless, the overall magnitude of the experimental cross section is quite well reproduced. Essential for this significant improvement over the results of Shor et al (87) is the inclusion of multiple collisions and of collisions between nuclear resonances.

3.6 *Dileptons*

As mentioned in Section 2.3.4 and discussed in detail by Cassing et al (12), in the energy range above about 300 MeV/u, direct, real photons can probably not be seen with current experimental equipment, but the spectroscopy of virtual, massive photons offers an interesting alternative.

Data on dilepton production in the energy regime of about 1 up to 5 GeV/u, obtained at the Bevalac with the dilepton spectrometer (DLS) have been available for about two years (61, 100, 101). While first theoretical descriptions of dilepton production were based on thermal models (63, 93, 94), Ko and coworkers (35, 95) and Wolf et al (34, 96, 97) have now also performed BUU calculations; the calculations take pions and nucleon resonances explicitly into account. This treatment gives an excellent description of measured proton spectra, and it agrees within about 20% with the measured pion yields (34). The model thus seems to give a realistic simulation of nucleon and pion dynamics during a heavy-ion collision in the energy range between 400 MeV/u and 2 GeV/u. The calculations also include dileptons produced by $\pi^+\pi^-$ annihilation, neutron-proton and π-nucleon bremsstrahlung, as well as those coming from the Dalitz decays of the Δ ($\Delta \rightarrow N\gamma^* \rightarrow Ne^+e^-$) and N*(1440). For the comparison with the experimental data, the experimental acceptance filter has been used.

In Figure 7 I show the results of calculations made by Wolf et al for the reaction $^{40}Ca + {}^{40}Ca$ at 1 and at 2 GeV/u. The various contributions are clearly shown; bremsstrahlung and Δ Dalitz decay contribute about equally to the total yield, which agrees reasonably well with experiment. The spectrum at 1 GeV/u, in particular, shows a dominant contribution of $\pi^+\pi^-$ annihilation for invariant masses larger than about 500 MeV. The annihilation signal is less pronounced at the higher energy because the bremsstrahlung spectrum hardens there and partly masks the annihilation component; also, going to a heavier system does not increase the relative importance of the pion annihilation signal (34, 97).

It will therefore not be easy to detect evidence for a nontrivial pion dispersion relation in the $\pi^-\pi^+$ events, i.e. a shift of the annihilation threshold to lower invariant masses, or an increase by about a factor of 3 of the annihilation component close to threshold (98, 99) [note that the

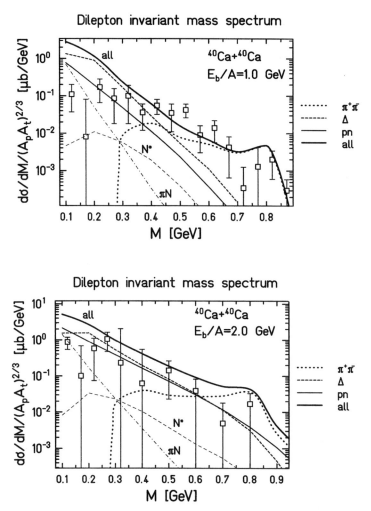

Figure 7 Invariant mass spectrum of dileptons for the reaction $^{40}Ca + {}^{40}Ca$ at 1 and 2 GeV/u bombarding energy. The solid line gives the contribution of massive photons produced in pn bremsstrahlung, the short-dashed curve that of the Δ Dalitz decay ($\Delta \rightarrow N\gamma^* \rightarrow Ne^+e^-$). The dotted curve shows the contribution from $\pi^+\pi^-$ annihilation and the dash-dotted curve that of πN bremsstrahlung (from 34).

predicted effects are now much smaller than those originally expected (63, 90, 93)]. At 1 Gev/u bombarding energy, the annihilation contribution near threshold is much smaller than the bremsstrahlung component so that the threshold shift and the increase at threshold will probably not be observable.

More promising seems to be a detailed, high statistics study of the high energy end of the $\pi^+\pi^-$ annihilation component. Since pion annihilation in vacuo is well known to proceed through the ρ meson [vector-meson dominance (66)], a detailed determination of the peak position around 700–800 MeV in the mass spectrum could give information on the properties of the ρ in dense nuclear matter. This is particularly interesting since there exist theoretical predictions, based on quark models, that call for a significant drop of the masses of pseudoscalar mesons with increasing density (1, 102). Also the in-medium properties of the Δ may become accessible by studying its electromagnetic Dalitz decay, which makes an important contribution to the measured dilepton spectrum.

All the calculations discussed up to now have been performed without using electromagnetic form factors for the nucleons (recall the discussion in Section 2.3.4). In order to see the possible effect of such form factors Wolf et al have performed the calculation for p + ^{9}Be with an electromagnetic form factor for the baryons of the form of Equation 5. Whereas this has virtually no effect at 1 GeV, it raises the cross section for 2.1 GeV above the data by up to a factor of three at $M \simeq 700$ MeV (see Figure 8). For the ^{40}Ca + ^{40}Ca reaction, for which the calculations already give quite high cross sections, the inclusion of the form factor raises the calculations

Effect of the formfactor

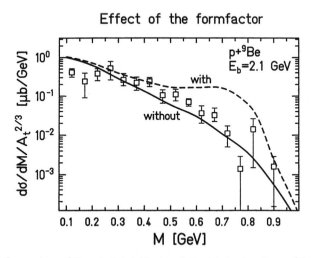

Figure 8 Comparison of the calculated dilepton yield with the data for p + ^{9}Be at 2.1 GeV (100). Shown are the results of a calculation without (*solid line*) and with (*dashed curve*) an electromagnetic form factor for the protons obtained from the vector-meson dominance model (from 34).

above the data by up to a factor of 20 at $M \simeq 800$ MeV. Based on the reasonable agreement obtained in the calculation without form factors, Wolf et al (34) speculated that the ρ meson could be suppressed in dense nuclear matter; subsequently several speculations have appeared in the literature about the modification of the ρ meson in the nuclear medium (102, 103). A final conclusion on this point has to wait, however, until the effects of the electromagnetic form factors of the nucleon, discussed in Section 2.3.4, on the elementary dilepton spectra have been taken into account.

4. SUMMARY

The existing data on hard-photon production in the energy range up to about 100 MeV/u, as well as measurements of source velocities, can all be understood in a model that assumes an incoherent superposition of bremsstrahlung emitted in proton-neutron collisions. This same mechanism has thus been assumed to be applicable to the production of other particles as well. I have shown that pion yields in this energy range can be described by the same mechanism; at the lowest energies the pion yields become sensitive to the details of the momentum distribution.

At the higher energies, around 1 GeV/u, the most promising probes are either dileptons, because of the absence of final-state interactions, or alternatively, heavy mesons, which because of their finite rest mass can be produced only if the available energy in NN collisions exceeds the production threshold. I have previously developed criteria for particles and energies that select probes sensitive to the hot, that is, equilibrated, parts of the reaction volume (72). Heavy mesons, such as the η meson or the K^+, seem to be best suited for this purpose; the K^+ has the additional important advantage of small final-state interactions. Finally, dileptons can potentially give very important information on hadronic interactions and the behavior of meson and baryon resonances in dense nuclear matter.

The unresolved problems in the theoretical descriptions, requiring intensive studies, are all related to the medium corrections to the nucleon-nucleon interactions that enter both into the mean-field part of the description and into the collision cross sections. Progress in this area requires a joint theoretical and experimental effort. Before we can hope to obtain reliable information on the EOS in heavy-ion reactions, we must be sure that we can control these in-medium effects in a less complicated dynamical environment, namely that of nucleon+nucleus reactions, where the nuclear phase-space density gets much less distorted during the course of the reaction. Experiments determining the inclusive particle production cross sections for photons, dileptons, and heavy mesons (η, K^+) in proton + nucleus reactions for proton energies from about 200 MeV up to 2 GeV would be most helpful for this purpose.

Finally, it is important to realize that the development of realistic transport theories and their numerical implementation represents a major step forward in our understanding of nuclear collisions. There is no longer the need to use different models for preequilibrium processes and for the final, thermalized stages of a nucleus-nucleus collision.

ACKNOWLEDGMENTS

This article is based on the work of many of my colleagues. I wish to acknowledge in particular many detailed discussions on the topics treated here with W. Bauer, G. F. Bertsch, G. E. Brown, P. Kienle, C. M. Ko, R. Malfliet, J. Randrup, and last but not least with my colleagues at Giessen, particularly T. Biro, W. Cassing, V. Metag, K. Niita, and Gy. Wolf. I am particularly indebted to W. Bauer and W. Cassing for a careful reading of the manuscript and many helpful comments. This work was supported by BMFT and GSI Darmstadt.

Literature Cited

1. Brown, G. E., In *Proc. 3rd Int. Conf. on Nucleus-Nucleus Collisions*, St. Malo, France, 1988; *Nucl. Phys.* A488: 689c (1988)
2. Blaizot, J. P., *Phys. Rep.* 64: 171 (1980)
3. Sharma, M. M., et al., *Phys. Rev.* C38: 2562 (1988)
4. Brown, G. E., In *Proc. Int. Nuclear Physics Conf.*, Sao Paulo, Brazil, 1989. Singapore: World Scientific (1990), p. 3
5. Satz, H., Specht, H. J., Stock, R., eds., *Proc. 6th Int. Conf. on Ultra-Relativistic Nucleus-Nucleus Collisions—Quark Matter*, 1987, Nordkirchen, Germany; *Z. Phys.* A38 (1988)
6. Mosel, U., *Fields, Symmetries, and Quarks.* Hamburg, New York: McGraw-Hill (1989)
7. Lang, A., et al., *Phys. Lett.* B245: 147 (1990)
8. Stöcker, H., Greiner, W., *Phys. Rep.* 137: 277 (1986)
9. Gutbrod, H. H., Poskanzer, A. M., Ritter, H. G., *Rep. Prog. Phys.* 52: 1267 (1989)
10. Braun-Munzinger, P., Stachel, J., *Annu. Rev. Nucl. Part. Sci.* 37: 1 (1987)
11. Bertsch, G. F., Das Gupta, S., *Phys. Rep.* 160: 189 (1988)
12. Cassing, W., et al., *Phys. Rep.* 188: 363 (1990)
13. Cassing, W., Mosel, U., *Prog. Part. Nucl. Phys.* 25: 235 (1990)
14. Botermans, W., Malfliet, R., *Phys. Rep.* 198: 115 (1990)
15. Vasak, D., Müller, B., Greiner, W., *J. Phys.* G11: 1309 (1985)
16. Ko, C. M., Bertsch, G. F., Aichelin, J., *Phys. Rev.* C31: 2324 (1985)
17. Bauer, W., et al., *Nucl. Phys.* A456: 159 (1986)
18. Heuer, R., et al., *Z. Phys.* A330: 315 (1988)
19. Bauer, W., et al., *Phys. Rev.* C34: 2127 (1986)
20. Metag, V., See Ref. 1; *Nucl. Phys.* A488: 483c (1988)
21. Lippert, Th., et al., *J. Mod. Phys.* In press (1992)
22. Koch, V., et al., *Phys. Lett.* B236: 135 (1990)
23. Botermans, W., Malfliet, R., *Phys. Lett.* B171: 22 (1986)
24. Cassing, W., *Z. Phys.* A327: 447 (1987)
25. Cassing, W., Niita, K., Wang, S. J., *Z. Phys.* A331: 439 (1988)
26. Aichelin, J., Stöcker, H., *Phys. Lett.* B176: 14 (1986)
27. Aichelin, J., *Phys. Rep.* In press (1991)
28. Cassing, W., Wang, S. J., *Z. Phys.* A328: 423 (1987)
29. Blättel, B., et al., *Phys. Rev.* C38: 1767 (1988)

30. Blättel, B., et al., *Nucl. Phys.* A495: 381c (1989)
31. Ko, C. M., Li, Q., *Phys. Rev.* C37: 2270 (1988)
32. Vasak, D., Gyulassy, M., Elze, H.-Th., *Ann. Phys.* NY 173: 462 (1987)
33. Weber, K., et al., *Nucl. Phys.* A515: 747 (1990)
34. Wolf, Gy., et al., *Nucl. Phys.* A517: 615 (1990)
35. Xiong, L., et al., *Nucl. Phys.* A512: 772 (1990)
36. Wang, S. J., et al., *MSU Rep. MSUCL-752*, to be published (1991)
37. McMillan, W. G., Teller, E., *Phys. Rev.* 72: 1 (1947)
38. Bertsch, G. F., *Phys. Rev.* C15: 713 (1977)
39. Cassing, W., *Z. Phys.* A327: 87 (1987)
40. Cugnon, J., Lejeune, A., Grange, P., *Phys. Rev.* C35: 861 (1987)
41. Ter Haar, T., Malfliet, R., *Phys. Rev.* C36: 1611 (1987)
42. Faessler, A., *Nucl. Phys.* A495: 269c (1989)
43. Bohnet, A., et al., *Nucl. Phys.* A494: 349 (1989)
44. Jaenicke, J., et al., *Univ. Heidelberg preprint*, to be published (1990)
45. Bertsch, G. F., et al., *Nucl. Phys.* A490: 745 (1989)
46. de Jong, F., ter Haar, B., Malfliet, R., *Phys. Lett.* B22: 485 (1989)
47. VerWest, B. J., Arndt, R. A., *Phys. Rev.* C25: 1979 (1982)
48. Bauer, W., *Phys. Rev.* C40: 715 (1989)
49. Bhalerao, R. S., Liu, L. C., *Phys. Rev. Lett.* 54: 865 (1985)
50. Liu, L.-C., Londergan, J. T., Walker, G. E., *Phys. Rev.* C40: 832 (1989)
51. Vetter, T., et al., *Phys. Lett. B.* In press (1991)
52. Laget, J. M., Wellers, F., Lecolley, J. F., *Phys. Lett.* B257: 254 (1991)
53. Cassing, W., et al., *Univ. Giessen preprint*, submitted to *Phys. Lett.* (1990)
54. Flaminio, V., et al., *CERN-HERA* 84-01 (1984)
55. Randrup, J., Ko, C. M., *Nucl. Phys.* A343: 519 (1980)
56. Jackson, J. D., *Classical Electrodynamics*. New York: Wiley. 2nd ed. (1975)
57. Nifenecker, H., Bondorf, J. P., *Nucl. Phys.* A442: 478 (1985)
58. Schäfer, M., et al., *Z. Phys. A.* In press (1991)
59. Nakayama, K., *Phys. Rev.* C39: 1475 (1989); Herrmann, V., Speth, J., Nakayama, K., *Phys. Rev. C*, submitted (1990)
60. Nifenecker, H., Pinston, J. A., *ISN Grenoble preprint* (1990)
61. Roche, G., et al., See Ref. 1; *Nucl. Phys.* A488: 477c (1988)
62. Deleted in proof
63. Gale, C., Kapusta, J., *Phys. Rev.* C35: 2107 (1987)
64. Schäfer, M., et al., *Phys. Lett.* B221: 1 (1989)
65. Haglin, K., Kapusta, J., Gale, C., *Phys. Lett.* B224: 433 (1989)
66. Perkins, D. H., *Introduction to High Energy Physics*. Reading: Addison-Wesley. 2nd ed. (1982)
67. Ericson, T., Weise, W., *Pions and Nuclei*. Oxford: Clarendon (1988)
68. Brown, G. E., Rho, M., Weise, W., *Nucl. Phys.* A454: 669 (1986)
69. Randrup, R., Vandenbosch, R., *Nucl. Phys.* A490: 418 (1988)
70. Niita, K., Cassing, W., Mosel, U., *Nucl. Phys.* A504: 391 (1989)
71. Nifenecker, H., Pinston, J. A., *Prog. Part. Nucl. Phys.* 23: 271 (1989)
72. Mosel, U., See Ref. 4, p. 617
73. Benenson, W., et al., *Phys. Rev. Lett.* 43: 683 (1979)
74. Noll, H., et al., *Phys. Rev. Lett.* 52: 1284 (1984)
75. Niita, K., *Univ. Giessen preprint*, to be published (1990)
76. Braun-Munzinger, P., See Ref. 4, p. 641
77. Suzuki, T., et al., *Phys. Lett. B.* In press (1991)
78. Metag, V., In *Proc. Int. School of Physics*, Varenna, 1989, In press (1991); Metag, V., *Ann. Phys.* 48: In press (1991)
79. Ashery, D., Schiffer, J. P., *Annu. Rev. Nucl. Part. Sci.* 36: 207 (1986)
80. Schürmann, B., Zwermann, W., *Mod. Phys. Lett.* A3: 1441 (1988)
81. DePaoli, A. L., et al., *Phys. Lett.* B219: 194 (1989)
82. Aichelin, J., Ko, C. M., *Phys. Rev. Lett.* 55: 2661 (1985)
83. Dorfan, D. E., et al., *Phys. Rev. Lett.* 14: 995 (1965)
84. Baldin, A. A., et al., *JETP Lett.* 48: 137 (1988)
85. Carroll, J. B., et al., *Phys. Rev. Lett.* 62: 1829 (1989)
86. Shor, A., et al., *Phys. Rev. Lett.* 63: 2192 (1989)
87. Shor, A., Perez-Mendez, V., Ganezer, K., *Nucl. Phys.* A514: 717 (1990)
88. Ko, C. M., Ge, X., *Phys. Lett.* B205: 195 (1988)
89. Koch, P., Dover, C. B., *Phys. Rev.* C40: 145 (1989)
90. Ko, C. M., Xia, L. H., *Phys. Rev.* C40: R1118 (1989)

91. Danielewicz, P., *Phys. Rev.* C42: 1564 (1990)
92. Batko, G., et al., *Phys. Lett. B.* In press (1991)
93. Xia, L. H., et al., *Nucl. Phys.* A485: 721 (1988)
94. Ko, C. M., Xia, L. H., Siemens, P. J., *Phys. Lett.* B231: 16 (1989)
95. Xiong, L., et al., *Phys. Rev.* C41: R1355 (1990)
96. Batko, G., et al., In *Proc. NATO Adv. Sci. Inst. on the Nuclear Equation of State*, Pensicola, Spain, 1989. New York: Plenum (1990), p. 353
97. Wolf, Gy., et al., In *Proc. Corinne, Nantes*, 1990, ed. D. Ardouin. Singapore: World Scientific (1990), p. 406
98. Korpa, C. L., Pratt, S., *Phys. Rev. Lett.* 64: 1502 (1990)
99. Korpa, C. L., et al., *Phys. Lett.* B246: 333 (1990)
100. Naudet, C., et al., *Phys. Rev. Lett.* 62: 2652 (1989)
101. Roche, G., et al., *Phys. Lett.* B226: 228 (1989)
102. Weise, W., See Ref. 4, p. 211
103. Schuck, P., et al., See Ref. 97, p. 323

Annu. Rev. Nucl. Part. Sci. 1991. 41: 55–96

TESTS OF THE ELECTROWEAK THEORY AT THE Z RESONANCE

H. Burkhardt

CERN, Geneva, Switzerland

Jack Steinberger

CERN, Geneva, Switzerland, and Scuola Normale Superiore, Pisa, Italy

CONTENTS

1. LEP

The Large Electron-Positron collider (LEP) at CERN came into operation in the summer of 1989 to produce its first Z's. The present center-of-mass energy is limited to 105 GeV. Energy upgrading is in progress, and by 1994

55

0163–8998/91/1201–0055$02.00

the available energy is expected to exceed the W^+W^- pair production threshold of 160 GeV by at least 10 GeV, and perhaps even more. LEP was conceived in the 1970s, following the prediction of the Z and $W^\pm$ by the new electroweak theory (1), a theory that was experimentally established in 1973 by the discovery of the neutral current it had predicted (2). The construction of such a machine may first have been suggested by Richter (3). It was hoped that it might enable discovery of the Z, but fortunately for physics the Z as well as the W were found earlier, in $p\bar{p}$ collisions (4).

The design luminosity of the LEP collider is 1.6×10^{31} cm^{-2} s^{-1} at the Z mass, and it has almost been reached (within a factor of 2): in 1990 about three quarters of a million Z's were observed, hundreds of times more than had been seen in $p\bar{p}$ colliders, and under much cleaner conditions. The analysis of these Z decays is proceeding along several lines:

1. The study of the electroweak interaction at a higher energy and with higher precision than was possible before;
2. Searches for new particles, such as different varieties of Higgs bosons, supersymmetric particles, new quarks or leptons, excited leptons, and others;
3. Studies of QCD in hadronic Z decay; and
4. Studies of heavy quarks, in particular the b quark, which is produced in interesting quantities in Z decay.

Substantial results in all four of these fields have already occasioned of the order of one hundred publications. Here we limit ourselves to the first line of investigation, which has provided excellent evidence that there are just three fermion families, given an accurate value of the electroweak mixing angle through the precise measurement of the Z mass, allowed substantially more precise checks on the electroweak theory than were possible previously, and through its virtual effects, set limits on the mass of the top quark.

2.   THE FOUR DETECTORS

Four of the eight collision points in LEP are equipped with detectors. These are constructed to measure (a) the momenta of charged particles by magnetic tracking, and (b) the energies of both charged and neutral particles in electromagnetic followed by hadronic calorimeters, over as much of the 4π solid angle as possible. All four detectors emphasize the identification of electrons and muons. With respect to previous 4π detectors, they provide much more complete coverage and more precision in energy and angular measurement, especially in the calorimetry. The four collaborations each number several hundred physicists from dozens of institutions around the world.

2.1 *ALEPH*

The ALEPH detector is shown in Figure 1 (5). Tracking is done in three locations: a silicon strip microvertex chamber to reconstruct secondary vertices (this was not yet operational when the data reported here were taken), an inner drift chamber with an important event trigger function, and the main device, a time projection chamber, which is 3.6 m in diameter and 4.4 m long and which furnishes ionization density information for particle separation. The electromagnetic calorimeter is a lead–proportional wire plane sandwich with 45 layers. The signal is read out from 72,000 three-storey towers that project to the collision zone, as well as from each of the 45 wire planes in each of the 36 modules. Space resolution is $\sim 0.8° \times 0.8°$; energy resolution is $0.19/\sqrt{E(\text{GeV})} + 1.1\%$. The superconducting coil produces a field of 1.5 T. The return yoke serves as a hadron calorimeter. Sampling is done in streamer tubes, placed every

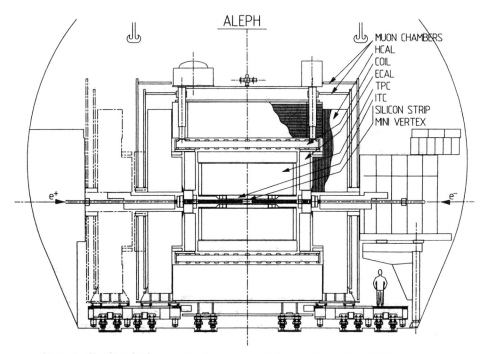

Figure 1 The ALEPH detector. The scale can be set from the fact that the TPC is 4.4 meters long and has a diameter of 3.6 meters.

5 cm of iron, with a total iron thickness of 1.2 m. The signal is read out in 4500 projective towers as well as on strips, every centimeter. The strip readout of the hadron calorimeter, together with two outer double layers of streamer tubes that are read out every 5 mm, serves to identify muons. A reconstructed Z decay into tau leptons is shown in Figure 2. The main emphasis in ALEPH has been on the precision of momentum measurement, with $\Delta p/p = 0.08\% \cdot p(\text{GeV})$, corresponding to a resolution of 3.6% for 46-GeV muons (see Figure 3), and also on the fine granularity of the electromagnetic calorimeter, essential in the identification of electrons as well as photons immersed in hadronic jets (see Figure 4).

2.2 DELPHI

The DELPHI detector is shown in Figure 5 (6). The inner tracking utilizes a silicon microstrip detector to reconstruct secondary vertices, an inner jet drift chamber, and a time projection chamber, 2.4 m in diameter and 3 m in length, which also furnishes ionization density information. A special feature of the DELPHI detector is its ring-imaging Čerenkov system, pioneered by that collaboration. When it becomes fully operational, this system, using both liquid and gas radiators, will permit e-π separation in the interval 0 to 4 GeV and π-K separation in the intervals 0.5 to 7 and 9

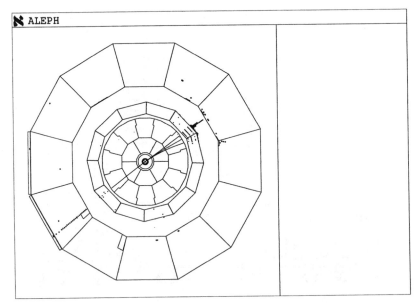

Figure 2 An $e^+e^- \rightarrow \tau^+\tau^-$ event in ALEPH. The single track is due to the decay $\tau^- \rightarrow \mu^- + v_\mu + v_\tau$, the three-track decay to $\tau^+ \rightarrow 2\pi^+ + \pi^- + v_\tau$.

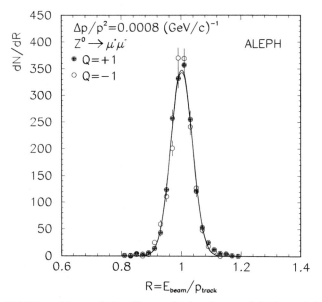

Figure 3 ALEPH tracking resolution illustrated using the muonic Z decay; $\Delta p/p = 3.6\%$ at 45 GeV.

to 25 GeV (see Figure 6). There follows an outer tracking of five layers of drift tubes. The momentum resolution achieved for the muons from Z decay is 7%. The barrel electromagnetic calorimeter is a heavy projection chamber, in which the ionization produced in the gas between lead layers is drifted to the ends of the 90 cm long modules. This technique, also pioneered by DELPHI, offers excellent spatial resolution. The superconducting coil produces a field of 1.2 T. It is surrounded by a scintillator hodoscope for time-of-flight information. The return yoke is instrumented as a hadron calorimeter with streamer tube layers every 5 cm of iron, for a total iron thickness of 1 m. The signals are read out in ~ 4000 projective towers. The whole setup is surrounded by three layers of muon detectors. An exceptional hadronic event is shown in Figure 7.

2.3   *L3*

The L3 detector is shown in Figure 8 (7). It is remarkable in its sheer physical size, and in the fact that the general tracking is minimized in favor of precise outer tracking for muons only. The central tracking chamber is a jet-type time expansion chamber with 18 cm inner diameter, 80 cm outer diameter, and 105 cm length, designed to measure the direction and the sign of the electric charge of charged particles up to 50 GeV. The elec-

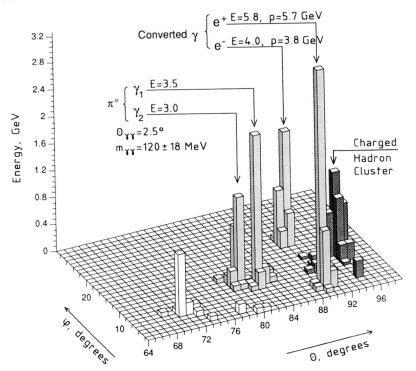

Figure 4 Spatial resolution of the ALEPH electromagnetic calorimeter and its usefulness in the identification of electrons and photons. Part of a typical hadron jet is shown. Photons and electrons are identified by the lateral and longitudinal (not shown) development of the shower. Electrons must have, in addition, an associated track whose momentum matches the energy.

tromagnetic calorimeter is constructed of 10,752 bismuth germanium oxide scintillator prisms, corresponding to an angular resolution of $2° \times 2°$. The energy resolution is exceptional, particularly at low energy. At 45 GeV it is 1.4%, entirely limited by systematics (see Figure 9). For the results reported here, the barrel part was in place, but not the two endcaps. The hadron calorimeter is a 58-layer uranium–proportional wire plane sandwich, arranged in 144 modules, read out on the wires with a high granularity. The momenta of the penetrating muons in the polar angle interval 40°–140° are measured with high precision in the very large (inner diameter = 4.8 m, outer diameter = 11 m) solenoidal magnet with a field of 0.5 T. The momentum resolution for the 45-GeV muons of Z decay is 2.2 to 3%. A hadronic event is shown in Figure 10.

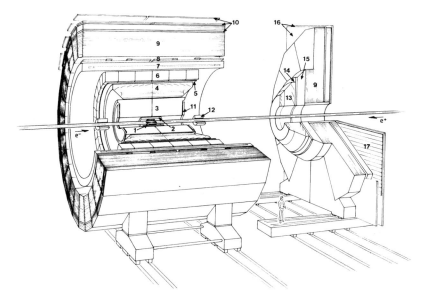

Figure 5 The DELPHI detector. *1* = microvertex detector. *2* = inner detector. *3* = time projection chamber (TPC). *4* = barrel ring-imaging Čerenkov counter (RICH). *5* = outer detector. *6* = high-density projection chamber (HPC). *7* = superconducting solenoid. *8* = time-of-flight counters (TOF). *9* = hadron calorimeter. *10* = barrel muon chambers. *11* = forward chamber A. *12* = small-angle tagger (SAT) = luminosity monitor. *13* = forward RICH. *14* = forward chamber B. *15* = forward electromagnetic calorimeter. *16* = forward muon chambers. *17* = forward scintillator hodoscope.

2.4 *OPAL*

The OPAL detector is shown in Figure 11 (8). Particles are tracked by means of jet-type drift chambers, at a pressure of 4 atm, in order to increase both space and ionization density resolution. The inner "vertex" part covers the radial interval between 9 and 22 cm and is 1 m long; the outer, main chamber has an outer diameter of 3.7 m and is 4 m long. There are 177 wires in the combined radial interval. The jet chamber is surrounded by a layer of chambers that drift the ionization along the z direction, in order to measure the z coordinates with higher precision than is possible by the charge division used in the jet chamber. The warm solenoid provides the axial field of 0.44 T. The momentum resolution for 45-GeV muons is 6.8%. A layer of scintillation counters is used to measure the time of flight and in the trigger. An electromagnetic "presampler," a double layer of streamer tubes, is used to measure the positions of photons and electrons that have showered in the two radiation lengths of the coil. The electromagnetic calorimeter is made of 11,700 projective lead-glass blocks, cor-

Particles in jets

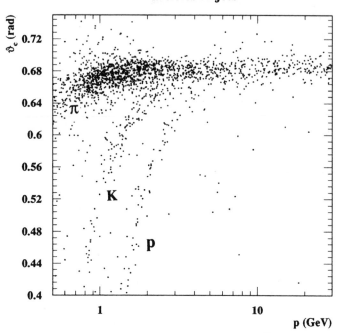

Figure 6   Particle separation in the DELPHI barrel ring-imaging Čerenkov counter, liquid part. Čerenkov angle vs particle momentum.

responding to a granularity of $1.9° \times 1.9°$. Energy resolution is 3% for 45-GeV electrons. The return yoke consists of eight 10-cm slabs of iron, with layers of streamer tubes in between, serving as a hadron calorimeter. These are read out in 976 projective towers and on strips every centimeter. The whole apparatus is surrounded by four layers of muon detectors. Figure 12 shows a typical hadronic event. One of the great successes of OPAL is the tracking precision. Figure 13 illustrates the precision achieved in reconstructing the impact parameter of a track near the vertex, using the muons of Z decay as an example. The result, 40 μm, is truly excellent. Figure 14 shows the particle separation obtained in the ionization measurement.

3. PHYSICS AT THE Z RESONANCE

The center-of-mass energy available at LEP, 105 GeV at present, is clearly much less than the 2000 GeV at the Tevatron p$\bar{\text{p}}$ collider, even after

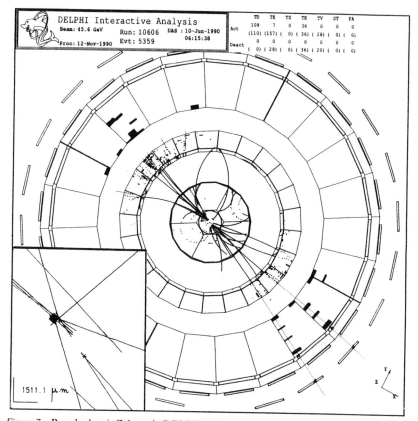

Figure 7 Rare hadronic Z decay in DELPHI. The inset shows the decay of a neutral particle after a trajectory of 1.5 mm, on the basis of the very precise silicon strip measurements. The two secondaries are identified as muons. From their momenta and included angles, it is identified as a J/ψ, probably from B decay.

allowance is made for the fact that the quark-antiquark center-of-mass energy is only a fraction of the p$\bar{\text{p}}$ energy. The strength at the e^+e^- collider is the simplicity of the initial state, which is carried over to the final state so that, in general, the rates for different channels can be calculated in the framework of present theories and can be experimentally distinguished (except that the different quark-flavor channels can only rarely be identified).

At collision energies near the Z mass, the cross section is enhanced through the Breit-Wigner resonance denominator by a factor of the order of $(m_Z/\Gamma_Z)^2 \approx 10^3$. Since the nonresonant cross section is very small, of the order of $4\alpha^2/s$, the resonance presents a very important experimental

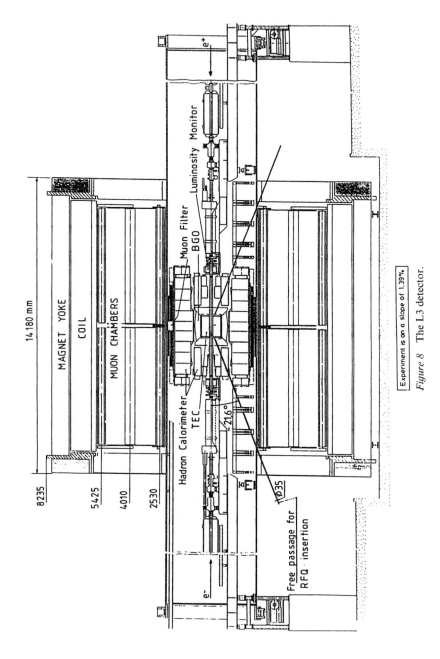

Experiment is on a slope of 1.39%.

Figure 8   The L3 detector.

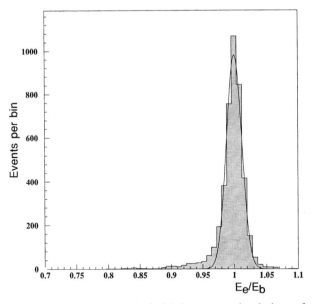

Figure 9 The energy resolution observed in the L3 electromagnetic calorimeter for electrons of Bhabha events, $\sigma(E_e)/E_{beam}$ is 1.4%,

opportunity, by its magnitude alone. For this reason, all the work at LEP has been on or near the Z peak.

3.1 *Model-Independent Analysis of the Z Line Shape*

The cross-section studies we discuss here are processes in which, in Born approximation, either a photon on a Z is exchanged and a fermion pair is emitted:

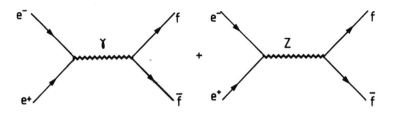

The fermion f may be a charged lepton e, μ, or τ; a neutrino ν_e, ν_μ, or ν_τ; or one of five quark flavors u, d, s, c, or b. No novel channels have been observed.

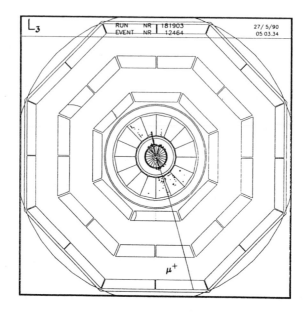

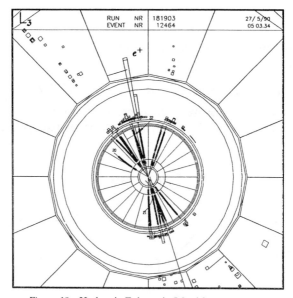

Figure 10   Hadronic Z decay in L3 with muon.

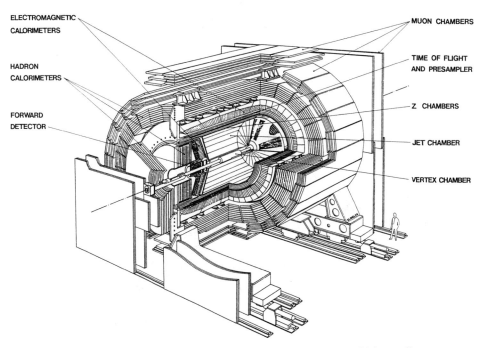

ELECTROMAGNETIC
CALORIMETERS

HADRON
CALORIMETERS

FORWARD
DETECTOR

MUON CHAMBERS

TIME OF FLIGHT
AND PRESAMPLER

Z CHAMBERS

JET CHAMBER

VERTEX CHAMBER

Figure 11   The OPAL detector. Muon chambers are 11 meters high overall.

The energy dependence of Z production is described by a relativistic Breit-Wigner line shape, which must, however, be corrected for initial-state radiation (9):

$$\sigma_f(s) = \frac{12\pi\Gamma_e\Gamma_f}{m_Z^2}\frac{s}{|s - m_Z^2 + is\Gamma_Z/m_Z|^2}[1 + \delta_{rad}(s)]. \qquad 1.$$

Here m_Z is the Z mass, Γ_e, Γ_f are the partial decay widths of the Z to the electron and final-state fermion respectively, $s = E_{cm}^2$ is the square of the center-of-mass energy, and $\delta_{rad}(s)$ is the initial-state radiation correction. The contribution of small terms due to photon exchange and its interference with Z exchange have been omitted from Equation 1 for clarity. In the analysis of the experimental results these terms are always included. The correction for the initial-state radiation, $\delta_{rad}(s)$, is large, as can be seen in Figure 15. It is 40% on the Z peak, but is known with a precision adequate for the present levels of experimental accuracy. Fits of Equation 1 to the experimental line shapes of the different channels furnish model-independent values for m_Z, Γ_Z, Γ_f, and σ_f^{peak}, where σ_f^{peak} is defined as the

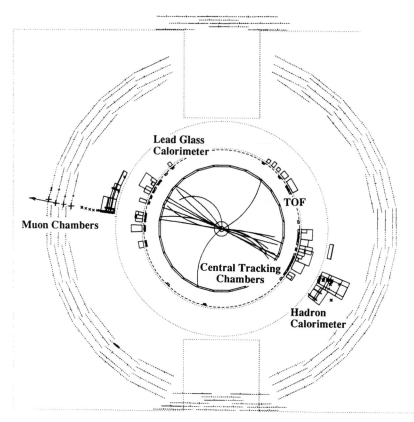

Figure 12 Hadronic Z decay in OPAL.

cross section at the peak of the resonance for channel f after unfolding the initial-state radiation:

$$\sigma_f^{\text{peak}} = \frac{12\pi}{m_Z^2} \frac{\Gamma_e \Gamma_f}{\Gamma_Z^2}.$$

2.

3.2 *Electroweak Born Approximation and Radiative Corrections*

The three parameters of the electroweak theory can be fixed at low energies through measurements of the electromagnetic fine structure constant α, the Fermi constant G_F obtained from the muon lifetime, and the weak

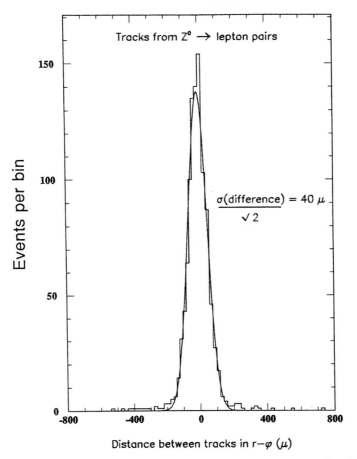

Figure 13 Illustration of the precision of track reconstruction in OPAL: the distance between the two tracks of muonic Z decay, at the vertex. The error in the impact parameter is 40 μm.

mixing angle $\sin^2 \theta_w$ obtained by measuring the ratio of charged to neutral current cross sections in neutrino scattering. The masses of the fermions and of the Higgs particle are not predicted by the theory and can be considered as additional parameters. The masses of the $W^{\pm}$ and Z are instead directly related to the three coupling parameters of the theory through

$$\frac{m_W^2}{m_Z^2} = \cos^2 \theta_w \qquad\qquad 3.$$

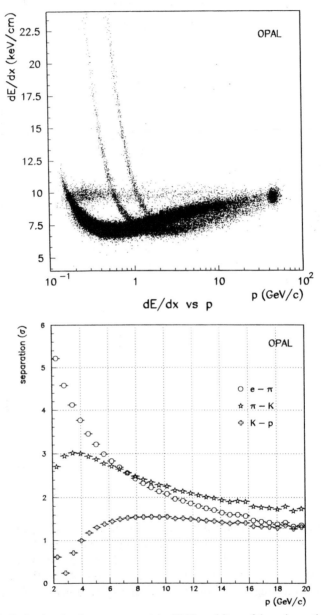

Figure 14   Ionization density measurement in OPAL and its usefulness in particle identification. (*Top*) dE/dx versus p. (*Bottom*) Reliability of identification (σ) versus p.

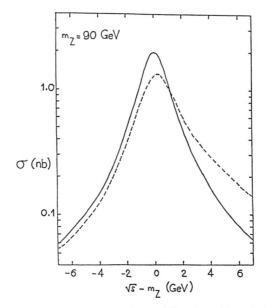

Figure 15 Z line shape for the decays $Z \rightarrow \mu^{+}\mu^{-}$ and $Z \rightarrow \tau^{+}\tau^{-}$ without (*solid line*) and with (*dashed line*) initial-state radiative correction.

and

$$\frac{\pi\alpha}{\sqrt{2G_F}} = m_Z^2 \cos^2 \theta_w \sin^2 \theta_w = A_0^2 = (37.2802 \pm 0.0003 \text{ GeV})^2. \qquad 4.$$

In the Born approximation the differential cross section for the reaction $e^+e^- \rightarrow f\bar{f}$ is

$$\frac{d\sigma_t}{d\Omega} = \frac{\alpha^2 N_c^f}{4s}[F_1(s)(1+\cos^2 \theta)+2F_2(s)\cos \theta]. \qquad 5.$$

Here

θ is the center-of-mass production angle,

N_c^f is the color factor: 3 for quarks, otherwise 1,

$F_1(s) = Q_f^2 - 2v_e v_f Q_f \operatorname{Re}(\chi) + (v_e^2 + a_e^2)(v_f^2 + a_f^2)|\chi|^2,$

$F_2(s) = -2a_e a_f Q_f \operatorname{Re}(\chi) + 4v_e a_a v_f a_f |\chi|^2,$

Q_f is the electric charge in units of the charge of the positron,

$\chi = s/(s - m_Z^2 + im_Z\Gamma_Z)$ is the Breit-Wigner resonance denominator,

$v_f = (I_3^f - 2Q_f\sin^2\theta_w)/(2\sin\theta_w\cos\theta_w)$ is the weak vector coupling constant,

$a_f = I_3^f/(2\sin\theta_w\cos\theta_w)$ is the weak axial-vector coupling constant,

I_3 is the third component of weak isospin, and

$\Gamma_Z = \Sigma_f\Gamma_f$ is the total width of the Z.

The first term in F_1 is pure photon exchange, the last term is pure Z exchange, and the central term is the interference between the two. The integrated cross section at and near the peak is dominated by the Z exchange, $|\chi|^2$ term:

$$\sigma_f = N_c^f\frac{4\pi}{3}\frac{\alpha^2}{s}(v_e^2 + a_e^2)(v_f^2 + a_f^2)|\chi|^2. \qquad 6.$$

Since the partial widths are

$$\Gamma_f = \frac{N_c^f}{3}\alpha m_Z(v_f^2 + a_f^2), \qquad 7.$$

the Born approximation cross section (Equation 5), when integrated over the solid angle near the Z pole, has the Breit-Wigner form for spin-1 exchange:

$$\sigma_f = \frac{12\pi\Gamma_e\Gamma_f}{m_Z^2}\frac{s}{|s - m_Z^2 + im_Z\Gamma_Z|^2}. \qquad 8.$$

The Born approximation is not adequate for the analysis of the LEP experiments; it is essential to include higher order radiative corrections. These include initial-state photon, final-state photon, and final-state gluon radiation corrections, plus vertex and propagator corrections and other even smaller corrections. They have been very much studied, and a summary of information and references to other papers are given in the *Proceedings of the 1989 CERN Workshop on Z physics at LEP 1* (10). The initial-state radiation can be factored out as shown in Equation 1.

The other corrections are small and calculable in the Standard Model. However, they depend on the unknown Higgs boson and top quark masses, as well as on the value of the strong coupling constant α_s, which is not yet very precisely known. The dominant effects of these corrections can be summarized by replacing $\sin^2\theta_w$ with an "effective mixing angle" $\overline{\sin^2\theta_w}$

at the Z mass, by replacing α with its renormalized value at the Z mass (10):

$$\alpha(m_Z^2) = \frac{\alpha}{1 - \Delta\alpha} = 1.064\alpha,$$

by inserting the factor $\rho = 1/(1 - \Delta\rho)$ in the relationships between m_W and m_Z and between G_F and m_Z:

$$\frac{m_W^2}{m_Z^2} = \rho \overline{\cos^2\theta_w} \qquad\qquad 9.$$

$$\frac{\pi\alpha(m_Z^2)}{\sqrt{2G_F}} = \rho m_Z^2 \overline{\sin^2\theta_w} \, \overline{\cos^2\theta_w}, \qquad\qquad 10.$$

with

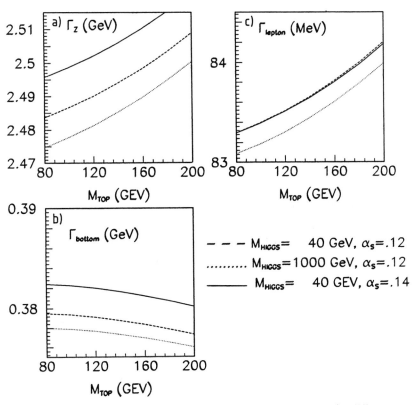

Figure 16   Dependence of (*a*) Γ_Z, (*b*) Γ_b, and (*c*) Γ_ℓ on m_t, m_H, and α_s (10).

$$\Delta\rho \simeq 3G_F m_t^2/8\sqrt{2}\pi^2 - 11 G_F m_Z^2 \sin^2\theta_w \ln(m_H/m_W)/12\pi^2\sqrt{2}$$
$$= 0.0026 m_t^2/m_Z^2 - 0.0015 \ln(m_H/m_W),$$

and, for the quark channels, by including the second-order QCD correction and by replacing N^f with (11)

$$N_c^f(1 + \alpha_s/\pi + 1.4\alpha_s^2/\pi^2). \qquad\qquad 11.$$

Finally, radiative corrections to the Z propagator lead to an s dependence in the imaginary part that is taken into account by replacing Z in Equation 5 by

$$\chi = \frac{s}{(s - m_Z^2) + is\Gamma_Z/m_Z}. \qquad\qquad 12.$$

These changes give the major radiative corrections for all channels $e^+e^- \rightarrow f\bar{f}$ (12), except the $b\bar{b}$ channel, which requires an additional term taking into account relatively large t-b amplitudes in the vertex corrections. To give the reader some feeling for the magnitude of the corrections, we show in Figure 16 the dependences of Γ_ℓ, Γ_b, and Γ_Z on the top quark mass for different choices of the Higgs mass and α_s.

3.3   Asymmetries

The forward-backward asymmetry in the angular distribution (Equation 5), as well as the polarization asymmetries, measures the vector–axial-vector coupling ratios for the different channels.

3.3.1   FORWARD-BACKWARD ASYMMETRIES   The forward-backward asymmetry A_{FB} is defined as the difference between forward and backward cross sections divided by their sum:

$$A_{FB}^f = \frac{\int_0^1 (d\sigma_f/d\cos\theta)\,d\cos\theta - \int_{-1}^0 (d\sigma_f/d\cos\theta)\,d\cos\theta}{\int_0^1 (d\sigma_f/d\cos\theta)\,d\cos\theta + \int_{-1}^0 (d\sigma_f/d\cos\theta)\,d\cos\theta}.$$

From Equation 5, we find $A_{FB} = \frac{3}{4}F_2/F_1$. Well above and below the Z peak, the asymmetry is dominated by the interference term in F_2,

$$A_{FB}^f \simeq \frac{3}{2}\frac{a_e a_f Q_f \,\mathrm{Re}(\chi)}{(v_e^2 + a_e^2)(v_f^2 + a_f^2)},$$

but exactly on the peak this term vanishes, and

$$A_{FB}^f = 3 \frac{v_e a_e v_f a_f}{(v_e^2 + a_e^2)(v_f^2 + a_f^2)}. \qquad 13.$$

For leptons, $v_\ell/a_\ell = 1 - 4\sin^2\theta_w$ is very small, leading to the small asymmetry at the peak $A_{FB}^{\ell,peak} \simeq 3(v_\ell/a_\ell)^2$. This asymmetry at the peak furnishes a direct measure of v_ℓ/a_ℓ, but the sensitivity is compromised by the smallness of v_ℓ/a_ℓ together with the quadratic dependence.

For quarks,

$$A_{FB}^q = 3 \frac{v_e}{a_e} \frac{v_q}{a_q(1 + v_q^2/a_q^2)},$$

where

$$v_q/a_q = 1 - \tfrac{8}{3}\sin^2\theta^w \qquad \text{for up-type quarks,}$$

$$v_q/a_q = 1 - \tfrac{4}{3}\sin^2\theta_w \qquad \text{for down-type quarks.}$$

In contrast to leptons, v_q/a_q is not very small compared to one. This itself would facilitate the measurement of v/a in the quark asymmetries, but at least with our current technology, this advantage is outweighed by the difficulty of separating the individual quark channels and identifying the quark jet relative to the antiquark jet in hadronic decays.

3.3.2 POLARIZATION ASYMMETRIES   We consider here the case in which the polarization of one of the four particles is measured. If that particle were the incident electron or positron, we would obtain some potentially very interesting results, but that is not yet possible at LEP.

At present the only measurement possibility is offered by the τ, which can signal its polarization in its decay. The polarization dependence of the cross section is

$$\frac{d\sigma}{d\cos\theta}(\cos\theta, p) \propto (1 + \cos^2\theta)F_1(s) + 2\cos\theta F_2(s)$$

$$+ p[(1 + \cos^2\theta)F_3(s) + 2\cos\theta F_4(s)], \qquad 14.$$

where p is twice the helicity of the τ.

At the Z pole,

$$F_1(m_Z^2) = (v_e^2 + a_e^2)(v_f^2 + a_f^2)$$

$$F_2(m_Z^2) = 4v_e a_e v_f a_f$$

$$F_3(m_Z^2) = 2(v_e^2 + a_e^2)v_f a_f$$

$$F_4(m_Z^2) = 2v_e a_e(v_f^2 + a_f^2).$$

The polarization asymmetry A_{pol} is defined by

$$A_{\text{pol}} = \langle p \rangle \equiv -\frac{\sigma_{p=1} - \sigma_{p=-1}}{\sigma_{p=1} + \sigma_{p=-1}} = -\frac{F_3}{F_1}$$

$$= -\frac{2v_f a_f}{v_f^2 + a_f^2},$$ 15.

and the forward-backward polarization asymmetry $A_{\text{pol}}^{\text{FB}}$ is defined by

$$A_{\text{pol}}^{\text{FB}} \equiv$$

$$-\frac{\int_0^1 d\cos\theta \left[\dfrac{d\sigma_{p=1}}{d\cos\theta} - \dfrac{d\sigma_{p=-1}}{d\cos\theta}\right] - \int_{-1}^0 d\cos\theta \left[\dfrac{d\sigma_{p=1}}{d\cos\theta} - \dfrac{d\sigma_{p=-1}}{d\cos\theta}\right]}{\int_{-1}^1 d\cos\theta \left[\dfrac{d\sigma_{p=1}}{d\cos\theta} + \dfrac{d\sigma_{p=-1}}{d\cos\theta}\right]}$$

$$= -\frac{3}{4}\frac{F_4}{F_1} = -\frac{3}{4}\frac{2v_e a_e}{v_e^2 + a_e^2}.$$ 16.

4. CROSS SECTIONS AT THE Z PEAK

The experimental results reviewed here are the cross sections for the four channels: $q\bar{q}$, e^+e^-, $\mu^+\mu^-$, and $\tau^+\tau^-$, expressed as a function of the center-of-mass energy, the production angle, and in the case of the τ, its polarization. The experiments are designed to measure the Z mass, to check the predictions of the electroweak theory, and to detect the effects of the top quark and/or Higgs boson in the small radiative corrections. Since nearly one million Z decays have been observed so far, and many more are expected, elimination of systematic errors to a level well below 1% becomes necessary. This is a major consideration in the triggering of events, in their selection and acceptance, in the measurement of the luminosities, and in the calibration of the beam energy. During this first period, LEP was operated in cycles lasting about one week; about one half of the luminosity of each cycle was at the energy of the Z peak; the rest was divided roughly equally among the six energies, ± 1, ± 2, and ± 3 GeV from the peak energy, for a total luminosity of ~ 10 pb^{-1}. The data-taking efficiencies of the four experiments varied between 60 and 80%.

In the following discussions on triggering, event selection, acceptance, and luminosity determination, it would be too lengthy and tedious to give the details for any, much less all four, of the experiments. We limit ourselves to some more general remarks that we illustrate using ALEPH. Results are given for each of the collaborations and then combined to give a LEP result.

4.1 *Event Triggers*

Highly efficient triggering on all four event classes was achieved relatively easily at LEP because of low background rates. All detectors use multiple ORed triggers. The trigger redundancy (25 in the case of ALEPH) not only increases efficiency, but more importantly makes it possible to *measure* the efficiency. Triggering efficiencies of very nearly 100% for each of the four event types are obtained, with uncertainties of less than 0.1%.

4.2 *Event Selection*

The four channels have very distinct characteristics (Figure 17). The dominant channel is the qq̄ channel, with 88% of the observed events. These events typically have many charged tracks (on the average 20) and large energy deposits in both calorimeters, with total energy near the center-of-

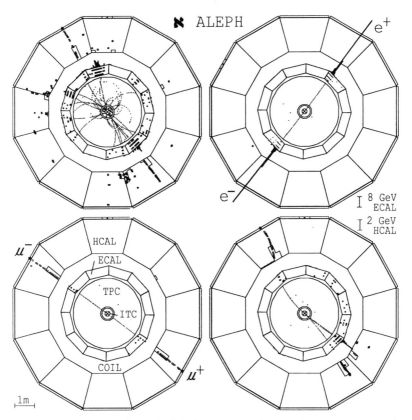

Figure 17   Typical events for each of the four channels (from ALEPH): $Z \rightarrow q\bar{q}$, $Z \rightarrow e^{+}e^{-}$, $Z \rightarrow \mu^{+}\mu^{-}$, and $Z \rightarrow \tau^{+}\tau^{-}$.

mass energy. The e^+e^- events are characterized by two back-to-back tracks, each with the momentum equal to the beam energy, with all the energy in the electromagnetic calorimeter. The $\mu^+\mu^-$ events have two back-to-back tracks that penetrate the hadron calorimeter, with very little calorimetric energy. The τ's, depending on their decay, each have one or three tracks, typically with substantial missing energy, and are also back-to-back. These characteristics can be modified by initial- or final-state radiation and by the interaction of the outgoing particles in the beam pipe and detector, but it is generally possible to separate the channels with purities exceeding 99% with little loss in efficiency. The dominant background, the e^+e^- inelastic scattering (the two-photon channel) with the production of additional particles, can be dealt with on the basis of the typically small energy and the forward-backward energy imbalance of these events.

The selection efficiencies and purities are determined using Monte Carlo simulations of the different event channels and backgrounds. The acceptances are close to unity, with a small uncertainty for hadronic ($q\bar{q}$) events ($99.1 \pm 0.2\%$ in ALEPH) and somewhat larger inefficiencies and uncertainties for leptons, especially τ's ($98.4 \pm 0.3\%$ for the combined ALEPH lepton sample).

Event generators for all channels exist with a high precision and ability to reproduce the observed events, and all four detectors have been simulated in full detail (13). As an example, Figure 18 reproduces data and simulations for the thrust axis of hadronic events observed in ALEPH.

4.3   *Luminosity Measurement*

As has always been the case at e^+e^- colliders, the luminosity is measured with the help of Bhabha scattering, at angles small enough so that the cross section is dominated by the t channel, and the contribution of Z-exchange interference is a small correction. The luminosity detectors of the four collaborations work in the angular region between 40 and 120 mrad. It is interesting that the theoretical uncertainties in the Bhabha cross section contribute nonnegligibly to the error as the experimental precision improves (14). The experimental challenge is to define the geometrical acceptance with high accuracy. In ALEPH, the luminosity detector is a lead–proportional wire sandwich calorimeter covering the small-angle region on each side. It is read out with high angular resolution in 754 projective towers (see Figure 19). One side, which alternates from event to event, is geometrically more restrictive and defines the fiducial acceptance. The angular uncertainty in the fiducial boundary, event by event, is about 0.08 mrad, and the total systematic uncertainties in the luminosity are 0.4%. The L3 experiment also uses a segmented calorimeter, but it is not

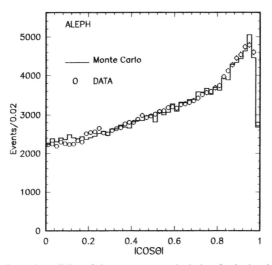

Figure 18 Check on the validity of the acceptance calculation for hadronic events. Comparison of simulation and data for the thrust axis distribution (from ALEPH).

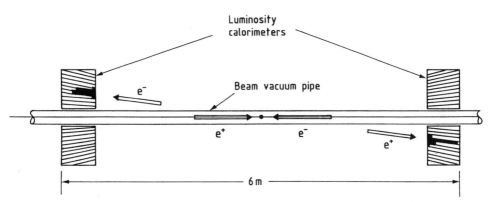

Figure 19 Disposition of the luminosity calorimeter in ALEPH. The center planes of the calorimeter are 2.7 m from the interaction point, and the diameters are 80 cm.

projective and gives an experimental uncertainty of 0.7%. Figure 20 shows the precision with which the angular acceptance is understood in this detector. The DELPHI experiment uses masked calorimeters and quotes an error of 0.8%. The OPAL experiment uses proportional tubes inserted into a segmented calorimeter after the first four radiation lengths to define the geometrical acceptance for the absolute cross-section measurement with a precision of 0.7%. The theoretical uncertainty in the most recent Bhabha calculations is believed to be ~0.5% (15).

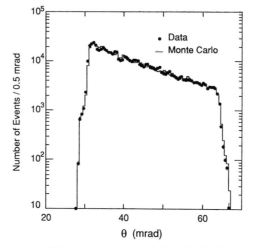

Figure 20   Check on the validity of the acceptance calculation for luminosity events. Comparison of simulation and data for the polar angle distribution (from L3).

4.4   *LEP Beam Energy*

One of the important results of these first measurements at LEP is the measurement of the Z mass. A precise knowledge of the beam energy is therefore essential. The calibration of the LEP beam energy is at present based on the measurement of the frequency difference between protons and positrons circulating in the same LEP orbit at ~ 20 GeV. The relative increase in beam energy from 20 to 45 GeV is established using flux loops on each of the LEP magnets. The present uncertainty is 20 MeV in the center-of-mass energy (0.22 parts per thousand) (16). It is expected that even higher precision will be achieved by using the precession frequency of transversely polarized beam electrons.

4.5   *Results on the Z Line Shape and the Number of Fermion Families*

The most comprehensive results that can be included in this review are those presented by the four collaborations at a recent conference in Aspen (17). The results are based on all data obtained in 1989 and 1990. They are preliminary, but substantial changes are not expected before publication. They are based on the event numbers and acceptances shown in Table 1. Major parts of these results have been published previously (18).

The basic results on the line shape are the cross sections for the four channels q$\bar{\text{q}}$, e$^+$e$^-$, $\mu^+\mu^-$, and $\tau^+\tau^-$ expressed as a function of the center-of-mass energy. Figure 21 shows the OPAL result for the hadronic channel,

Table 1 Numbers of selected events ($N_{ev.}$), angular acceptances, as well as acceptance efficiencies for the four experiments and the four dominant Z-decay channels

Channel	Parameters	ALEPH	DELPHI	L3	OPAL
q$\bar{q}$	$N_{ev.}$	175,000	120,000	115,000	166,000
	Ang. accept.	4π	4π	4π	4π
	Efficiency (%)	99.1$\pm$0.2	96.9$\pm$0.5	99.0$\pm$0.4	98.4$\pm$0.6
e$^+$e$^-$	$N_{ev.}$	6942	2615	4175	5415
	Ang. accept.	$\cos\theta \, {}^{-\ 0.9}_{+\ 0.7}$	$\|\cos\theta\| < 0.7$	$\|\cos\theta\| < 0.7$	$\|\cos\theta\| < 0.7$
	Efficiency (%)	98.8$\pm$0.4	97.3$\pm$0.7	99.2$\pm$0.6	99.3$\pm$0.7
$\mu^+\mu^-$	$N_{ev.}$	6691	2489	3245	7240
	Ang. accept.	$\|\cos\theta\| < 0.9$	$\|\cos\theta\| < 0.93$	$\|\cos\theta\| < 0.8$	$\|\cos\theta\| < 0.95$
	Efficiency (%)	99.6$\pm$0.6	91.5$\pm$0.8	78.3$\pm$0.6	91.6$\pm$0.5
$\tau^+\tau^-$	$N_{ev.}$	6260	2039	2540	5559
	Ang. accept.	$\|\cos\theta\| < 0.9$	$\|\cos\theta\| < 0.73$	$\|\cos\theta\| < 0.7$	$\|\cos\theta\| < 0.9$
	Efficiency (%)	85.4$\pm$0.7	87$\pm$1.4	75.4$\pm$1.6	86.1$\pm$ 1.1
$\ell^+\ell^-$	$N_{ev.}$	24757	9676		
	Ang. accept.	$\|\cos\theta\| < 0.9$	$\|\cos\theta\| < 0.69$		
	Efficiency (%)	98.4$\pm$0.3	90.1$\pm$0.6		

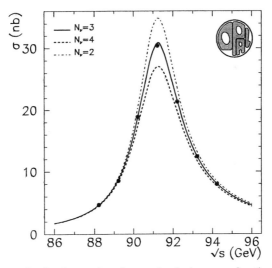

Figure 21 Cross section for the reaction e$^+$e$^-$ → q$\bar{q}$ → hadrons as a function of the center-of-mass energy. The electroweak expectations for 2, 3, and 4 families are shown, with one fitted free parameter, the Z mass (from OPAL).

and Figure 22 the ALEPH results for the leptonic channels. From these, the parameters m_Z, Γ_Z, and σ^{peak} can be found by fitting the model-independent resonance expression (Equation 1) (19). It is common practice now to obtain m_Z and Γ_Z from a combined fit to all channels. The partial widths are derived from Γ_Z and the peak cross sections as follows (see Equation 2):

$$\Gamma_h = m_Z\Gamma_Z\sqrt{R\sigma_h^{\text{peak}}/12\pi},$$

$$\Gamma_{e,\ell} = m_Z\Gamma_Z\sqrt{\sigma_{e,\ell}^{\text{peak}}/12\pi},$$

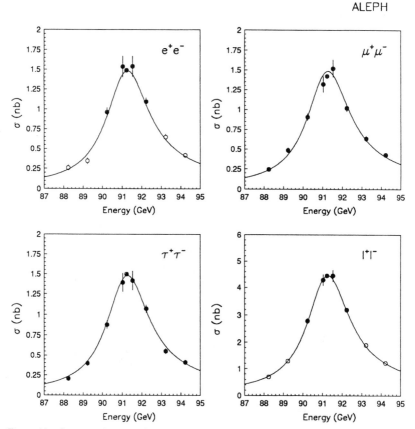

Figure 22 Cross section as a function of the center-of-mass energy for each of the three lepton channels, as well as the combined channel. The *t*-channel contribution for the electron channel has been subtracted (from ALEPH).

$$\Gamma_{\mu,\tau} = m_Z\Gamma_Z\sigma_{\mu,\tau}^{peak}/\sqrt{12\pi\sigma_e^{peak}}$$

and

$$\Gamma_{inv} = \Gamma_Z - \Gamma_h - 3\Gamma_\ell = \Gamma_Z(1 - \sqrt{R\sigma_h^{peak}m_Z^2/12\pi} - 3\sqrt{\sigma_h^{peak}m_Z^2/12\pi R}),$$

where $R = \sigma_h^{peak}/\sigma_\ell^{peak}$. The subscript ℓ denotes the average of the three leptonic channels, assuming universality.

The results of the four collaborations are given in Table 2. There are no significant disagreements. The errors quoted include both statistical and systematic errors. In the combined results, the common theoretical uncertainty in the luminosity Bhabha cross section, taken to be 0.5%, has been included. Other possible sources of common systematic errors include the t-channel treatment in wide-angle Bhabha scattering and a point-to-point

Table 2 Results of the four LEP experiments on the Z line shape and partial decay widths

Result	ALEPH	DELPHI	L3	OPAL	Combined LEP result
m_Z, GeV	91.182 ±0.009	91.175 ±0.010	91.180 ±0.010	91.160 ±0.009	91.174 ±0.005 ± 0.020 LEP
Γ_Z, MeV	2488 ±17	2454 ±20	2500 ±17	2497 ±17	2487 ±9
σ_h^{peak}, nb	41.76 ±0.39	41.98 ±0.63	40.92 ±0.47	41.23 ±0.47	41.46 ± 0.29
Γ_h, MeV	1756 ±15	1718 ±22	1739 ±19	1747 ±19	1744 ±10
Γ_e, MeV	84.2 ±0.9	81.6 ±1.3	83.0 ±1.0	83.5 ±1.0	83.3 ± 0.5
Γ_μ, MeV	80.9 ±1.4	88.4 ±2.4	84.3 ±2.0	83.5 ±1.5	83.3 ±0.9
Γ_τ, MeV	82.9 ±1.6	84.9 ±2.7	83.3 ±2.6	83.1 ±1.9	83.3 ±1.0
Γ_ℓ, MeV	83.3 ±0.7	83.4 ±1.0	83.3 ±0.8	83.4 ±0.7	83.3 ±0.4
$R = \Gamma_h/\Gamma_\ell$	21.07 ±0.19	20.61 ±0.33	20.88 ±0.28	20.94 ±0.24	20.94 ±0.12
$\Gamma_{inv.}$, MeV	481 ±14	486 ±21	511 ±18	499 ±17	493 ±9.5
N_ν, MeV	2.90 ±0.08	2.93 ±0.13	3.08 ±0.10	2.99 ±0.10	2.96 ±0.06

error in the LEP energy calibration affecting the width measurement. We believe that, at present, these common errors can still be neglected. The three lepton partial widths are equal at their level of precision of $\sim 1\%$, in agreement with the universality of the weak interaction.

One of the important consequences of these results concerns the number of fermion families. In the Standard Model, each light neutrino family contributes 166.5 ± 0.5 MeV (assuming the top quark mass to be $90 < m_t < 160$ GeV and the Higgs boson mass to be $50 < m_H < 1000$ GeV) to the "invisible width" $\Gamma_{inv} = \Gamma_Z - \Gamma_h - 3\Gamma_\ell$. There is no evidence for fermions other than those of the three known families, but fermions with masses beyond the reach of existing accelerators can be postulated. However, for the known families, the neutrino masses are, if not zero, much less than the masses of their charged counterparts, at least by a factor of a hundred. In the absence of a theory of the masses, it is at least very plausible to assume that the neutrinos of higher families would still have masses well below one half the Z mass. They would then contribute to the invisible width. So would any other nondetected neutral channels, such as other weakly interacting particles emitted in the decay of the Z.

From the experimental result $\Gamma_{inv} = 493 \pm 9.5$ MeV, and with the above theoretical value of the neutrino partial width, Γ_ν, we obtain

$$N_\nu = 2.96 \pm 0.06.$$

The result leaves no room for a fourth fermion family and little room for any other new neutral weakly interacting particle. If one wants to imagine a fourth neutrino family with mass only slightly less than $m_Z/2$, the error on the result for Γ_{inv} corresponds to a lower mass limit of 45.5 GeV for a Dirac type of neutrino, and of 42.3 GeV for a Majorana type, at the 95% confidence level.

The fact that the result is 3 and not 2 is striking confirmation of the separate identity of the τ neutrino, although strong evidence for this already exists in the experimental results on τ decay, which are in good agreement with the electroweak theory. Finally, the fact that the value of N_ν is, with good precision, compatible with an integer is an interesting confirmation of the Standard Model.

4.6   Lepton and b Quark Forward-Backward and τ Polarization Asymmetries

In principle, the forward-backward asymmetries can be measured for all eight types of charged fermions, and so the v^2/a^2 ratio can be measured independently for every one. Measurements of each of the three lepton channels have been reported by all four collaborations (18); the statistical significance is still limited, however. The results are in agreement with universality. Here we content ourselves with reporting only the combined lepton results. The quark flavors are not, in general, separable; however, with limited efficiency and purity, the b$\bar{\text{b}}$ channel can be separated. The b-quark asymmetry has been reported by the ALEPH and L3 collaborations (20). A global quark forward-backward asymmetry, based on the overall charges assigned to the jets, has been reported by ALEPH (21).

In the determination of polarization asymmetries, only the effects of initial beam polarization and the τ polarization are measurable. The former, which offers the possibility of very precise measurements of v/a, waits for the development of longitudinally polarized beams. However, the τ polarization can be measured using the energy distributions of the decay products in various decay channels. A first result has been reported by ALEPH (22).

A typical result for the lepton angular distribution is shown in Figure 23. It illustrates also a particular problem of the e^+e^- channel: the important contribution of the t-channel photon exchange at small angles. This is a quantum electrodynamic process, calculated (23) and subtracted. Figure 24 shows the dependence of the asymmetry on the center-of-mass energy, together with a two-parameter (v, a) electroweak fit. The interesting point

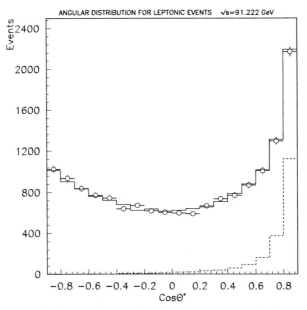

Figure 23 Angular distribution of the combined leptonic channels on the Z resonance peak. The calculated t-channel $e^+e^- \rightarrow e^+e^-$ contribution is also shown (from ALEPH).

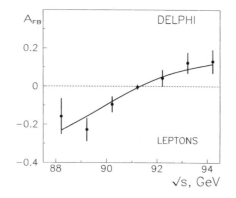

Figure 24 The leptonic forward-backward asymmetry, A^{ℓ}_{FB}, as a function of the center-of-mass energy. Only the very small asymmetry at the Z peak is dominated by v_ℓ^2/a_ℓ^2. The rest is dominated by Q_ℓ^2/a_ℓ^2 (from DELPHI).

is the very small asymmetry on the resonance peak. The much larger asymmetries away from the peak are due to the interference of the axial weak current with the electromagnetic (vector) current. As noted in Section 3.3.1, the asymmetry on the peak measures the ratio of vector to axial-vector current strengths: $A^{\ell}_{FB} \simeq 3v_\ell^2/a_\ell^2$.

This result can be combined with the measurement of the partial leptonic width to obtain both leptonic vector and axial-vector couplings independently of the details of the electroweak theory. For this purpose, it is customary to introduce the coupling constants $g_{V,f}$ and $g_{A,f}$, incorporating the Fermi constant, by using Equation 10 to rewrite Equation 7:

$$\Gamma_f = \frac{\sqrt{2}G_F m_Z^3 N_c^f}{12\pi}(g_{V,f}^2 + g_{A,f}^2).$$

In the electroweak theory g_V and g_A are related to the mixing angle:

$$g_{V,f} = \sqrt{\rho}(I_3^f - 2Q_f \overline{\sin^2\theta_w}) = \sqrt{\rho}v_f \cdot 2\overline{\sin\theta_w \cos\theta_w}$$

$$g_{A,f} = \sqrt{\rho}I_3^f = \sqrt{\rho}a_f \cdot 2\overline{\sin\theta_w \cos\theta_w}.$$

Of course, $g_{V,f}/g_{A,f} = v_F/a_f$. In the electroweak theory $v_\ell/a_\ell = 1 - 4\sin^2\theta_w$, so that the leptonic forward-backward asymmetry (for that matter, the other asymmetries as well) provides a measure of the "effective" mixing angle without the complication of top quark or Higgs boson corrections. The results for the lepton asymmetry, v_ℓ/a_ℓ, and $g_{V,\ell}$ and $g_{A,\ell}$ are given in Table 3 for the combined lepton sample.

For the measurement of the b-quark forward-backward asymmetry, the b$\overline{\text{b}}$ channel is selected using electronic and muonic b decay, which together

Table 3 Result of the combined lepton sample for the asymmetry A_{FB}^{ℓ} on the Z peak, the ratio of vector to axial vector coupling strengths, and $g_{V,\ell}^2$ and $g_{A,\ell}^2$

	ALEPH	DELPHI	L3	OPAL	Combined LEP result
A_{FB}^{ℓ}	0.024 ±0.008	0.008 ±0.013	0.024 ±0.014	0.007 ±0.008	0.016 ± 0.005
v_{ℓ}^2/a_{ℓ}^2	0.0082 ±0.0026	0.0028 ±0.0044	0.0080 ±0.0048	0.0023 ±0.0028	0.0054 ± 0.0016
$g_{V,\ell}^2$	0.0020 ±0.0007	0.0007 ±0.0014	0.0020 ±0.0012	0.0006 ±0.0007	0.0014 ± 0.0004
$g_{A,\ell}^2$	0.2484 ±0.0022	0.2510 ±0.0028	0.2490 ±0.0030	0.2509 ±0.0022	0.2498 ± 0.0012
$\overline{\sin^2\theta_w}$					$0.2317 \, {}^{+\,0.0030}_{-\,0.0026}$

account for 20% of b decay. The main backgrounds are due to the leptonic decay of charm quarks and to hadrons misidentified as leptons. The requirement of high lepton momentum as well as transverse momentum increases the purity of the sample at the expense of efficiency (see Figure 25). Typical values are 80% and 10% respectively for $p > 3$ and $p_T > 2$ GeV.

As seen in Section 3.3.1, on the Z pole

$$A_{FB}^b = \frac{3[(v_\ell/a_\ell)(v_b/a_b)]}{[1+(v_\ell/a_\ell)^2][1+(v_b/a_b)^2]}.$$

Since $v_b/a_b = 1 - \frac{4}{3}\sin^2\theta_w \simeq 0.69$ is about ten times larger than v_ℓ/a_ℓ, the b asymmetry is potentially more sensitive to $\sin^2\theta_w$ than the leptonic asymmetry. But this is more than compensated, at present, by the combined effect of selection inefficiency and impurity of the b sample.

In addition, there is the effect of $b\bar{b}$ mixing. In first approximation the b ($\bar{b}$) is identified by a positive (negative) lepton, and this is always valid for the charged B mesons. However, the neutral B^0 ($\bar{B}^0$) meson can transform into its charge conjugate before decay, and so falsify the signature. The $B\bar{B}$ mixing parameter, $\chi \equiv$ number of charge conjugate decays divided by the total number of b decays, is measured using events with two leptons, one associated with each jet. After background corrections, χ is obtained from the number of like-sign double-lepton events, normalized to the total number of double-lepton events.

The observed asymmetry must be corrected for this mixing. The experi-

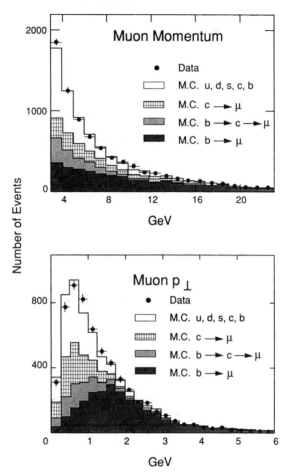

Figure 25   Purity of the b-quark channel as a function of the momentum and transverse momentum of the muon used to tag the channel (from L3).

mental results for the mixing (24) are included in Table 4, which gives the b-asymmetry results for the ALEPH and L3 collaborations (20).

The τ polarization on the Z peak is a measure of v_τ/a_τ (see Equation 15):

$$\langle P_\tau \rangle = A_{\text{pol}}^\tau = \frac{-2v_\tau/a_\tau}{1+(v_\tau/a_\tau)^2}.$$

It can be inferred from the energies of the decay products, which in turn reflect the angle of decay in the τ rest system.

The sensitivity to the polarization varies substantially from one decay

Table 4 The observed b̄b forward-backward asymmetries A_{FB}^b, the observed mixing parameter χ, the corrected forward-backward asymmetry $A_{FB,corr.}^b$, and the corresponding weak mixing angle, for the ALEPH and L3 collaborations, as well as the combined result

	ALEPH	L3	Combined result
A_{FB}^b	0.104 ± 0.025	0.084 ± 0.025	0.094 ± 0.018
χ	$0.132 \begin{smallmatrix} +0.027 \\ -0.026 \end{smallmatrix}$	$0.178 \begin{smallmatrix} +0.049 \\ -0.040 \end{smallmatrix}$	$0.144 \begin{smallmatrix} +0.25 \\ -0.22 \end{smallmatrix}$
$A_{FB,corr.}^b$	0.141 ± 0.044	$0.130 \begin{smallmatrix} +0.044 \\ -0.042 \end{smallmatrix}$	0.135 ± 0.031
$\overline{\sin^2 \theta_w}$	0.225 ± 0.008	0.226 ± 0.008	0.226 ± 0.006

channel to another. The hadronic decay channels $\tau \to \nu_\tau + \pi$ and $\tau \to \nu_\tau + \rho$ are more useful here than the leptonic channels. The only result at present is that of ALEPH (22). Figure 26 shows the energy spectrum of the pion in the channel $\tau \to \nu_\tau + \pi$. Positive helicity corresponds to a linear

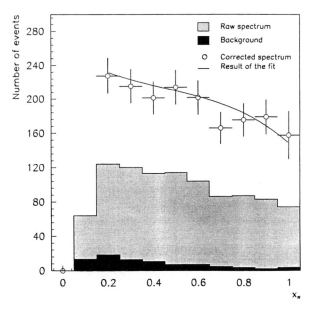

Figure 26 Energy spectrum of pions in the channel $e^+e^- \to \tau^+\tau^-$, $\tau \to \pi + \nu_\tau$, used to measure the τ polarization. A flat spectrum corresponds to no polarization. Helicity $+1$ (-1) corresponds to a linear spectrum with zero at the maximum (minimum) pion energy (from ALEPH).

distribution that is zero at $E_\pi = E_{max}$, negative to a distribution going to zero at $E_\pi = 0$. The slope of the observed distribution gives the τ polarization. In addition the $\rho\nu$, $a_1\nu$, and leptonic decay channels have been studied, with the result (22)

$$A^\tau_{pol} = \langle P_\tau \rangle = -0.152 \pm 0.045.$$

This corresponds to $\sin^2\theta_w = 0.230 \pm 0.006$.

Some measure of the forward-backward asymmetry of the sum of all quarks can be obtained from the asymmetry in the momentum-weighted charges of the jets. Such a study has been performed by ALEPH (21). The hadronic events are divided into two jets by a plane perpendicular to the thrust axis. The "charge" of each jet is determined by summing the charges of individual tracks multiplied by a power of their momentum. By simulation it is found that the power 1 maximizes the sensitivity. With this choice the charge asymmetries expected for each of the five quark flavors are determined. These differ substantially from flavor to flavor, negative for up-type and positive for down-type quarks. Only a small net charge asymmetry remains, but the statistical error is small because of the large number of hadronic events:

$$A_{FB,Q} \equiv \langle Q_{forward} - Q_{backward} \rangle = 0.0084 \pm 0.0016.$$

This net asymmetry is the flavor average of the forward-backward asymmetries multiplied by the product of the production fraction $p_f(\Sigma p_f = 1)$, times the average jet charge expected for a given quark, q_f^{jet}:

$$A_{FB,Q} = 3 \sum_f \frac{v_e}{a_e} \frac{v_f/a_f}{(1 - v_f^2/a_f^2)} p_f q_f^{jet}.$$

An uncertainty of 18% is assigned to the determination of the q_f^{jet} on the basis of quark hadronization programs. The result, in terms of $\sin^2\theta_w$, is

$$\overline{\sin^2\theta_w} = 0.2300 \pm 0.0053.$$

Combining all asymmetry measurements available at present, one obtains $\sin^2\theta_w = 0.2306 \pm 0.0022$.

5.   COMPARISON WITH THE STANDARD MODEL

The electroweak theory is defined by two coupling constants, the mixing angle, the masses of the fermions, and the mass of the Higgs particle. The coupling constants can be taken to be the fine structure constant α and the Fermi constant G_F. Instead of the mixing angle, it is convenient to use the

Z mass m_Z. One of the important results of these experiments is therefore the precise value of $m_Z = 91.174 \pm 0.021$ GeV as compared with the best previous results:

UA2 $m_Z = 91.49 \pm 0.99$ GeV   (Ref. 25)

CDF $m_Z = 90.9 \pm 0.4$ GeV   (Ref. 26)

SLAC $m_Z = 91.14 \pm 0.12$ GeV   (Ref. 27).

Fermion masses and the Higgs boson mass enter through the radiative corrections to the partial widths, cross sections, and asymmetries. The precise measurements of these quantities at the Z peak therefore permit some insight into the as-yet-unknown t quark and Higgs boson masses. As pointed out in Section 3.2, the dominant but not exclusive correction involving the t quark and Higgs boson masses is through $\Delta\rho$, where they occur approximately in the combination:

$$\Delta\rho \simeq 0.0026 m_t^2/m_Z^2 - 0.0015 \ln{(m_H/m_W)}.$$

In order to separate the effects of m_t and m_H it is therefore necessary, at least until the top is seen and its mass measured, to use the fact that some parameters, such as Γ_b, have m_t dependences not contained in ρ. This separation is not possible at present levels of precision at LEP. However, for the mass interval imagined as possible for the Higgs boson, $50 < m_H < 1000$ GeV [a lower bound of ~ 50 GeV has been established by the LEP experiments (28)], the t quark mass effect dominates. The main measured quantities relevant to a determination of this radiative correction and the t quark mass are m_Z, m_W, Γ_Z, Γ_ℓ, and $\sin^2\theta_w$ as measured in the asymmetries. Four of these are measured at LEP; the fifth, m_W, is obtained from the ratio m_W/m_Z, measured in the p$\bar{\text{p}}$ collider experiments (29) and in the ratio of neutral to charged current deep-inelastic neutrino scattering (30). The combined result for m_W, given the LEP result for m_Z, is $m_W = 80.05 \pm 0.22$ GeV.

For a particular Higgs boson mass, each of these quantities can be seen as a relationship between the top quark mass and the effective mixing angle, as shown in Figure 27 for $m_H = 200$ GeV. The lower limit on the t quark mass was obtained in the direct search by the CDF collaboration at Fermilab, $m_t > 89$ GeV, at the 90% confidence level (31). If this limit is included, Figure 27 shows everything currently known about the top quark mass as well as all the accurate information on the mixing angle. It also shows much of the relevant information about the consistency of the LEP results with the Standard Model, which is just the fact that the five experimental bands share a common area, or in other words, that best fits of $\sin^2\theta_w$ and m_t can be found with reasonable χ^2. These are

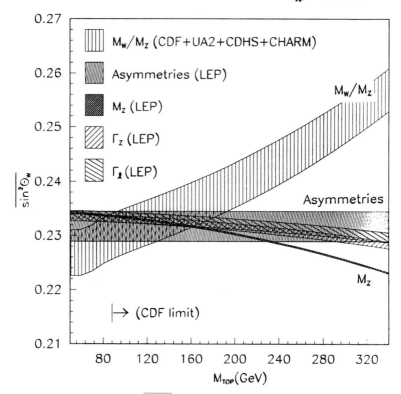

Figure 27 Correlation between $\overline{\sin^2 \theta_w}$ and m_t for m_Z, m_W, Γ_Z, Γ_ℓ, and the asymmetries, for a Higgs mass $m_H = 200$ GeV. This plot permits a visualization of what these measurements contribute to our knowledge of $\overline{\sin^2 \theta_w}$ and m_t through the radiative corrections. One can also see the compatibility of these measurements in the framework of the electroweak theory.

$$\overline{\sin^2 \theta_w} = 0.2327 \pm 0.0008 \pm 0.0003(m_H)$$

$$m_t = 126 \pm 30 \pm 18(m_H) \text{ GeV}$$

$$\chi^2 = 1.4/4 \text{ degrees of freedom,}$$

where m_H denotes the uncertainty corresponding to a Higgs boson in the mass range 50 to 1000 GeV. The precision of the new value of $\sin^2 \theta_w$ is primarily due to the precision in m_Z; Γ_ℓ, Γ_Z, and m_W contribute comparably to the limitation on the top mass.

In Table 5 the Standard Model expectations are given for the various measured quantities, as well as the value of $\overline{\sin^2 \theta_w}$ that follows from them.

Table 5 Combined LEP results and Standard Model predictions for m_Z, σ_h^{peak}, Γ_Z, Γ_h, Γ_ℓ, R, N_ν, and asymmetries[a]

	Experiment	Standard Model	$\overline{\sin^2\theta_w}$
m_Z, GeV	91.174 ± 0.021		0.2327 ± 0.0012
Γ_Z, MeV	2487 ± 9	2486 ± 10	0.2327 ± 0.0010
σ_h^{peak}, nb	41.46 ± 29	41.42 ± 0.05	0.2328 ± 0.0026
Γ_h, MeV	1744 ± 10	1737 ± 8	0.2320 ± 0.0012
Γ_ℓ, MeV	83.5 ± 0.4	83.6 ± 0.3	0.2331 ± 0.0014
$R = \Gamma_h/\Gamma_\ell$	20.94 ± 0.12	20.80 ± 0.06	0.225 ± 0.010
Γ_{inv}	493 ± 9.5	499 ± 2	0.2356 ± 0.0037
N_ν	2.96 ± 0.06	Integer	
$(v/a)_\ell^2$ from A_{FB}^ℓ	0.0054 ± 0.0016	0.0048 ± 0.0007	$0.2317 \; ^{+\,0.0030}_{-\,0.0026}$
$(v/a)_\tau$ from $A_{\tau,pol}$	0.079 ± 0.028	0.069 ± 0.005	0.231 ± 0.007
A_{FB}^b	0.135 ± 0.031	0.097 ± 0.007	0.226 ± 0.006
$A_{FB,Q}$	0.0084 ± 0.0016		0.2300 ± 0.0053

[a] In the predictions the measured values of m_Z are used, as well as $\alpha_s = 0.120 \pm 0.008$, $m_t = 125 \pm 35$ GeV, and $50 < m_H < 1000$ GeV. The column in $\sin^2\theta_w$ lists the value that follows from the particular experiment.

The Standard Model expectations assume $90 < m_t < 160$ GeV, $50 < m_H < 1000$ GeV, and where relevant, $\alpha_s = 0.120 \pm 0.008$. The agreement with the Standard Model can be judged both in comparing the experimental and theoretical values and in comparing the results for $\sin^2\theta_w$ with one another and with the value of 0.2327 ± 0.0012 that follows from m_Z. The results for Γ_Z, Γ_ℓ, and the asymmetries are largely independent of each other, and each checks the model at a level of $\Delta\sin^2\theta_w \approx 0.001$–$0.002$.

The predictions for σ_h^{peak} and R are approximately independent of the top and Higgs radiative corrections, since they are ratios of widths, and in the ratio these corrections nearly cancel. They reflect, however, the uncertainty in α_s through the QCD radiative correction of Γ_h. This introduces the main theoretical uncertainty, $\sim 0.3\%$ in R and $\sim 0.1\%$ in σ_h^{peak}. The two measurements confirm the structure of the theory at the 0.5% level.

The strong coupling constant at the Z mass, $\alpha_s(m_Z^2)$, can be deduced from the measured values of Γ_Z and R, because of the QCD correction $(1 + \alpha_s/\pi + 1.4\alpha_s^2/\pi^2)$ of the hadronic partial widths. With $m_t = 125 \pm 35$ GeV and $50 < m_H < 1000$ GeV,

$$\alpha_s = 0.119 \pm 0.028 \text{ from } \Gamma_Z,$$

$\alpha_s = 0.146 \pm 0.022$ from R,

and, combined,

$\alpha_s = 0.136 \pm 0.017$ (statistical error only).

This result is not statistically as precise as those derived from the shape parameters of hadronic events, but is useful because it has different systematic uncertainties.

6. SUMMARY AND FUTURE

The results of the cross-section measurements on and near the Z peak during this first year of LEP operation, based on the analysis of $\sim 700,000$ Z-decay events by the four experimental collaborations, have shown that there are just three neutrino families with mass less than $m_Z/2$, and therefore very probably just three fermion families. The mass of the Z has been measured with a precision of 1 part in 4000, and the predictions of the Standard Model have been verified at a level of approximately one half of a per cent for the total Z widths, for partial Z-decay widths, and for asymmetries in the angular distributions. No evidence for inadequacy of the Standard Model (which might indicate new directions in physics) has been found. A rough value for the mass of the top is obtained through the contributions of the top to radiative corrections. Other results of this first operation, such as new particle searches, QCD studies, and heavy-quark studies, are not reported here.

A program to increase the LEP energy above the W^+W^- production threshold should be completed in 1994; this will make it possible to study other electroweak processes, such as the coupling of the Z to a W^+W^- pair. Given sufficient luminosity and time, it will also permit a measurement of the W mass with an error of about 75 MeV, three times better than the present value.

In the meantime, before the higher energy is available, it is expected that the integrated luminosity at and near the Z peak will be increased by a factor of 10 or more, with consequent improvement in sensitivity to deviations from the Standard Model. It is also hoped that it will be possible to implement the longitudinal polarization of the electron beam in order to measure the polarization asymmetry with high precision. One might hope to achieve here a precision in v_e/a_e of $\sim 10^{-3}$, corresponding to an error in $\sin^2 \theta_w$ of 2.5×10^{-4}. Together, the accurate W^- mass measurement, the polarization asymmetry, and the improved precision in the cross-section and partial-width measurements will represent a new level of sensitivity to possible new phenomena in the electroweak sector.

Here we have mentioned only the expected impact of the planned LEP improvements on the electroweak theory. Of course, the increased luminosity and energy will also open up new possibilities in the areas of particle searches and B physics.

ACKNOWLEDGMENTS

It is a pleasure to thank all members of the four LEP experiments for making their results available for this review. In particular we wish to thank Drs. Ugo Amaldi, Edward Blucher, Albrecht Böhm, Marcello Mannelli, Manel Martinez, Aldo Michelini, and David Stickland for their generous collaboration.

Literature Cited

1. Glashow, S. L., *Nucl. Phys.* B22: 579 (1961); Weinberg, S., *Phys. Rev. Lett.* 19: 1264 (1967); Salam, A., in *Proc. 8th Nobel Symp.*, ed. N. Svartholm. Stockholm: Almqvist & Wiksell (1968), p. 367; Glashow, S. L., Iliopoulos, J., Maiani, L., *Phys. Rev.* D2: 1285 (1970); 't Hooft, G., *Nucl. Phys.* B33: 173 (1971) and B35: 167 (1971); Passarino, G., Veltman, M., *Nucl. Phys.* B160: 151 (1979)
2. Hasert, F. J., et al., *Phys. Lett.* B46: 138 (1973)
3. Richter, B., *Nucl. Instrum. Methods* 136: 47 (1976)
4. Arnison, G., et al. (UA1 Collab.), *Phys. Lett.* B122: 103 (1983); Banner, M., et al. (UA2 Collab), *Phys. Lett.* B122: 476 (1983); Arnison, G., et al. (UA1 Collab.), *Phys. Lett.* B126: 398 (1983); Bagnaia, P., et al. (UA2 Collab.), *Phys. Lett.* B129: 130 (1983)
5. Decamp, D., et al., *Nucl. Instrum. Methods* A294: 121 (1990)
6. Aarnio, P., et al. (DELPHI Collab.), preprint CERN-PPE/90-128, submitted to *Nucl. Instrum. Methods* (1991)
7. Adeva, B., et al. (L3 Collab.), *Nucl. Instrum. Methods* A289: 35 (1990)
8. Ahmet, K., et al. (OPAL Collab.), preprint CERN-PPE/90-114, submitted to *Nucl. Instrum. Methods* (1991)
9. Greco, M., Pancheri-Srivastava, G., Srivastava, Y., *Nucl. Phys.* B171: 118 (1980); Kuraev, E. A., Fadin, V. S., *Sov. J. Nucl. Phys.* 41: 466 (1985); Altarelli, G., Martinelli, G., in *Physics at LEP*, ed. J. Ellis, R. Peccei (Rep. CERN 86-02). Geneva: CERN (1986), p. 46; Berends, F. A., Burgers, G., van Neerven, W. L., *Nucl. Phys.* B297: 429 (1988)
10. Altarelli, G., Kleiss, R., Verzegnassi, C., eds., *Z. Physics at LEP 1* (CERN 89-08). Geneva: CERN (1989), Vol. 1, in particular the contributions of: Consoli, M., Hollik, W., pp. 7–54; Burgers, G., Jegerlehner, F., pp. 55–88; Bardin, D., et al., pp. 89–128
11. Bardeen, W. A., et al., *Phys. Rev.* D18: 3998 (1978); Chetyrkin, K. G., et al., *Phys. Lett.* B85: 277 (1979); Dine, M., Sapirstein, J., *Phys. Rev. Lett.* 43: 668 (1979); Celemaster, W., Gonsalves, R. J., *Phys. Rev. Lett.* 44: 560 (1979); *Phys. Rev.* D21: 3112 (1980)
12. Lynn, B. W., Stuart, R. G., *Nucl. Phys.* B253: 216 (1985); Borelli, A., et al., *Nucl. Phys.* B333: 357 (1990); Berends, F. A., et al., *Phys. Lett.* B203: 177 (1988)
13. Altarelli, G., Kleiss, R., Verzegnassi, C., eds., *Z Physics at LEP 1* (CERN 89-08). Geneva: CERN (1989), Vol. 3
14. Böhm, M., Denner, A., Hollik, W., *Nucl. Phys.* B304: 687 (1988); Berends, F. A., Kleiss, R., Hollik, W., *Nucl. Phys.* B304: 712 (1988); Burkhardt, H., et al., *Z. Phys.* C43: 497 (1989); Jadach, S., et al., *Phys. Lett.* B253: 469 (1991)
15. Beenakker, W., Berends, F. A., van der Marck, S. C., Small-angle Bhabha scattering, Preprint-90-0695, Leiden Univ. (1990), submitted to *Nucl. Phys. B* (1991)
16. Bailey, R., et al., LEP energy calibration, preprint CERN-SL/90-95 (1990); Hatton, V., et al., LEP absolute energy in 1990, LEP Performance Note 12, Dec. 1990
17. Presented at the Aspen Winter Conference on Elementary Particle Physics, Aspen, Colo., 1991, and in the following CERN internal notes: OPAL Physics Note 90-21 (1990); DELPHI Note 90-62 PHYS 80 (1990)

18. Decamp, D., et al. (ALEPH Collab.), *Phys. Lett.* B231: 519 (1989); B234: 399 (1990); B235: 399 (1990); *Z. Phys.* C48: 365 (1990); Aarnio, P., et al. (DELPHI Collab.), *Phys. Lett.* B231: 539 (1989); B241: 425 (1990); Abreu, P., et al., *Phys. Lett.* B241: 435 (1990); preprint CERN-PPE/90-119 (1990); Adeva, B., et al. (L3 Collab.), *Phys. Lett.* B231: 509 (1990); B236: 109 (1990); B237: 136 (1990); B238: 122 (1990); B247: 173 (1990); B249: 341 (1990); B250: 183 (1990); Akrawy, M. K., et al. (OPAL Collab.), *Phys. Lett.* B231: 530 (1989); B235: 379 (1990); B240: 497 (1990); B247: 458 (1990)
19. Computer program ZFITTER/ZBIZON: Bardin, D., et al. (Dubna-Zeuthen radiative correction group), *Z. Phys.* C44: 493 (1989); *Comp. Phys. Commun.* 59: 303 (1990); Computer program ZAPP: Berends, F. A., Burgers, G., van Neerven, W. L., *Nucl. Phys.* B297: 429 (1988); B304: 921 (1988); Computer program MIZA: Martinez, M., et al., preprint CERN-PPE/90-109, submitted to *Z. Phys.* (1991)
20. Decamp, D., et al. (ALEPH Collab.), "A measurement of the Z → b$\bar{\text{b}}$ forward-backward asymmetry." See Ref. 17; Adeva, B., et al. (L3 Collab.), *Phys. Lett.* B252: 713 (1990)
21. ALEPH Collab., *Phys. Lett.* B259: 377 (1991)
22. ALEPH Collab., See Ref. 17
23. Beenakker, W., Berends, F. A., van der Marck, S. C., *Nucl. Phys.* B349: 323 (1991)
24. Decamp, D., et al. (ALEPH Collab.), *Phys. Lett.* B258: 236 (1991); Adeva, B., et al. (L3 Collab.), *Phys. Lett.* B252: 703 (1990)
25. Alitti, J., et al. (UA2 Collab.), *Phys. Lett.* B241: 150 (1990)
26. Abe, F., et al. (CDF Collab.), *Phys. Rev. Lett.* 63: 720 (1989)
27. Abrams, G. S., et al. (Mark 2 Collab.), *Phys. Rev. Lett.* 63: 2173 (1989)
28. Decamp, D., et al. (ALEPH Collab.), *Phys. Lett.* B246: 306 (1990); Abreu, P., et al. (DELPHI Collab.), preprint CERN-PPE/90-163 (1990); Adeva, B., et al. (L3 Collab.), *Phys. Lett.* B248: 203 (1990); Akrawy, M. Z., et al. (OPAL Collab.), *Phys. Lett.* B253: 511 (1991)
29. Abe, F., et al. (CDF Collab.), *Phys. Rev. Lett.* 65: 2243 (1990); Alitti, J., et al. (UA2 Collab.), *Phys. Lett.* B241: 150 (1990)
30. Abramowicz, H., et al. (CDHS Collab.), *Phys. Lett.* 57: 298 (1986); Blondel, A., et al., *Z. Phys.* C45: 361 (1990); Allaby, J. V., et al. (CHARM Collab.), *Phys. Lett.* B177: 446 (1986); *Z. Phys.* C36: 611 (1987)
31. Abe, F., et al. (CDF Collab.), Argonne preprint ANL-HEP-PR-90-109, submitted to *Phys. Rev. D* (1991)

Annu. Rev. Nucl. Part. Sci. 1991. 41: 97–132

HADRON COLLIDER PHYSICS[1]

Marjorie D. Shapiro

Lawrence Berkeley Laboratory, Berkeley, California 94720

James L. Siegrist

SSC Laboratory, Dallas, Texas 75237

KEY WORDS: QCD, electroweak, heavy flavor, jet production, W and Z bosons, top quark

CONTENTS

[1] This work was supported by the Office of Energy Research, Office of High Energy and Nuclear Physics, Division of High Energy Physics of the US Department of Energy under Contract DE-AC03-76SF00098.

0163–8998/91/1201–0097$02.00

Hadron colliders provide an important laboratory for testing the Standard Model of strong and electroweak interactions. Because such colliders have the highest available center-of-mass energy ($\sqrt{s}$), they probe the shortest accessible length scales and hence provide a unique opportunity both to study the fundamental fields of the Standard Model and to search for deviations from the predictions of the Standard Model. We present here recent results in the field of experimental hadron collider physics.

1.   PHENOMENOLOGICAL OVERVIEW

1.1   *Particle Production in Soft Processes*

The total $p\bar{p}$ cross section, σ_{tot}, is dominated by soft processes. Because most $p\bar{p}$ interactions involve low momentum transfers, it is not possible to calculate the total cross section using perturbative quantum chromo-dynamics (QCD). Instead, it is necessary to parametrize σ_{tot} phenom-enologically. The models used to describe the features of soft hadronic interactions share two common features (1). First, because the momentum transfer in most collisions is small, particles are produced with limited transverse momentum (p_t) with respect to the incoming $p\bar{p}$ direction. Second, because there are no severe dynamical constraints in the problem, the particles have a distribution of longitudinal momenta with respect to the beamline ($p_\parallel$) that is determined chiefly by the available phase space.

The three-dimensional phase-space element can be expressed in terms of p_t and $p_\parallel$:

$$\frac{d^3p}{E} = d\phi \frac{dp_t^2}{2} \frac{dp_\parallel}{E},$$

1.

where ϕ is the azimuthal angle. Hence the invariant single-particle cross section can be written

$$E\frac{d\sigma}{d^3p} = \frac{1}{\pi}\frac{d\sigma}{dp_t^2\,dy},$$

2.

where we have integrated over azimuth and where the rapidity y is defined as

$$y \equiv \frac{1}{2}\ln\left(\frac{E+p_\parallel}{E-p_\parallel}\right)$$

3.

so that $dy = dp_\parallel/E$. In the case where particle masses can be neglected, $y \approx -\ln[\tan(\theta/2)]$. Here θ is the angle between the particle's momentum vector and the beam line. This angular variable is called pseudorapidity (η):

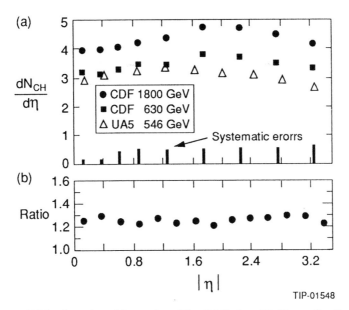

Figure 1 (*a*) The charged particle pseudorapidity distribution $dN_{ch}/d\eta$ as a function of the pseudorapidity η as measured by CDF ($\sqrt{s}$ = 1800, 630 GeV) and by UA5 ($\sqrt{s}$ = 546 GeV) (55). In all cases, the statistical uncertainty is smaller than the plotted point. An estimate of the systematic uncertainty for the CDF data is shown on the lower edge of the plot. (*b*) The ratio of $dN_{ch}/d\eta$ at 1800 GeV to that at 630 GeV.

$$\eta \equiv -\ln[\tan(\theta/2)].$$ 4.

Because it is independent of mass and therefore requires only an angular measurement, η rather than y is the variable most commonly used at hadron colliders.

Rapidity is a natural phase-space element and the distribution of particles is expected to be roughly flat in this variable. This fact is demonstrated in Figure 1, which shows the charged particle multiplicity $dN_{ch}/d\eta$ as a function of pseudorapidity for several center-of-mass energies. These data were taken using "minimum bias" triggers, triggers sensitive to the complete nondiffractive cross section. At all center-of-mass energies ($\sqrt{s}$), the cross section is nearly independent of η. The value of $dN_{ch}/d\eta$ grows slowly with increasing center-of-mass energy.[2]

---

[2] Note that the flat rapidity "plateau" must end at some value η_{max}. This value is set by the kinematic limit $\eta_{max} \approx \ln(2E/m)$, where m is the mass of the produced particle, usually a pion.

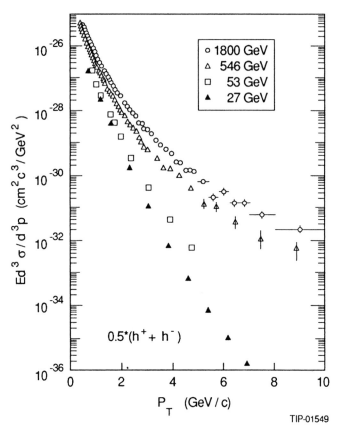

Figure 2   The energy dependence of the single-particle invariant cross section $Ed^3\sigma/d^3p$ as measured by the CDF ($\sqrt{s} = 1800$ GeV) (56), UA1 ($\sqrt{s} = 546$ GeV) (57), British-Scandinavian ($\sqrt{s} = 53$ GeV) (58), and Chicago-Princeton ($\sqrt{s} = 27$ GeV) (59) collaborations.

As can be seen in Figure 2, the single-particle p_t spectrum falls rapidly for minimum bias events. However, as the center-of-mass energy increases, a high p_t tail becomes apparent in the data. The effect is also observed (see Figure 3) in the behavior of the cross section $d\sigma/d\Sigma E_t$, where ΣE_t is the total transverse energy observed in the event:

$$\Sigma E_t \equiv \sum_i E_i \sin \theta_i. \qquad\qquad 5.$$

Here E_i is the energy in detector cell i with center at position θ_i and the sum is taken over all detector cells. The nonexponential tail at high

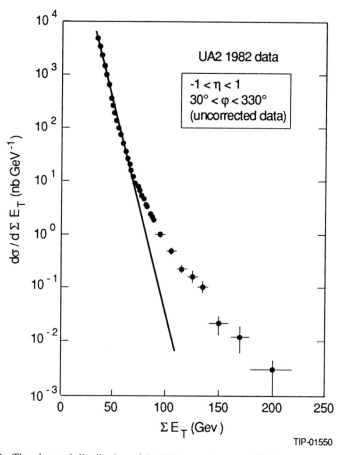

Figure 3 The observed distribution of $d\sigma/d\Sigma E_t$ as a function of ΣE_t as measured by the UA2 experiment. The solid line shows the exponential falloff at low ΣE_t.

transverse energy indicates the presence of a component of the cross section that does not result from the soft physics described above. This is the onset of hard scattering, which is the focus of the remainder of this article. [The high summed E_t tail in the E_t spectrum is a particular example of high p_t phenomena in hadron colliders (for more discussion, see 2).]

1.2 *Large Momentum Transfer Processes*

Because the strong coupling constant α_s becomes small for large momentum transfers, high p_t scattering is well described by perturbative QCD (for a review of the QCD parton model, see 3). Initial hadrons are treated as a set of quasi-free partons (quarks and gluons) that scatter elastically to produce large p_t partons. The momentum distributions of the initial partons are described by a set of distribution functions $f_i(x, \mu)$ that specify the probability for finding a parton of type i in the proton carrying a fraction of the proton's total momentum that is between x and $x + dx$. Here μ is an arbitrary scale at which the distribution functions are evaluated and should be taken to be of the order of the hard momentum scale Q. The hard scattering process is represented by the following parton model formula:

$$\sigma \approx \sum_{ij} \int dx_1 \, dx_2 \hat{\sigma}_{ij} f_i(x_1, \mu) f_j(x_2, \mu). \qquad 6.$$

Here i and j label the types of incoming partons (gluons and the various flavors of quarks and antiquarks) and $f_i(x, \mu)$ is the parton structure function for parton species i. The invariant mass of the parton-parton system ($\sqrt{\hat{s}}$) is related to the hadron-hadron center-of-mass energy ($\sqrt{s}$) by $\hat{s} = x_1 x_2 s$. The parton cross section σ_{ij} can be calculated perturbatively and is expressed as an expansion in $\alpha_s(\mu)$.

At collider energies, the hard scattering cross section is dominated by gluon-gluon scattering. This is true because color factors enhance the gluon cross section relative to quarks and because the gluon structure functions dominate at low x, where the cross section is largest. Independent of Q^2, parton elastic scattering is dominated by t-channel gluon exchange. Thus, the angular distribution in the center of mass is similar to Rutherford scattering:

$$\frac{d\sigma}{d\hat{t}} = \frac{|M|^2}{16\pi\hat{s}^2}, \qquad 7.$$

where $\hat{t}$ and $\hat{s}$ are the normal Mandelstam variables, evaluated in the parton center-of-mass system. High p_t leptons are produced by the weak decays of heavy quarks and electroweak bosons. Their production rates are therefore reduced relative to elastic parton scattering by several orders of magnitude.

The lowest-order QCD calculations provide reasonable descriptions of the inclusive jet and boson cross sections. There is always an ambiguity, however, in the normalization of such calculations because the calculated rate depends on the choice of momentum scale μ used for evaluation of α_s and for the evolution of the quark and gluon distribution functions. This theoretical uncertainty is in general reduced if a next-to-leading-order calculation is done. The contribution of the next-to-leading-order term has been calculated in several cases: the total cross section for W/Z production and for Drell-Yan scattering ($p\bar{p} \rightarrow e^+e^-$, $\mu^+\mu^-$) (4), the cross section for producing an isolated high transverse momentum photon (5), the total cross section for producing a heavy quark–antiquark pair (6), and the single-jet inclusive cross section (7).

1.3 Experimental Considerations

To study hard scattering phenomena, one must measure the properties of quarks and gluons (jets), electroweak bosons (photons, W's, Z's), and neutral, noninteracting particles (neutrinos, supersymmetric particles). Because the total inelastic cross section is so large relative to the hard scattering rate, significant event selection must be accomplished at the trigger level. These considerations place several requirements on any multipurpose detector designed to run at a hadron collider. We review here the general considerations for detector design. Descriptions of the UA1 (8) and UA2 (9) detectors at CERN and the CDF (10 and references therein) and D0 (11) detectors at Fermilab are available in the literature.

The high energies reached in hadron colliders necessitate the use of calorimetric detectors. The high multiplicity environment demands that the detector have good segmentation. The fact that inclusive production is generally flat in rapidity and uniform in ϕ (for a constant E_t cut) means that pseudorapidity and ϕ are natural segmentation variables. A large solid-angle coverage is highly desirable. Because the jet rate dominates all other processes, a high level of rejection against jet events is necessary when studying electrons, muons, and missing-energy signals. For muons and electrons, this means that high quality tracking information is important. In a high rate environment, it is advantagous to have this information available in the trigger. Good calorimeter resolution and the absence of cracks are also necessary to eliminate mismeasured jets as a major source of missing transverse energy.

Typically, collider detectors employ large sampling calorimeters. These detectors have good resolution at high energy and are sensitive to both charged and neutral particles. In most cases many longitudinal samples are summed in depth to form projective "towers" in η-ϕ. Some longitudinal segmentation, however, is essential. Calorimeters are typically divided into "electromagnetic" and "hadronic" sections, often constructed with different materials. The electromagnetic and hadronic segments can also be further subdivided to give additional information about the longitudinal shower development.

Tracking chambers are an essential ingredient of collider detectors, providing a necessary tool for lepton and photon identification. The high overall multiplicity of particles produced in hadronic collisions means that tracking detectors must have good two-track resolution and must provide high quality extrapolation to calorimeters and muon detectors. While a momentum measurement can aid in background rejection and is necessary for some physics studies (such as the measurement of jet fragmentation functions or the reconstruction of final-state particles such as K^0's, D^0's, B's), it is not essential for a collider detector. Two of the four large $p\bar{p}$ collider detectors (UA2 and D0) have no magnetic field.

The large collider detectors all operate with a number of different triggers. A prescaled minimum bias trigger provides a representative sample of nondiffractive events. Jet triggers select hard scattering events either by requiring a minimum ΣE_t in the calorimeters or requiring a localized cluster of energy above a specified E_t. Electron triggers require an electromagnetic cluster with little hadronic energy and often incorporate a tracking requirement as well. Muon triggers require a set of hits in the muon chamber that point back to the interaction region. Here again, a track requirement can also be imposed.

2.   JET PHYSICS

Jet production is the dominant high p_t process at the CERN and Tevatron colliders. The study of such jet events allows high statistics tests of the QCD model of strong interactions. The basic assumption of these measurements is that observed jet cross sections and angular distributions closely follow those of the partonic processes. This assumption relies on our ability to define a prescription for finding jets that is both experimentally straightforward and well matched to the theoretical calculation of interest (for a discussion of jet definition, see 12).

Hadronization is a soft process. Hadrons are therefore produced with limited p_t with respect to the initial parton direction, forming collimated "jets" of particles. In general, hard scattering events appear as two "beam jets" and two or more high p_t scattered jets. The beam jets are remnants of the initial protons and antiprotons after the hard scattering has occurred. The resulting beam jet particle distributions look a great deal like the soft minimum bias events discussed in Section 1.1. The presence of a hard scatter in the event can result in a higher overall multiplicity, but the "underlying event" in hard scattering processes is well described by a flat rapidity distribution of low p_t particles.

In contrast, high p_t jets can be observed as a cluster of energy. Such jets were first unambiguously observed in $p\bar{p}$ collisions by the UA2 experiment in 1982 (13). Using a simple "cluster algorithm" that combined neighboring calorimeter cells, the UA2 group found that for large ΣE_t most of the total energy observed in their calorimeter was deposited in two back-to-back clusters. Figure 4 shows the fraction of the total observed transverse energy found in the highest (h_1) and two highest (h_2) clusters as a function of the total transverse energy in the event. At high transverse energies, most of the transverse energy E_t is found in the two highest clusters. Figure 5 shows the distribution of the difference in azimuth of the two highest E_t clusters in events with $\Sigma E_t \geq 60$ GeV. The large enhancement at $\phi = 180°$ is what one expects from a hard $2 \rightarrow 2$ scattering process.

The longitudinal momentum distribution of the hadrons in the high p_t jets is governed by phase-space factors. The mathematical formalism developed in Section 1.1 holds in this case as well. Hadrons in jet events have a roughly flat distribution in rapidity that is measured relative to the jet axis. In the laboratory frame, contours of equal particle density form circles in η-ϕ space (14). It is therefore natural to define a jet in terms of the energy flow within a fixed η-ϕ cone. Fixed-cone algorithms have been used by both the UA1 (15) and CDF (16) collaborations and have been incorporated into next-to-leading-order QCD calculations (17).

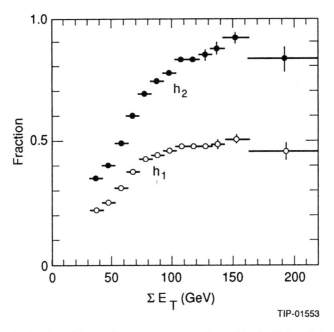

TIP-01553

Figure 4   The fraction of the total transverse energy observed in the highest (h_1) and two highest (h_2) clusters as a function of the total transverse energy of the event, as measured by the UA2 experiment.

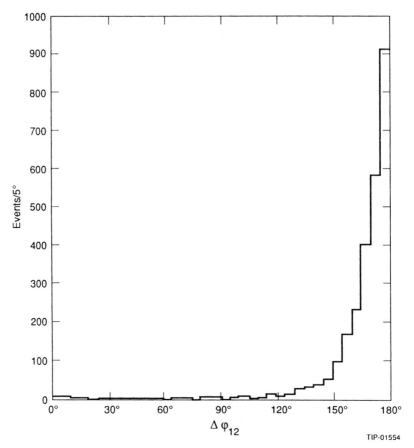

Figure 5   The distribution of the difference in azimuth between the two highest E_t clusters in events with ($\Sigma E_t \geq 60$ GeV), as measured by the UA2 experiment.

2.1 *Single-Jet Cross Section and Compositeness Limits*

At all values of p_t, the jet cross section is dominated by the t-channel exchange of a gluon. Because the matrix elements for all the dominant diagrams are similar, the relative rates of quark-quark, quark-gluon, and gluon-gluon scattering are determined by the parton distribution functions and by color factors. At low $\hat{s}$ gluon scattering dominates, while quark diagrams become important at high $\hat{s}$.

The similarity of the t-channel matrix elements allows us to write the jet cross section using a single effective subprocess (SES) approximation (18):

$$\frac{d\sigma}{dp_t\,dy_1\,dy_2} = F(x_A)F(x_B)\hat{\sigma}_{SES}(AB \to 1, 2),$$ 8.

where

$$F(x) = G(x) + \tfrac{4}{9}\Sigma_i[Q_i(X) + \bar{Q}_i(x)],$$ 9.

and the sum is taken over all quark species. UA1 and UA2 have extracted $F(x)$ using inclusive jet data and compared the resulting values to QCD predictions. The results of such a comparison are shown in Figure 6 and demonstrate that the jet cross section at low x cannot be explained by quark-antiquark scattering alone but is in good agreement with the full QCD calculation. This plot provides clear evidence for the non-Abelian nature of QCD and the existence of a three-gluon coupling.

Figure 7 shows the inclusive jet cross section measured by the CDF collaboration. The error bars plotted on the data points include both statistical errors and the portion of the systematic uncertainty that depends on E_t. In addition, the size of the overall normalization uncertainty is indicated in the insert. The curve shown with the data represents the predictions of a next-to-leading-order QCD calculation (17). The normalization uncertainty in this calculation is of order 10%. Both the overall rate and the shape of the data are in good agreement with theory.

The measurement of $d\sigma_{Jet}/dE_t$ can be used to study models of quark compositeness. If quarks are made of more fundamental objects, their strong coupling will be modified. Based on a simple assumption of color-singlet isoscalar exchange between left-handed quarks (19), the Lagrangian for this interaction is

$$\mathscr{L} = \pm \frac{g^2}{2\Lambda_c^2}(\bar{u}_L\gamma^\mu u_L + \bar{d}_L\gamma^\mu d_l)(\bar{u}_L\gamma_\mu u_L + \bar{d}_L\gamma_\mu d_l),$$ 10.

where $g^2/4\pi \equiv 1$. For energies far below Λ_c, this term acts like an effective local four-fermion interaction. The inclusive cross section will contain a term that is independent of $\hat{s}$, and will cause a flattening of the cross section as a function of E_t, that is, an excess of events in the high E_t region.

Figure 8 shows a preliminary measurement of $d\sigma/dE_t$ from the CDF experiment, along with the predictions of QCD and predictions for the composite model described above with a value of $\Lambda_c = 950$ GeV. Although a compositeness limit has not yet been set using this data, it is clear that values of Λ_c below about a TeV are excluded (20). Recent results from the UA2 collaboration give a lower limit for Λ_c of 825 GeV at a 95% confidence level (21).

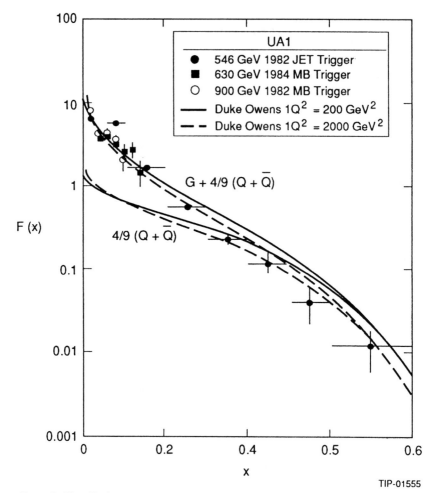

TIP-01555

Figure 6   The effective structure function as measured by the UA1 experiment. The curves show the QCD predictions at two values of scale, with and without the gluon contribution.

2.2 *Two-Jet Angular Distribution*

The majority of hard scattering events contain two back-to-back jets. This dijet system can be described in terms of six independent variables, three boost variables that transform to the hard scattering center of mass (β_x, β_y, and β_z) plus three center-of-mass variables ($\hat{s}$, the invariant mass of the hard scattering system, ϕ the azimuthal position of one of the jets, and

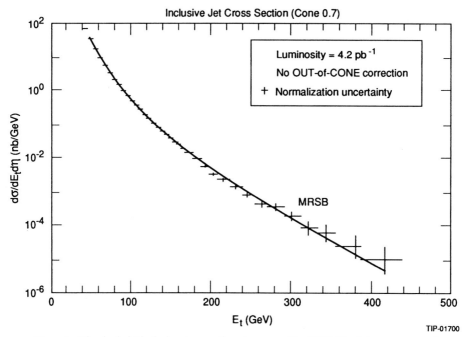

Figure 7   The single-jet inclusive cross section as measured by CDF. The data are compared to a calculation of order(α_s^3).

$\cos\theta^*$, the scattering angle of one of the jets with respect to the beam line). The distributions in two of these six variables, β_z and $\hat{s}$, are determined primarily by the parton distribution functions. The azimuth, ϕ, shows no dynamical structure for unpolarized beams.

The transverse boosts β_x and β_y result from higher-order QCD processes. These boosts are often described by the phrase "intrinsic k_t" and are caused by the emission of additional gluons during the hard scattering process. In all collider experiments, the observed dijet k_t results from two sources, the intrinsic k_t caused by gluon emission and experimental effects such as finite energy resolution. The UA2 group has developed a technique for separating these two effects. The mean value of k_t is about 5 GeV (22).

Because there is a t-channel pole in the cross section, the distribution in $\cos\theta^*$ takes on a Rutherford-like shape:

$$\frac{d\sigma}{d\cos\theta^*} \approx \alpha_s^2(\mu)\hat{s}\frac{1}{1-\cos^2\theta^*}. \qquad 11.$$

This form represents the expected angular distribution for a fixed cutoff

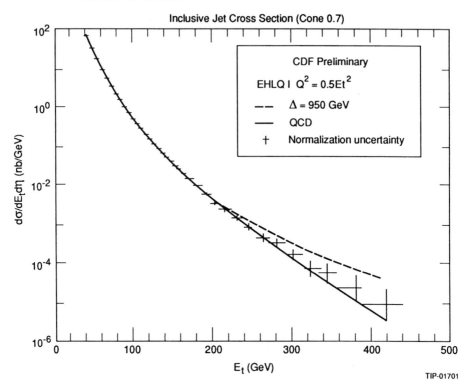

Figure 8   The inclusive cross section as measured by CDF. The predictions of QCD (EHLQ, Set I structure functions with $Q^2 = 0.5E_t^2$) and QCD modified by a compositeness term with $\Delta = \Lambda_c = 950$ MeV.

in parton invariant mass and for a fixed range in the boost parameter β_z. It is common to plot the dijet angular distribution in terms of the variable χ:

$$\chi \equiv \hat{u}/\hat{t} = \frac{1+\cos\theta^*}{1-\cos\theta^*} \approx e^{|\eta_1 - \eta_2|}. \qquad 12.$$

The distribution in χ is approximately constant for $\chi > 2$:

$$\frac{d\sigma}{d\chi} \approx \frac{\pi\alpha_s^2(Q^2)}{\hat{s}} \frac{(\chi^{-2}+\chi^{-1}+1+\chi+\chi^2)}{(1+\chi)^2}. \qquad 13.$$

Figure 9 shows the χ distribution as measured by UA1, along with QCD predictions (23). The figure shows that the parton model description pro-

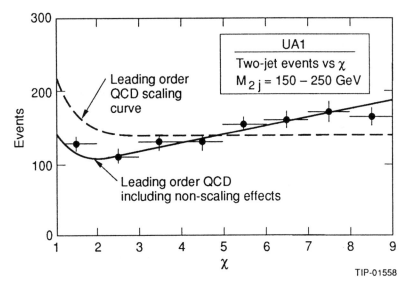

Figure 9 The distribution of χ for two-jet events as measured by the UA1 collaboration. The curve shows the predictions of a lowest-order two-parton scattering QCD calculation, with and without contributions due to QCD scaling violations.

vides a good description of the data if one includes the effect of scaling violations in the calculation.

2.3 Three-Jet Angular Distributions

The production of three-jet events is common at collider energies. These events result from the production of a hard gluon via initial- or final-state bremsstrahlung (24). The three-jet fraction is a strong function of the minimum p_t cut on the third jet; typically ~ 20% of all jet events show a third jet.

The scattering of three massless partons can be described by nine independent variables. As in the two-jet case, there are three boost variables (β_x, β_y, and β_z). The distribution of energy in the center-of-mass system is described by three internal variables: $\hat{s}$, the invariant mass of the three-jet system, and x_3 and x_4, the energy fractions of the two leading jets.[3] In addition, the orientation of the three-jet system can be described by three Euler-like angles: θ^*, the angle between the leading jet and the beam line,

---

[3] These variables are scaled to the subprocess center-of-mass energy such that $x_3 + x_4 + x_5 = 2$.

ϕ^*, the azimuthal position of the leading jet, and ψ^*, the angle of rotation about the leading jet axis (ψ^* is the angle between the plane formed by the leading jet and the beam line and the plane formed by the two other jets).

The cos θ^* distribution for three-jet events is nearly identical in shape to that for the two-jet system (25). As in the two-jet case, the dominant three-jet diagrams involve the t-channel exchange of a gluon propagator. The angular distribution in ϕ is flat since the initial partons are unpolarized, but the ψ^* distribution peaks at 0° and 180°. This structure is the result of singularities in the cross section for gluon radiation along the beam line. The regions of ψ^* near 0° and 180° are experimentally difficult to measure because the p_t of the softest jet decreases rapidly as the three-jet system is rotated into a configuration where the three jets are planar with the beam line.

Figures 10 and 11 show the distributions of the variables x_3 and x_4 as measured by the CDF experiment (26). The solid lines show the shapes these distributions would have in a phase-space model, while the diamonds

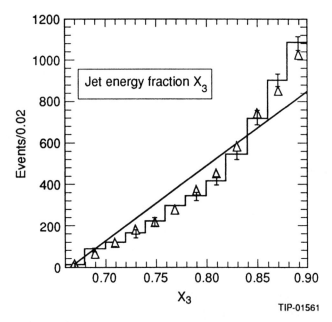

TIP-01561

Figure 10 The distributions of the variable x_3 as measured by CDF (histogram). The solid line is the prediction of a phase-space model while the diamonds show the QCD predictions.

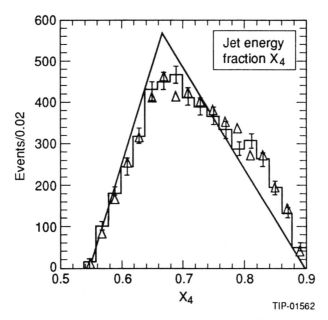

Figure 11   The distributions of the variable x_4 as measured by CDF (histogram). The solid line is the prediction of a phase-space model while the diamonds show the QCD predictions.

show the predictions of a QCD calculation. For both variables, QCD provides a good description of the data. Significant deviations from the phase-space model are seen. These plots provide evidence that three-jet events result from a bremsstrahl process.

In summary, the dynamics of jet production is well described by perturbative QCD. Quantitative agreement between theory and experiment exists over a wide kinematic range, not only for the inclusive cross section but for angular and energy variables as well. This agreement is essential to our understanding of collider physics. Since the hard scattering cross section is dominated by QCD processes, a thorough understanding of QCD phenomena is important for interpreting all collider results.

3.   LEPTON IDENTIFICATION

Leptonic decays provide an important tag of electroweak processes at hadron colliders. The presence of leptons can be used to select relatively pure samples of W and Z bosons and heavy quarks. At collider energies, however, leptons from the decay of bottom quarks are produced with low to moderate p_t; even leptons from W's and Z's have transverse momenta well below those for typical QCD jets. Jet production presents a large background to the study of low p_t leptons and must be rejected at the trigger level.

Both electron and muon identification can be enhanced with isolation cuts. Such cuts remove events with significant additional energy deposition near the lepton. Isolation criteria are sensitive to the production process and the mass of the decaying particle. Standard Model processes that tend to produce isolated leptons include Drell-Yan and weak Drell-Yan processes, and heavy quark decay for very massive top quarks. Processes that produce nonisolated leptons include b and c quark decays.

3.1 *Electron Identification*

The CDF experiment provides an example of electron identification in a magnetic detector (27). The CDF central electron trigger requires an electromagnetic energy deposit of $E_t > 12$ GeV within a "trigger tower" ($\delta\eta = 0.2, \delta\phi = 15°$) in association with a track of $p_t > 6$ GeV/c. The ratio of hadronic to electromagnetic energy in the calorimeter is required to be less than 0.125 in the trigger. This sample contains significant background from π^0-π^+ overlap, early-showering charged pions, and photon conversions and Dalitz pairs. Such backgrounds are rejected in the offline analysis in the following manner:

1. A requirement is made that the measured track momentum (p) and calorimeter energy deposit (E) be consistent (a typical cut is $E/p < 1.5$).

2. The requirement that the leakage into the hadronic section of the calorimeter be small is tightened. The ratio of hadronic to electromagnetic energy must be less than 0.05.

3. Gas proportional chambers with cathode strip readout ("strip chambers") imbedded in the calorimeter near shower maximum provide an accurate measurement of the shower position. This position can be compared to the extrapolated track position measured by the Central Tracking Chamber.

4. Events containing a single charged track and multiple π^0's are rejected by requiring the transverse spread of the electromagnetic cluster to be consistent with that expected for an electron. The lateral shape is measured in the calorimeter by studying the fraction of the energy deposited in the towers surrounding the central one hit by the electron candidate. A measurement of this shape is also made in the strip chambers, where a χ^2 test to the electron hypothesis is performed in both the wire and the strip projections.

5. Conversion electrons and Dalitz pairs are identified in the tracking chambers. The CDF algorithm is estimated to be $\sim 80\%$ efficient at finding conversions.

The UA2 detector does not have a magnetic field and hence cannot use E/p as a tool for selecting electrons. Nevertheless, the experiment has excellent electron identification (28). As in the magnetic detectors, the major methods for rejecting background are (a) requiring the electron candidate have both longitudinal and transverse shower development consistent with that expected for a single electromagnetic shower, and (b) requiring a good match between the position of the electromagnetic cluster and the extrapolated track position at the face of the calorimeter. In the UA2 detector this track position is measured using a preshower converter consisting of 1.5 radiation lengths of tungsten followed by a proportional chamber to provide a good position measurement. In an effort to improve the background rejection for lower energy and less isolated electrons from top and bottom quark decay, a major upgrade of UA2 added the following features:

1. A cylindrical drift chamber (Jet Vertex Detector). Its purpose was to measure tracks close to the beam interaction point.
2. A highly segmented silicon hodoscope. The hodoscope rejected conversion pairs by measuring dE/dx losses.
3. A scintillating fiber detector (SFD). It provided a measurement of the track position immediately in front of the central calorimeter and served as a preshower counter.
4. A transition radiation detector (TRD). This provided an independent method for separating electrons and pions.

The additional background rejection from the combination of added detector components was about a factor of 20.

3.2   Muon Identification

The principles of muon identification are presented here, using the UA1 experiment as an example (29). Muons in the UA1 detector are measured with two sets of chambers separated by 60 cm. Each set contains planes of drift chambers and limited steamer tubes. The coverage is $\approx 70\%$ of the full solid angle and the detector is placed behind approximately nine interaction lengths of iron.

The UA1 muon trigger requires a muon track "stub" consisting of at least three out of four possible hits in the chamber. The track must point back to the interaction region within a cone of ± 150 mrad. The thickness of the absorber and the pointing requirement translate to an effective minimum p_t cut of about 2 GeV on the muon trigger.

About 40% of the events written to tape by the UA1 collaboration are triggered by the muon system (29). These data are analyzed in the following manner. A filter is applied to remove cosmic rays, events with shower leakage through the calorimeter cracks and central detector tracks with obvious kinks. Muon candidates are selected by requiring a good match between the muon stub and the track measured in the central drift chamber. Single-muon events are selected by requiring a track candidate with $p_t > 6$ GeV and $|\eta| < 1.5$. A dimuon sample is selected by requiring $p_t > 3$ GeV and $|\eta| < 2.0$ for each muon (with at least one muon satisfying $|\eta| < 1.3$), and the mass of the muon pair to exceed 6 GeV.

The residual background in these samples is dominated by decays of pions and kaons and the background from misassociation of tracks in the central detector and the muon detector (≈ 0.07 per muon). Other backgrounds include noninteracting hadrons, shower leakage, particles penetrating cracks, and residual cosmic rays (total ≈ 0.025 per muon). The π and K decay background is reliably estimated by folding the measured single-hadron p_t spectrum with the probability that a hadron of a given p_t will decay to a muon with p_t^μ. That is,

$$\frac{d\sigma^{\text{background}}}{dp_t^\mu} = \int \frac{d\sigma}{dp_t^{\text{hadrons}}} \text{Prob}(p_t^{\text{hadron}} \to p_t^\mu)\, dp_t^{\text{hadron}}, \qquad 14.$$

where the probability $\text{Prob}(p_t^{\text{hadron}} \to p_t^\mu)$ is estimated using Monte Carlo techniques. For isolated muons, a requirement that the energy deposited in the calorimeter be consistent with minimum ionizing deposition can also be applied.

The size of the background in the muon sample is quite dependent on the physics process being studied. The background decreases rapidly with p_t. In addition, since most hadrons are produced within jets, an isolation cut will significantly improve the signal-to-noise ratio.

3.3   Neutrino Identification

Electroweak decays often involve the production of neutrinos. Since these particles cannot be detected directly, their presence must be inferred from the presence of a large momentum imbalance in the event. Because all collider detectors have holes in the forward and backward region to allow the beam to enter and exit the apparatus, no detector is capable of measuring the energy flow in the beam direction. Instead, the technique for finding noninteracting neutral particles involves the search for large missing transverse momentum.

Calorimetric detectors have the advantage that they are sensitive both to charged and neutral particles. They are therefore the most suitable for missing-momentum measurements. Because a calorimeter measures energy rather than momentum, the term "missing E_t" (E_t) is usually used to describe the magnitude of the missing transverse momentum. We define the missing transverse energy by the relation

$$E_t \equiv -\sum_i \mathbf{E}_t = -\sum_i E_t \hat{\mathbf{n}}_i, \qquad\qquad 15.$$

where the sum is over towers in the calorimeter and where $\hat{\mathbf{n}}_i$ is the outward-pointing normal to the tower center.

An E_t analysis is sensitive to all types of detector imperfections. Backgrounds for E_t analyses include mismeasurement of jet events due to finite detector resolution, loss of energy in cracks, and loss of jets in the beam direction. For both the UA1 and CDF experiments, and for the upgraded UA2 detector, mismeasured jets are the primary source of E_t.

For sampling calorimeters, the resolution in general scales with the square root of the incident energy. If the missing E_t resolution is dominated by calorimeter effects, then the fractional E_t resolution should scale as $1/\sqrt{\Sigma E_t}$. The UA1 group and later the CDF group have studied the E_t resolution and have found that this form holds. They therefore define the "missing E_t significance" to be

$$S \equiv \frac{E_t}{\sqrt{\Sigma E_t}}. \qquad\qquad 16.$$

4.   ELECTROWEAK PHYSICS

The study of electroweak gauge bosons is a major element of collider physics. The $W^\pm$ and Z^0 were first discovered in hadron collisions. Such collisions remain as yet the only way of producing charged vector bosons. The large QCD background inhibits the study of hadronic decays of the W and Z. Although a combined W/Z signal has been observed by UA2 in the dijet invariant mass distribution (31), most studies at hadron colliders have been limited to leptonic decays.

Collider studies of the W and Z have two major thrusts. First, measurements of the boson production properties provide tests of perturbative QCD. Second, measurements of the boson masses and decay distributions provide information on the electroweak structure of nature. In particular, precision measurements of the W mass, in conjunction with results from LEP and from deep inelastic neutrino scattering, provide an important check of electroweak radiative corrections (e.g. 32).

4.1   W Boson Production and Decay

The lowest-order process for producing W or Z bosons in a $p\bar{p}$ collider is quark-antiquark annihilation (33). This process produces no transverse momentum, and hence for leptonic decays the two leptons must balance p_t. Higher-order calculations of the W production cross section have been completed, with the following results (34): The total cross section changes by an overall factor $K \approx 1 + 8\pi/9\alpha_s(M_W^2)$. In addition, the W does not have zero p_t. A correct calculation of the transverse momentum spectrum for W bosons requires a nonperturbative treatment of multiple soft gluon emission, which is handled via resummation techniques (35). The mean transverse momentum of the W is of order 10 GeV. The measured W production cross section (multiplied by the leptonic branching ratio, B) at SPS and Tevatron energies, $\sigma B = 2.6 \pm 0.6 \pm 0.5$ nb at the Tevatron, is in good agreement with these predictions (36). Figure 12a shows the transverse momentum distribution of W candidates, as measured by the UA2 collaboration (37). The curves show QCD predictions (including soft gluon resummation) as a function of the value of Λ_{QCD} used to evaluate α_s. Figure 12b shows the predictions at high p_t in more detail. The agreement with theory is excellent.

The angular distribution of W decays is determined by helicity conservation and the spin-1 nature of the W. For $W^+ \rightarrow e^+ v$, the e^+ is preferentially produced along the $\bar{p}$ direction, and the angular distribution in the center-of-mass system is

$$\frac{d\sigma}{d\cos\theta} \approx \frac{\hat{s}(1+\cos\theta)^2}{(\hat{s}-M_W^2)^2+(\Gamma_W M_W)^2}, \tag{17.}$$

where M_W and Γ_W are the mass and decay width of the W, respectively. A transformation of variables allows us to find the electron p_t distribution (here we use the lowest-order calculation, where the W transverse momentum is constrained to be zero). In the center-of-mass frame, the e and v are back to back and balance p_t. Thus $p_t^2 = \frac{1}{4}\hat{s}\sin\theta^2$. Evaluating the Jacobian

$$\frac{d\cos\theta}{dp_t^2} = -\frac{2}{\hat{s}}\left(1-4\frac{p_t^2}{\hat{s}}\right)^{-1/2} = -\frac{2}{\hat{s}\cos\theta} \tag{18.}$$

gives the result

$$\frac{d\sigma}{dp_t} \approx \frac{1+\cos^2\theta}{\cos\theta} \sim \frac{1-2p_t^2/\hat{s}}{(1-4p_t^4/\hat{s})^{1/2}}. \tag{19.}$$

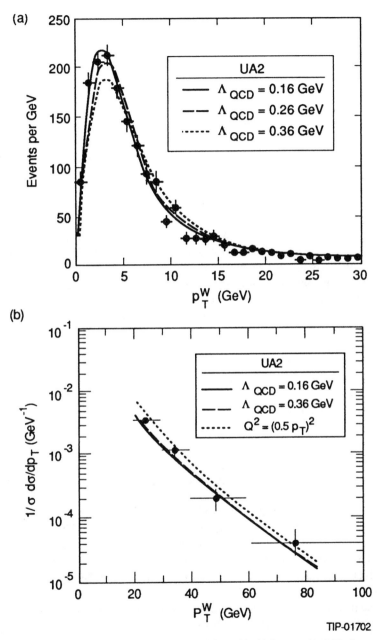

Figure 12   The W boson p_t distribution observed by UA2: (*a*) for $p_t < 30$ GeV, the curves show three different values of Λ_{QCD}; (*b*) the differential cross section for high p_t W production along with the QCD predictions for various Λ_{QCD} and scales Q^2.

The cross section diverges for $\theta = \pi/2$ (and $p_t = \sqrt{\hat{s}}/2$). This divergence is called the Jacobian peak. When the above expression is integrated over all values of $\hat{s}$, the presence of the Breit-Wigner removes the singularity but leaves a sharp peak at $p_t = \sqrt{\hat{s}}/2$. Thus, the lepton p_t spectrum is a good estimator of the W mass.

The nonzero transverse momentum spectrum of the W boson smears the electron and neutrino Jacobian peaks. In order to determine the mass of the W most accurately, one should find a variable that is less sensitive to the smearing. The natural choice is the transverse mass. If a W is produced with transverse momentum, its decay products must both be boosted. The e-v transverse mass is defined to be

$$m_T^2 = (|p_{t_e}| + |p_{t_v}|)^2 - (\vec{p}_{t_e} + \vec{p}_{t_v})^2 \qquad\qquad 20.$$

and depends on the p_t of the W only to order $(p_t/M_W)^2$. Figure 13 shows the W transverse mass distribution as measured by CDF (38). When fitting for the W mass, CDF has chosen to select a clean W sample by requiring that the event have no additional clusters (in addition to the lepton candidate) with $E_t > 7$ GeV. The curves are a fit to the data using a model that includes the predicted QCD cross section and angular distribution and the effect of finite detector resolution. The value of the mass obtained in this way is $M_W = 79.97 \pm 0.35 \pm 0.24$ GeV for decays to electrons, and $79.90 \pm 0.53 \pm 0.32$ GeV for muons, where the first error is statistical and the second is systematic. Results are in good agreement with and have similar errors to those obtained by the UA2 group: $M_W = 80.49 \pm 0.43 \pm 0.24$ GeV (39).

4.2 Z Boson Production

The production of a Z and its subsequent decay into two leptons provides an extremely clean signal. Figure 14 shows the dilepton mass distribution as measured by CDF for (a) muons and (b) electrons (40). The mass value derived from a combined fit for both decay modes is $M_Z = 90.9 \pm 0.3 \pm 0.2$ GeV. This result should be compared to the LEP result of $M_Z = 91.16 \pm 0.03$ GeV (41). Although the error is larger for CDF than for the LEP experiments, the measurement is of remarkable precision. Z production cross secions and p_t spectra both agree with QCD expectations (42).

The measurement of the ratio of the W and Z production cross sections with subsequent decays into electron(s),

$$R = \frac{\sigma(W \to ev)}{\sigma(Z \to ee)}, \qquad\qquad 21.$$

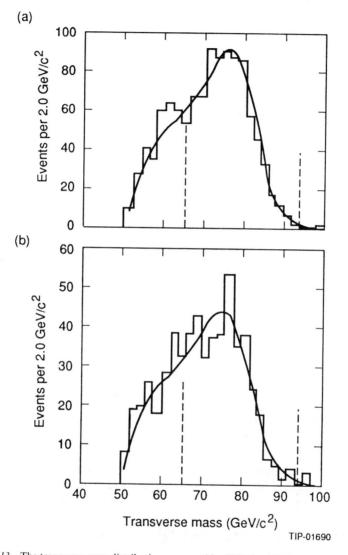

Figure 13 The transverse mass distribution measured by CDF for (*a*) W → e*v* and (*b*) W → *μv* candidates, along with the best mass fit to the data. The range of transverse mass used in the fit is indicated by the dashes.

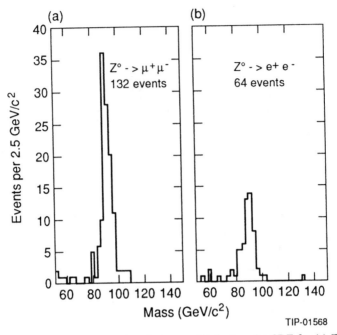

Figure 14   Number of events vs invariant mass distributions by CDF for (*a*) Z → μ⁺μ⁻ candidates and (*b*) Z → e⁺e⁻ candidates as measured with the track detector.

together with the LEP measurement of the total decay with the Z (Γ_Z), allows the total decay width of the W (Γ_W) to be calculated with greater precision than that obtained by direct measurement. The CDF measurement of $R = 10.2 \pm 0.8 \pm 0.4$ yields a result $\Gamma_W = 2.19 \pm 0.20$ GeV (43). The measurement of R provides a method for placing limits on the top quark mass that are independent of the top decay mode. The CDF R measurement excludes M_t below 41 GeV (35 GeV) at the 90% (95%) confidence level. With the existing statistical and systematic uncertainties, this limit is less stringent than those obtained at LEP (44).

5. HEAVY FLAVOR PRODUCTION

5.1 *Bottom Production*

The production of b quarks can be studied in hadron colliders by tagging their semileptonic decays. At CERN and Tevatron energies, b quarks are the most copious source of prompt leptons in the range $5 \leq p_t \leq 20$ GeV. At higher transverse momenta, the dominant mechanism for producing leptons is the decay of weak vector bosons. Figure 15 shows the inclusive

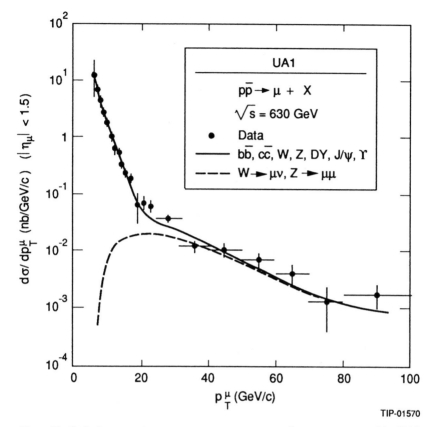

TIP-01570

Figure 15 Inclusive muon transverse momentum spectrum for muons measured by UA1. The data are compared with Monte Carlo predictions including the production of $b\bar{b}$, $c\bar{c}$, W, Z, Drell-Yan, J/ψ, and Υ. The data have been corrected for background and acceptance, but not for momentum resolution errors.

muon spectrum as measured by the UA1 collaboration (45). This spectrum is well described by Standard Model lepton sources. The combined contribution of $W \rightarrow \mu\nu$ and $Z \rightarrow \mu\mu$ decays is indicated by the dashed line. The remainder of the spectrum is dominated by b quark decays.

UA1 has studied the relative contributions of b quark, c quark, and π and K decays to the muon spectrum both in single-muon and dimuon events. Figure 16 shows several results from this study for event samples containing like-sign and unlike-sign muon pairs. Figures 16*a* and *b* show the distribution of muons as a function of the muon transverse momentum measured with respect to the direction of the nearest jet (p_t^{rel}).

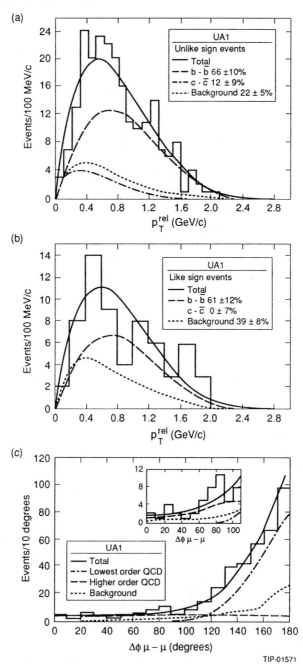

Figure 16 The distribution of p_t relative to the jet axis for muons in dimuon events, as measured by UA1.

Since b quarks are massive, they tend to produce leptons with higher p_t^{rel} than do c quarks. Heavier quarks, such as top, produce even more isolated leptons. The subject of b quark production is an area of active work in CDF (46). Preliminary results on exclusive states such as $B^+ \to J/\psi K^+$ show that b physics studies at hadron machines hold future promise in their own right. The earlier UA1 work was largely aimed at setting top quark mass limits, by studying lepton isolation distributions. More recent direct searches by CDF and UA2 based on larger data samples have since set more interesting limits.

5.2 Top Searches

The Standard Model requires the existence of a top quark, t, the SU(2) partner to the bottom quark. The nonobservation of the t at LEP and SLC places a model-independent lower limit on m_t of 45.8 GeV (44). Upper limits on the top quark mass arise from consideration of the consistency of radiative corrections and the various measurements of the Standard Model parameters. Kennedy & Langacker find $m_t < 190$ GeV at 95% confidence level for Higgs masses in the range $M_Z \leq M_H \leq 1$ TeV (47).

Next-to-leading-order calculations of the top quark production cross section in hadron colliders have been performed (48). For the CERN experiments, the $W \to t\bar{b}$ process dominates for 40 GeV $< m_t < M_W$, while at the Tevatron the dominant process is $p\bar{p} \to t\bar{t}$ throughout the mass range. In order to set limits on top quark production, semileptonic decay modes are used to reduce background from Standard Model processes. The transverse momentum of the leptons from such decays can be rather low, depending on the top quark mass: These leptons must be separated from b and c quark semileptonic decays. In addition, since the derived limit depends on the assumed t → lepton decay rate, the limit applies only to a Standard Model top quark. In models containing more than one Higgs doublet, the decay of the t to a charged Higgs boson will reduce the sensitivity of the leptonic search.

To search directly for leptons originating from top quark decay, the CDF collaboration uses both electron and muon decay modes (49). For events in which two heavy quarks decay semileptonically, the signature is two oppositely charged leptons among the final-state decay products. The cleanest combination from the point of view of other Standard Model backgrounds is the presence of one e and one μ in the event. This mode has no Drell-Yell background or direct Z decay contamination and, in spite of the small branching fraction, can be used to set interesting limits. Events are selected that contain oppositely charged electron-muon pairs above a p_t threshold. Figure 17 shows the predicted number of events above thresholds between 5 and 40 GeV for the signal with various top

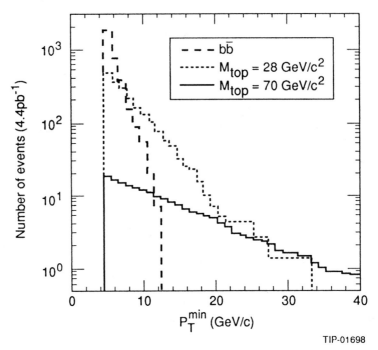

Figure 17   The predicted number of $\mu^{\pm}e^{\mp}$ events with both the E_t of the electron and the P_t of the muon greater than P_t^{min}. The histograms are for b$\bar{b}$ and t$\bar{t}$ production as indicated.

masses and the residual b$\bar{b}$ background. In addition to the electron and muon identification cuts discussed previously, "isolation" cuts are made that eliminate candidate leptons with any extra E_t (> 5 GeV) in a cone (radius $R = 0.4$ in η-ϕ) about the lepton direction. The isolation cut serves to reduce the b$\bar{b}$ background so that by setting the lepton p_t threshold to 15 GeV, the b$\bar{b}$ background is largely eliminated while retaining good efficiency for the top signal. One event is observed by CDF with both leptons above 15 GeV p_t, resulting in a lower limit on the top mass of 72 GeV (95% confidence level) (Figure 18) (49). This limit has since been raised to 89 GeV by combining the above data with a sample of dielectron candidates and by lowering the minimum p_t cut on the muon (50).

Another event topology with higher branching fraction, but significant background from W+jets production, is an event with one semileptonic decay mode, and the second top quark decays to jets. Both UA2 (51) and CDF (52) have used this channel to set limits. The transverse mass variable was used to search for an excess of events below the W Jacobian peak. The event selection required an electron, missing $E_t > 20$ GeV, and two

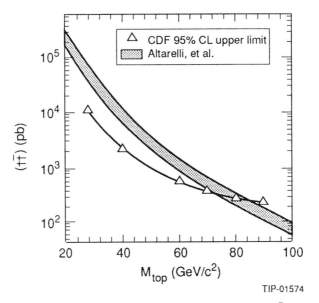

Figure 18   CDF results on the upper limit (95% confidence level) for the $t\bar{t}$ production cross section as a function of the top quark mass. The shaded band shows the result of a theoretical calculation of the $t\bar{t}$ production cross section.

or more jets with $E_t > 10$ GeV (Figure 19). The 95% confidence level lower limits obtained by this method are 77 GeV for CDF (52) and 69 GeV for UA2 (51), see Figure 20. The method based on the transverse mass distribution becomes less effective as the top quark mass limit approaches the W mass, since the changes in shape of the distribution become smaller, and the normalization of the distribution, as calculated in QCD, is somewhat uncertain.

For top quark masses above the W mass, the decay of the top leads to the production of a real W. The search strategy for the high mass range includes either a direct search for the W or a search for evidence of b or $\bar{b}$ decay fragments in the event. The search for the W decay to jets will be difficult because of the poor mass resolution and the large QCD background of lepton+jets events. The CDF group hopes to employ soft muons from the b decay, or information from the recently installed silicon vertex detectors to help enhance the search. With these improvements to the apparatus, and the expected fivefold increase in the integrated luminosity, the sensitivity is expected to extend well above 120 GeV in top quark mass.

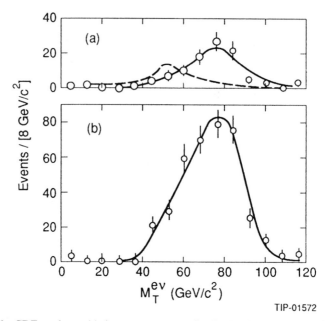

TIP-01572

Figure 19 CDF results on (*a*) the transverse mass distribution for the electron + ≥ two-jet data (*points*), and predictions for W + two-jet (*solid curve*), and t t̄ production with $M_{\text{top}} = 70$ GeV (*dashed curve*). (*b*) The measured transverse mass distribution for the electron + ≥ one-jet data together with W + one-jet prediction, normalized to equal number of events.

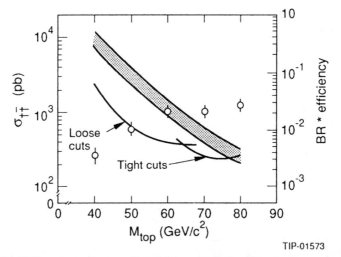

TIP-01573

Figure 20 CDF results on the upper limit (95% confidence level) for the t t̄ production cross section (*solid curve*). The predicted cross section is given by the shaded area. Plotted points show the t t̄ branching ratio multiplied by the efficiency as a function of the top quark mass (*right-hand scale*).

6. SUPERSYMMETRIC STANDARD MODEL EXTENSIONS

Supersymmetric extensions of the Standard Model hypothesize an additional global symmetry that connects particles of differing spin (for a general review, see 54). The minimal model ($N = 1$ supersymmetry) has a single generator Q_x that transforms as spin-$\frac{1}{2}$ under the Lorentz group. Q_x acting on an ordinary massless state of helicity generates a superpartner of helicity $h-\frac{1}{2}$. Because of various commutation rules satisfied by Q_x, when Q_x is applied again, the result vanishes. For this reason, the supermultiplets are doublets, with two particles that differ by half a unit of spin angular momentum. The superparticles corresponding to the spin-1 gauge bosons are the spin-$\frac{1}{2}$ gauginos—gluino, wino, zino, and photino. The superparticles corresponding to the spin-$\frac{1}{2}$ fermions are spin-0 partners to the quarks and leptons named squarks and sleptons.

Since Q_x commutes with the other operators, the quantum numbers of the superpartners are the same as for the original particles, but they carry an additional quantum number R that is absolutely conserved. Since supersymmetry is not seen readily in nature, the symmetry must be broken. The conservation of R implies that the lightest superpartners must be stable. The expected scale of supersymmetric scale breaking should be of order M_W, so such particles might be expected to appear in the fragments of hadron-hadron collisions. In p$\bar{\text{p}}$ collisions, the gluinos ($\tilde{\text{g}}$) and squarks ($\tilde{\text{q}}$) are expected to be pair produced via QCD processes with a rather high rate.

In the simplest models, the photino ($\tilde{\gamma}$) is assumed to be the lightest superparticle and hence is stable. The decay modes for $m_{\tilde{\text{q}}} > m_{\tilde{\text{g}}}$ are $\tilde{\text{q}} \rightarrow \tilde{\text{g}}\text{q}$ and $\tilde{\text{g}} \rightarrow \text{q}\bar{\text{q}}\tilde{\gamma}$, and for $m_{\tilde{\text{g}}} > m_{\tilde{\text{q}}}$, they are $\tilde{\text{g}} \rightarrow \tilde{\text{q}}\text{g}$ and $\tilde{\text{q}} \rightarrow \text{q}\tilde{\gamma}$. The final states are composed of normal quarks and gluons, along with photinos. Since the photinos do not interact in the detector, the signature for supersymmetric (SUSY) particles is the production of some number of jets along with missing transverse momentum. The Standard Model backgrounds to this signature are dominated by the $\ell\nu$ decay mode of the W, and by Z+jet events, where the Z decays to $\nu\bar{\nu}$.

The CDF group has made a search for SUSY particles in the decay modes discussed above using a data sample of 25.3 nb^{-1}. To isolate events with the expected signature, events with $E_t > 30$ GeV and one cluster with $E_t > 15$ GeV are selected. The significance of the measured missing p_t (see Equation 16) is required to be above 2.8 to eliminate dijet fluctuations. This corresponds to roughly a 4σ cut on the E_t.

To further eliminate surviving dijet background, events containing two clusters with $E_t > 5$ GeV that are back to back in ϕ to within $\pm 30°$ are removed from the sample. Cosmic rays are removed if more than 3 GeV of the energy are deposited out of time in the central hadron calorimeters. There are 115 events that survive these cuts. Since cosmic rays deposit large amounts of energy in the electromagnetic calorimeters where no timing information is available, the ratio of transverse momentum in tracks to the cluster transverse momentum is required to be above 0.2. In addition, to remove $W \rightarrow e\nu$ events, CDF requires that the leading cluster deposit at least 10% of its energy in the hadron calorimeter. After correcting the missing p_t for the presence of tracks passing through cracks in the calorimetry, two events remain. They have $E_t = 35.2$ and 36.1 GeV.

To set a limit on SUSY particle production, the systematic errors on the luminosity, event selection, and jet energy scale are taken into account. Limits are set using the most conservative set of structure functions, namely those that give the weakest limit. Figure 21 shows the result along with limits from UA1 and UA2. The discontinuity along the line $m_{\tilde{q}} = m_{\tilde{g}}$ is due to the differing acceptances for the allowed decay modes. Searches

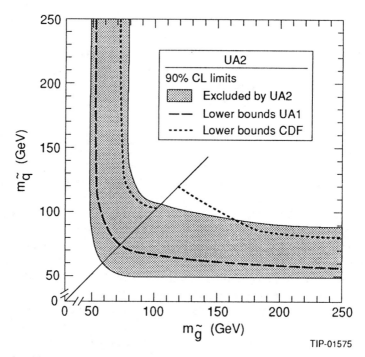

TIP-01575

Figure 21   The 90% confidence level excluded region in the $m_{\tilde{q}}$ and $m_{\tilde{g}}$ plane, as measured by the UA1, UA2, and CDF collaborations.

made in the latest data samples thus far show no evidence for SUSY particle production.

7. CONCLUSIONS

Experience with multipurpose detectors at hadron-hadron colliders is growing rapidly. The calculability of the event rates for known and hypothetical processes makes hadron-hadron colliders an interesting laboratory for both new particle searches and precision parameter measurements. This review focused mainly on measurements of particles and processes predicted by the Standard Model. A wealth of recent results shows that these detectors are also sensitive to non–Standard Model physics (53). So far, all results are in agreement with Standard Model predictions. Large data sets should be collected by the CDF and D0 experiments in the near future and should provide a wealth of new information, especially on heavy flavor (bottom and top quark) production.

Literature Cited

1. Abramovskii, V. A., Kanchelli, O. V., Gribov, V. N., In *Proc. 16th Int. Conf. on High Energy Physics*, ed. J. D. Jackson, A. Roberts. Batavia, Ill: Fermi Natl. Lab. (1972), 1: 363
2. Feynman, R. P., *Photon-Hadron Interactions*. Reading, Mass: Benjamin (1972); Berman, S. M., Jacob, M., *Phys. Rev. Lett.* 25: 1683 (1970); Berman, S. M., Bjorken, J. D., Kogut, J. B., *Phys. Rev.* D4: 3388 (1971)
3. Altarelli, G., *Phys. Rev.* 81C: 1 (1982)
4. Altarelli, G., et al., *Nucl. Phys.* B143: 521 (1978); Erratum, *Nucl. Phys.* B146: 544 (1978)
5. Aurenche, P., et al., *Phys. Lett.* B169: 441 (1986); B140: 87 (1984)
6. Dawson, S., Ellis, R. K., Nason, P., *Nucl. Phys.* B303: 607 (1988)
7. Ellis, R. K., Sexton, J. C., *Nucl. Phys.* B269: 445 (1986); Ellis, S. D., Kunszt, Z., Soper, D., Zurich preprint ETH-TH/90-3 (1990)
8. UA1 Collaboration, *Nucl. Instrum. Methods* B147: 392 (1984); *Phys. Scr.* 23: 397 (1981)
9. Beer, A., et al., *Nucl. Instrum. Methods* 224: 360 (1984)
10. Abe, F., et al., *Nucl. Instrum. Methods* 271: 487 (1988)
11. D0 Collaboration, *D0 Design Report.* Batavia, Ill: Fermi Natl. Lab. (1983)
12. Hinchliffe, I., Shapiro, M., *Rep. QCD Working Group*, in *Proc. Summer Study on High Energy Physics in the 1990's*, ed. S. Jensen. Singapore: World Scientific (1989)
13. Banner, M., et al., *Phys. Lett.* B118: 203 (1982)
14. Schwitters, R., in *Proc. 1983 SLAC Summer Inst.*, SLAC Rep. No. 267, ed. P. McDonough. Stanford, Calif: SLAC (1983)
15. Arnison, G., et al., *Phys. Lett.* B118: 185 (1982)
16. Abe, F., et al., *Phys. Rev. Lett.* 62: 613 (1989)
17. Ellis, S. D., et al., See Ref. 7
18. Owens, J. F., *Rev. Mod. Phys.* 59: 465 (1987)
19. Eichten, E., Lane, K., Peskin, M., *Phys. Rev. Lett.* 50: 811 (1983)
20. Hessing, T. L., for the CDF Collaboration see Ref. 50
21. UA2 Collaboration, CERN preprint CERN-EP/90-188 (1990); *Phys. Lett.* B257: 232 (1991)
22. UA2 Collaboration, *Z. Phys.* C30: 341 (1986)
23. UA1 Collaboration, *Phys. Lett.* B158: 494 (1985)
24. Berends, F. A., et al., *Phys. Lett.* B103: 124 (1982)
25. Appel, J., et al., *Z. Phys.* C30: 341 (1986)
26. Plunkett, R., for the CDF Collaboration, Jet Dynamics at the Tevatron Col-

lider, in *Proc. 8th Topical Workshop on Proton-Antiproton Collider Physics.* Castiglione Della Pescaia, Italy. Singapore: World Scientific (1989)

27. Abe, F., et al., *Nucl. Instrum. Methods* A271: 387 (1988); *Phys. Rev. Lett.* 62: 1005 (1989)
28. UA2 Collaboration, *Z. Phys.* C47: 11 (1990); Ansorge et al., *Nucl. Instrum. Methods* A265: 33 (1988)
29. UA1 Collaboration, *Phys. Lett.* B186: 237 (1987); *Z. Phys.* C37: 505 (1988)
30. Deleted in proof
31. UA2 Collaboration, *Phys. Lett.* B186: 152 (1987)
32. Amaldi, U., et al., *Phys. Rev.* D36: 1385 (1987); Kennedy, D. C., Langacker, P., *Phys. Rev. Lett.* 65: 2967 (1990); Altarelli, G., Barbieri, R., CERN TH.5863/90 (1990)
33. Yamaguchi, Y., *Nuovo Cimento* A43: 193 (1966); Drell, S. D., Yan, T. M., *Phys. Rev. Lett.* 25: 316 (1970)
34. Altarelli, G., et al., *Nucl. Phys.* B246: 12 (1984); Altarelli, G., Ellis, K., Martinelli, G., *Z. Phys.* C27: 617 (1985)
35. Altarelli, G., et al., *Phys. Rev. Lett.* 63: 720 (1989); Arnold, P. B., Kauffman, R. P., ANL-HEP-PR-90-70; *Nucl. Phys. B.* In press (1991)
36. UA2 Collaboration, *Z. Phys.* C47: 11 (1990); Abe, F., et al., *Phys. Rev. Lett.* 62: 1005 (1989)
37. UA2 Collaboration, *Z. Phys.* C47: 523 (1990)
38. Abe, F., et al., *Phys. Rev. Lett.* 65: 2243 (1990)
39. UA2 Collaboration, *Phys. Lett.* B241: 150 (1990)
40. Abe, F., et al. *Phys. Rev. Lett.* 63: 720 (1989)
41. Particle Data Group, *Phys. Lett.* B239, Vol. 2 (1990)
42. Abe, F., et al., Fermilab PUB-90-229-E, *Phys. Rev. D.* In press (1991)
43. Abe, F., et al., *Phys. Rev. Lett.* 64: 152 (1990)
44. ALEPH Collaboration, *Phys. Lett.* B236: 511 (1990)
45. UA1 Collaboration, see Ref. 29
46. Baden, D., for the CDF Collaboration, *B Physics at CDF*, in *Proc. SLAC Summer School Topical Conf. on Gauge Bosons and Heavy Quarks*, ed. E. Brennan. Stanford, Calif: SLAC (1990)
47. Kennedy, D. C., Langacker, P., See Ref. 32
48. Dawson, S., et al., See Ref. 6
49. Abe, F., *Phys. Rev. Lett.* 64: 147 (1990)
50. Sliwa, K., for the CDF Collaboration, *Proc. 25th Rencontre de Moriond*, ed. J. Tran Thanh Van. Savoie, France: Les Arcs (1990)
51. UA2 Collaboration, *Z. Phys.* C46: 179 (1990)
52. Abe, F., *Phys. Rev. Lett.* 64: 142 (1990)
53. UA2 Collaboration, *Phys. Lett.* B235: 363 (1990); UA2 Collaboration, *Phys. Lett.* B238: 442 (1990); Abe, F., et al., *Phys. Rev. Lett.* 62: 1825 (1989)
54. Hinchliffe, I., *Annu. Rev. Nucl. Part. Sci.* 36: 505–43 (1986)
55. Alner, C. G., et al., *Phys. Rev.* 154: 247 (1987)
56. Abe, F., et al., *Phys. Rev. Lett.* 61: 1819 (1988)
57. Arnison, G., et al., *Phys. Lett.* B118: 167 (1982)
58. Alper, B., et al., *Nucl. Phys.* B87: 19 (1975)
59. Anreasyan, D., et al., *Phys. Rev.* D19: 764 (1979)

Annu. Rev. Nucl. Part. Sci. 1991. 41: 133–85

ADVANCES IN HADRON CALORIMETRY

Richard Wigmans[1]

PPE Division, CERN, 1211 Geneva 23, Switzerland

KEY WORDS: shower development, compensation, particle identification

CONTENTS

1. INTRODUCTION

Particle detectors are the experimental tools of physicists who study the fundamental constituents of matter and the elementary forces that govern its behavior. In the past decades, the rapid increase in our knowledge in this field had gone hand in hand with a continuous refinement of the experimental techniques that were used to study the increasingly complicated subnuclear processes. The relation between experimental particle physics and research and development in the field of particle detection has

---

[1] On leave of absence from NIKHEF/H, Amsterdam, The Netherlands.

0163–8998/91/1201–0133$02.00

developed into a symbiotic one. The increasingly higher energies at which experiments could be performed required at each step different approaches for particle detection, while the new possibilities offered by more sophisticated detectors led to new discoveries that urged the need for higher energy accelerators. Obviously, this interaction process was catalyzed by the revolutions in electronics and computing.

This article reviews a class of detectors that provide a perfect example of this symbiosis. Hadron calorimeters were used for the first time in accelerator-based experiments in the 1970s, when the 400 GeV proton synchrotrons at FNAL and CERN and the proton-proton collider ISR became available. In 1983, hadron calorimeters played an important role in the detection of the intermediate vector bosons, one of the most fundamental physics discoveries in this century. This discovery demonstrated the powerful possibilities of these (at that time rather crude) instruments and motivated further development work. At present, hadron calorimeters play an important role in many ongoing experiments, and future projects at the multi-TeV proton-proton colliders LHC and SSC would be meaningless without them.

Calorimeters exist in a wide variety, but the underlying principle is simple. Basically, a calorimeter is a block of matter in which the particle to be measured interacts and transforms (part of) its energy into a measurable quantity. The resulting signal may be electrical, optical, thermal, or acoustical. It is, of course, important that the signal be proportional to the energy that one wants to measure, a situation not always easy to achieve.

The reasons why calorimeters have emerged as the key detectors in almost any experiment in particle physics can be divided into two classes. First, there are reasons related to the calorimeter properties:

1. Calorimeters are sensitive to both charged and neutral particles.
2. Owing to differences in the characteristic shower patterns, some crucial particle identification is possible.
3. Since calorimetry is based on statistical processes, the precision of the measurements improves with increasing energy, in contrast to other detectors.
4. The calorimeter dimensions needed to contain showers increase only logarithmically with the energy, so that even at the highest energies envisaged it is possible to work with rather compact instruments.
5. Calorimeters do not need a magnetic field for energy measurements.
6. They can be segmented to a high degree, which allows precise measurements of the direction of the incoming particles.
7. They can be fast—response times better than 50 ns are achievable—which is important in a high-rate environment.

8. The energy information can be used to trigger on interesting events with very high selectivity.

Second, there are reasons related to the physics to be studied. Here the emphasis has clearly shifted from the classical tracking devices, which aimed at reconstructing the 4-vectors of all individual particles produced in the interaction, to measuring more global event characteristics, indicative of interesting processes at the constituent level. These characteristics include missing (transverse) energy, total transverse energy, jet production, and multijet spectroscopy. Calorimeters are extremely well suited for this purpose.

As the name suggests, hadron calorimeters are intended for detecting hadrons. Traditionally, many experiments employ a separate electromagnetic (e.m.) calorimeter, intended for precise measurements of electrons and γ's. This e.m. calorimeter is backed up by a hadron calorimeter, which should absorb pions, kaons, protons, etc. However, most frequently these particles will interact with a nucleus in the preceding matter in this setup and may then deposit a considerable fraction of their initial energy in the e.m. calorimeter. This causes problems if the response to electrons and hadrons is different, which is usually the case. There is a growing tendency to combine both functions in one instrument, divided into an e.m. and a hadronic section. In such detectors, the e.m. section has superior properties for electron detection, for example a finer granularity. In one very recent approach, even this subdivision has been abandoned.

The title of this chapter should, therefore, not be interpreted too strictly. The instruments called hadron calorimeters may in practice also be used for the detection of other particles, such as electrons, photons, muons, and even neutrinos. Since they are being used for that purpose, I do not refrain from mentioning their properties in this respect.

The scope of this article is necessarily limited. It concentrates on the recent improvements in our understanding of the fundamental aspects of hadron calorimetry: How do these devices work, what are the fundamental limitations to their performance, and how does the calorimeter performance relate to the needs of current and future experiments? This approach is taken at the expense of the practical aspects. I do not elaborate on the technical problems that arise when a device with a mass measured in kilotons, requiring (sub)millimetric precision and containing more than 10^4 readout channels of spectroscopic quality has to be designed, built, operated, and maintained. Of course, these problems are crucial to realizing the powerful possibilities offered by this detection technique, and a lot of ingenuity and creativity have gone and are going into finding adequate or optimal solutions. For these aspects I suggest the reader consult a recent review (1), which also contains an extensive list of references to the specialized literature.

The article is also limited in that it is restricted to hadron calorimeter applications in accelerator-based experiments. Historically, the first large-scale detectors of this type were used in cosmic-ray experiments. They have found a wide application in underground experiments looking for nucleon decay, cosmic neutrinos, etc. In these experiments, the calorimeter is not only a detector but also a target and (for nucleon decay) a source. Therefore, the total instrumented mass is a very important parameter. Because of the completely different boundary conditions (low event rates, low energies, extremely rare processes), the emphasis in the development of these detectors has been on cheap, reliable technology with high signal-to-background separation capability. For a review of these detectors, I refer to a recent article by Davis et al (2).

In Section 2, I describe the various processes by which particles lose their energy when traversing dense matter and by which they are eventually absorbed. I discuss shower development phenomena, the effects of the electromagnetic and strong interactions, and the consequences of differences between these interactions for the calorimetric energy measurement of electrons and hadrons, respectively.

Section 3, which deals with the performance of calorimeter systems, starts with a discussion of the so-called compensation mechanism, its relevance for the performance of hadron calorimeters, and the methods to achieve compensation in calorimeters. The factors that determine and limit the energy, position, and time resolution are discussed, as well as the possibilities for particle identification and two-particle separation.

Section 4 begins with an overview of the different tasks performed by calorimeters in existing or planned experiments. Topics include event selection, particle identification, and multijet spectroscopy. The important points are illustrated by examples. The rest of this section is devoted to future experiments at multi-TeV pp colliders, in which calorimetry will play a determining role. The various requirements on the calorimeter performance are discussed and compared with the state of the art in calorimeter technology.

2.   THE ABSORPTION OF PARTICLES IN DENSE MATTER

When a particle traverses matter, it will generally interact and lose a fraction of its energy. The medium is excited in this process, or is heated and hence the word calorimeter. The interaction processes that play a role depend on the energy and the nature of the particle. They are the result of the electromagnetic, the strong, and more rarely, the weak forces between the particle and the medium constituents. In this section, I describe the various mechanisms by which particles may lose their energy and eventually be absorbed.

2.1   *Electromagnetic Absorption*

The best-known energy loss mechanism contributing to the absorption process is the electromagnetic interaction experienced by charged particles that traverse matter. The charged particles ionize the medium. This process forms the basis for many detectors since the liberated electrons may be collected by means of an electric field and thus yield an electric signal.

The e.m. interaction may, however, manifest itself in many other ways. Charged particles may excite atoms or molecules without ionizing them. The deexcitation from these metastable states may yield (scintillation) light, which can be used as a source of calorimeter signals. Charged particles traveling faster than the speed of light characteristic for the traversed medium lose energy by emitting Čerenkov light. At high energies, knock-on electrons (δ rays) and bremsstrahlung are produced; and even nuclear reactions induced by the e.m. interaction may occur.

The e.m. field quantum, the photon, interacts via three different processes: the photoelectric effect, Compton scattering, and electron-positron pair production. The relative importance of these three processes depends strongly on the photon energy and the electron density ($\sim Z$) of the medium.

Except at the lowest energies, the absorption of electrons and photons is a multistep process, in which particle multiplication may occur (shower development). This phenomenon, which leads to the absorption of high energy particles in relatively small volumes, is discussed in the next subsections. Muons, subject to only the e.m. interaction, do not show such behavior up to very high energies (100 GeV). They lose their energy primarily through ionization and δ rays. These mechanisms account for an energy loss of typically 2 MeV/g·cm^{-2}, and therefore it takes very substantial amounts of material to absorb high energy muons (1 TeV muons may penetrate several kilometers of the Earth's crust).

2.1.1   BELOW 10 MEV   Already at fairly low energies, relatively simple showers may develop. Let us consider as an example the γ's of a few MeV characteristic of nuclear deexcitation. Such a γ may create an electron-positron pair. These charged particles lose their kinetic energy through ionization of the medium. When the positron comes to rest, it annihilates with an electron, and two γ's of 511 keV are created. These γ's are absorbed in a series of Compton scattering processes, ending with the photoelectric effect. The Compton electrons and photoelectrons lose their energy through ionization. In the sketched process, the energy of the original γ is absorbed through ionization of the detector medium by a number of charged particles, one positron and several electrons.

At these low energies, a modest role may also be played by photonuclear reactions, e.g. γn, γp, or photoinduced nuclear fission. However, the cross sections for these processes usually do not exceed 1% of the cross sections for the processes mentioned above and may therefore, in general, be neglected.

2.1.2 ABOVE 10 MEV   Most of the energy loss mechanisms relevant to high energy shower development were mentioned above: ionization for electrons and positrons, pair production, Compton scattering, and the photoelectric effect for photons. There is one more, crucial mechanism that contributes at higher energies: bremsstrahlung.

In their passage through matter, electrons and positrons may radiate photons as a result of the Coulomb interaction with the nuclear electric field. In this process, the electron itself undergoes a change in direction (usually small) (multiple or Coulomb scattering).

Bremsstrahlung is by far the principal source of energy loss by electrons and positrons at high energies. As a consequence, high energy e.m. showers are quite different from the ones discussed in the previous subsection, since an important multiplication of shower particles occurs. A primary high energy electron (> 1 GeV) may radiate thousands of photons on its way through the detector. The photons carrying more energy than 5–10 MeV will create e^+e^- pairs. The fast electrons and positrons from this process may in turn lose their energy by radiation, and so forth. The result is a shower that may consist of thousands of different particles, electrons, positrons, and photons. The overwhelming majority of these particles is very soft. The average energy of the shower particles is obviously a function of the depth inside the detector: the further the shower has developed, the softer the spectrum of its constituents becomes.

The energy loss mechanisms are governed by the laws of quantum electrodynamics (QED) (3). They depend primarily on the electron density of the medium in which the shower develops. This density is roughly proportional to the average Z of the medium, since the number of atoms per unit volume is within a factor of about two for all solids.

The results of calculations on the energy loss mechanisms for photons and electrons are shown in Figure 1, as a function of energy, in three materials with very different Z values: carbon ($Z = 6$), iron ($Z = 26$), and uranium ($Z = 92$) (4, 5). At high energies, above 100 MeV, pair production by photons and energy loss by radiation dominate in all cases, but at low

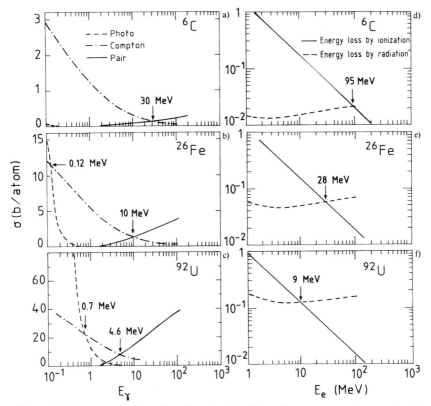

Figure 1 The cross sections for photoelectric effect, Compton scattering, and pair production, as a function of the photon energy in carbon (*a*), iron (*b*), and uranium (*c*). The fractional energy loss by radiation and ionization, as a function of the electron energy in carbon (*d*), iron (*e*), and uranium (*f*).

energies the differences among the various materials are considerable. Both the energy at which Compton scattering starts dominating pair production and the energy at which ionization losses become more important than bremsstrahlung are strongly dependent on the material and are roughly inversely proportional to Z.

These conditions determine the "critical energy" (ε_c), the point at which no further particle multiplication occurs in the shower. Above this energy, γ's produce on average more than one charged particle (pair production), and electrons lose their energy predominantly by creating new γ's. Below ε_c, γ's produce only one electron each, and these electrons do not produce new γ's themselves.

Figure 1 also shows that the contribution of the photoelectric effect is extremely Z dependent ($\sigma_{pe} \propto Z^5$). In carbon, it plays a role only at energies below a few keV, while in uranium it is the dominating process up to 0.7 MeV.

The approximate shape of the longitudinal shower profile can be deduced from Figure 1. If the number of electrons and positrons were to be measured as a function of depth in the detector, one would first find a rather steep rise due to the multiplication. This rise continues up to the depth at which the average particle energy equals ε_c. Beyond that point no further multiplication will take place, and since more and more electrons are stopped, the total number of remaining particles decreases slowly.

The positrons will be found predominantly in the early shower part, i.e. before the maximum is reached. Showers in high-Z materials will contain more positrons than in low-Z materials, because positron production continues to lower energies. The average energy of the shower particles is also lower in high-Z materials, since radiation losses dominate to lower energies. These effects will have interesting consequences.

Because the underlying physics is simple and well understood, e.m. shower development can be simulated in great detail by Monte Carlo techniques. One program, EGS4 (6), has emerged as the world-wide standard for this purpose. It is extremely reliable, and in the following sections several of its results are shown.

2.1.3 ELECTROMAGNETIC SHOWER CHARACTERISTICS   Since the e.m. shower development is primarily determined by the electron density in the absorber medium, it is to some extent possible, and in any case convenient, to describe the shower characteristics in a material-independent way. The units that are frequently used to describe the characteristic shower dimensions are the radiation length (X_0) for the longitudinal development and the Molière radius (ρ_M) for the transverse development.

The radiation length is defined as the distance over which a high energy (>1 GeV) electron loses on average 63.2% ($1-1/e$) of its energy to bremsstrahlung. The average distance that a very high energy photon travels before converting to an e^+e^- pair equals $9/7\ X_0$. The Molière radius is defined by the ratio of X_0 and ε_c, where ε_c is the electron energy at which the losses through radiation and ionization are the same. For approximate calculations, the following relations hold:

$$X_0 \approx 180 A/Z^2 \text{ (g/cm}^2\text{)} \quad \text{and} \quad \rho_M \approx 7A/Z \text{ (g/cm}^2\text{)}.$$

Expressed in these quantities, the shower development is approximately material independent. Figure 2 shows the longitudinal development of a 10-GeV electron shower in Al, Fe, and Pb, as obtained with EGS4 simu-

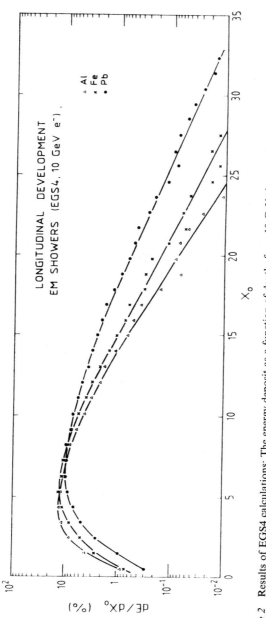

Figure 2 Results of EGS4 calculations: The energy deposit as a function of depth, for a 10 GeV electron shower developing in aluminum, iron, and lead, showing approximate scaling of the longitudinal shower profile when expressed in units of radiation length, X_0.

lations. The profile is as expected from the discussion in Section 2.1.2. Globally, it indeed scales with X_0. The differences among the various materials can be understood as well. The radiation length is defined for particles with energies in the GeV region, and therefore does not take into account the peculiarities occurring in the MeV region. The shift of the shower maximum to greater depth for high-Z absorbers is a consequence of the fact that particle multiplication continues to lower energies; the slower decay beyond this maximum is due to the fact that lower energy electrons still radiate.

The figure shows that it takes roughly $25X_0$ to absorb at least 99% of the shower energy. This corresponds to 14 cm of Pb, 44 cm of Fe, or 220 cm of Al. If the energy is increased, only very little extra material is needed to achieve the same containment. A 20 GeV photon will travel on average $1.3X_0$ before converting into an e^+e^- pair of 10 GeV each. It therefore takes only an extra $1.3X_0$ to contain twice as much energy.

The radiation length is, strictly speaking, defined for infinite energy and has no meaning in the MeV energy range. Figure 1 shows that the total cross section around the region where Compton scattering takes over from pair production is considerably lower than at very high energies, particularly in high-Z materials. As a consequence, the mean free path of photons of a few MeV is ~ 3 cm in lead or $5X_0$!

The lateral spread of an e.m. shower is caused by two effects: (*a*) electrons diverge from the axis by multiple scattering; and (*b*) in the energy region where the total cross section reaches a minimum, bremsstrahlung photons may travel quite far from the shower axis, especially if they are emitted by electrons that themselves travel under a considerable angle to this axis.

The first process dominates in the early stages of the shower development, while the second process is predominant beyond the shower maximum, particularly in high-Z media. Figure 3 shows the lateral distribution of the energy deposited by an e.m. shower in lead at various depths (7). The two components can be clearly distinguished (note the logarithmic ordinate). The radial profile shows a pronounced central core surrounded by a halo. The central core disappears beyond the shower maximum. Similar calculations in aluminum have shown that the radial profile, expressed in ρ_M units, is indeed narrower than in lead. Like the radiation length, the Molière radius does not take into account the peculiarities occurring in the MeV region.

Figure 3 shows that e.m. showers are very narrow, especially in the first few radiation lengths. The Molière radius of lead is ~ 1.5 cm. With a sufficiently fine-grained calorimeter, the showering particle can therefore be localized with a precision of ~ 1 mm (see Section 3.3).

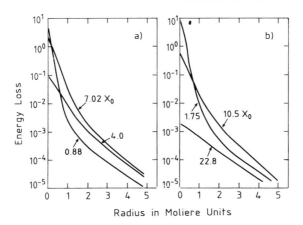

Figure 3   Results of EGS4 calculations: The lateral distribution of the energy deposited by a 1 GeV e.m. shower in lead at various depths.

2.1.4   ENERGY LOSS BY MUONS   Muons passing through matter also lose their energy through electromagnetic processes. Compared to electrons, however, the cross sections for higher-order QED processes, such as bremsstrahlung or e^+e^- pair production, are suppressed by a factor of $(m_\mu/m_e)^2 \approx 40,000$. The critical energy is, for example, at least 200 GeV. At energies below 100 GeV, the energy loss by muons will therefore be dominated by ionization processes.

The mean energy loss per unit path length for these processes, $\langle dE/dx \rangle$, is given by the well-known Bethe-Bloch formula (8). For relativistic muons, $\langle dE/dx \rangle$ falls rapidly with increasing β, reaches a minimum value near $\beta = 0.96$ (minimum-ionizing particles), then undergoes what is called the relativistic rise, leveling off at values of 2–3 MeV/g · cm^{-2} in most materials.

In practical calorimeters, the total energy loss $\Delta E/\Delta x$ may differ quite a bit from the value calculated from $\langle dE/dx \rangle$. This is because of the relatively small number of collisions with atomic electrons and the very large fluctuations in energy transfer that may occur in such collisions. Therefore, the energy loss distribution will in general be peaked at values below the ones calculated from $\langle dE/dx \rangle$ and have a long tail toward large energy losses, the so-called Landau tail (9). Only for very substantial amounts of matter, the equivalent of at least 100 m of water, will the energy loss distribution become approximately Gaussian.

At muon energies above 100 GeV, higher-order QED processes such as bremsstrahlung rapidly start dominating over ionization losses and may lead to very large energy losses in the calorimeter. This was experimentally demonstrated by the HELIOS Collaboration (10).

2.2   Strong-Interaction Processes

The absorption of particles subject to the strong interaction (hadrons) in a block of matter proceeds in a way that is very similar in many respects to the one described for electromagnetically interacting particles, although the particle production mechanisms are substantially more complicated. When a high energy hadron penetrates a block of matter, it will at some point interact with one of its nuclei. In this process, mesons are usually produced (π, K, etc). Some fraction of the initial particle energy is transferred to the nucleus. The excited nucleus will release this energy by emitting a certain number of nucleons (and at a later stage, low energy γ's) and it will lose its kinetic (recoil) energy by ionization. The particles produced in this reaction (mesons, nucleons, γ's) may in turn lose their kinetic energy by ionization and/or induce new reactions, thus causing a shower to develop.

Some of the particles (e.g. π^0, η) produced in this cascade process decay via the electromagnetic interaction. Therefore, hadron showers generally contain a component that propagates electromagnetically. The fraction of the initial hadron energy converted into π^0's and η's varies strongly from event to event, depending on the detailed processes occurring in the early phase of the shower development, i.e. the phase during which production of these particles is energetically possible.

On average, approximately one third of the mesons produced in the first interaction will be π^0's. In the second generation of interactions, the remaining hadrons may also produce π^0's if they are sufficiently energetic, and so on. And since production of π^0's by hadronically interacting mesons is an irreversible process, the average fraction of the initial hadron energy converted into π^0's increases approximately logarithmically with the energy.

Although the showers produced by hadrons and by electrons show many similarities, there exist some characteristic differences that have crucial consequences.

2.2.1   SHOWER DIMENSIONS   The hadronic shower development is based largely on nuclear interactions, and therefore the shower dimensions are governed by the nuclear interaction length, λ_{int}, which approximately scales with the nuclear radius as $A^{1/3}$ (g/cm^2).

Figure 4 shows the results of measurements that give a good impression of the longitudinal and lateral development of 300 GeV π^- showers in uranium (11). The profiles look very similar to e.m. showers (Figures 2 and 3), albeit on a very different scale. It takes about 80 cm of uranium to contain the 300 GeV π^- showers at the 95% level, while 10 cm would be sufficient for electrons at the same energy.

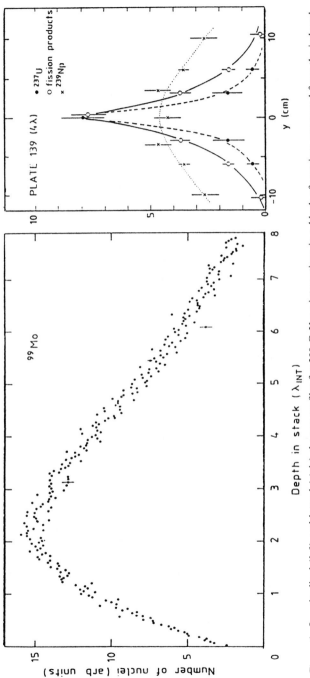

Figure 4 Longitudinal (*left*) and lateral (*right*) shower profiles for 300 GeV π^- interactions in a block of uranium, measured from the induced radioactivity. The ordinates indicate the number of radioactive decays of a particular nuclide, produced in the absorption process of the high energy pion. Since the different nuclides are produced by different types of shower particles, such experimental data may yield valuable information on details of the shower development process (from 11).

The leakage as a function of the detector depth was measured by several groups (12, 13) for hadron energies ranging from 5 to 210 GeV. It turns out that the detector size needed to contain, for instance, more than 99% of the shower energy increases only very slowly with the energy, from $6\lambda_{int}$ at 5 GeV to $9\lambda_{int}$ at 210 GeV (12).

One may use the differences in the characteristic pattern of energy deposition for particle identification. Since λ_{int} scales with $A^{1/3}$ and X_0 with A/Z^2, the separation between electromagnetically interacting particles (e, γ, π^0) and hadrons works best for high-Z materials, for which the ratio λ_{int}/X_0 may reach values larger than 30 (see Section 3.5).

2.2.2 INVISIBLE ENERGY   A second crucial difference between the shower development for high energy electrons and that for hadrons is that, in the latter case, a certain fraction of the dissipated energy is undetectable ("invisible"). Apart from neutrinos and high energy muons, which may be generated in the hadronic showers and which will generally escape the detector, I refer mainly to the energy needed to release nucleons from the nuclear field that keeps them together. Some fraction of this nuclear binding energy loss may be recovered when neutrons are captured by other nuclei. Protons, α particles, and heavier nucleon aggregates released in nuclear reactions will, however, lose only their kinetic energy, by ionization. The fraction of invisible energy can be quite substantial, up to 40% of the energy dissipated in nonelectromagnetic form (14).

At low energies (<2 GeV) the probability that charged hadrons lose their kinetic energy without undergoing nuclear interactions, i.e. by ionization alone, increases rapidly. In this case, as for muons and e.m. showers, there are no invisible energy losses. As a consequence, hadron calorimeters suffer in general from signal nonlinearities at low energy (15; see Section 3.2.2).

2.2.3 NONRELATIVISTIC SHOWER PARTICLES   A third difference, which has important consequences for the calorimetric energy measurement of elementary particles, results from the fact that a large fraction of the energy deposited in hadronic showers is carried by nonrelativistic particles, protons and neutrons. I mention three consequences:

1. Many protons produced in the shower development process have a specific ionization $\langle dE/dx \rangle$ that is 10 to 100 times the minimum-ionizing value, depending on the Z of the traversed medium. As a consequence, the fraction of the energy of such particles detected by sampling calorimeters, consisting of alternating layers of absorber and active material with very different Z values, may be considerably different from the fraction detected for minimum-ionizing particles (14).

2. Some frequently used active calorimeter media show a strongly non-

linear response to densely ionizing particles. They suffer from saturation [scintillator (e.g. 16)] or recombination effects [liquid argon (17), room-temperature liquids (18)]. Such effects may suppress the response, i.e. the signal per unit deposited energy, by as much as a factor of five (19). These effects are much smaller or are absent when gases or silicon are used as the active calorimeter medium.

3. Neutrons, which lose their kinetic energy exclusively through strong interactions, may travel quite long distances before being absorbed. In calorimeters where neutrons contribute significantly to the signal, this may lead to a considerable prolongation of the pulse duration for hadronic signals, compared with e.m. signals. Typical time constants for the neutron contribution to the calorimeter signal amount to 10 ns for the release of kinetic energy and 1 μs for the γ's created in the thermal neutron capture process. These phenomena may be exploited for particle identification, and in particular for electron-pion separation (see Section 3.5.1).

2.2.4 THE ROLE OF NEUTRONS   Regarding calorimetric applications, perhaps the most crucial difference between e.m. and hadronic shower development comes from the fact that a considerable fraction of the energy is carried by nonionizing particles, i.e. the soft (few MeV) neutrons from nuclear evaporation processes.

Since these neutrons lose their kinetic energy exclusively through collisions with atomic nuclei, their contribution to the signal of sampling calorimeters is completely dependent on the cross sections for neutron scattering in the materials composing the calorimeter, and on the kinetic energy loss of the neutrons in such reactions. It is well known that hydrogen in particular is very efficient in slowing down neutrons.

It was shown both experimentally (11) and theoretically (14, 20, 21) that in calorimeters with hydrogenous active material the neutrons generated in the shower development may deposit a large fraction of their kinetic energy in the active layers, while charged particles are sampled only at the level of a few percent. This effect is an important tool for "compensating" calorimeters.

3.   CALORIMETER PERFORMANCE

In this section, I discuss the performance of calorimeters, the instrumented blocks of dense matter in which the particles interact and are absorbed through the processes described in the previous section and thus yield signals from which the particle properties (energy, direction, type) can be derived.

Historically, one distinguishes between electromagnetic and hadronic calorimeters. There is now a growing tendency to combine these two functions into one instrument. Another distinction that may be made concerns their composition: homogeneous, fully sensitive devices as opposed to sampling calorimeters. The latter consists of a passive absorber with active material embedded in it, most frequently in the form of a sandwiched layer structure. In this way only a small fraction of the initial particle energy, ranging from 10^{-5} for gas calorimeters to a few percent for solid or liquid readout media, is deposited in the active layers.

Although additional fluctuations, affecting the energy resolution, arise because only a fraction of the energy is deposited in the active material, the sampling technique is becoming more popular, particularly in accelerator-based experiments. There are several reasons for this popularity. (a) Since very dense absorber materials can be used, calorimeters can be made extremely compact. Even at the highest energies envisaged today, 2 m of lead or uranium are sufficient to contain all showers at the 99% level. (b) At increasing energies, the energy resolution tends to be dominated by systematic effects; therefore, the effects of sampling fluctuations become less important. (c) Contrary to fully sensitive devices, sampling calorimeters can be made to be compensating. Before discussing actual devices in detail, I first elaborate on this latter point, which is crucial for the performance of hadron calorimeters.

3.1 Compensation

3.1.1 THE ROLE OF THE e/h RATIO   In a given calorimeter, hadron showers are detected with an energy resolution that is worse than that for e.m. showers. This is mainly because in hadronic showers fluctuations occur in the fraction of the total energy carried by ionizing particles. Losses in nuclear binding energy (see Section 2.2.2) may consume up to 40% of the incident energy, with large fluctuations about this average.

As a consequence, the signal distribution for monoenergetic pions is wider than for electrons at the same energy and has, in general, a smaller mean value ($e/\pi > 1$). The calorimeter response to the electromagnetic (e) and nonelectromagnetic (h) components of hadron showers shows a similar difference ($e/h > 1$). Since the event-to-event fluctuations in the fraction of the energy spent on π^0 production (f_{em}) are large and non-Gaussian, and since the average fraction f_{em} increases (logarithmically) with energy, the following effects should be expected if $e/h \neq 1$:

1. The signal distribution for monoenergetic hadrons is non-Gaussian.
2. The fluctuations in f_{em} represent an additional contribution to the energy resolution.
3. The energy resolution σ/E does not improve as $E^{-1/2}$ with increasing energy.

4. The calorimeter signal is not a linear function of the hadron energy.

5. The measured e/π signal ratio is energy dependent.

Because of the latter effect, a comparison of e/π signal ratios measured with different calorimeters only makes sense if they were obtained at the same (preferably low) energy. In order to avoid this confusion, I prefer to use the energy-independent quantity e/h, which is not directly measurable but which can be derived from a fit to the e/π signal ratios measured at different energies (22, 23). In practice, the difference between e/h and e/π is small around 10 GeV and vanishes for e/h close to unity.

All these effects have been experimentally observed (10, 24, 25) (Figure 5) and can be reproduced with a simple Monte Carlo simulation. At increasing energies, deviations from the compensation condition, $e/h = 1$, rapidly become a dominating factor degrading the calorimeter performance, in particular for the energy resolution σ/E (Figure 5a). Over one order of magnitude in energy, signal nonlinearities of $\sim 20\%$ have been observed, both in overcompensating ($e/h < 1$) and undercompensating ($e/h > 1$) calorimeters (Figure 5b). But perhaps the most severe drawback of a noncompensating calorimeter, especially in an environment where high trigger selectivity is required, is the non-Gaussian response (Figure 5c). For example, if one wants to trigger on transverse energy, it will be very difficult to unfold a steeply falling $E_\perp$ distribution and a non-Gaussian response function. Moreover, severe trigger biases are likely to occur: if $e/h < 1$ (>1) one will predominantly select events that contain small (considerable) amounts of e.m. energy.

There is general agreement that for future applications at high energy hadron calorimeters should be compensating. It should be emphasized that other sources of experimental uncertainty, such as calibration errors, will produce effects similar to those caused by deviations from $e/h = 1$. Therefore, it is not necessary that e/h be exactly 1. It has been estimated that $e/h = 1 \pm 0.05$ is adequate to achieve energy resolutions at the 1% level (14).

3.1.2 METHODS TO ACHIEVE COMPENSATION   Because of the nuclear binding energy losses typical for (the nonelectromagnetic part of) hadronic showers, one might naively expect the e/h signal ratio to be larger than unity for all calorimeters. This is, however, a tremendous oversimplification. Based on our current understanding of hadron calorimetry (14, 20, 21), it may be expected that a large variety of very different structures can actually be made compensating. A wealth of available experimental data supports the framework of this understanding, and explicit predictions were experimentally confirmed.

The response of a sampling calorimeter to a showering particle is a

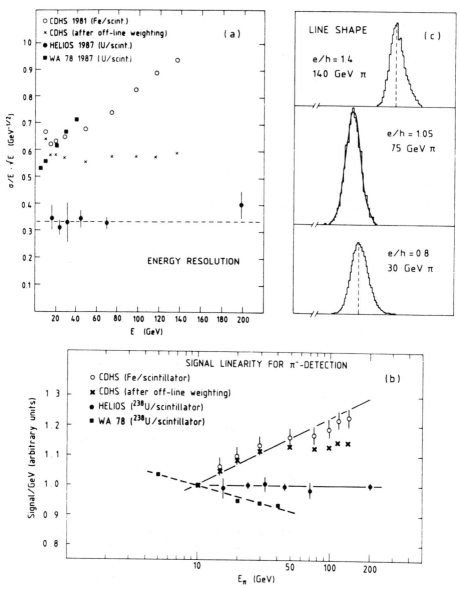

Figure 5 Measurements of pion absorption in undercompensating (24), compensating (10), and overcompensating (25) calorimeters. (*a*) The energy resolution $\sigma/E \cdot \sqrt{E}$ is given as a function of the pion energy. (*b*) The signal per GeV as a function of the pion energy, and (*c*) the signal distribution i.e. the number of events versus the pulse height in arbitrary units, for monoenergetic pions for calorimeters with different e/h ratios. The data show the advantages of $e/h \approx 1$; the energy resolution scales with $E^{-1/2}$, the signal is linear with energy, and the response is Gaussian.

complicated issue that depends on many details. This is particularly true for hadronic showers. It has become clear that showers can by no means be considered as a collection of minimum-ionizing particles with average dE/dx losses in absorber and active layers. The calorimeter signal is, to a very large extent, determined by very soft particles from the last stages of the shower development, simply because these particles are so numerous. Many observations support this statement.

Simulations of high energy e.m. showers in lead or uranium sampling calorimeters show that about 40% of the energy is deposited through ionization by electrons softer than 1 MeV (14). Measurements of pion signals in fine-sampling lead/plastic-scintillator calorimeters revealed that there is almost no correlation between the particles contributing to the signal of consecutive active layers (26, 27). This proves that the particles that dominate the signal travel on average only a very small fraction of a nuclear interaction length.

For a correct evaluation of the e/h ratio of a given calorimeter, the last stages of the shower development must therefore be understood in detail; i.e. the processes at the nuclear and even the atomic level must be analyzed. The particles that decisively determine the calorimeter response are soft photons in the case of e.m. showers and soft protons and neutrons from nuclear reactions in nonelectromagnetic showers. Since most of the protons contributing to the signal are highly nonrelativistic, the saturation properties of the active material for densely ionizing particles are of crucial importance (see Section 2.2.3).

There are many other factors that affect the signals from these shower components, and thus affect the e/h ratio. Among them are material properties, such as the Z values of the active and passive components, the hydrogen content of the active media (see Section 2.2.4), the nuclear level structure, and the cross section for thermal neutron capture by the absorber. There are also detector properties affecting the signals, such as its size, the signal integration time, the thickness of the active and passive layers, and the ratio of these thicknesses.

In order to achieve compensation, three different phenomena may be exploited:

1. The nonelectromagnetic response may be selectively boosted by using depleted uranium (^{238}U) absorber plates. The fission processes induced in the nonelectromagnetic part of the shower development yield extra energy, mainly in the form of soft γ's and neutrons (26). This phenomenon also led to the commonly used terminology, since the extra energy released in ^{238}U fission compensates for the nuclear binding energy losses.

2. One may selectively suppress the e.m. response by making use of the peculiarities of the energy deposition by the soft-photon component of

e.m. showers. Below 1 MeV, the photoelectric effect is an important energy loss mechanism. Since the cross section is proportional to Z^5, soft photons will interact almost exclusively in the absorber layers of high-Z sampling calorimeters. They will contribute to the signal only if the interaction takes place sufficiently close to the boundary layer, so that the photoelectron can escape into the active material. This effect may lead to a considerable suppression of the e.m. response (14, 21). It may be enhanced by shielding the active layers by thin sheets of passive low-Z material (19, 22, 28–30).

3. The most important handle on e/h is provided by the neutron response, in particular for calorimeters with hydrogenous active material (see Section 2.2.4). In this case, the fraction of the neutron's kinetic energy transferred to recoil protons in the active layers varies much more slowly with the relative amounts of passive and active material than does the fraction of the energy deposited by charged particles. Therefore, the relative contribution of neutrons to the calorimeter signal, and hence to e/h, can be varied by changing the sampling fraction (14, 20). A small sampling fraction enhances the relative contribution of neutrons. It is estimated that in compensated lead- or uranium-scintillator calorimeters, neutrons make up on average about 40% of the nonelectromagnetic signal (21, 22). The lever on the e/h ratio provided by this mechanism may be considerable. It depends on the energy fraction carried by soft neutrons (favoring high-Z absorbers), on the hydrogen contents in the active medium, and on the signal saturation for densely ionizing particles (Birk's constant for scintillators).

Apart from these methods, which aim at achieving $e/h = 1$ as an intrinsic detector property, a completely different approach has been applied, in order to reduce the mentioned disadvantages of an intrinsically non-compensating detector by means of off-line corrections to the measured data (24, 31). In this approach, which requires a very fine-grained detector, one tries to determine the π^0 content on a shower-by-shower basis, and a weighting scheme is used to correct for the different calorimeter responses to the π^0 and non-π^0 shower components.

An example of the results of calculations on e/h for uranium calorimeters is shown in Figure 6 (14). For hydrogenous readout materials (plastic-scintillator, warm liquids) the e/h value sensitively depends on the relative amount of active material, and configurations can be found with $e/h = 1$.

Experimental results clearly confirm the tendency predicted for plastic-scintillator readout (10, 25, 32). For nonhydrogenous readout materials, the neutron response, and hence the e/h ratio, can be affected through the signal integration time, taking advantage of the considerable energy released in the form of γ's when thermal neutrons are captured by nuclei, a process that occurs on a time scale of 1 μs. Mechanism 3 above cannot

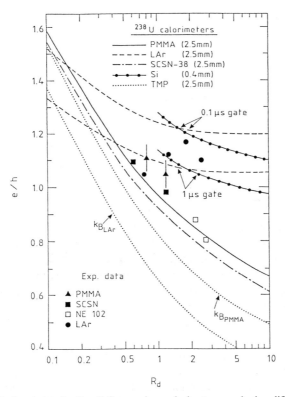

Figure 6 Calculated signal ratio e/h for uranium calorimeters employing different readout materials, as a function of the ratio R_d, the thicknesses of absorber and readout layers. Results of experimental measurements are included. The data concern calorimeters with the following readout materials: Plastic scintillator [polymethyl methacrylate (PMMA) or the polystyrene-based scintillator SCSN-38]; liquid argon (LAr); silicon (Si); or the warm liquid tetramethyl pentane (TMP). The saturation properties of the latter were assumed to obey Birk's law, with constants k_{BPMMA} and k_{BLAr} respectively. (From 14.)

be applied for such calorimeters. Experimental results obtained so far seem to confirm the prediction that it will be hard to achieve full compensation with liquid argon (LAr) readout (33–35). The data also confirm that e/h depends on the signal integration time Δt. The values range from 1.17 for $\Delta t = 100$ ns (23, 35) to 1.05 for $\Delta t = 2$ μs (23). The e/h values for Pb/LAr detectors are typically 10–15% larger, because fission energy is absent and because the e.m. shower response is less suppressed than in uranium. In U/Si detectors, a full exploitation of mechanisms 1 and 2 might yield a compensating calorimeter, since there are no saturation effects (19, 28, 30), but so far no calorimeter of this type has been constructed that was large enough to contain hadronic showers at a sufficient level to test this hypothesis. Detectors with gaseous readout media offer a convenient way to tune e/h to the desired value, i.e. through the hydrogen content of the gas mixture. This has been demonstrated experimentally by the L3 Collaboration (36). The curves for (the hydrogenous) TMP calorimeters given in Figure 6 are based on the assumption that the signal saturation in this liquid is equal either to LAr or to PMMA plastic scintillator. Experimental data indicate that the signal suppression in warm liquids is considerably larger, and that it depends on the electric field strength and on the particle's angle with the field vector (37, 38). No reliable experimental results on e/h exist so far for this type of calorimeter.

In summary, compensation is not a phenomenon restricted to uranium calorimeters, nor does the use of uranium absorber guarantee compensation. It has become clear that both the readout medium and the absorber material determine the e/h value. Compensation is easier to achieve with high-Z absorbers because of the large neutron production and the correspondingly large leverage on e/h. But even materials as light as iron allow compensation, if used in combination with, for instance, plastic scintillator, albeit with impractically thick absorber plates (14).

The neutron production in lead is considerably smaller than in uranium. In order to bring e/h to 1.0 for lead-scintillator detectors, the neutron signal must be more enhanced relative to charged particles than for uranium-scintillator calorimeters. As a consequence, the optimum sampling fraction is smaller for lead. The calculations (14) predicted e/h would become 1.0 for lead plates about four times as thick as the scintillator, while for uranium plates a thickness ratio of 1:1 is optimal. This prediction was experimentally confirmed by the ZEUS Collaboration (39). The authors found $e/h = 1.05 \pm 0.04$, an hadronic energy resolution scaling with $E^{-1/2}$ over the energy range 3–75 GeV, and no deviations from a Gaussian line shape.

The mechanisms described above, which make compensating calorimeters possible, apply only to sampling calorimeters. They are based on the fact that only a small fraction of the shower energy is deposited in the active part of the calorimeter; thus, by carefully choosing the parameter values, one may equalize the response to the electromagnetic and non-electromagnetic shower components. This does not work for homogeneous devices, because losses will inevitably occur that cannot be compensated for in the nonelectromagnetic shower part. Measurements performed so far with homogeneous hadron detectors support this conclusion. These detectors are found to be strongly undercompensating ($e/h \gg 1$). The measurements also show the resulting nonlinearity, the non-Gaussian response, and the poor energy resolution (40, 41).

3.2 *Energy Response and Resolution*

3.2.1 FLUCTUATIONS IN THE ENERGY MEASUREMENT   The detection of particle showers with calorimeters is based on statistical processes: the production of ionization charge, scintillation or Čerenkov photons, phonons, or electron-hole pairs in semiconductors. The energy resolution for particle detection is therefore determined, among other factors, by fluctuations in the number n of primary, uncorrelated processes. The relative width of the signal distribution σ_S/S for the detection of monoenergetic particles with energy E will therefore relate to n as $\sigma_S/S \approx \sqrt{n}/n$, which for linear calorimeters leads to the familiar relation $\sigma_E/E = c/\sqrt{E}$.

It has become customary to quote a value of c to express calorimetric energy resolutions, where E is given in units of GeV. Because of the statistical nature of calorimetry, the relative energy resolution σ_E/E improves with increasing energy. This very attractive feature is one of the reasons why these instruments have become so popular.

Fluctuations in the number of primary processes constituting the calorimeter signal form the ultimate limit for the energy resolution. However, in most detectors the energy resolution is dominated by other factors. These factors may concern statistical processes with a Gaussian probability distribution, or they may be of a different nature. In the latter case, their contribution to the energy resolution will cause deviations from the $E^{-1/2}$ scaling law. Such deviations are, of course, most apparent at high energies. Among the factors that will cause deviations from $E^{-1/2}$ scaling are instrumental ones: noise and pedestal contributions to the signal, uncertainties due to calibration and nonuniformities, or incomplete shower containment.

The energy resolution of sampling calorimeters is frequently dominated by the very fact that the shower is sampled (42). The nature of these sampling fluctuations is purely statistical and, therefore, they contribute as $c/\sqrt{E}$ to the final energy resolution. It is not possible to calculate the contribution of sampling fluctuations to the energy resolution with one single formula. Much depends on specific details of the particular combination of active and passive materials of which the calorimeter is composed. On the other hand, the contribution of sampling fluctuations to the energy resolution can be measured in a straightforward way, by comparing energy resolutions measured with different fractions of the active calorimeter channels (26, 27). From such measurements and from Monte Carlo simulations of shower development (43), the following picture has emerged. A major contribution to the sampling fluctuations comes from fluctuations in the number of different shower particles contributing to the calorimeter signal. In some devices (e.g. gas or Si readout), the fluctuations in the energy that the individual shower particles deposit in the active calorimeter layers also have to be taken into account. Many of the charged particles produced in hadronic showers have a dE/dx of 100–1000 times the minimum-ionizing value (protons, nuclear fragments). And since the signals from gas and Si detectors do not saturate for these densely ionizing particles, local "hot spots" may form at random places in the shower development, a phenomenon known as "Texas towers."

The contribution of sampling fluctuations to the energy resolution depends on the sampling fraction, i.e. the fraction of the energy that is deposited in the active calorimeter layers, and on the sampling frequency, i.e. the number of readout active layers in a given detector volume, or to be more precise, the total boundary surface between active and passive material in this volume. It has been established that, for a given sampling fraction, the contribution of sampling fluctuations to the energy resolution scales as $\sigma_{samp}/E = 1/\sqrt{f}$ with the sampling frequency f (44). In calorimeters with dense active material (plastic-scintillator, LAr), the contribution of sampling fluctuations to the e.m. energy resolution tends to scale as $\sigma_{samp}/E = \sqrt{t_{abs}/E}$ for a particular combination of passive and active material, a fixed thickness of the active planes, and a thickness t_{abs} of the passive planes. The sampling fluctuations also depend on the thickness t_{act} of the active planes. Photon conversions in these planes contribute a term that scales as $c\sqrt{1/t_{act}}$ for fixed t_{abs} (15). The relative contribution of this term depends on the Z values of the active and passive calorimeter layers. For Fe/LAr, one finds that σ_{samp}/E scales approximately as $t_{act}^{-1/4}$ for fixed t_{abs}.

When detecting electrons and hadrons with the same calorimeter, sampling fluctuations for the latter particles are considerably larger. First of all, the number of different shower particles contributing to the hadronic signal is smaller, because (a) individual shower particles may traverse many planes, and (b) the average energy deposited by individual particles in the active layers is larger (soft protons!). Moreover, the spread in the dE/dx loss of individual shower particles in the active layers is much larger than for e.m. showers. In hadronic shower detection, two additional sources of fluctuation play a role, sources that have no equivalent in e.m. calorimeters and that tend to dominate the energy resolution of practically all hadron calorimeters constructed up to now. First, there are the effects of the non-Gaussian fluctuations in the π^0 shower component f_{em} (Section 3.1.1); these contribute a constant term to the energy resolution, which only vanishes for compensating detector structures.

Second, there are intrinsic fluctuations, in the fraction of the initial energy that is transformed into ionizing shower particles (the visible energy; see Section 2.2.2). These fluctuations form the ultimate limit for the energy resolution achievable with hadron calorimeters. In general one may therefore write (ignoring instrumental contributions such as shower leakage, noise, calibration, etc)

$$\sigma_{had}/E = \sqrt{\frac{c_{int}^2 + c_{samp}^2}{E}} + a. \qquad\qquad 1.$$

This formula shows that at high energy one will want to have a as small as possible, which requires compensation; moreover, it is useless to make the sampling much finer than the limit set by the intrinsic fluctuations.

It turns out that calorimeters with hydrogenous readout are not only advantageous for achieving compensation: they may also yield considerably lower values for c_{int} than other detectors (14). The intrinsic resolution is largely dominated by fluctuations in the nuclear binding energy losses. Since most of the released nucleons are neutrons in the case of high-Z target material, there is a correlation between this invisible energy and the kinetic energy carried away by neutrons. Efficient neutron detection therefore reduces the effect of the intrinsic fluctuations on the energy resolution.

The ZEUS Collaboration measured the intrinsic resolution limit for compensating uranium- and lead-scintillator calorimeters to be $20\%/\sqrt{E}$ and $13\%/\sqrt{E}$, respectively (27). This difference, which means that in principle better energy resolutions can be achieved with lead calorimeters than with uranium ones, can be explained as follows. The extent to which the mechanism described above works depends on the degree of correlation between the nuclear binding energy losses and the kinetic neutron energy. This correlation is expected to be better in lead than in uranium, since many of the neutrons in uranium come from fission processes. These fission neutrons are less strongly correlated to the nuclear binding energy losses.

3.2.2   THE ENERGY RESOLUTION AND RESPONSE OF HADRON CALORIMETERS
As pointed out in Section 3.1.1, the energy resolution σ/E does not in general scale as $E^{-1/2}$ for hadron calorimeters. Only for devices with e/h sufficiently close to 1.0 is such scaling observed down to values of $\sigma/E \approx 2\%$, where instrumental effects usually start dominating the results.

The energy resolution of homogeneous detectors is dominated by their noncompensating nature. The value of σ/E does not improve below $\sim 10\%$, even at energies as high as 150 GeV (40, 41).

Most sampling calorimeters currently employed as hadron detectors use iron as the absorber material, with active layers consisting of plastic scintillator, LAr, or wire chambers. None of these (undercompensating,

$e/h > 1$) detectors has achieved energy resolutions better than $\sim 50\%/\sqrt{E}$ (at 10 GeV), while rapid deviations from $E^{-1/2}$ scaling occur at higher energies. Similar results were obtained with the overcompensating ($e/h < 1$) uranium-scintillator calorimeter operated by the WA78 experiment (25).

Efforts to determine the π^0 content on a shower-by-shower basis and to correct the signal by means of a weighting scheme (24, 31) did result in a restoration of the $E^{-1/2}$ scaling for detection of single pions of known energy; these weighting algorithms may, however, introduce signal nonlinearities, and doubts remain about the applicability of such a scheme for detecting jets of unknown composition and energy (22, 31).

The deviations from $E^{-1/2}$ scaling are less dramatic in U/LAr calorimeters, which have e/h values closer to unity. The energy resolution obtained by the D0 collaboration amounts to about $60\%/\sqrt{E}$ at 10 GeV (33); SLD found a similar value in their initial prototype studies (34).

With compensating calorimeters, considerably better energy resolutions have been obtained. For uranium/plastic-scintillator detectors, values of $\sim 35\%/\sqrt{E}$ were reported by the HELIOS (10) and ZEUS Collaborations (32), scaling with $E^{-1/2}$ up to ~ 100 GeV. ZEUS also measured $44\%/\sqrt{E}$ for a compensating Pb/plastic-scintillator sandwich detector (10 mm Pb/2.5 mm scintillator) (39). Given that the contribution of intrinsic fluctuations to the resolution is only $13\%/\sqrt{E}$ (27), the latter result is very strongly dominated by sampling fluctuations (Equation 1). This is a consequence of the small sampling fraction ($\sim 2\%$) needed to make lead-scintillator calorimeters compensating. The only way to improve this result, while maintaining compensation, is therefore to increase the sampling frequency. The SPACAL Collaboration (44) achieved this by using 1 mm thick scintillating plastic fibers as active material (Figure 7). Compared to the compensating lead/scintillating-plates calorimeter built by ZEUS, the sampling fluctuations were suppressed by a factor of 4 in this way. The energy resolutions σ/E for electrons and pions were measured to scale as $12.9\%/\sqrt{E} + 1.2\%$ and $27.7\%/\sqrt{E} + 2.5\%$, respectively. The energy-independent terms were shown to be mainly due to instrumental effects (e.g. light attenuation in the fibers) that could be eliminated using knowledge of the impact point of the particles. For single pions, an energy resolution of $30.6\%/\sqrt{E} + 1.0\%$ was obtained in that way (45).

Comparing energy resolutions measured with different calorimeters is a delicate matter since the results may have been obtained in very different ways. For example, in order to contain pion showers of more than 100 GeV to such a degree that the energy resolution is not dominated by fluctuations in longitudinal shower leakage, calorimeters should be at least $8\lambda_{int}$ deep, which in practice is rarely the case. In order to eliminate this

Figure 7   The SPACAL detector built at CERN as a research and development project for detectors at the multi-TeV pp colliders. The detector consists of scintillating plastic fibers embedded in a lead matrix at a volume ratio 4:1, needed for compensation. In total 176,855 fibers, bunched together in 155 hexagonal towers, were used to build this 13-ton, $9.5\lambda_{int}$ deep calorimeter. Each tower is read out by one photomultiplier. The fibers are running longitudinally, that is in the direction of the incoming particles. This leads to perfect hermeticity, but makes it difficult to achieve longitudinal segmentation. The SPACAL detector does not have separate e.m. and hadronic sections.

problem, one usually selects only those events that are sufficiently contained. In doing so, one creates a biased event sample, predominantly retaining events with a large fraction of e.m. energy (π^0 production). This leads to energy resolutions that are better than for an unbiased sample.

Good energy resolution is only meaningful if the signal uniformity is also good. Most frequently, the quoted energy resolution was measured for beam particles impacting the calorimeter at a single location. If the calorimeter response depends on the impact point, as it often does, this could lead to a measured resolution that is far too optimistic, in particular for jet detection. A better way of estimating a realistic performance figure is to measure the multiparticle resolution. Pions interacting in a target are measured in a calorimeter placed downstream. The HELIOS Collaboration found that the multiparticle resolution of their uranium-scintillator calorimeter was a factor two worse than the single-pion resolution, a consequence of signal nonuniformities (10). The multiparticle resolutions measured with the SPACAL detector, which is much more uniform (44), were found to be even better than the single-pion resolutions, because of the reduced sensitivity to the effects of light attenuation (45). At 150 GeV, values of $\sigma/E \approx 3\%$ were measured (Figure 8).

The best absolute values for σ/E were obtained for heavy ions. HELIOS reported 1.9% for 3.2 TeV ^{16}O ions (10); the WA80 group, which also operates a uranium-scintillator calorimeter, found 1.7% for 6.4 TeV ^{32}S ions (46). One should realize, however, that in this case a convolution of 16 or 32 independent 200 GeV nucleon showers, all starting in the same point, is measured; hence, strictly speaking, these results only tell something about the precision of the energy measurement for such a nucleon.

The effects of noncompensation on the signal linearity were discussed in Section 3.1.1 and experimental results were shown in Figure 5b. Signal

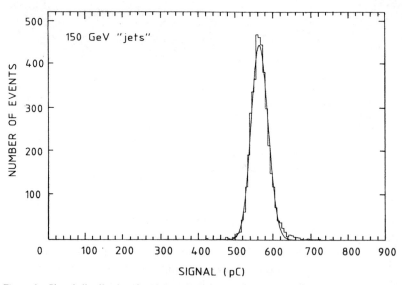

Figure 8 Signal distribution for the secondary particles produced by 150 GeV π^- in an upstream target, measured with the SPACAL detector (data from 45).

nonlinearities of ~20% over one order of magnitude in hadron energy are commonplace in noncompensating calorimeters such as those used by WA1 and WA78 and shown in Figure 5.

The weighting procedure mentioned above did not eliminate these non-linearities, particularly below 30 GeV (24). Therefore, a jet composed of 10 particles of 10 GeV would yield a signal considerably different from a 100 GeV pion signal, and the improvement in resolution is likely to be considerably smaller for jet detection.

Below 3 GeV, a different kind of nonlinearity occurs, even with compensating calorimeters. At these low energies, charged hadrons may lose their energy without undergoing strong interactions and the corresponding nuclear binding energy losses. They would then deposit their energy through ionization alone, like muons. The response to hadrons thus increases at low energies ($E_\pi < 3$ GeV), and the e/π signal ratio decreases. This was clearly confirmed by experimental data, as illustrated in Figure 9 (47).

3.3   Position and Angular Resolution

The position and the angle of incidence of particles can be obtained with calorimeters, using measurements of the transverse and longitudinal shower distributions. The localization can be derived from the center of gravity of the transverse distribution. The position resolution depends

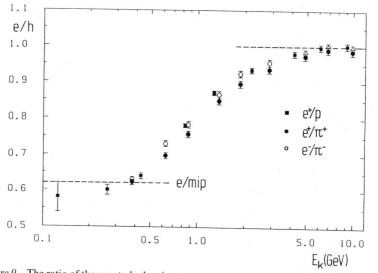

Figure 9   The ratio of the e.m. to hadronic response, e/h, as a function of the kinetic energy, E_k, showing the nonlinearity for hadron detection at low energy (data from 47).

on the characteristic width of the showers, as well as on the transverse granularity and on the signal-to-noise ratio of the calorimeter. The particles are best located in the early, narrow part of the shower. With the highly segmented Si detector planes that the ZEUS Collaboration installed at depths of $4X_0$ and $7X_0$ inside the calorimeter, one expects to be able to localize electrons to within a fraction of a millimeter (48). Submillimeter resolution was recently reported for a new type of Pb/LAr calorimeter, having an "accordion" structure with a fine transverse granularity $(1.4 \times 1.5X_0)$. For 125 GeV electrons, position resolutions were measured to be 0.5–0.7 mm (49). But even with a much cruder granularity, millimetric resolution can be readily obtained for these particles. Measurements with the e.m. calorimeter for the JETSET experiment (CERN-LEAR), which has a cell size of $3.4 \times 4.0X_0$, showed resolutions of 3–7.5 mm at 1 GeV (50), and the SPACAL group, whose detector has an effective cell size of $(9.7X_0)^2$, reported $\sigma_{em}(mm) = 17.1(mm)/\sqrt{E}$, where E is measured in GeV (51). Scaling with $1/\sqrt{E}$ is expected since the energy resolution for a single cell improves as $1/\sqrt{E_i}$ if the shower profile stays the same.

Since hadron showers are much broader than electromagnetic showers, the position resolution for detecting hadrons is worse, though a fine granularity and a good signal-to-noise ratio can improve the results. Especially at high energies, the showers are characterized by a very pronounced core, caused by the π^0 component, surrounded by an exponentially decreasing halo (Figure 10). For hadronic showers, the fine-grained SPACAL detector, with an effective cell size of $(0.33\lambda_{int})^2$, obtained position resolutions better than 5 mm for energies above 100 GeV (see Figure 11). There are strong indications that an even finer granularity would further improve these results (51, 52).

The angle of the incident particles is determined from longitudinal shower information. The measurement of the angular resolution for e.m. (σ_θ^e) and hadronic (σ_θ^h) showers has been carefully studied for several calorimeters used to investigate neutrino scattering (53–56). Typical results obtained in such experiments are

$$\sigma_\theta^e(mrad) \leq 20/\sqrt{E(GeV)} \quad (55, 56)$$
$$\sigma_\theta^e(mrad) \simeq 3.5 + 53/E(GeV) \quad (54).$$

For a hadron calorimeter that was carefully optimized by choosing a material in which e.m. and hadronic showers have approximately the same spatial dimensions, which is achieved if $\lambda \approx 3X_0$, a value $\sigma_\theta^h (mrad) \approx 160/\sqrt{E(GeV)} + 560/E(GeV)$ was reported (53, 54, 57).

3.4 The Time Response

The time response of a calorimeter is an important parameter when event rates are high. It may be of crucial importance, especially in future high

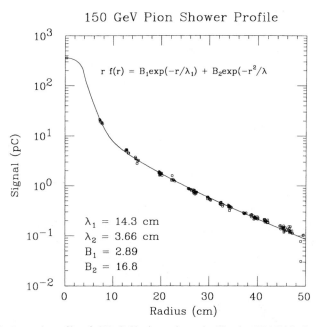

Figure 10   Lateral profile of 150 GeV pions showering in the SPACAL detector. The collected charge is given as a function of the radial distance to the impact point (data from 45).

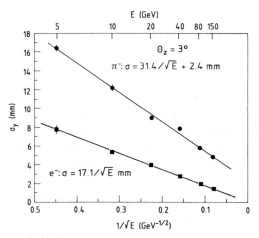

Figure 11   Position resolution for electrons and hadrons showering in the SPACAL detector, as a function of energy (data from 51).

luminosity multi-TeV pp colliders, where event rates will be measured in GHz. Two aspects should be distinguished: (*a*) the accuracy with which a signal can be attributed to a certain event or bunch crossing (timing); and (*b*) the signal duration, i.e. the time it takes before all the charge or light is collected. At high rates, charge remaining from previous events may affect the precision of the measured signals (pile-up).

The ultimate limit to the calorimeter properties in this respect is set by the physics of the shower development. However, the choice of calorimeter materials may make a big difference. In uranium calorimeters, a sizeable (10–15%) contribution to the hadronic signals comes from γ's produced in thermal neutron capture. This is a process on a time scale of ~ 1 μs. The Δt dependence of the e/h value found for U/LAr calorimeters may illustrate this point (Section 3.1.2). In calorimeters using lead as the absorber material, this process is negligible, first because of the small cross section for thermal neutron capture (a factor of 20 smaller than in ^{238}U) and second because the neutron production rate is about a factor of 3 smaller than in uranium (since there is no nuclear fission in lead).

Of course neutrons do contribute in a very essential way (compensation!) to the signal of compensating lead-scintillator calorimeters (see Section 3.1.2), but only by releasing their kinetic energy in elastic n-p scattering. This is a process with a time constant of ~ 10 ns. Figure 12 shows typical electron and pion signals, measured with a very fast digitizing oscilloscope. The pion signal clearly exhibits an exponential tail, which is absent for the electron signal (58). Other evidence for the intrinsic speed of this type of calorimeter comes from a measurement of the e/π signal ratio as a function of the time over which the signals are integrated (59). Figure 13 shows that this ratio is practically constant for gate widths exceeding 50 ns.

Apart from the physics of the shower development, instrumental effects may limit the time response of calorimeters. Detectors using gases (wire chambers) or liquids in the ionization chamber mode (LAr, TMP) to produce the signals are relatively slow since they rely on drifting charges. Using bipolar pulse-shaping techniques (60), rather fast signals (50–100 ns) can be obtained but charge collection will continue for another microsecond or so. This problem may be avoided in calorimeters using semiconductor or plastic scintillator as active material, although the wavelength shifters used to read out plastic-scintillator calorimeters may also introduce a considerable reduction of the signal speed (10). Scintillating fibers are definitely the fastest readout media to date (61). Calorimeter signals with a rise time of 1–2 ns and a full width at half maximum of 5 ns have been obtained (see Figure 12).

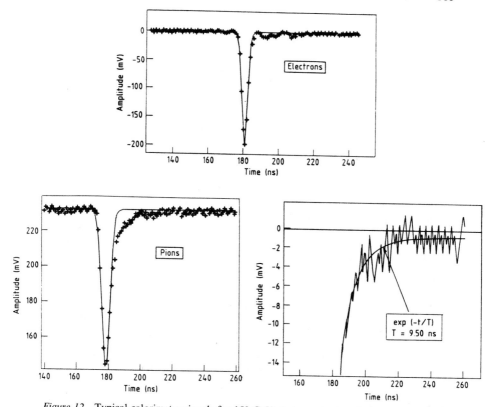

Figure 12 Typical calorimeter signals for 150 GeV electrons (*top*) and pions (*bottom left*) measured with the SPACAL detector. The pion signal exhibits a clear exponential tail with a time constant of ~10 ns, which is attributed to the neutron shower component (*bottom right*). The $t = 0$ point is arbitrary and the bin size is 1 ns (data from 58).

3.5   Particle Identification

Hadron calorimeters can identify a class of particles not readily identified by other methods. In the following, I briefly discuss the identification of electrons, muons, and neutrinos.

3.5.1 DISCRIMINATION BETWEEN ELECTRONS (PHOTONS) AND HADRONS Electron-pion discrimination can be achieved with calorimeters by exploiting the very different longitudinal and lateral energy deposition profiles of e.m. and hadronic showers. The separation works best in materials with very different radiation and nuclear interaction lengths, that is, at high Z (Figure 14).

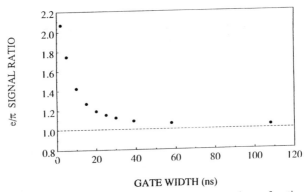

Figure 13   The e/π signal ratio for 80 GeV particles, measured as a function of the gate width with the SPACAL detector (data from 59).

Most frequently, the differences in longitudinal shower development are used for electron identification (62). This is the principal reason why all calorimeters operating in experiments at high energies are equipped with separate e.m. and hadronic sections. However, differences in lateral shower development may equally well be used for this purpose, and with fine-grained detectors excellent results can be obtained. With the longitudinally unsegmented SPACAL calorimeter, pion rejection factors of several hundred were obtained, with electron efficiencies better than 95% at energies in the range of 40–150 GeV, just by using the lateral shower information (58). When longitudinal shower information was incorporated, the rejection factor improved to more than 6000, at 80 GeV (51). This is illustrated in Figure 15.

Some experiments make use of a preshower counter to improve the electron identification (63). Such a detector exploits the fact that the energy deposition in a thin $(1–2X_0)$ sheet of high-Z material is very different for electrons and pions, and the difference readily permits an additional factor of 10 in e/π separation (58). When placed a few centimeters upstream of the calorimeter, the effect of such an absorber on the calorimeter performance is negligible, at least for energies above 10 GeV (45).

The ultimate limit to the applicability of this e/π separation technique is set by the charge exchange reaction $\pi^- p \to \pi^0 n$ (or $\pi^+ n \to \pi^0 p$), which may simulate an e.m. shower. For pions of a few GeV, the cross section for this reaction is at the 1% level of the total inelastic cross section, and it decreases logarithmically with increasing energy (64). Since the starting point of the pion shower generated by this process is determined by λ_{int}, a thin preshower counter may reduce the limit derived from the cross section by an order of magnitude.

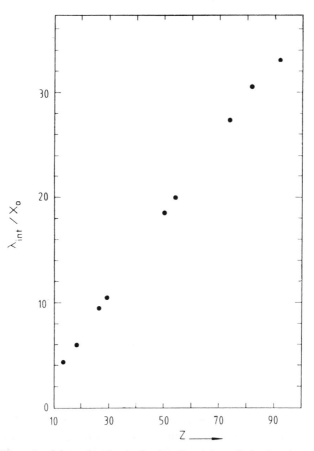

Figure 14 The ratio of the nuclear interaction length and the radiation length as a function of Z of the nuclei.

The SPACAL Collaboration developed a completely new method for e/π separation, which does not require any information on the energy deposition patterns. Exploiting the very high signal speed of the detector, the small, but very significant, differences in the time structure of electron and hadron signals were used in order to discriminate between these particles (65). At 80 GeV, a pion rejection factor of almost 1000 was obtained with this technique (see Figure 16), for an electron efficiency better than 99% (58). An interesting aspect of this technique is its speed, which holds promise for a very fast electron trigger. In the mentioned

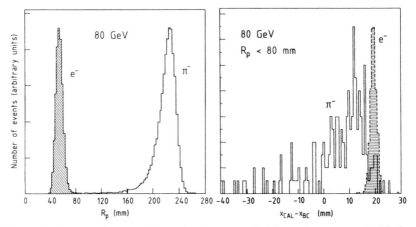

Figure 15 Electron-pion separation using shower profile information, measured with the SPACAL calorimeter. (*Left*) The diagram shows the distributions of the "effective width" R_p for electron and pion showers at 80 GeV, entering the detector at an angle of 2° with respect to the fiber axis. (*Right*) The diagram shows the displacement of the center of gravity of the showers with respect to the impact point of the particles, for those pions that could not be distinguished from electrons on the basis of their lateral shower characteristics (data from 51).

application, the discriminating information was available for the trigger logic after 85 ns.

While electron identification is relatively easy for isolated particles, the problems increase considerably for electrons near the core of jets, a situation that will occur frequently in heavy-flavor decay. Experimental data are lacking so far, but Monte Carlo studies indicate that very fine (three-dimensional) granularity is the key to solving this problem in the best possible way. I expect that the experience with the Si planes installed at depths of $4X_0$ and $7X_0$ inside the ZEUS calorimeter (66) will be very useful.

3.5.2 MUON IDENTIFICATION   There exist several calorimetric methods for discriminating between muons and hadrons:

1. In very deep calorimeters with fine longitudinal segmentation, high energy muons are clearly recognized as isolated, minimum-ionizing tracks, ranging far beyond the tracks produced in hadronic shower development. This is the technique used in experiments on incident neutrinos.

2. In calorimeters used in 4π experiments at colliders, which for practical reasons are barely deep enough to contain hadron showers, the lateral

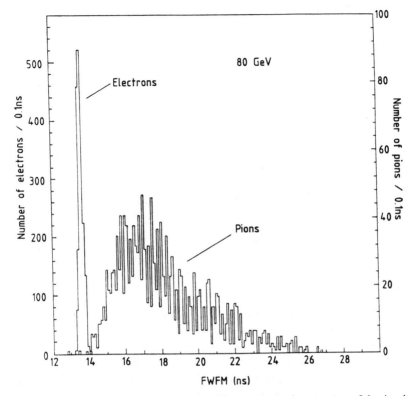

Figure 16 Electron-pion separation through differences in the time structure of the signals. The duration of the pulses, measured at 20% of the amplitude is compared for 80 GeV electrons and pions (data from 58).

profile of the energy deposition may be used to discriminate between muons and hadrons (Figure 17). The quality of the π/μ separation depends strongly on the lateral granularity of the calorimeter.

3. Calorimeters may simply absorb hadrons, so that tracks detected beyond the calorimeter are by definition muon candidates.

It is important to realize that the calorimeter that absorbs hadrons is also a source of muons. In hadronic showers, many charged pions and (particularly important for this problem) kaons are produced. Each of these secondary particles has a small but finite decay probability, thus producing a muon. The probability for muon production in hadronic showers was recently measured to be $\sim 10^{-4} E_\pi (\text{GeV})$ (67). For 80 GeV π^-, the muons generated in the showers and escaping the $9.5 \lambda_{\text{int}}$ deep

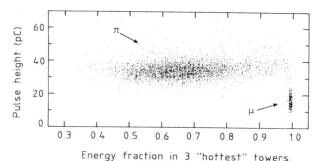

Figure 17   Calorimetric pion-muon separation through differences in the lateral energy deposition. The scatter plot shows the fraction of the total energy recorded in the three calorimeter towers with the highest signal versus the total calorimeter signal, for a mixed beam of 10 GeV pions and muons (data from 45).

detector had an average momentum of ~ 3 GeV/c. It was also shown that such muons could easily be distinguished from muons traversing the absorber if the calorimeter had a fine transverse granularity.

3.5.3   NEUTRINO IDENTIFICATION   Apparent missing energy or missing momentum has become a powerful means of inferring the presence of neutrinos among the collision products. Missing energy relies on a measurement of the total energy using 4π calorimetric coverage for all particles (charged, neutral, muons). This can in practice be achieved for collisions at e^+e^- storage rings or in fixed-target experiments. Neutrino production is inferred whenever the measured energy is lower than the total available energy and is incompatible with the resolution function of the detector. A total-energy measurement is not practical at hadron colliders, because a considerable fraction of the total energy is produced at angles too close to the colliding beams, and hence inaccessible for calorimetric measurements. In this case it is advantageous to implement a measurement of the imbalance in transverse energy, where neutrino production is signaled by $E_{\mathrm{Tmiss}} = \Sigma_i E_{\perp,i} \neq 0$, incommensurate with detector resolution. The intrinsic quality of such measurements was estimated to be $\sigma(E_{\perp,\mathrm{miss}}/E_{\mathrm{total}}) \approx 0.3/\sqrt{E_{\mathrm{total}}}$ (68), considerably better than the values achieved up to now: $\sigma(E_{\perp,\mathrm{miss}}/E_{\mathrm{total}}) \approx 0.7/\sqrt{E_{\mathrm{total}}}$ (69–71) or $\sigma[E_{\perp,\mathrm{miss}} \approx 0.6E_{\perp}^{0.43}(\mathrm{GeV})]$ (72).

3.6   *Particle-Particle Separation*

One tends to think of hadronic showers as phenomena that can be detected only by a detector with a very large volume. This is certainly true if one wants to contain the showers or do a precise energy measurement.

However, when it comes to localizing particles, the situation is quite different. Especially at high energies, more than half of the energy is on average deposited in a cylinder with a radius of a few centimeters ($\sim 0.2\lambda_{int}$, see Figure 10). This means that two particles hitting the detector and developing showers might be recognized as two separate particles even if they enter the detector at a relatively small distance from each other. The two-particle separation capability of a calorimeter depends on the effective density (λ_{int}) and on the transverse granularity. The SPACAL Collaboration found that two 80 GeV pions can be separated with more than 95% efficiency down to a distance of 8 cm (51). Figure 18 shows how a fine lateral segmentation may produce very detailed event information.

4. CALORIMETERS IN PARTICLE PHYSICS EXPERIMENTS

In this section, I briefly review the various tasks that calorimeters perform in current experiments. The requirements on the calorimeter performance in future experiments at the LHC and SSC are discussed.

4.1 *The Present Situation*

The calorimeter tasks in high energy physics experiments can be subdivided as follows: (*a*) event selection (triggering), (*b*) lepton identification, (*c*)

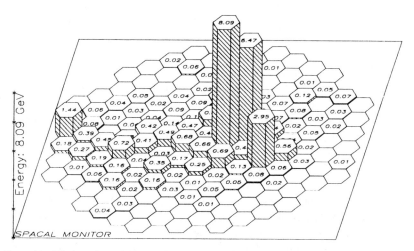

Figure 18   Display of a multiparticle event caused by a pion interacting in a target upstream of the calorimeter. The numbers correspond to the energy deposited in the individual towers of hexagonal shape, which have a diagonal width of 86 mm (data from 51).

energy measurement of jets and leptons, (d) hadron absorption, and (e) vertex reconstruction. The latter application is specific for v scattering experiments in which the detector serves as target at the same time. Therefore, experiments have given specific emphasis on angular resolution in their design (see Section 3.4).

Event selection is a prime task in all experiments, but especially in those in which the event rate is orders of magnitude beyond what can be recorded on tape, usually not more than 10 Hz. Experiments studying low-cross-section processes such as v scattering or e^+e^- collisions mainly use the calorimeter signals to distinguish beam interactions from background (cosmic rays, beam-gas scattering), for example through the total energy deposited. On the other hand, experiments studying strong-interaction processes often have to deal with primary event rates in the range from 10 kHz to 1 MHz. Thus the trigger not only has to determine that an interaction occurred, but also whether it was a potentially interesting event.

Usually, one selects at the first trigger level deep inelastic events through transverse energy patterns. Searches for events in which neutrinos or supersymmetric particles are produced are done using quantities like the missing energy ($E_{in} - E_{tot}$, only for fixed-target experiments), or the missing transverse momentum ($p_\perp^2$). A representative example is taken from HELIOS, a fixed-target experiment at the CERN SPS (10). The signals from each calorimeter cell i are split at the front end. Two thirds of the signal are digitized to determine the total integrated charge. The remaining one third is diverted into the Energy Flow Logic, which directly determines the physical quantities E_{tot}, $E_\perp$, and $p_\perp^2$ of the event, for triggering purposes. This is done as follows. The signal from each calorimeter cell i is further split into four parts, which are weighted with factors depending on the geometrical position of the cell. This weighting is done with precision resistors. Let θ_i and ϕ_i be the polar and azimuthal angles of the cell with respect to the vertex. The weighting factors for E_{tot} and $E_\perp$ are then 1 and $\sin \theta_i$, respectively. For the $p_\perp^2$ quantity, four sums are formed, labeled p_x^+, p_y^+, p_x^-, and p_y^-. Each cell contributes to two of those, depending on the hemisphere in which it is located, with weighting factors $\sin \theta_i \times \sin \phi_i$ and $\sin \theta_i \times \cos \phi_i$, respectively. By fast analog summing of the signals from all the calorimeter cells, signals proportional to E_{tot}, $E_\perp$, p_x^+, p_y^+, p_x^-, and p_y^- are formed. These are digitized with flash ADCs and by digital arithmetic the quantity $p_\perp^2 [= (p_x^+ - p_x^-)^2 + (p_y^+ - p_y^-)^2]$ is obtained. The digitized signals are compared with trigger thresholds and sent to a parallel-trigger processor. This whole process takes only 150 ns. Once the event rate is sufficiently reduced by such triggers, one can look for more detailed processes, which takes more time. For electron or jet production, the energy deposition pattern must be analyzed, a procedure that may easily take a good fraction of a millisecond, especially when information from other detectors is included in the analysis.

The methods used for the identification of leptons were discussed in Section 3.5. Usually, calorimeter information is combined with data from other detectors, for example a muon spectrometer or an upstream tracking system. As shown before, calorimeters can do an excellent job of identifying electromagnetic showers. However, in practice, most e.m. showers will be generated by γ's from π^0 decay, rather than by electrons. The UA2 Collaboration used their preshower detector in an elegant way to discriminate between these particles. With readout planes installed behind and at several distances in front of the $1.5X_0$ thick converter, they managed to distinguish electrons from γ's, which do not produce a track in the upstream planes and which after $\frac{9}{7}X_0$ on average convert into an electron-positron pair. Thanks to the good spatial resolution of the readout planes (composed of scintillating fibers), electrons could also be distinguished from γ's entering the detector close to a charged pion (63).

Accurate measurement of the electron energy has been given great emphasis in the LEP detectors. Because of the relatively low energy and the simplicity of the events, the charged hadrons produced in Z^0 decay can be accurately measured with a magnet tracking system. Therefore, hadronic calorimetry is relatively crude in these experiments. This may turn out to be a disadvantage when the process $e^+e^- \rightarrow W^+W^-$ will be studied in the future. Good hadronic calorimetry was given a high priority in the design of the p$\bar{\text{p}}$ experiments at CERN and FNAL, in particular in the experiments without a magnetic field, UA2 (73) and D0 (33). The UA2 Collaboration capitalized on the good energy resolution and granularity of their calorimeter, which allowed them to study the production of intermediate vector bosons based on the hadronic decay modes (74) (Figure 19). This is the first example of the use of a calorimeter as a jet spectrometer. Another example illustrating the advantages of accurate calorimetric energy measurements came from the heavy-ion experiments at CERN. Figure 20 shows the total energy distribution measured by the WA80 Collaboration when the 6.4 TeV ^{32}S beam from the CERN SPS was dumped in the U/plastic-scintillator calorimeter. It turned out that the beam did not consist of ^{32}S particles only. A few percent of the ions apparently dissociated underway and those with the proper $A/Z = 2$ ratio made it through the acceleration process (46). The good energy resolution of the calorimeter allows a detailed study of these contaminating lower-mass ions. Similar results were reported by HELIOS (10). Hadronic energy resolution will be crucial for the experiments at the ep collider HERA, since it is directly linked to the accuracy of the measurement of the proton

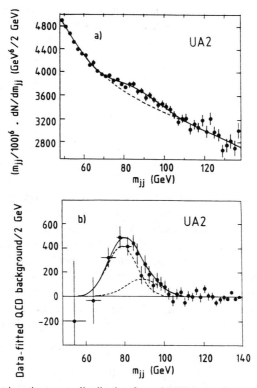

Figure 19 Two-jet invariant mass distribution from the UA2 experiment (74). The upper diagram shows the measured data points together with the results of best fits to the QCD background alone (*dashed curve*) or including the sum of two Gaussian functions describing $W, Z \to q\bar{q}$ decays (*full line*). The lower diagram shows the same data after subtracting the QCD background. The data are compatible with peaks at $m_W = 81$ GeV and $m_Z = 92.4$ GeV. The measured width of the bump is 8 GeV, of which 5 GeV can be attributed to nonideal calorimeter performance (71).

structure function at high values of Q^2, the region where evidence for eventual quark substructure will be sought (75).

Hadron absorbers have long been used for studying muon physics. The use of an instrumented absorber that allows π/μ discrimination and recognition of muons generated in hadronic showers (see Section 3.5.2) has not yet been as fully developed in collider experiments, as it has in ν scattering experiments. This is partly because of a lack of need. Muon production is usually studied by matching tracks measured upstream and downstream of the calorimeter system. Most experiments use a muon detection system based on thick layers of iron interleaved with wire cham-

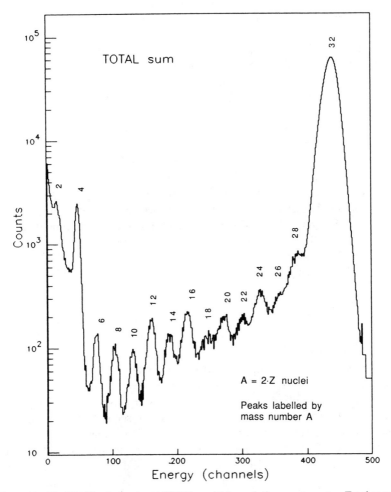

Figure 20 The WA80 calorimeter at CERN as a high resolution spectrometer: Total energy measured in the calorimeter for minimum-bias events, showing the composition of the momentum-selected heavy-ion beam (data from 46).

bers or streamer tubes, which at the same time serve as a backing calorimeter, indicating hadronic shower leakage. Except for L3, which operates with an external magnetic field, all experiments base their muon momentum measurements on deflection in a magnetic field on the inside of the calorimeter.

4.2 Requirements for Future Experiments

In the coming 8–10 years, a considerable fraction of the particle physics community will be preparing for experiments at the next generation of accelerators, the multi-TeV pp colliders LHC (8 + 8 TeV, at CERN) and SSC (20 + 20 TeV, near Dallas). These machines are characterized by very high luminosities, design figures range from 10^{33} to 4×10^{34} cm^{-2} s^{-1}. For an estimated total pp cross section of 100 mb, this translates into primary event rates of 100 MHz to 4 GHz. The bunch-to-bunch spacing will be about 15 ns, so that even for the lowest luminosity mentioned, more than one interaction per bunch crossing will occur on average. The luminosities must be very high because of the extremely small cross sections for the physics processes of interest. The cross section for hard quark and gluon scattering, the process through which new physics might be accessible, is proportional to E^{-2}, E being the center-of-mass energy of the collision. The cross sections for the processes envisaged (top quark, Higgs, SUSY particle production) are in most cases extremely small. Depending on the mass of these objects, they might be at the level of 1 pb (10^{-11} of the total cross section), and therefore unprecedented event rates will be needed. In a typical interaction, more than 100 particles will be produced.

With the possible exception of some very specific processes, calorimetry is the only way to perform meaningful experiments under these circumstances. The tasks of the calorimeter system will be basically the same as outlined in the previous subsection. However, the precision with which these tasks must be performed will be higher and the experimental conditions much more difficult than in the present generation of experiments. Let us, as an example, look at the event selection. Given the primary event rate and the fact that at most ten events per second can be recorded for off-line analysis, the calorimeter-driven trigger logic has to discard 99.999999% of the interactions, while retaining the rare interesting ones with close to 100% efficiency. This extremely high-quality selectivity has to be achieved at rates of 10^8–10^9 Hz. This is orders of magnitude beyond present capabilities.

What is needed for a calorimeter system to achieve this goal? Many workshops have been held, thousands of pages were filled, and tens of thousands of hours of computer time were spent in an attempt to answer this question. I do not even try to summarize this work here (see 76–82) but rather give a personal view from the perspective of this review. The usual approach to this problem is as follows. One takes a process (e.g. pp → H^0 → Z^{0}Z^0 → q$\bar{\text{q}}$e$^+$e$^-$, assuming a certain mass for the neutral Higgs boson, H^0) that might be detectable, and one generates Monte Carlo events for this process and for processes that contribute to the background for this signal. Most frequently, the conclusion reached is that the background from physics is so large that calorimeter properties like the energy resolution make only a marginal difference for the detectability of this process.

One of the few exceptions is the process $H^0 \to \gamma\gamma$, which is believed to be one of the most promising channels for detecting the Higgs particle if $m_W < m_H < 2m_W$. It is claimed that the e.m. energy resolution of the calorimeter may make a decisive difference (83–85). Since this process is a major source of inspiration for both Monte Carlo simulators and detector developers, I use it to illustrate my point of view.

At these relatively low values of the Higgs mass, the production rate is not the problem. The cross section for Higgs production is estimated to be of the order of 0.1 nb, so that, in one year of running at a luminosity of 10^{33} cm^{-2} s^{-1}, $\sim 10^6$ H^0's would be produced. Even though the branching ratio for the process $H^0 \to \gamma\gamma$ is only 10^{-3}, a reasonable number of events might therefore be expected. As for almost all other potential signals of new physics, the detectability depends crucially on the background. A certain and well-calculable background comes from direct photon production in $q\bar{q}$ and gg processes. An uncertain, and possibly strongly detector-dependent, contribution comes from jets mimicking γ's. It is estimated that at 100 GeV, two-jet events outnumber $H^0 \to \gamma\gamma$ events by eight orders of magnitude. Therefore, if the probability for a jet to be mistaken as a γ were even one per thousand, the consequences for the signal-to-background ratio could be disastrous. The most likely scenario for a jet to be mistaken as a γ is a leading π^0 that appears as an isolated e.m. shower. In general, there will also be other activity nearby in the calorimeter for such an event, but it is questionable to what extent this can be used as an argument to discard it. The $H^0 \to \gamma\gamma$ events will not be clean because of fragmenting remnants of the colliding protons, the possible occurrence of one or more other interactions in the same bunch crossing, and the signals in the calorimeter's memory of interactions in previous bunch crossings. However, the π^0 background could be unambiguously discarded if the energetic π^0 was recognized as such, that is, as two separate γ showers.

If the calorimeter were designed to optimize the suppression of background rather than to detect the signal (through extraordinary e.m. energy resolution), it would look quite different:

1. The calorimeter would be located as far from the interaction vertex as one could afford, so as to allow a maximum separation of the two γ's. At a distance of 2 m, the γ's from a decaying 100 GeV π^0 are separated by 4 mm, a distance that should be detectable.

2. The calorimeter would have a very high effective density, in order to limit the transverse shower size.

3. It would have a very fine lateral segmentation, for optimum "two-track" separation.
4. It would be as fast as possible, in order to make the real $H^0 \rightarrow \gamma\gamma$ events as clean as possible.

Some of these choices are orthogonal to extraordinary e.m. energy resolution. All detectors that are claimed to be able to achieve $\sigma_{em}/E \approx 3\%/\sqrt{E}$ or better have a few properties in common: they are very expensive and not very dense. This means that the showers will be considerably broader than needed. The effective Molière radius of a sampling calorimeter can be made at least a factor of three smaller than for the mentioned detectors, which moreover must be located close to the interaction point in order to be affordable.

If designed as described above, the calorimeter would also offer advantages in other types of analysis, perhaps including the study of the intermediate mass Higgs particle through its dominant $H^0 \rightarrow b\bar{b}$ decay mode (86). A large distance from the interaction point allows the particles from a jet to separate before they develop showers. Because of the high effective density, the particles would develop pencil-like showers in the calorimeter. A fine lateral segmentation would guarantee optimal information on the detailed structure of the jet (Figure 18), including information on its possible content of electrons, signatures of heavy flavor production.

A fine lateral segmentation is, in my opinion, very important. In the previous sections, several examples of its beneficial effects were shown. Among other things, a fine lateral segmentation would provide excellent position resolution (Figure 11), e/π separation (Figure 15), and π/μ separation (Figure 17). The possibility of recognizing muons generated in hadronic shower development might be crucial for another process that is frequently discussed, $pp \rightarrow H^0 \rightarrow Z^0Z^0 \rightarrow \mu^+\mu^-\mu^+\mu^-$, for a heavy Higgs particle. The high luminosity options that are being considered are mainly inspired by the small cross section and branching ratio for this process. At these high luminosities, muon production in the absorber will cause MHz event rates in the muon spectrometer.

The advantages of longitudinal segmentation are much less obvious. Traditionally, electron identification is used as an argument, but excellent e/π separation results were also obtained with a longitudinally unsegmented calorimeter (Figures 15 and 16). Therefore, given a certain number of electronics channels available, I would rather use these to make as fine a lateral segmentation as possible.

The high density will also limit the depth of the calorimeter needed for absorbing the hadrons to a sufficient extent. This is an important argument in view of the costs of an experiment, which often scale as R^{2-3}, R being the radius. The large distance from the interaction point will minimize the effects of radiation damage, which scale as R^{-2} with the distance (87, 88). Because of the high interaction rate, radiation damage to detector equipment is a major source of concern. For calorimeters it is, however, a local problem. At polar angles exceeding 15°, the radiation doses are manageable with current technology, even at the highest luminosities presently envisaged. At smaller polar angles, the doses increase rapidly. The main problem comes from e.m. showers generated in huge numbers by decaying π^0's and, when electronics are installed in the first few interaction lengths, from the neutron flux generated in the shower development. I am not so pessimistic in this respect. The fundamental understanding of radiation damage phenomena is continuously improving, and as a result, new products with better properties are becoming available. As an example, plastic scintillators may be mentioned. Until a few years ago, these materials were notorious for their sensitivity to ionizing radiation. Recently, it was shown that the performance of a lead/scintillating-fiber calorimeter would not deteriorate up to levels of the order of 0.1 MGy (89). This corresponds to four years of running at the SSC with a luminosity of 10^{33} cm^{-2} s^{-1}, for a detector placed 4 m from the interaction point at pseudorapidity $\eta = 3.4$ (polar angle of 4°).

The time response characteristics of the calorimeter will be very important at these high event rates. Given the choice, I would prefer to study individual pp collisions rather than events containing the charge produced in the previous 100 bunch crossings. Above a certain luminosity level (and depending on the experimental goals one may argue what this level will be), detector speed will be a decisive advantage.

I think two calorimeter properties will prove essential at the trigger level: compensation and hermeticity. As discussed in Section 3.1, noncompensation will lead to biased event samples. The events selected with a non-compensating calorimeter are more likely to be characterized by their (anomalous) π^0 content than by anything else. Lack of hermeticity will bias event samples selected on missing $E_\perp$ signatures. These problems are, of course, especially serious when a very high selectivity is required.

As was shown in Section 3.2, lack of compensation may also severely deteriorate the hadronic energy resolution. To illustrate this point, Figure 21 shows the total energy distribution for hadronically decaying Z^0's, measured by the calorimeters of the ALEPH experiment at LEP. The rms width of this distribution amounts to 14.3%, or 13.6%/$\sqrt{E}$. Similar results were obtained by other LEP experiments. In these experiments, measuring the momentum of individual tracks may significantly improve the Z^0 mass resolution, but at LHC/SSC this option will not be available, especially not at the trigger level.

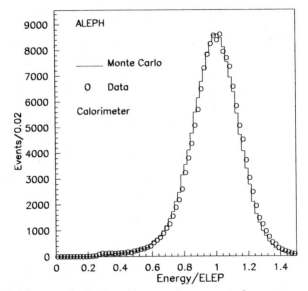

Figure 21 Total energy distribution of hadronically decaying Z^0 particles, measured with the calorimeters of the ALEPH experiment at LEP (ALEPH Collaboration, 1991, private communication). The distribution is asymmetric because of noninteracting particles, acceptance losses, nonlinearities in response, etc. The energy scale is normalized to the nominal machine energy. A Gaussian fit in the region ~ 0.85–1.5 gave a σ of 14.3%, or $136\%/\sqrt{E}$. This result improved to 13.1% if Z^0s were selected that produced jets only in the central detector region (θ between $45°$ and $135°$ from the beam line). When the calorimetric information on the charged decay products was replaced by the measured momenta of the tracks, the resolution was further improved to $\sigma = 10.3\%$ for all Z^0s, and to $\sigma = 9.3\%$ for jets in the central detector region.

Calorimeters consisting of a separate e.m. and hadronic section with very different e/h values, will have poor energy resolution for hadron detection. This is because of the large longitudinal shower fluctuations, the large fluctuations in the fraction of the shower energy carried by π^0s, and the large fluctuations in the fraction of the shower energy deposited in the e.m. calorimeter section. It should be noted that the dedicated calorimeters proposed in the framework of the $H^0 \rightarrow \gamma\gamma$ search are expected to have e/h values larger than 1.5. This is also the case for the "accordion" calorimeter considered at CERN (49). The reason is basically simple. Compensation requires a small sampling fraction in order to maximize the contribution of neutrons to the signal (see Section 3.1, Figure 6); extraordinary e.m. resolution requires a large sampling fraction in order to minimize the sampling fluctuations. The e.m. energy resolution of a compensating calorimeter can only be improved through the sampling frequency (Section 3.2.1), and it will be very hard to do better than $10\%/\sqrt{E}$.

With state-of-the-art compensating calorimeters, mass resolutions for intermediate vector bosons on the order of 5% are within reach, and once reached it will be possible to distinguish between hadronically decaying W and Z particles (see also Figure 19). Because of the important role played by these particles in many processes, the prospect of triggering on events in which they are produced is exciting. It should be emphasized that in pp colliders the mass resolution of hadronically decaying particles is not only determined by the hadronic energy resolution, but also by factors coming from the algorithms used to define jets, from event pile-up, etc. In ep or e^+e^- machines such factors are less important. A distribution as shown in Figure 21 could probably be measured with a resolution of $\sim 3\%$ with state-of-the-art technology.

When entering the world of multi-TeV physics, there are two aspects in which nature works in our favor, and one should take maximum advantage of these features:

1. Jets become much more "jet-like," a nicely collimated bunch of particles resulting from the fragmentation of ultrarelativistic quarks or gluons. When one compares the hadronic events at the e^+e^- colliders PETRA ($\sqrt{s} = 29$ GeV) and LEP (91 GeV), the difference is very striking.
2. The properties of jets (energy, direction, etc) can be measured much more precisely than at current energies.

Calorimeters are instruments for measuring energy. The history of nuclear and particle physics is filled with examples proving that precision in this matter pays off. Let me illustrate this with one example taken from nuclear physics, where more than half a century ago calorimeters were used for the first time in nuclear science. Figure 22 shows the γ-ray spectrum of decaying uranium nuclei, measured with a scintillation counter and with a high purity Ge detector.[2] The advent of the latter technology caused a revolution in experimental nuclear physics. We are now entering an era in which hadron calorimetry will permit measurements of fragmenting quarks with nuclear spectroscopic quality. If nature is kind to us, a new world might open up.

5. SUMMARY AND OUTLOOK

The use of calorimetric detection methods in high energy physics has rapidly evolved from a technique employed for some rather specialized

[2] I thank G. Roubaud for helping me in obtaining this data.

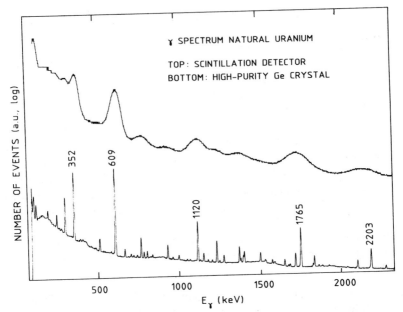

Figure 22   Nuclear gamma-ray spectrum of decaying uranium nuclei, measured with a BGO
scintillation counter (*upper curve*) and with a high purity Ge crystal (*lower curve*). Courtesy
of G. Roubaud, CERN.

applications (neutrino scattering) to the prime experimental tool in modern
experiments. This evolution, inspired by the physics goals of the experiments concerned, has paralleled crucial developments in the calorimeter
technology itself. The sophistication of the instrumentation, as well as the
understanding of the basic principles of operation and of the limitations of
the technique, has reached a very mature level and offers many fascinating
possibilities.

Nevertheless, in view of the strongly increasing demands on calorimeter
performance for future experiments, particularly at the proposed multi-
TeV pp colliders, further research and development are needed, focusing
on those calorimeter features that are likely to be essential in the next
decade: hermeticity; energy resolution; rate capability; calibration and
monitoring; radiation resistance; and the ability to discriminate among
electrons, pions, photons, and muons.

I see two trends that are likely to shape detectors for future experiments.
First, compensating calorimeters will become the standard. As was pointed
out in Section 3.1, the compensation requirement does not necessarily
imply the use of uranium absorber. When hydrogenous readout material

is used, compensation can be achieved with a variety of absorber materials, of which lead is probably the most attractive. Second, I expect further development toward integrated calorimetry, that is, instruments that combine the functions of electromagnetic and hadronic shower measurement with electron and muon identification. The traditional subdivision into an electromagnetic calorimeter and a hadron calorimeter will disappear.

Research and development and prototype studies will become even more essential than in the present generation of experiments. The time scales involved and the size and cost of the calorimeters simply do not permit design errors.

I hope that the information provided in this review may guide and encourage those who want to contribute to the further development of this very powerful and elegant experimental technique.

Acknowledgments

I thank P. Jenni and C. Scheel for their critical reading of the manuscript and for helpful discussions. I am grateful to many colleagues who made their results available to me for this article. This work was supported by the LAA project, which the author gratefully acknowledges.

Literature Cited

1. Fabjan, C. W., Wigmans, R., *Rep. Prog. Phys.* 52: 1519 (1989)
2. Davis, R. Jr., Mann, A. K., Wolfenstein, L., *Annu. Rev. Nucl. Part. Sci.* 39: 467 (1989)
3. Tsai, Y. S., *Rev. Mod. Phys.* 46: 815 (1974)
4. Storm, E., Israel, H. I., *Nucl. Data Tables* 7: 565 (1970)
5. Pages, L., et al., *Atomic Data* 4: 1 (1972)
6. Nelson, W. R., Hirayama, H., Rogers, D. W. O., *SLAC Rep.* 165. Stanford: SLAC (1985)
7. Yuda, T., *Nucl. Instrum. Methods* 73: 301 (1969)
8. Rossi, B., *High-Energy Particles.* Englewood Cliffs, NJ: Prentice Hall (1964)
9. Kopp, K., et al., *Z. Phys.* C28: 171 (1985)
10. Åkesson, T., et al., *Nucl. Instrum. Methods* A262: 243 (1987)
11. Leroy, C., et al., *Nucl. Instrum. Methods* A252: 4 (1986)
12. Catanesi, M. G., et al., *Nucl. Instrum. Methods* A260: 43 (1987); Catanesi, M. G., et al., *Nucl. Instrum. Methods* A292: 97 (1990)
13. Barreiro, F., et al., *Nucl. Instrum. Methods* A292: 259 (1990)
14. Wigmans, R., *Nucl. Instrum. Methods* A259: 389 (1987)
15. Wigmans, R., in *Proc. ICFA Sch. Instrumentation in Elementary Particle Physics*, Trieste, ed. C. W. Fabjan, J. E. Pilcher. Singapore: World Scientific (1987)
16. Akimov, Y. K., *Scintillator Counters in High Energy Physics.* New York: Academic (1965)
17. Anderson, D. F., Lamb, D. C., *Nucl. Instrum. Methods* A265: 440 (1988)
18. Munoz, R. C., et al., *J. Chem. Phys.* 85: 1104 (1986)
19. Wigmans, R., see Ref. 81, p. 608
20. Brückmann, H., et al., *Nucl. Instrum. Methods* A263: 136 (1988)
21. Brau, J. E., Gabriel, T. A., *Nucl. Instrum. Methods* A238: 489 (1985)
22. Wigmans, R., *Nucl. Instrum. Methods* A265: 273 (1988)
23. Groom, D. E., see Ref. 79, p. 59
24. Abramowicz, H., et al., *Nucl. Instrum. Methods* 180: 429 (1981)
25. de Vincenzi, M., et al., *Nucl. Instrum. Methods* A243: 348 (1986)
26. Fabjan, C. W., Willis, W. J., in *Proc. Calorimeter Workshop*, FNAL, Batavia, Ill., ed. M. Atač. Batavia, IL: FNAL

(1975), p. 1; Fabjan, C. W., et al., *Nucl. Instrum. Methods* 141: 61 (1977)
27. Drews, G., et al., *Nucl. Instrum. Methods* A290: 335 (1990); Tiecke, H. (The ZEUS Calorimeter Group), *Nucl. Instrum. Methods* A277: 42 (1989)
28. Wigmans, R., *Rep. NIKHEF-H* 87-13. Amsterdam: NIKHEF-H (1987)
29. Brau, J. E., Gabriel, T. A., *Nucl. Instrum. Methods* A279: 40 (1989)
30. Borchi, E., et al., *Phys. Lett.* B222: 525 (1989); Angelis, A. L. S., et al., *Phys. Lett.* B242: 293 (1990); Borchi, E., et al., *Preprint CERN-PPE/91-18.* Geneva: CERN (1991)
31. Braunschweig, W., et al., *Nucl. Instrum. Methods* A275: 246 (1989); A265: 419 (1988)
32. Behrens, U., et al., *Nucl. Instrum. Methods* A289: 115 (1990); d'Agostini, G., et al., *Nucl. Instrum. Methods* A274: 134 (1989)
33. Abolins, M., et al., *Nucl. Instrum. Methods* A280: 36 (1989)
34. Hitlin, D., *Rep. CALT* 68-1305. Pasadena: Caltech (1985)
35. Yu, B., Radeka, V., eds., *Rep. BNL* 52244. Upton, NY: Brookhaven Natl. Lab. (1990), p. 14
36. Galaktionov, Y., et al., *Nucl. Instrum. Methods* A251: 258 (1986)
37. Aubert, B., et al., *Nucl. Instrum. Methods* A286: 147 (1990)
38. Bacci, C., et al., *Nucl. Instrum. Methods* A292: 113 (1990)
39. Bernardi, E., et al., *Nucl. Instrum. Methods* A262: 229 (1987)
40. Hughes, E. B., et al., *Nucl. Instrum. Methods* 75: 130 (1969)
41. Benvenuti, A., et al., *Nucl. Instrum. Methods* 125: 447 (1975)
42. Amaldi, U., *Phys. Scripta* 23: 409 (1981)
43. del Peso, J., Ros, E., *Nucl. Instrum. Methods* A276: 456 (1989)
44. Acosta, D., et al., *Nucl. Instrum. Methods* A294: 193 (1990)
45. Acosta, D., et al., *Preprint CERN-PPE/91-85.* Geneva: CERN (1991)
46. Young, G. R., et al., *Nucl. Instrum. Methods* A279: 503 (1989)
47. Andresen, A., et al., *Nucl. Instrum. Methods* A290: 95 (1990)
48. Gössling, C., in *Proc. 24th Int. Conf. on High-Energy Physics*, Munich. Berlin: Springer-Verlag (1988), p. 1208
49. Aubert, B., et al., *Proposal CERN/DRDC/90-31.* Geneva: CERN (1990); Aubert, B., et al., see Ref. 76, 3: 368
50. Hertzog, D., et al., *Nucl. Instrum. Methods* A294: 446 (1990)
51. Acosta, D., et al., *Preprint CERN-PPE/91-11.* Geneva: CERN (1991)
52. Binon, F., et al., *Nucl. Instrum. Methods* 188: 507 (1981)
53. Diddens, A. N., et al., *Nucl. Instrum. Methods* 178: 27 (1980)
54. Bogert, D., et al., *IEEE Trans. Nucl. Sci.* NS-29: 336 (1982)
55. DeWulf, J. P., et al., *Nucl. Instrum. Methods* A252: 443 (1986)
56. DeWinter, C., et al., *Rep. CERN-EP/88-81.* Geneva: CERN (1988)
57. Abt, I., et al., *Nucl. Instrum. Methods* 217: 377 (1983)
58. Acosta, D., et al., *Nucl. Instrum. Methods* A302: 36 (1991)
59. Paar, H. P., see Ref. 77, ed. V. Kelly, T. Dombeck, G. Yost. Singapore: World Scientific (1991); Acosta, D., et al., In preparation, Geneva: CERN (1991)
60. Willis, W. J., Radeka, V., *Nucl. Instrum. Methods* 120: 221 (1974); Radeka, V., *Annu. Rev. Nucl. Part. Sci.* 38: 217 (1988); Radeka, V., Rescia, S., *Nucl. Instrum. Methods* A265: 228 (1988)
61. Hartjes, F. G., Wigmans, R., *Nucl. Instrum. Methods* A277: 379 (1989)
62. Baumgart, R., et al., *Nucl. Instrum. Methods* A272: 722 (1988)
63. Alitti, J., et al., *Nucl. Instrum. Methods* A279: 364 (1989); see also Munday, D. J., et al., *Proposal CERN/DRDC/90-27.* Geneva: CERN (1990)
64. Barns, A. V., et al., *Phys. Rev. Lett.* 37: 76 (1976)
65. DeSalvo, R., et al., *Nucl. Instrum. Methods* A279: 467 (1990)
66. Krüger, J., ed., *DESY Rep. PRC* 87-02. Hamburg: DESY (1987)
67. Acosta, D., et al., In preparation. Geneva: CERN (1991)
68. Willis, W. J., Winter, K., *Rep. CERN* 76-18. Geneva: CERN (1976), p. 131
69. Arnison, G., et al., *Phys. Lett.* B139: 115 (1984)
70. Bagnaia, P., et al., *Z. Phys.* C24: 1 (1984)
71. Jenni, P., *Nucl. Phys. Proc. Suppl.* B3: 341 (1988)
72. Alitti, J., et al., *Phys. Lett.* B235: 363 (1990)
73. Beer, A., et al., *Nucl. Instrum. Methods* A224: 360 (1984)
74. Alitti, J., et al., *Z. Phys.* C49: 17 (1991)
75. Feltesse, J., in *Proc. Workshop on Physics at HERA*, Hamburg, ed. R. Peccei. Hamburg: DESY (1988), 1: 33
76. Jarlskog, G., Rein, D., eds., *Proc. Large Hadron Collider Workshop*, Aachen, CERN 90-10, Vols. 1–3. Geneva: CERN (1990)
77. Kelly, V., Dombeck, T., Yost, G., eds., *Proc. SSC Symp. on Detectors for SSC*, Ft. Worth. Singapore: World Scientific (1991)
78. Fernandez, E., Jarlskog, G., eds., *Proc.*

HADRON CALORIMETRY 185

ECFA Study Week on Instrumentation Technology for High-Luminosity Hadron Colliders, Barcelona, CERN 89-10, Vols. 1 and 2. Geneva: CERN (1989)
79. Donaldson, R., Gilchriese, M. G. D., eds., *Proc. Workshop on Calorimetry for the Supercollider*, Tuscaloosa, 1989. Singapore: World Scientific (1990)
80. Jensen, S., ed., *Proc. Summer Study on High Energy Physics in the 1990's*, Snowmass. Singapore: World Scientific (1989)
81. Donaldson, R., Gilchriese, M. G. D., eds., *Proc. Workshop on Experiments, Detectors and Experimental Areas for the Supercollider*, Berkeley. Singapore: World Scientific (1987)
82. Mulvey, J. H., ed., *Proc. Workshop on Physics at Future Accelerators*, La Thuile and Geneva, CERN 87-07. Geneva: CERN (1987)
83. Atwood, D. M., et al., see Ref. 81, p. 728
84. Barter, C., et al., see Ref. 80, p. 98
85. Seez, C., et al., see Ref. 76, 2: 474
86. Brau, J., Pitts, K. T., Price, L. E., see Ref. 80, p. 103
87. Groom, D. E., ed., *Rep. SSC-SR-1003*. Berkeley: SSC (1988); *Erratum SSCL-285*. Dallas: SSC (1990)
88. Stevenson, G. R., see Ref. 76, 3: 556
89. Acosta, D., et al., *Preprint CERN-PPE/91-45*. Geneva: CERN (1991)

Annu. Rev. Nucl. Part. Sci. 1991. 41: 187–218

ISOBAR EXCITATIONS IN NUCLEI

Carl Gaarde

The Niels Bohr Institute, University of Copenhagen,
DK-2100 Copenhagen, Denmark

KEY WORDS: spin response in delta region

CONTENTS

1. INTRODUCTION

The isobar at 1232 MeV is the simplest excitation of the nucleon. In a quark model it is a spin-isospin flip of a quark with no change of the orbital motion, a Gamow-Teller excitation. The Δ isobar, however, is also

187

0163–8998/91/1201–0187$02.00

a resonance in the pion-plus-nucleon system; the study of the Δ isobar in the nuclear medium is therefore closely related to the in-medium behavior of the π, which in turn is the carrier of the strong force. We are therefore examining the coexistence of the nucleon, the pion, and the Δ isobar in the nucleus.

Dramatic effects have been predicted for pionic modes in general (1–3). The experimentally observed attraction in the spin-isospin channel for free nucleons, originating from one-pion exchange, was expected to lead to an attractive particle-hole interaction in the nucleus. This interaction could result in collective modes, and eventually in a pion condensate. It has long been known that this condensate is not realized in nuclei. On the contrary, recent $(\vec{p}, \vec{p}')$ and $(\vec{p}, \vec{n})$ data seem to show that correlations for pionic modes in the quasi-elastic region are weak and that the attraction in the spin-isospin channel is gone. Why do we expect correlations in the Δ region? The situation looks similar. The pion exchange is again the dominant interaction between N and Δ isobar, and it is attractive at finite momentum transfer. The main differences could be that the attraction is larger to begin with and also that medium effects, for example Pauli blocking, could be different.

The experimental facts can be summarized as follows. The Δ energy in nuclei as observed in charge exchange reactions is lower than that on a proton target. This is not seen with electromagnetic probes, for example, where only a broadening of the Δ peak is observed. This shift in Δ energy is ascribed to correlations in the spin-longitudinal channel, or using other terminology, the coupling of the Δ-hole state and the pion results in a mixed state, the pisobar.

The shift in Δ energy is around 30 MeV, and on a nuclear structure scale it is a very large effect. The question is, however, whether the effect we see is real or an artifact. The charge exchange reactions probe the nucleus at low densities, but what would be the effect at normal or even higher densities (4–6). Several publications have recently discussed this problem in connection with heavy-ion collisions at relativistic energies (7–9). Here one could expect Δ isobars to form at high densities and the signals in detectors would strongly depend on possible medium effects on pions and Δ isobars.

The experimental problem is specificity. In this review we present data obtained with different probes, and we show that the information from charge exchange reactions is the most specific on these medium effects. The emphasis is on data from inclusive charge exchange reactions, where the experimental challenge is to separate the spin-longitudinal from the spin-transverse response. This can be done by using different probes to enhance one or the other response, or by using polarized beams to measure spin observables. We discuss recent data in which such a separation has been attempted.

The disadvantage to using such simple experiments is that only the nuclear surface is probed. We briefly discuss data from exclusive reactions, which in principle are more specific. Details on the propagation of pions and Δ isobars, before they reach the detectors, are needed to analyze such data. The result is often less specificity.

2. THE NUCLEAR SPIN RESPONSE

We emphasized above that the Δ is a spin excitation and should be discussed in conjunction with the general spin response of the nucleus. Figure 1 shows the regions of the particle-hole and Δ-hole states in a plot of energy transfer (ω) versus three-momentum transfer (q). Also shown are lines along which the system can be probed. The pion line $\omega_\pi = \sqrt{(q^2+m_\pi^2)}$ corresponds to pion absorption, the photon line $\omega = q$ to photon absorption. The region below the photon line is accessible to inelastic scattering. In Figure 1 are shown two lines corresponding to the (^{3}He, t) reaction at two angles at 2 GeV. The response along these lines is discussed below.

What kind of correlations or collective effects can we expect at finite momentum transfers in nuclei? Most of our knowledge on correlations comes from small momentum transfer phenomena. Specifically, for the spin channel strong correlations are observed leading to collective states for

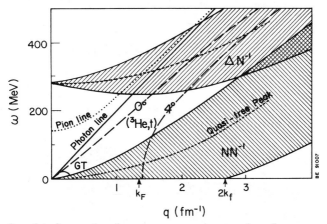

Figure 1 In a plot of energy transfer ω versus momentum transfer q, four curves are given, along which the nuclear response is probed in different reactions. The quasi-free regions for NN^{-1} and ΔN^{-1} excitations are indicated.

Gamow-Teller ($\Delta l = 0$) and spin dipole ($\Delta l = 1$) transitions (10). For the spin quadrupole strength at somewhat larger momentum transfer, the collectivity is already smaller (11). At even larger q where well-developed quasi-elastic peaks are observed in the spectra ($q \geq 1.2$ fm^{-1}) the correlations in the spin channels seem to be small. Data from a ($\vec{p}, \vec{p}'$) experiment at 500 MeV—where spin observables were obtained at one angle, corresponding to $q \approx 1.75$ fm^{-1}—showed no enhancement in the spin-longitudinal channel (12). This result is confirmed in a recent ($\vec{p}, \vec{n}$) experiment, where spin transfer data were measured (T. Taddeucci, private communication).

In this paper we concentrate on the response in the Δ-hole region as measured with different probes. We refer to a schematic two-level model first proposed by Guichon & Delorme (14). In this model, the π in the medium and the Δ-hole state are treated as a two-level system, and as can be seen from Figure 1, these two levels would cross at some q. Any interaction attractive or repulsive between the two would, however, make them repel each other and the two states would mix. This is illustrated in Figure 2. At small q, the pionic branch is dominantly a pion state and in the crossing region it is an equal mixture of a pion and a Δ-hole state. We note, however, that in such a model the pionic branch with increasing q

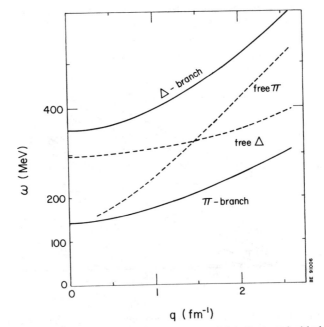

Figure 2 The two-level model by Guichon & Delorme (14) is illustrated with the Landau-Migdal parameter $g' = 0.5$. The figure is adapted from an article by Ericson (6).

becomes more of a Δ-hole state: at $q \approx 3$ fm^{-1} it is a 70% Δ-hole and 30% pion state (6, 15). We see from Figure 1 that the charge exchange reactions cover these regions of the ω-q plane very well.

The model above is schematic as infinite matter is assumed and the width of the Δ is neglected. A more realistic model has been developed by Delorme & Guichon (16), in which the finiteness of the nucleus is included as well as effects from the Δ width, both decay and spreading width. The essential features from the schematic model survive in this more realistic treatment. Strength is removed from the pion line and the bare Δ-hole region and is shifted up and down as indicated in Figure 2. The details depend on values for the parameters used such as coupling constants and g', the Landau-Migdal parameter.

The dependence on the density is especially important because of the experimental conditions for testing these models. The hadronic probes are all strongly absorbed, which means that the response is measured at low densities at the surface of the nucleus. It seems that only inelastic neutrino scattering would measure the pion-like response at full density (6). The two-level model and the following discussion refer to the pion-like response, i.e. only the spin-longitudinal part of the response. The spin-transverse part of the Δ-hole states has nothing to couple to.

The properties of the response function can also be discussed in terms of Δ-hole interactions. These interactions are usually taken as boson exchange potentials, for example a prescription like "$\pi + \rho + g'$" is often used. The interaction is a sum of π and ρ exchange and a Landau-Migdal term. In such a description the interaction is also very different in the spin-longitudinal and spin-transverse channels. The problem is the same, however. The interaction is strongly dependent on density and the current data on the spin-longitudinal response is dominated by signals from low density regions. In the following sections we discuss data on Δ excitations obtained with different probes.

3.   ELECTROMAGNETIC PROBES

First we discuss the study of the Δ excitation with electromagnetic probes and distinguish between photon absorption where $\omega = q$ and inelastic electron scattering with $t = \omega^2 - q^2$ negative and the exchange of a virtual photon. In both cases the spin-transverse response is probed since the spin part of the electromagnetic interaction is $\vec{\sigma} \times \hat{q}$.

3.1   *Photon Absorption*

Figure 3 shows data for photon absorption for a number of nuclei from ^{9}Be to ^{238}U (17–21). The total absorption cross section per nucleon is given

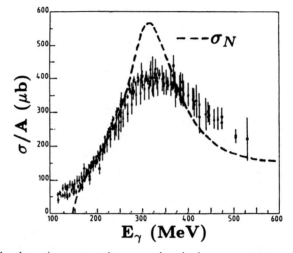

Figure 3 The absorption cross section per nucleon is given versus γ energy. The nucleon cross section is taken as $\sigma_N = 1/2(\sigma_p + \sigma_n)$ (17–21). The figure is adapted from a paper by Ghedira (21).

versus γ energy, and within the experimental uncertainties the response is universal. Also shown is the cross section on the nucleon taken as $\sigma_N = 1/2(\sigma_p + \sigma_n)$. A figure similar to this one was given by Ahrens et al (22). One can see that the width of the Δ is larger for the nuclear targets, combined with a small upward shift. For ^{12}C the width (FWHM) is $\Gamma = 240$ MeV and a shift in energy of $\delta = 19 \pm 5$ MeV is deduced.

A less model-dependent analysis of the energy shift can be obtained from a dispersion relation analysis of nuclear Compton scattering (22). The conclusion is that the resonance energy is shifted less than 5 MeV in nuclei relative to the free nucleon value.

The Δ-hole model (23, 24) has successfully been applied to photon absorption. The ingredients in this model are an effective Δ-hole interaction and potentials for nucleons and Δ isobars in the nucleus. The real part of the latter potential is around 30 MeV and includes various effects. It is usually difficult to distinguish between this real part and the effective Δ-hole interaction. The imaginary potential is taken to be around 40 MeV and should describe the combined effect of the narrowing of the Δ width from the Pauli blocking of the $\Delta \rightarrow N + \pi$ decay and the widening from two-nucleon ($N\Delta \rightarrow NN$) decay. The model describes the data fairly well and restrains specifically the parameters of the effective Δ-hole interactions in the spin-transverse channel.

3.2   *Electron Scattering*

With inelastic electron scattering the region below the photon line can be studied. Data have been obtained at ALS (Saclay) (25) and MIT at energies between 620 and 750 MeV (26, 27) and at SLAC at energies up to 1650 MeV (28, 29). In some cases the response has been separated in a longitudinal (a nonspin or Coulomb) part and a transverse or magnetic contribution. It is found that the Δ region is completely dominated by the transverse response.

The data at the lower energies show a behavior similar to that shown by the photon absorption data. The cross section per nucleon is almost the same and an increase of the apparent width of the Δ for nuclear targets is observed.

With increasing four-momentum transfer, the Δ peak becomes less and less pronounced in the spectrum and the analysis becomes more difficult. It seems, however, that the energy of the Δ peak increases with increasing four-momentum transfer. Part of this effect could be due to problems in defining the background. The dependence on q_μ^2 is not inherent in the Δ-hole model.

4.   PION-NUCLEAR REACTIONS

In the resonance region the formation of the Δ plays a dominant role for many properties in pion-nucleus scattering, and the Δ-hole model has successfully been used in this energy region though only for π data on light nuclei (24).

The pion absorption cross section is very large and the forward angle scattering is strongly modified by effects of the medium. The spin-longitudinal nature of the $\pi N\Delta$ vertex results in coherent pion scattering in the forward direction. This is seen in a partial wave analysis of elastic pion-nucleus scattering. The low partial waves scatter as if from a black disk. This pattern is also seen in the total cross section. The optical theorem relates the forward elastic amplitude to the total cross section. In Figure 4 total cross sections for photon and pion scattering on ^{12}C are compared. The figure is from a book by Ericson & Weise (30). A downward shift is seen for the pion curve, and the partial wave analysis shows this in more detail (31). The shifts for the low partial waves are quite large. We note that this feature emerges from the Δ-hole model with pion exchange as the Δ-hole interaction, but we should emphasize again that detailed calculations have only been performed for light nuclei. For heavy nuclei the signature of the Δ almost disappears from the total reaction cross section versus energy (32). We also note that this is not in conflict with the two-

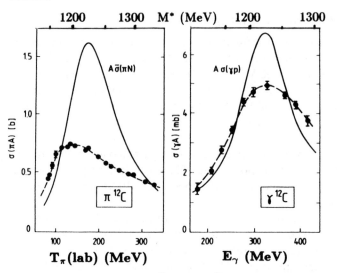

Figure 4 The total cross sections for $\pi + {}^{12}C$ and $\gamma + {}^{12}C$ reactions are compared. Also shown are the corresponding cross sections on the nucleon times $A (= 12)$. The figure is adapted from the book by Ericson & Weise (30).

level model referred to above nor to the effects seen in charge exchange reactions and discussed in the following sections. There we discuss correlations at larger momentum transfers or higher partial waves, and in the two-level model strength is removed from the pion line. Pion-induced reactions in such a picture are not the ideal way to study pionic modes even if the probing interaction, the driving force, has the correct spin structure.

We do not further discuss pion reactions here. Instead, we refer the reader to the book mentioned above and a number of review articles (3, 23, 31, 33–35).

5.   BARYON CHARGE EXCHANGE REACTIONS

The cross section for the elementary process $NN \rightarrow N\Delta$ is very large and the charge exchange reaction at intermediate energies is completely dominated by this process in the Δ energy region. The non-spin-transfer part of the cross section is much smaller than the $\Delta S = 1$ part involved in forming the isobar in nuclei. The spectra from charge exchange reactions are therefore much simpler to analyze than inelastic spectra at the same energy. We discuss in the following only data obtained with charge exchange reactions.

5.1 $NN \to \Delta N$

The data on the elementary $NN \to N\Delta$ transition are summarized in Figures 5, 6, and 7. In Figure 5 is shown the cross section for the $p+p \to n+\Delta^{++}$ reaction versus four-momentum transfer at four energies (36). In Figure 6 the total cross section for $pp \to p\pi^+n$ is given as a function of energy with emphasis on the threshold region (37). We see a cross section almost independent of energy above this threshold, and from Figure 5 one can see that this is so also for the t dependence. This is similar to the situation for the $pn \to np$ reaction, where the forward angles cross section is independent of energy from 50 MeV to as high as it is measured. Such a behavior would come from a one-pion-exchange model for the two

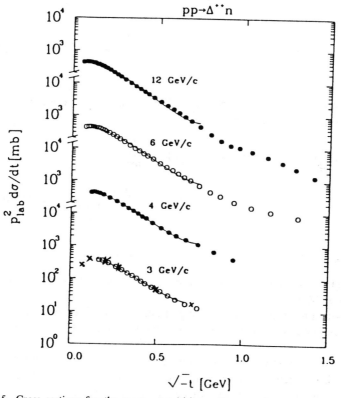

Figure 5 Cross sections for the $p+p \to n+\Delta^{++}$ reaction are given versus momentum transfer (36). The thin drawn curves correspond to an "absorbed" one-pion exchange model (36). The crosses in the 3-GeV/c data correspond to the simple expression given in the text with $\Lambda_\pi = 630$ MeV/c.

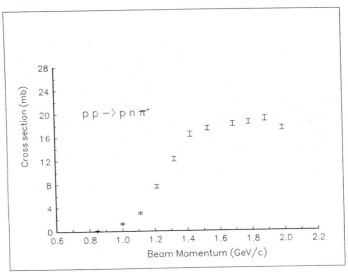

Figure 6 The total cross section for the $p+p \rightarrow n+p+\pi$ reaction is given as a function of energy (37).

processes. For the $pp \rightarrow n\Delta^{++}$ reaction, we can write the square of the transition matrix element as (38, 39)

$$M^2 = \frac{1}{4}\left(-\sqrt{2}g_\pi f_\pi^*/m_\pi\right)^2 \left(\frac{\Lambda_\pi^2 - m_\pi^2}{\Lambda_\pi^2 - t}\right)^4 \frac{-2t}{(t-m_\pi^2)^2}$$
$$\times \frac{[(m^*+m)^2 - t]^2}{3m^{*2}}[(m^*-m)^2 - t]$$

with

$$g_\pi = 13.61, \qquad f_\pi/m_\pi = g_\pi/2m, \qquad f_\pi^*/f_\pi = 2.0.$$

In Figure 5 the curve given corresponds to the cut-off parameter equal to $\Lambda = 650$ MeV/c.

In this model the cross section is independent of energy. The observed t dependence is also described by this expression, with the value for Λ as given.

The spin structure for the elastic and the $NN \rightarrow N\Delta$ amplitudes does not follow such a simple prescription. In Figure 7 the squares of the spin amplitudes for the two processes are given. The spin-longitudinal amplitude is denoted δ, and ε and β are the transverse amplitudes in and out of the scattering plane, respectively. The results for the elastic process may be summarized as follows (40; D. V. Bugg, private communication).

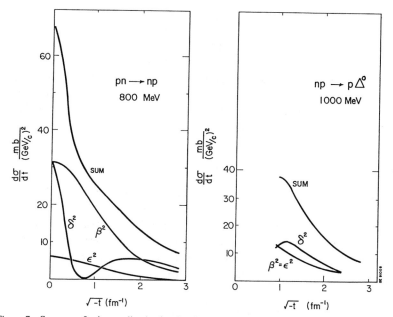

Figure 7   Squares of spin amplitudes for the elementary charge exchange reactions at similar energies are compared. The data for the np → pn reaction are from Dubois et al (40) and D. V. Bugg (private communication). For the np → pΔ^0 reaction the sum of squares corresponds to Figure 5 (corrected for isospin by a factor 1/3). The ratio of the spin-longitudinal (δ^2) and spin-transverse terms (β^2 and ε^2) is taken from data on the p($\bar{d}$, 2p[$^1S_0$])Δ^0 reaction (42). The two transverse amplitudes are assumed to be equal since a separation has not yet been achieved.

The δ and β amplitudes are rather independent of energy, whereas ε is larger than β at 100 MeV, but decreases to become rather small at 800 MeV (Figure 7). We note that such a behavior is not consistent with a boson exchange potential, be it π or ($\pi + \rho$) exchange. In such models the two transverse amplitudes would be equal (in Born approximation). The numbers for the N → Δ transition are deduced from the p($\bar{d}$, ^{2}He)Δ^0 reaction (42) and refer to the N-N vertex; they are therefore directly comparable to the elastic amplitudes.

This experiment has thus far determined only the ratio between longitudinal and transverse cross section; in the plot we assumed the two transverse amplitudes to be the same.

We see from Figure 7 a strong similarity between the two reactions, an indication that the same basic process is in play. A description in terms of one-pion exchange modified by short-range absorption accounts reasonably well for the available data. At low energy, the absorption can be parametrized as a δ force in r space leading to $\delta = \beta = \varepsilon$ at $q = 0$ for the elastic amplitudes. In the high energy limit, the absorption is an impact parameter and we find $\delta = \beta$ and $\varepsilon = 0$ for $q = 0$ (43, 44). Such a model also describes essential features of the data on spin observables from Wicklund et al (36). The cross-section data in Figure 5 are taken from that paper.

We conclude that the elementary NN $\rightarrow$ NΔ reaction has a very large cross section with a complicated spin structure. The process is a mixed spin-transverse and spin-longitudinal transition in a ratio of about 2:1 around an energy of 1 GeV.

5.2   Composite Projectiles

In this section we discuss form factors for a number of projectile-ejectile systems used in the study of Δ excitations in nuclei. The simplest one is the d-2p[^{1}S$_0$] form factor, the most complicated is from ^{40}Ar-induced reactions. In the plane wave impulse approximation (PWIA) we can write the cross section as a product of the elementary cross section, the projectile-ejectile form factor, and the target response function:

$$\frac{d\sigma}{dt} = N\left\{F_T^2(t)\frac{d\sigma_T}{dt}(t)R_T + F_L^2(t)\frac{d\sigma_L}{dt}(t)R_L\right\}.$$

For the nucleon target the response functions R would be equal to unity. N is the absorption factor. In this approximation the transverse and longitudinal part also separates and by proper choice of projectile-ejectile systems one or the other of the two responses can be emphasized. In Figure 8 longitudinal and transverse form factors are shown for a number of projectile-ejectile systems. The d-2p[^{1}S$_0$] form factor is calculated from wave functions for the two states. The d state in the deuteron has a very large effect in enhancing the longitudinal form factor with increasing momentum transfer. The ratio of longitudinal to transverse form factor can be parametrized as $(F_L/F_T)^2 = \exp(-0.26t)$ and the ratio is then almost 3 at $\sqrt{-t} = 2$ fm^{-1}. The calculated form factors are found to describe the data very well for the p($\vec{d}$, 2p[^{1}S$_0$])n elastic case (45).

For the ^{3}He-t form factor the d state plays a similar role and gives in fact an even larger ratio $(F_L/F_T)^2 = \exp(-0.31t)$. In this case the transverse form factor can be deduced from magnetic electron scattering. The spin-longitudinal part is based on calculations and verified with the data for the p(^{3}He, t)Δ^{++} reaction (38). We see from Figure 8 that the (^{3}He, t) reaction is really a very good reaction for the study of the spin-longitudinal response at finite q. The form factor is not very steep, so it gives a large cross section for the Δ excitation and puts a strong emphasis on the spin-longitudinal cross section.

Formfactors

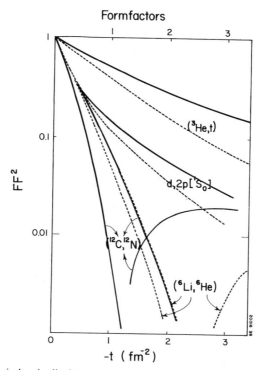

Figure 8   The spin-longitudinal (*solid*) and spin-transverse (*dashed*) form factors squared (FF^2) are given for a number of projectile-ejectile systems. The spin-longitudinal form factor for ^{6}Li-^{6}He, happens to almost coincide with the transverse form factor for ^{12}C-^{12}N. Only the larger is shown beyond the first minimum for these two reactions.

In Figure 9 we demonstrate that the expression above is a very good approximation for the p(^{3}He, t)Δ^{++} reaction. The elementary cross section is taken as the one-pion-exchange expression proven in Figure 5 to describe the pp $\rightarrow$ nΔ^{++} data. The extracted form factor is then shown to be consistent with measured transverse and calculated spin-longitudinal form factors.

The ^{6}Li-^{6}He and ^{12}C-^{12}N form factors are again based on calculations checked against data on the transverse form factors from electron scattering. We see a very different behavior for the two cases; the ratio of the form factors changes dramatically with momentum transfer. We also see rather steep form factors leading to much smaller cross sections for Δ formation than for the (^{3}He, t) reaction, for example. In Figure 10 we show the t dependence for the p(^{12}C, ^{12}N)Δ^0 reaction. We see a very impressive agreement with the simple approach taken here, the plane wave

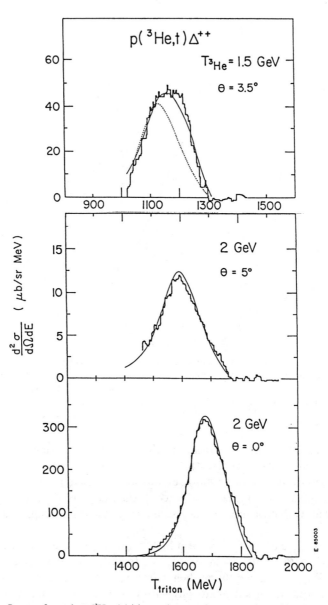

Figure 9   Spectra from the p(^{3}He, t)Δ^{++} reaction are shown together with calculated cross sections (38).

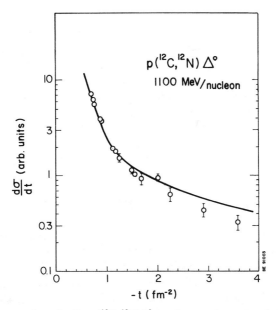

Figure 10 Cross sections for the $p(^{12}C, ^{12}N)\Delta^0$ reaction are shown together with the calculated dependence on momentum transfer. The curve is based on the form factors given in Figure 8 and the elementary cross section given in Figure 5 and an assumed ratio of transverse to longitudinal cross section of 2 to 1.

approximation multiplied by a constant attenuation factor. We note that the $(^{12}C, ^{12}N)$ reaction between 1 and 3 fm^{-2} is an extremely selective probe for spin-longitudinal modes. Distortion effects could complicate the analysis for heavier targets.

5.3   *Spin Observables*

The most direct way to separate the spin response into its components is the use of polarized beams. The $(\vec{p}, \tilde{n})$ reaction would be the simplest to analyze and such experiments are under way. Data have been obtained for two other reactions in which spin transfer variables are involved. At Laboratoire National Saturne, tensor-polarized beams of d and ^{6}Li have been used to measure the tensor-analyzing power in the $(\vec{d}, 2p[^1S_0])$ (42) and $(^6\vec{Li}, ^6He)$ reactions. In both cases the spin-1 and isospin-0 projectile is transformed into a spin-0 and isospin-1 ejectile. A measurement of the cross section for the different polarization states in the beam is enough to determine tensor-analyzing power. The measured quantity is

$$P = \rho_{20}(\tfrac{1}{2}T_{20} + \sqrt{\tfrac{3}{2}}T_{22}\cos 2\phi).$$

The term ρ_{20} is the beam polarization and ϕ is the angle between the polarization axis and the normal to the scattering plane. At zero degrees, where T_{22} is zero, one then measures T_{20}, which in the plane wave limit directly determines the ratio between longitudinal and transverse cross section.

In Figure 11 the results are compared for the deuteron and ^{12}C as targets (42). The data on the proton (which within error bars is equal to the deuteron results) provide the most direct and model-independent information on the ratio of longitudinal to transverse cross section for the elementary NN → NΔ process. We comment on the data for ^{12}C below.

6.   CHARGE EXCHANGE REACTIONS ON NUCLEI

The triton spectra from the (^{3}He, t) reaction on ^{12}C are shown in Figure 12. The spectra correspond to different cuts in the ω-q plane and we see that the raw spectra directly give essential features of the nuclear response. At small momentum transfer, the collective spin-isospin excitations are observed and at larger energy transfer and momentum transfer the quasi-free peaks from particle-hole and Δ-hole excitations dominate the spectra. The charge exchange reaction is an efficient tool for the study of the nuclear spin response.

6.1   *(p, n) Data*

Data for the (p, n) reaction have been obtained at Los Alamos (46, 47) at 800 MeV and at Gatchina (48) at 1 GeV. At both energies the energy of the Δ peak is around 30 MeV lower for nuclear targets than for the proton.

In Figure 13 a typical zero-degree spectrum from Lind (49) is shown together with model calculations performed by Udagawa et al (50). The calculated cross section is a sum of a longitudinal and a transverse part. The ΔN^{-1} interaction, parametrized as $g' + (\pi + \rho)$ exchange, shifts the longitudinal cross section downward but has little effect on the transverse cross section. In this model the shift of the Δ energy is then due to correlations. Also shown in the figure are the contributions from different J^π transfers. The residual interaction has a larger effect on the low angular momentum states, but the surface character of the reaction tends to emphasize the ΔN^{-1} states with larger J. The Δ is moving in a complex potential. This potential, together with the natural width of the Δ, is supposed to describe medium effects such as narrowing of width from Pauli blocking and spreading width from two-nucleon decay. The final shape of the Δ peak is then an envelope of ΔN^{-1} states, whose energies and widths depend on the ΔN^{-1} interaction and on this Δ potential. It is obviously very

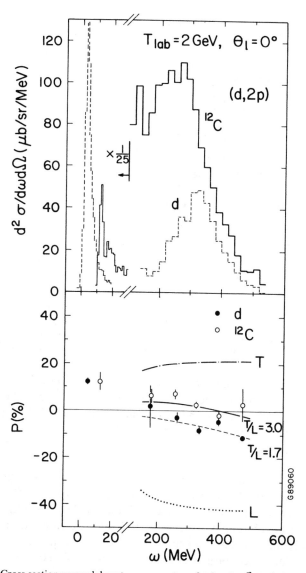

Figure 11   Cross section versus laboratory energy transfer for the $(\vec{d}, 2p[^1S_0])$ reaction on d and ^{12}C targets (42). The lower half of the figure gives the tensor-analyzing power (see text). The dotted curve corresponds to a pure spin-longitudinal response, e.g. one-pion exchange. The dot-dashed curve is for a pure spin-transverse transition. The two other curves correspond to best fits assuming constant ratio of transverse to longitudinal cross section.

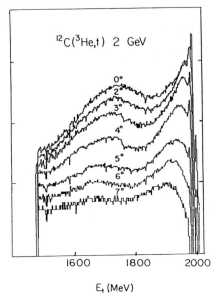

$^{12}C(^3He,t)$ 2 GeV

0°
2°
3°
4°
5°
6°
7°

1600   1800   2000

E_t (MeV)

Figure 12 Triton spectra from ^{12}C on a logarithmic scale for a range of angles.

difficult to disentangle these effects and it is essential to get more experimental information on the response in the Δ region. Spin observables would be helpful.

The model does, however, give a consistent picture of the available data. This is demonstrated in Figure 13 for the (p, n) data, but the γ and π absorption data are also reasonably well accounted for with the parameters used in the calculations by Udagawa et al (50).

6.2 ($^3He, t$) Data

Rather systematic data on the Δ formation in nuclei with the ($^3He, t$) reaction have been obtained at Saturne at energies around 2 GeV (38, 51, 52) and at Dubna at energies between 2.5 and 16 GeV (53, 54).

In Figure 14 zero-degree spectra for ^{12}C and the proton are compared. It was really these spectra that showed in such a clear manner the shift of the Δ energy in nuclear targets that started the discussion of medium effects for charge exchange reactions.

We first discuss the A dependence of the cross section given in Figure 15 for the quasi-elastic and Δ peak at four angles (11). This dependence is consistent with quasi-free excitation. The integrated cross section can be written as a product of the elementary cross section and the number of nucleons effectively contributing:

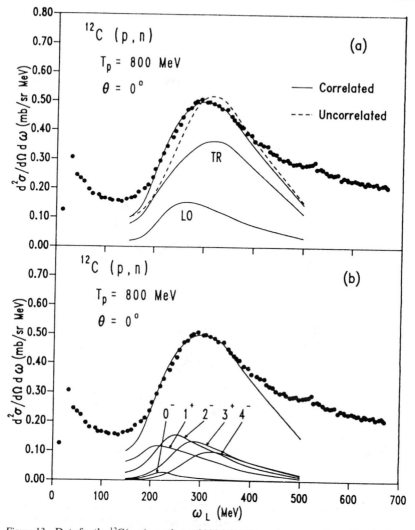

Figure 13 Data for the ^{12}C(p, n) reaction at 800 MeV (49) are shown together with calculated cross sections (50). In (*a*), the transverse and longitudinal cross sections are shown separately, and in (*b*) the contributions from different multipoles are given.

$$\frac{d\sigma}{d\Omega}(t) = \int \frac{d^2\sigma}{dE\,d\Omega}dE = N_{eff}\frac{d\sigma^{NN}}{d\Omega}(t)$$

at the momentum transfer, t, in question. N_{eff} is calculated as an attenuation factor times N, the number of neutrons for the quasi-elastic peak, or $(Z+$

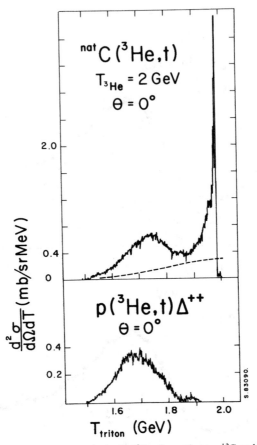

Figure 14   Zero-degree spectra from the (^{3}He, t) reaction on ^{12}C and the proton.

$\frac{1}{3}N$) for the Δ peak. The incoming and outgoing particles are attenuated along straight lines through densities of nuclear matter determined experimentally from electron scattering. The attenuation is taken as $\exp[x(b)]$ with

$$x(b) = \int dz\, \rho(\sqrt{b^2+z^2})\sigma_{NN},$$

a function of the impact parameter b. The term σ_{NN} usually represents the nucleon-nucleon total cross section, but in this case it is the projectile-target-nucleon total cross section. Here ρ is the density at $r = \sqrt{b^2+z^2}$, where z is the position along the trajectory. In Figure 15 the

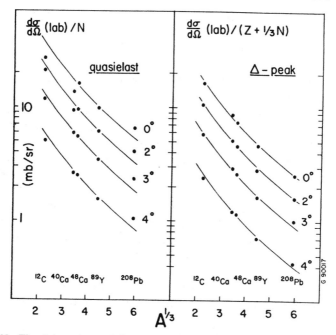

Figure 15 The A dependence of the cross sections at different angles for the quasi-elastic and the Δ peak (11). The spectra for ^{12}C are shown in Figure 12. The cross sections are given by N or $(Z + \frac{1}{3}N)$. The curves give the calculated A dependence of the attenuation factor and are all parallel in this logarithmic plot.

cross sections are divided by N or $(Z + \frac{1}{3}N)$ and the curves give the calculated attenuation versus A. In this model the attenuation should be independent of angle. The agreement with the data implies that the simple model developed above describes the essential features of the A dependence. The discrepancy for the quasi-elastic peak at small angles arises because the relevant parameter in this case is not the total number of neutrons, and nuclear structure effects determine the cross sections.

This model for the attenuation can also be used to estimate the average densities at which the reactions take place. The σ_{NN} in the above formula is adjusted to give the measured absolute cross sections. With σ_{NN} equal to 20, 37, and 55 mb for (p, n) at 800 MeV, (d, 2p) at 2 GeV, and (^{3}He, t) at 2 GeV, we find average densities of 0.5, 0.35, and $0.2\rho_0$ respectively. We note that these average densities are quite small, but that the distributions are rather broad. A more detailed calculation of the distortion effects has been performed by Dmitriev (55). The three nucleons in ^{3}He or t are distorted individually with the constraint from the internal wave functions.

It is shown that the cross section now receives contributions from higher density regions of the target.

In Figure 16 is given the centroid energy of the Δ peak in the (^{3}He, t) reaction on ^{12}C and the proton. The zero-degree spectra are shown in Figure 14. We see that the energy shift increases with increasing momentum transfer in line with the two-level model (Figure 2).

In Figure 17 we show the results of a detailed calculation by Delorme & Guichon (16). In this calculation the specific spin structure of the (^{3}He, t) probe is properly taken into account, as discussed in a previous section. The response function is calculated in a semiclassical approximation with the ΔN^{-1} interaction included in a random phase approximation (RPA). The width of the Δ in the medium is also considered. If the distortion effects are treated in an eikonal approximation, one can conclude that the distortion does not shift the centroid of the Δ peak. We see from the figure that the calculation accounts very well for the data and the authors conclude that part of the energy shift is a result of Δ-hole correlations in the nucleus.

The (^{3}He, t) data have also been analyzed by Udagawa et al (50). The distortions are treated in a full distorted wave calculation that includes the

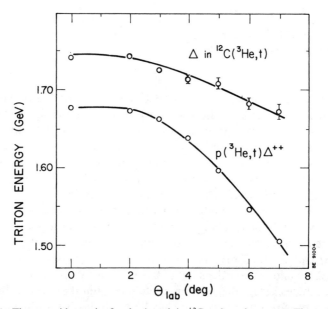

Figure 16 The centroid energies for the Δ peak in ^{12}C and on the proton. The zero-degree spectra are given in Figure 14. The spectra for ^{12}C are shown in Figure 12 for all the angles. The curves are to guide the eye.

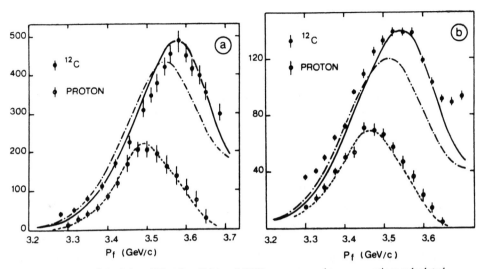

Figure 17 (³He, t) data (52) at $\theta = 0°$ (*a*) and 3° (*b*) are compared to cross sections calculated by Delorme & Guichon (16) with (*solid line*) and without (*dot-dashed line*) correlations in the ¹²C response function.

finite size of ³He and t. A distorted wave approach also seems to show that the distortion effects do not shift the Δ peak. The agreement with the data is quite good and would probably be improved by including the D state in the ³He and t wave functions. A somewhat simpler approach has been used by Esbensen & Lee (56), who assumed a surface response. They point out that a broadening of the Δ in the medium also leads to an apparent energy shift, because of the steep form factor.

We conclude this section by noting that the detailed models suggest effects from Δ-hole correlations consistent with the experimental observation, and that the shift increases with increasing momentum transfer. The (³He, t) form factor emphasizes very strongly the spin-longitudinal response with increasing momentum transfer.

The (³He, t) data at the higher energies from Dubna show at least two new features. At the highest energies the heavier Δ resonances at 1620 and 1700 MeV are excited, but with rather small cross sections. The other interesting effect is that the apparent shift in Δ energy for proton and ¹²C targets seems to become smaller with increasing energy. This is in line with predictions from the two-level model discussed above. The higher bombarding energy means that the response is probed at smaller momentum transfer, and the coupling between the Δ-hole states and the pion is smaller. The smaller momentum transfer also means that the enhancement of the spin-longitudinal part of the ³He-t form factor is less.

The contribution from projectile excitations in the (^{3}He, t) reaction was discussed by Oset et al (57). The question is how large this contribution is. It is seen from Figure 9 that it is small in the p(^{3}He, t)Δ^{++} case. The calculated curves in that figure describe the shape in detail without any need for projectile excitation. For nuclear targets the neutrons would enhance the contribution from projectile excitation more than the protons would. This comes from isospin coupling coefficients. Oset et al conjecture that the apparent energy shift between proton and nuclear targets is due to this projectile excitation. This explanation is not consistent with the (p, n) and ($\vec{d}$, 2p[^{1}S$_0$]) data.

6.3   ($\vec{d}$, 2p[^{1}S$_0$]) Data

In Figure 10 the zero-degree spectra from the ($\vec{d}$, 2p[^{1}S$_0$]) reaction on ^{12}C and the proton were compared (42). An energy shift between the two Δ peaks is observed, and an analysis based on quasi-free Δ excitation and an eikonal approximation for the distortion effects can only account for half of the shift. Data for the deuteron and ^{12}C have been obtained out to 7.2° at 2 GeV bombarding energy (44). This corresponds to $\sqrt{-t} = 2.3$ fm^{-1}.

The (d, 2p) data can also be used to comment on the contribution from projectile Δ excitation. At larger angles, where the experimental background conditions for the (d, 2p) studies are well under control, we notice a strict proportionality between the spectra from proton and deuteron target. Projectile excitation should in this case play a much larger role for the proton target.

The angular distributions for the Δ peaks are consistent with the d-2p[^{1}S$_0$] form factors as given in Figure 8. The (d, 2p) reaction on nuclei is therefore expected to emphasize the spin-longitudinal response. This is not the immediate conclusion from the data on the tensor-analyzing power for ^{12}C. The data seem to show the opposite, namely that the ratio of the spin-transverse to the spin-longitudinal cross section is larger for ^{12}C than for the proton or deuteron targets. This is true at all angles between 0° and 7.2°. We ascribe this to a distortion effect. The conclusion from the direct comparison of the raw data for ^{12}C and the proton assumes a plane wave approximation, and we have strong indications from an analysis of the corresponding data for the quasi-elastic peak that distortions are quite large for the spin observables in the ($\vec{d}$, 2p) reaction (44). For the quasi-elastic case, we can compare to the recent ($\vec{p}$, $\vec{p}'$) and ($\vec{p}$, $\vec{n}$) data, where the distortion effects are smaller.

The apparent discrepancy between the tensor-analyzing power and the energy shift as a result of spin-longitudinal correlations is then ascribed to an effect of distortion. A detailed analysis of this effect has not yet been performed.

6.4   *Heavy-Ion-Induced Reactions*

Isobar excitations have also been studied in heavy-ion-induced reactions. The inclusive reactions in these cases are extreme surface reactions. The density at which the nucleon-Δ transition takes place and its effect on the subsequent propagation of the Δ could still be dependent on the combined density in the overlap region of target and projectile. The analysis of the data is often difficult because of complex form factors and, not least, distortion effects. Data have been obtained at Saturne for reactions with projectiles of ^{6}Li, ^{12}C, ^{16}O, ^{20}Ne, and ^{40}Ar at energies around 1 GeV per nucleon (58, 59). The outgoing particles are momentum analyzed in a magnetic spectrometer, and all the bound states in the ejectile contribute to the cross section. The angular distributions are extremely forward peaked, consistent with a very simple model for the reaction, namely a factorization of a form factor for the projectile-ejectile system, a response function for the target, an elementary $NN \rightarrow N\Delta$ cross section, and an attenuation factor:

$$\frac{d^2\sigma}{dt\,d\omega} = NFF^2(t)\frac{d\sigma}{dt}(N \rightarrow \Delta)R(t, \omega).$$

This is similar to the approach by Delorme & Guichon (16) for the (^{3}He, t) reaction.

Angular distributions have been measured in some cases. We discussed the p(^{12}C, ^{12}N)Δ^0 reaction above, where the simple approach certainly works very well.

In most of the heavy-ion-induced reactions the ejectile has several bound states. The measured cross section is then a sum of cross sections to these states. A quantitative analysis would need all the form factors and the relative contributions for these ejectile states. Only in two cases do we know the form factors reasonably well. These are the (^{6}Li, ^{6}He) and (^{12}C, ^{12}N) reactions, in which the ground states are the only bound states. The form factors were given in Figure 8 and were found to be rather steep, partly because they correspond to Gamow-Teller transitions with transferred angular momentum $l = 0$. In addition, the larger size of the projectiles, compared to ^{3}He, for example, gives a steeper form factor (in q space).

In Figure 18 spectra from ^{20}Ne-induced reactions are shown. The opening angle of the spectrometer is so large that in fact the spectra give the total cross sections. The ^{20}Ne-induced reactions are examples of many ejectile states contributing to the total yield. In the nucleon sector, the cross section is increasing with the neutron excess in the (^{20}Ne, ^{20}F) reaction, the (p, n)-like reaction. The opposite is seen in the (n, p)-like channel. This dependence on neutron excess shows that the low angular momentum

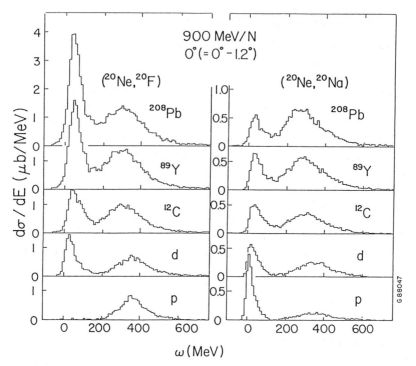

Figure 18 Zero-degree spectra for charge exchange reactions with a ^{20}Ne beam at 900 MeV per nucleon. The finite solid angle in the spectrometer together with the extreme forward peaking of the cross sections means that the spectra give the total cross sections.

transfers contribute significantly to this part of spectra. The A dependence of the Δ cross section is consistent with the model we used for the $(^3\text{He}, t)$ reaction. The cross section is proportional to N_{eff}, the number of effective nucleons contributing. N_{eff} is then equal to $D(Z + \frac{1}{3}N)$ or to $D(\frac{1}{2}Z + N)$ for (p, n)- or (n, p)-like reactions, respectively. D is the attenuation factor. In Figure 19 we give the numbers for the $(^{12}\text{C}, {}^{12}\text{N})$ reaction, where the cross sections are divided by DN in the nucleon sector and by $D(\frac{1}{3}Z + N)$ in the Δ sector.

The heavy-ion-induced reactions also show the shift of the energy of the Δ peak discussed above. We see a rather different behavior for the two channels. For the (n, p)-like channel the shift seems to increase with A, whereas for the (p, n)-like reaction the energy of the Δ peak is the same for nuclei with $A \geq 12$. This was also seen for (p, n) and $(^3\text{He}, t)$ reactions. This difference between the (p, n)- and (n, p)-like reactions is also seen for reactions with ^{12}C.

In Figure 20 spectra from the $(^{40}\text{Ar}, {}^{40}\text{K})$ reaction are shown. A very

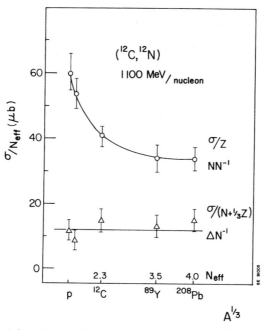

Figure 19   The *A* dependence of the total cross sections for the (^{12}C, ^{12}N) reaction in the nucleon sector and the Δ sector of the spectra. The cross sections are divided by the effective number of nucleons contributing. In Figure 15 the corresponding curve for (^{3}He, t) was not divided by the attenuation factor, but otherwise it is directly comparable.

large shift between the Δ energy for the two targets is observed. This is a case in which very many ejectile states contribute to the spectrum and a quantitative analysis will thus be very difficult.

The (^{12}C, ^{12}N) reaction should be simpler. Only the ground state in ^{12}N is a bound state, and the form factor for the ^{12}C-^{12}N transition was given in Figure 8. It was also demonstrated that the p(^{12}C, ^{12}N)Δ^0 reaction could be described fairly well in the plane wave limit. Preliminary data for the ^{208}Pb(^{12}C, ^{12}N) reaction, where angular distributions were obtained, show an effect similar to that given in Figure 16. The shift in energy between the Δ peak in the proton and ^{208}Pb increases with momentum transfer. The form factor for the ^{12}C-^{12}N transition as given in Figure 8 will enhance the spin-longitudinal response with increasing momentum transfer. The conclusion would therefore be that this is another confirmation of the ideas on the origin of the energy shift. In the ^{12}C-^{12}N transition, the distortion effects could be very important, so that the measured q dependence (the external momentum transfer) is not the in-medium dependence.

Guet et al (41) analyzed the total spectrum in the ^{12}C-induced

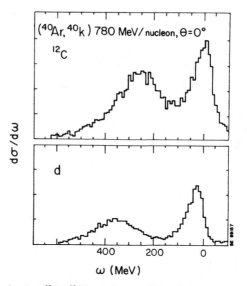

Figure 20 Spectra for the (^{40}Ar, ^{40}K) reaction on ^{12}C and the deuteron with a solid angle similar to that in Figure 18.

reaction. In an eikonal approximation for the absorption, they found that the impact parameter distribution is quite narrow but also that the combined density of target and projectile is around $0.5\rho_0$.

We conclude this section on heavy-ion Δ excitations with the following observations.

1. The Δ peak is a significant part of the spectra.
2. The A dependence of the cross section is consistent with quasi-free Δ formation modified by the form factor of the projectile-ejectile system and a dependence on absorption as described in an eikonal approximation.
3. Large energy shifts between the Δ in the nucleon and nuclear targets are found.
4. A detailed analysis of these data is complicated. For two reactions, ($^6\vec{\text{Li}}$, ^{6}He) and (^{12}C, ^{12}N), the form factors are fairly well known and model calculations should be made. Data have been obtained on cross section and tensor-analyzing power with the polarized $^6\vec{\text{Li}}$ beam at Saturne. Preliminary data on the momentum transfer dependence of the Δ energy in nuclei and the nucleon seem to confirm the findings for the (^{3}He, t) reaction.

7. EXCLUSIVE SPECTRA

In the preceding sections we discussed inclusive spectra, where only energy transfer and momentum transfer are measured. In some cases spin observables have also been obtained. The data from inclusive reactions are much simpler to analyze in terms of models. Response function technique can be applied, and integration over intermediate states or channels can often lead to rather model-independent predictions in terms of sums of strengths or cross sections.

For exclusive reactions this is usually not the case, and detailed models are needed. We refer very briefly to three different types of exclusive experiments relating to properties of the Δ isobar in nuclei.

7.1 Δ Decay Following ($^3He, t$) Reaction

Extensive data have been obtained in a coincidence experiment performed at Saturne. The momentum vector for charged pions and protons are measured in a drum-shaped detector, Diogène, in coincidence with forward-scattered tritons, which are analyzed in a separate dipole magnet (60). Data from proton and deuteron targets are compared to data from ^{12}C and ^{208}Pb (61).

In a recent experiment the Diogène detector was used to measure momenta of charged particles following proton-induced reactions (62). The decay of the Δ isobar in such a reaction also plays a dominant role. The selection of the Δ formation as the initial step is less specific, however.

The basic question is whether a collective ΔN^{-1} state will decay differently than a quasi-free ΔN state. The models presented in Figures 13 and 17 have definite numbers for the decay width for two-nucleon decay and "free" decay. The nuclear medium can, however, strongly affect the propagation of the decay particles and change the distributions seen in the detectors. Specific channels such as coherent pion propagation can be examined in such coincidence experiments. The contribution from Δ excitation in the projectile can also be tested. The analysis is in progress.

7.2   Quasi-Free Δ Decay

In an experiment by Nagae et al (63) the decay of the Δ isobar formed in a (p, p') reaction at 3 GeV has been studied. The decay is measured in an arrangement that determines the complete kinematics. All the momenta of light particles in the reaction

$$A(p, p)A^*(\Delta^0)$$
$$\hookrightarrow p + \pi^-,$$

that is, the momenta of two protons and a π^- are determined. Nagae et al conclude that the data are consistent with quasi-free Δ decay and that the energy of the Δ peak, as measured with the $p + \pi^-$ condition for Δ^0 decay, is not shifted relative to the expected position of the quasi-free Δ peak (63). This is an interesting result and may be summarized as follows: a condition on quasi-free decay leads to quasi-free Δ formation. The result is still consistent with the inclusive spectra presented in the previous sections and with the model calculation as given in Figure 13, for example.

7.3 Δ Decay Following π Absorption

Particle emission following π absorption has been studied at Los Alamos for π energies in the Δ resonance region (64). Data have been obtained for a number of nuclei throughout the periodic table. For some of the lighter nuclei it has been possible to follow all possible decay channels. The results confirm that channels other than $N + \pi$ and $2N$ contribute to the decay. The data are analyzed using Monte Carlo simulations of the distributions and, like the analysis of the $(^3He, t)$ Δ decay, a complicated procedure.

8.   SUMMARY

We have presented data for Δ isobar excitations in nuclei as studied with different probes. The electromagnetic excitation is the cleanest way to excite the isobar, but the photon is blind to correlations in the spin-longitudinal channel. The energy of the isobar excited with an electromagnetic probe is found to be equal to the free value, whereas the width is increased to as much as 240 MeV. The cross section per nucleon is almost constant from ^9Be to ^{235}U. Pion-induced reactions are dominated in the resonance energy region by the formation of the Δ isobar, and strong medium effects are observed in various channels.

The emphasis in this review is on the data from charge exchange reactions. These reactions involve the pion in a different way. The exchanged quantum between the probing field and the nucleus is dominantly the pion. This does not mean that the driving force is purely spin longitudinal. Any short-range correlations will bring in spin-transverse components. In all the charge exchange reactions a downward shift of the Δ peak is observed for nuclear targets. Model calculations for (p, n) and $(^3He, t)$ data ascribe this shift to Δ-hole correlations in the spin-longitudinal channel. The $(^3He, t)$ and $(\vec{d}, 2p[^1S_0])$ reactions are interesting probes because the d state in the wave functions enhances the spin-longitudinal form factor very much at larger momentum transfer. A two-level model for the pion and Δ-hole states predicts increasing effects with increasing momentum transfers. This seems to be confirmed by experiment. Data have also been obtained with tensor-polarized beams of d and ^{6}Li. The tensor-analyzing power in the Δ region with the $(\vec{d}, 2p[^1S_0])$ reaction on ^{12}C, interpreted in a plane wave approximation, is not consistent, however, with a shift caused by spin-longitudinal correlations. Strong distortion effects are seen in the corresponding case in the quasi-elastic region.

An extensive set of data has been obtained with heavy-ion-induced reactions. The Δ-isobar excitation in these reactions is also a prominent part of the spectra. Very large energy shifts for the Δ peak are observed, but the quantitative analysis of the data is complex and has not yet been done.

The central question is whether or not the shift discussed above really is due to Δ-hole correlations (or as described by another formulation, to the formation of pisobars). If so, we could expect much stronger medium effects at higher densities, as in heavy-ion collisions for example. Further experiments with polarized beams may clarify the nature of the correlations, and exclusive experiments following the decay of the Δ isobar in the nucleus could give new information on the density dependence of the correlations.

ACKNOWLEDGMENTS

Much of the data and analysis presented here was obtained in the collaboration around Laboratoire National Saturne. In particular I thank C. Ellegaard and J. Syrak Larsen at NBI; D. Bachelier, J. L. Boyard, M. Roy-Stephan, and T. Hennino from IPN, Orsay; P. Radvanyi and P. Zupranski from LNS; C. Goodman from IUCF; A. Brockstedt and M. Østerlund from Lund; and J. Y. Grossiord, A. Guichard, and J. R. Pizzi from IPN, Lyon.

I am grateful for helpful discussions with Michèle Roy-Stephan and Thomas Sams. I would like specifically to express my gratitude for the interest and long lasting support I have received from Gerry Brown, including at times his moral support as well.

Literature Cited

1. Migdal, A. B. *Rev. Mod. Phys.* 50: 107 (1978)
2. Brown, G. E., Weise, W., *Phys. Rep.* 27: 1 (1976)
3. Oset, E., Toki, H., Weise, W., *Phys. Rep.* 83: 281 (1982)
4. Chanfray, G., Ericson, M., *Phys. Lett.* B141: 163 (1984)
5. Dmitriev, V., Suzuki, T., *Nucl. Phys.* A438: 697 (1985)
6. Ericson, M., *Nucl. Phys.* A518: 116 (1990)
7. Bertsch, G., Brown, G. E., Koch, V., *Nucl. Phys.* A490: 745 (1988)
8. Gyvlassy, M., Greiner, W., *Ann. Phys.* 109: 485 (1977)
9. Brown, G. E., et al., *Nucl. Phys.* A505: 823 (1989)
10. Gaarde, C., et al., *Nucl. Phys.* A369: 258 (1981)
11. Brockstedt, A., et al., *Nucl. Phys. A.* In press (1991)
12. Rees, L. B., et al., *Phys. Rev.* C34: 627 (1986)

13. Deleted in proof
14. Guichon, P. A. M., Delorme, J., *Proc. Journees d'etudes Saturne, Piriac.* Gif sur Yvette: LNS (1989), p. 53
15. Chanfray, G., *Ann. Phys.* C2.15: 151 (1990)
16. Delorme, J., Guichon, P. A. M., *Proc. 10th Sess. d'etudes d'Aussois*, in *Preprint Lycen.* Lyon: IPN (1989)
17. Ahrens, J., et al., *Nucl. Phys.* A251: 479 (1975)
18. Arends, J., et al., *Phys. Lett.* B98: 423 (1981)
19. Ahrens, J., et al., *Phys. Lett.* B146: 303 (1984)
20. Carlos, P., et al., *Nucl. Phys.* A431: 573 (1984)
21. Ghedira, L., These, Univ. Paris XI (1984)
22. Ahrens, J., Ferreira, L. S., Weise, W., *Nucl. Phys.* A485: 621 (1988)
23. Koch, J. H., Moniz, E. J., Otsuka, N. *Ann. Phys.* 154: 99 (1984)
24. Horikawa, Y., Thies, M., Lenz, F., *Nucl. Phys.* A345: 386 (1980)
25. Barreau, P., et al., *Nucl. Phys.* A402: 515 (1983)
26. O'Connell, J. S., et al., *Phys. Rev. Lett.* 53: 1627 (1984)
27. O'Connell, J. S., et al., *Phys. Rev.* C35: 1063 (1987)
28. Sealock, R. M., et al., *Phys. Rev. Lett.* 62: 1350 (1989)
29. Barab, D. T., et al., *Phys. Rev. Lett.* 61: 400 (1988)
30. Ericson, T., Weise, W., *Pions and Nuclei.* Oxford: Clarendon (1988)
31. Hirata, M., et al., *Ann. Phys.* 120: 205 (1979)
32. Carroll, A. S., et al., *Phys. Rev.* C14: 635 (1976)
33. Ashery, D., Schiffer, J. P., *Annu. Rev. Nucl. Part. Sci.* 36: 253 (1986)
34. Schiffer, J. P., *Comments Nucl. Part. Phys.* 10: 243 (1981)
35. Schiffer, J. P., *Comments Nucl. Part. Phys.* 14: 15 (1985)
36. Wicklund, A. B., et al., *Phys. Rev.* D34: 19 (1986)
38. Shimizu, F., et al., *Nucl. Phys.* A386: 571 (1982)
38. Ellegaard, C., et al., *Phys. Lett.* B154: 110 (1985)
39. Dmitriev, V., Sushkov, O., Gaarde, C., *Nucl. Phys.* A459: 503 (1986)
40. Dubois, R., et al., *Nucl. Phys.* A377: 554 (1982)
41. Guet, C., et al., *Nucl. Phys.* A494: 558 (1989)
42. Ellegaard, C., et al., *Phys. Lett.* B231: 365 (1989)
43. Williams, P. K., *Phys. Rev.* D1: 1312 (1970)
44. Sams, T., Thesis. Niels Bohr Inst., Univ. Copenhagen (1990)
45. Ellegaard, C., et al., *Phys. Rev. Lett.* 59: 974 (1987)
46. King, N. S. P., et al., *Phys. Lett.* B175: 279 (1986)
47. Jeppesen, R. G., et al., *Tech. Prog. Rep.*, Univ. Colo. at Boulder, NPL 987: 73 (1984)
48. Baturin, V. N., et al., Gatchina Preprint LNPHI 1322 (1987)
49. Lind, D. A., *Can. J. Phys.* 65: 637 (1987)
50. Udagawa, T., Hong, S. W., Osterfeld, F., *Phys. Lett.* B245: 1 (1990)
51. Ellegaard, C., et al., *Phys. Rev. Lett.* 50: 1745 (1983)
52. Contardo, D., et al., *Phys. Lett.* B168: 331 (1986)
53. Ableev, V. G., et al., *JETP Lett.* 40: 763 (1984)
54. Ableev, V. G., et al., *Yad. Fiz.* 46: 549 (1987), 48: 27 (1988)
55. Dmitriev, V., *Phys. Lett.* B226: 219 (1989)
56. Esbensen, H., Lee, T. S. H., *Phys. Rev.* C12: 1966 (1985)
57. Oset, E., Shiino, E., Toki, H., *Phys. Lett.* B224: 249 (1989)
58. Bachelier, D., et al., *Phys. Lett.* B172: 23 (1986)
59. Roy-Stephan, M., *Nucl. Phys.* A488: 187c (1988)
60. Alard, J. P., et al., *Nucl. Instrum. Methods* A261: 379 (1987)
61. Hennino, T., *Nucl. Phys.* A527: 399c (1991)
62. Lemaire, M. C., et al., *Phys. Rev.* C43: 2711 (1991)
63. Nagae, T., et al., *Phys. Lett.* B191: 31 (1987)
64. Ransome, R. D., et al., *Phys. Rev. Lett.* 64: 372 (1990)

Annu. Rev. Nucl. Part. Sci. 1991. 41: 219–67

LOW ENERGY ANTIPROTON PHYSICS

Claude Amsler

Physik-Institut der Universität Zürich, Schönberggasse 9,
CH-8001 Zürich, Switzerland

Fred Myhrer

Department of Physics, University of South Carolina, Columbia,
South Carolina 29208, USA

KEY WORDS: antiproton scattering, antiproton-proton annihilation, meson
             spectroscopy, antiproton-nucleus

CONTENTS

1. INTRODUCTION

This report summarizes the present experimental and theoretical status of antinucleon-nucleon ($\overline{N}N$) interactions at low energy (below 2 GeV/c). We discuss elastic scattering, hyperon pair production, and annihilation into mesonic channels, including the search for exotic meson states in the annihila-

219

0163–8998/91/1201–0219$02.00

tion debris. New results from the Low Energy Antiproton Ring (LEAR) at CERN, from KEK in Japan, and from Brookhaven National Laboratory (BNL) are now available. Some of the experiments and their comparison to theoretical predictions have been reviewed earlier (1–3). We extend these reviews to more recent data and theoretical calculations.

Before discussing low energy $\bar{N}N$ interactions, it should be emphasized that the physics is based on models, as is most of strong interaction physics. Presumably, the theory of quantum chromodynamics (QCD) describes the strong force between quarks and gluons. At low energy and low momentum transfer, the QCD coupling constant becomes large and could lead to color confinement of quarks. A picture of the baryon structure has emerged from our present knowledge of QCD. At large distances the quarks feel the confinement forces, while at short distances they are almost free; thus their interactions can be treated perturbatively. This is the basis for phenomenological models like the MIT bag model (4) or the nonrelativistic quark model in which confinement is simulated by an harmonic oscillator potential (5, 6). Both models successfully describe the baryon mass differences, the excited baryon mass spectra, the baryon magnetic moments, etc, where the few model parameters are fixed by reproducing the proton mass, the nucleon-Δ mass difference, and the Λ mass. Low energy strong interactions are guided by the requirements of chiral symmetry, which, when incorporated into the bag model, describes the nucleon as a core of three confined valence quarks surrounded by a cloud of pions (7, 8). The pions are distributed around the quark core in a manner dictated by chiral symmetry. As a consequence, the observed root mean square (rms) charge radius of the proton (0.9 fm) is larger than that of the quark core because of the corrections from the pion cloud.

When using this model to describe the $\bar{N}N$ interactions, it becomes clear that the quark core and the pion cloud participate differently in $\bar{N}N$ reactions. When only the pion clouds overlap (large $\bar{N}N$ impact parameters), long-range meson exchanges are expected to describe the process. However, for S-wave scattering (zero-impact parameter), both meson exchanges and quark-antiquark interactions contribute to the scattering process as a result of the overlapping cores. In this picture, annihilation occurs only when the quark cores overlap, which means that annihilation takes place at $\bar{N}N$ distances somewhat shorter than the pion Compton wavelength.

For $\bar{N}N$ scattering (elastic $\bar{p}p$, $\bar{p}n$, and charge exchange $\bar{p}p \rightarrow \bar{n}n$), all annihilation channels contribute to the inelasticity. This means that only the dominant bulk properties of annihilation are needed to treat the loss of flux from the $\bar{N}N$ scattering channels. The long-range part of the $\bar{N}N$ potential is determined from the NN meson exchange potential. The only requirement is that the exchanged mesons of odd G parity (π and ω) must have opposite sign in the $\bar{N}N$ potential. In some $\bar{N}N$ models a complex short-range potential is added to simulate the annihilation processes (9, 10) and the sizes of the hadrons themselves. Another model assumes meson exchanges at large distances and an absorptive boundary condition at short distances ($R \approx 0.5$ fm) (11). In all these models the long-range $\bar{N}N$ meson exchange potential (MEP) is very attractive and pulls the $\bar{N}N$ wave function into the annihilation region. The net effect is that the effective absorption radius (R_{eff}) is larger than the radius of the absorptive boundary condition. As a consequence, the forward slope of the elastic differential cross section (and its energy dependence) is given by the MEP (12). In other words, the forward low energy $\bar{N}N$ scattering ($p_{lab} < 800$ MeV/c) is dominated by MEP. Thus the longer range (real part) of the $\bar{N}N$ potential is playing a significant role, which can be studied in $\bar{p}p$ scattering. These models describe the forward differential cross sections well. They do not, however, fit the angular and energy dependence of recent $\bar{p}p$ polarization data from LEAR. The new scattering data are so precise that one can test some basic aspects of these models and possibly gain further insight into the scattering process. One should keep in mind that NN scattering above $p_{lab} = 1.5$ GeV/c is mainly diffractive (13), so we expect $\bar{N}N$ to become diffractive at a somewhat lower energy. Hence the MEP description is not useful above about 1 GeV/c.

It should also be stressed that, even at very low momenta, elastic $\bar{p}p$ scattering requires more partial waves than its pp counterpart. Because of the strong absorption of the $\bar{p}p$ S wave (quark cores overlap), the dominant partial waves are the P waves, and some D waves are already required at 300 MeV/c (14). This is in contrast to pp scattering, which is dominated by S waves at the same momentum. Furthermore, all phase shifts are complex because of annihilation, and both isospin 0 and 1 contribute in each partial wave. Hence a treatment of $\bar{p}p$ scattering is intrinsically more complex than for the familiar NN system. The experimental information at low energy is still quite fragmentary, with only differential cross sections being measured down to 180 MeV/c (14).

At low momentum, the annihilation cross section is very large and exceeds the elastic cross section. The study of the annihilation process might give new insight into the physics of hadronization. At high energy, multiparticle production dominates the total cross section. Thus high energy multiparticle production and low energy annihilation processes can be considered complementary studies of hadronization. The high energy production seems to be dominated by the dynamics of stretching color-flux tubes, which hadronize into jets, as in the Lund model (15). In contrast, low energy $\bar{N}N$ annihilation can be thought of as generating a hot, concentrated quark gas of energy ~ 2 GeV that subsequently evaporates into an average of five pions. In fact, some aspects of $\bar{p}p$ annihilation seem to follow from a statistical thermodynamical description. In this respect, $\bar{p}p$ low energy studies are small-scale versions of high energy heavy-ion experiments searching for the quark-gluon plasma, which many anticipate will be observed at the planned relativistic heavy-ion collider (RHIC) at BNL.

While the statistical models successfully describe the final-state pion multiplicity distribution and branching ratios of in-flight $\bar{p}p$ annihilation, they are not designed to fit the branching ratios for annihilation at rest into two or three meson resonances from specific $\bar{p}p$ atomic states. Some two- and three-body meson final states have recently been investigated at LEAR and KEK for $\bar{p}p$ annihilation at rest. The angular momentum of the initial $\bar{p}p$ atomic states, following $\bar{p}$ capture in hydrogen, can be determined by measuring the atomic x rays in coincidence with the final mesons. Here the annihilation process is studied under controlled circumstances that contrast with the dynamical hadronization process associated with jet production. A strong dependence on the initial angular momentum is observed, and current QCD-inspired model calculations fail to reproduce the measured branching ratios satisfactorily.

Antiproton-proton annihilation at low energy is also a tool to investigate the production of meson resonances with masses below 2 GeV. Apart from the standard $\bar{q}q$ states, one can produce exotic mesons such as a two-quark–two-antiquark $\bar{q}^2q^2$ or other multiquark mesons as well as hybrid mesons, for example $\bar{q}qg$, or glueballs (mesons made exclusively of gluons, g). Speculations based on various models predict the existence of such exotic mesons, but none has been convincingly observed in any hadronic reaction, although several candidates exist (for reviews, see 16, 17). In $\bar{p}p$ annihilation, the LEAR experiments have not observed any narrow baryonium (quasi-nuclear $\bar{N}N$) states, but broad states are not excluded (18). Relevant data on exotic mesons have been hindered by the lack of statistics and by the lack of detectors capable of precisely reconstructing exclusive final states involving kaons and neutral mesons like π^0's and η's. With the Crystal Barrel, Jetset, and Obelix detectors, which are on the floor at LEAR, there is for the first time a reasonable chance of observing an exotic meson. We review in this report the LEAR results pertaining to the existence of such exotic states.

2. ANTINUCLEON-NUCLEON SCATTERING

At low energies ($p_{lab} < 1$ GeV/c), the forward $\bar{p}p$ scattering processes (elastic and charge exchange $\bar{p}p \rightarrow \bar{n}n$) are described by the long-range meson exchanges, supplemented by models or parametrizations for p̄p annihilation (which is effective at N̄N relative distances of 0.6 to 1.4 fm in most models). The Schrödinger equation with the meson exchange potential (MEP) and a parametrization for the annihilation (often an optical potential) is solved to ensure that unitarity is satisfied when the observables are calculated. We present these ideas and then examine the experimental evidence that supports them.

2.1 *Theoretical Considerations*

2.1.1 THE MESON EXCHANGE FORCES The long-range meson exchange potentials include one-pion exchange, two-pion exchange, ω exchange, and the short-distance nuclear forces.

The one-pion-exchange (OPE) potential is well tested in nuclear physics (see Figure 1). It describes the higher NN scattering partial waves (20) and the deuteron asymptotic D/S ratio (21). In both cases the pion and the nucleons are treated as pointlike particles. The pion-nucleon coupling constant is well known, $g_{\pi NN}^2/4\pi \simeq 14.4$. Since the tensor part of the OPE potential behaves as r^{-3} at short distances, a cutoff is needed in practical calculations (a natural cutoff is the structure of the nucleon and pion).

The two-pion-exchange (TPE) potential (Figure 1) is calculated using dispersion theory by the Paris (22) and the Stony Brook (23) groups. Part of the TPE potential is also calculated by the Bonn group in an effective field theory (24). This TPE potential is simulated in one-boson exchange potentials (OBEP) by isoscalar-scalar and isovector-vector (ρ meson) ex-

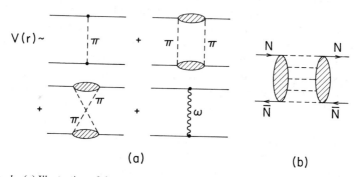

(a)   (b)

Figure 1 (a) Illustration of the one-pion, two-pion, and ω exchange potentials. The shaded ellipses for two-pion exchange are calculated in dispersion theory using as input pion-nucleon and pion-pion scattering data. For a recent review see (19). (b) The process N̄N → intermediate meson states → N̄N, illustrating that many intermediate meson states contribute to N̄N scattering via unitarity.

changes. Because of its large width, the ρ is included in TPE in a more natural way as a "bare" $\bar{q}q$ state embedded in the two-pion continuum, which has a range of $(2m_\pi)^{-1}$ or less. The TPE potential is not tested to high accuracy in NN scattering.

The ω exchange potential is assumed to give the NN repulsion in most meson exchange models. However, the value of the omega-nucleon coupling constant g_ω is very uncertain. We expect from SU(3) and vector-meson dominance $g_\omega^2/4\pi$ to be 4.5 (25), but in meson exchange models, like the Paris model, one needs a value of 10 to 12 to fit NN data. Be aware that cutoffs (or short-distance parametrizations) play a role in determining the latter values, as we discuss next.

The short-distance nuclear forces are often parametrized with (arbitrary) form factors (cutoffs). For short distances, it is difficult to separate the physics of a form factor from two or more meson exchanges. Many NN potential models introduce a short distance cut-off parametrization instead of form factors (22, 26). Some OBEP models also include heavier meson exchanges that we do not consider here. During the last ten years, model calculations have shown that most of the NN repulsion can be explained by using the Pauli principle on the quark level (with some help from the color-magnetic quark forces, which contribute to the nucleon-Δ mass difference). The quark degrees of freedom give S waves and $^1P_1$ NN phase shifts similar to hard-core phase shifts as a function of energy (19, 27). Model agreement with data is impressive. Including an ω exchange with a modest $g_\omega^2/4\pi = 4.5$ does not change the results of these phase-shift calculations. The quark model results are still being debated.

An $\bar{N}N$ potential can be generated from the one-pion, two-pion, and ω NN meson exchange potentials described above if one uses G-parity arguments (9, 28). This gives a good description for large $\bar{N}N$ separations. It generates a strongly attractive long-range force and, as we discuss below, it describes the $\bar{N}N$ forward differential cross-section data. However, the meson exchanges between nucleons implicitly assume pointlike nucleons and mesons. In NN OBEP, a form factor of $\sim (q^2+\Lambda^2)^{-1}$ is used with $\Lambda \sim 1.0$–1.5 GeV to parametrize the short-distance potentials (24), whereas the chiral or cloudy bag models of the nucleons give a "softer" πNN form factor (7, 8). This parametrization of the short-distance NN potentials should not be directly transferred to $\bar{N}N$ and there is a good reason for using quark models to describe these short-distance potentials: for NN and $\bar{N}N$ separations less than 1 fm, the quark cores are expected to overlap. Quark degrees of freedom for short-distance forces naturally lead in $\bar{N}N$ to annihilation processes, discussed in the next subsection.

A final comment on meson exchange forces: Theoretical calculations of NN or $\bar{N}N$ interactions should show how the observables change when the values of the cutoff parameters in the various potentials are changed. In this way one may better judge which parts of the calculations reflect the physics input.

2.1.2   THE ANNIHILATION AND $\bar{N}N$ SCATTERING   To describe $\bar{N}N$ scattering observables we need, as alluded to above, a model for the annihilation process (see Sections 3.3 and 3.4). In $\bar{p}p$ scattering, annihilation occurs when the $\bar{N}$ and N quark cores overlap. For chiral bag models, the nucleon quark core has an rms radius of 0.65 to 0.75 fm; we thus expect annihilation to be effective at $\bar{N}N$ distances of 1 fm or less. The annihilation model is therefore intimately tied to the short-distance forces discussed above. For scattering processes, only the bulk properties of annihilation contribute through unitarity to $\bar{N}N$ scattering, where we have to sum over all intermediate meson channels. The processes shown in Figure 1b can be described by optical potentials constructed so that unitarity is satisfied in the scattering process. We have only a primitive model-dependent understanding of the annihilation process itself, but since the many competing annihilation channels have to be added, it is reasonable to assume that the sum will have no strong spin nor isospin dependence.

The following models support our belief that the spin and isospin dependences indeed average out. The simplest model to describe the bulk properties of annihilation is the hot gas model (29). The energy spectrum of the pions in the annihilation reaction $\bar{p}p \rightarrow \pi^{\pm} +$ anything is well described by a hot gas with a temperature of $T \sim 100$ MeV. As a consequence of this model, we expect quantum numbers to be distributed in a statistical manner, as reinforced by the isospin statistical model (30). This model, which has been confirmed experimentally for $\bar{p}p$ annihilating into pions (31), says that for a given number of pions n in the final state, the cross-section ratios of the different charged channels (for instance for $n = 5$, the channels $2\pi^+2\pi^-\pi^0$ or $\pi^+\pi^-3\pi^0$ or $5\pi^0$) are determined by a statistical distribution in the pion charges according to the weight $(n_{\pi^+}!n_{\pi^-}!n_{\pi^0}!)^{-1}$, where $n = n_{\pi^+} + n_{\pi^-} + n_{\pi^0}$. Finally, in the threshold dominance model, recently elaborated by Vandermeulen (32), the energy dependence of the cross sections for $\bar{p}p$ into two or more pions is fitted with two parameters. The model averages over spin and isospin of the intermediate two-meson channels leading to the same final state. We discuss these models in some detail in Section 3.3.

Another feature of annihilation is based on dispersion arguments (33), which say that the annihilation forces should be of short range, approximately $\exp(-2Mr)/r$, where M is the nucleon mass and r is the relative $\bar{N}N$ distance (9, 10, 34, 35) (see Figure 1b). On the quark level, model calculations also find a rapid damping of the annihilation strength with increasing r for distances larger than 0.7 fm (36). In this model the absorption is so strong that for $r < 0.7$ fm $\bar{N}N$ scattering becomes insensitive to the short-distance ($r < 0.7$ fm) potential behavior and hence absorption is complete (the wave function becomes zero) at very short $\bar{N}N$ distances. An extreme model simulates annihilation by a black sphere of radius R, which, together with MEP, fits data for $R \approx 0.5$ fm (11). The black sphere effectively absorbs all low impact parameter $\bar{N}N$ scattering.

Other types of models for annihilation used in $\bar{N}N$ scattering are based on an effective meson-baryon theory that implicitly assumes pointlike particles. Annihilation is described by baryon exchanges in which $\bar{N}N$ couples to two mesons (37–39). These baryon exchanges correspond to Yukawa forces with a range of the order of $(2M)^{-1}$, which is small compared to the rms radii of the baryons themselves, a point we elaborate on in the next subsection. In a coupled-channel calculation the $\bar{N}N$ wave function is nonzero at very small distances, contrary to the optical models (39). These approaches introduce vertex form factors that also have ranges of the order of the inverse nucleon mass. One question is whether this coupled-channel description can be a reasonable parametrization of the successful Vandermeulen model (32) in which heavy intermediate mesons (ω, ρ, a_0, a_1, f_0, K*, ϕ, etc) are needed and dominate (see Section 3.3). Some of these intermediate states have been incorporated in recent distorted wave Born approximation (DWBA) calculations by the Jülich group (40).

2.1.3   BARYONIUM AND ANNIHILATION   In the original work of Shapiro and collaborators, $\bar{N}N$ resonances and bound states were predicted on the basis of the strongly attractive meson exchange potential (41). We refer to these molecular $\bar{N}N$ states as quasi-nuclear or baryonium states. On the other hand, $\bar{q}^2 q^2$ Regge trajectories are also predicted, for which the intercepts and slopes have been determined (42). The $\bar{N}N$ potential trajectories (which are not straight lines) are very sensitive to the short-range parametrization of the potential, which introduces uncertainties large enough for the $q^2 \bar{q}^2$ and the isospin zero $\bar{N}N$ trajectories to coincide around the $\bar{N}N$ threshold (43). Hence baryonium states are difficult to distinguish from pure $\bar{q}^2 q^2$ states.

It was argued by many that the coupling to the annihilation channels was weak. This implies that one should observe narrow $\bar{N}N$ states. However, any $\bar{N}N$ model that described scattering cross sections required annihilation to be effective at 1 fm $\bar{N}N$ separation, despite having a short Yukawa range (44–47); this requirement made the $\bar{N}N$ states very broad.

The crucial argument of Shapiro leading to narrow baryonium states was based on meson-baryon dispersion theory (48, 49) (see Figure 1b). He and his collaborators argued that annihilation due to baryon exchange had to be of very short range and they therefore concluded that annihilation is weak at $\bar{N}N$ separation of 1.0–1.5 fm, which is the typical size of an $\bar{N}N$ state calculated from the many meson exchange potentials, neglecting annihilation. The flaw in this line of argument is that a nucleon has a sizeable quark core and these quarks are confined. Quark confinement implies that dispersion theory arguments do not readily apply to quark degrees of freedom. If, as indicated by quark models, the baryon Compton wavelength is much smaller than the quark confinement radius, the latter determines the shortest $\bar{N}N$ distance down to which the dispersion arguments apply.

2.2   Scattering Experiments

2.2.1   INTEGRATED CROSS SECTIONS   The $\bar{p}p$ cross sections—total (50, 51), elastic (14), annihilation (52), and charge exchange (53)—have all been measured at LEAR energies down to ~ 200 MeV/c. The Obelix collaboration has even measured the annihilation cross section at 70 MeV/c (54). At low energy (~ 300 MeV/c), the annihilation cross section exceeds the elastic cross section by a factor of two.

As is evident from the measured elastic differential cross section at $p_{\text{lab}} = 180$ MeV/c, the P wave is important. Because of the strong S-wave absorption, the imaginary parts of the S-wave amplitudes are close to their unitarity limit $\sim 1/2k$ (where $k \approx p_{\text{lab}}/2$ is the center-of-mass momentum). For $p_{\text{lab}} \approx 180$ MeV/c, P waves already contribute approximately 40% of the total cross section (14) in contrast to $\sim 10\%$ in pp. (The strong P wave is caused by the strongly attractive MEP, which pulls the $\bar{N}N$ wave function into the shorter-range absorption region). As discussed recently by Kroll & Schweiger (55), to reflect the importance of the P waves, the total cross section at LEAR energies should be parametrized as

$$\sigma_{\text{tot}}(\text{mb}) = 108.4/k + 13.5 + 187k + 495k^2 - 784k^3 + \cdots, \qquad 1.$$

where the center-of-mass momentum k is in units of fm^{-1}. At $p_{\text{lab}} = 180$ MeV/c the center-of-mass momentum is $k \simeq 0.46$ fm^{-1}. For much lower values of k, the two last terms in Equation 1 can be dropped. However, it is obviously a very bad approximation to drop both the third term, which comes from the P waves, and the last two terms in Equation 1 for $p_{\text{lab}} \geq 180$ MeV/c. Unfortunately, the total cross section has been approximated by the empirical formula

$$\sigma_{\text{tot}}(\text{mb}) = 54/p_{\text{lab}}(\text{GeV}/c) + 66 \qquad 2.$$

as quoted in the review by Walcher (1). This last formula should no longer be used. The annihilation cross section is empirically given by (52)

$$\sigma_a(\text{mb}) = 46.6/p_{\text{lab}}(\text{GeV}/c) - 30.1 + 60.3p_{\text{lab}}(\text{GeV}/c). \qquad 3.$$

The annihilation cross section exceeds the unitarity limit for an S wave at 200 MeV/c by a factor 1.7, requiring a strong P wave.

Originally, these cross sections were measured to search for resonances of the $\bar{p}p$ system. The controversy on the existence of N$\bar{\text{N}}$ resonances has been settled at LEAR (for a review, see 18). The $\bar{p}p$ total cross section does not show any evidence for either broad or narrow states below 1 GeV/c. For narrow states ($\Gamma < 3.5$ MeV) the upper limit for the integrated cross section is 2 mb MeV above 400 MeV/c (50) and 8 mb MeV between 220 and 400 MeV/c (51). No structure (with an upper limit of 5 mb MeV) is observed in the inclusive annihilation cross section between 400 and 600 MeV/c (56). For limits on $\bar{\text{N}}$N bound states, see Section 3.1.5.

2.2.2 ELASTIC SCATTERING   The elastic $\bar{p}p$ cross section has been measured at LEAR between 180 and 1500 MeV/c (14, 57, 58). The main contribution to the OBEP potential is OPE, with some influence from heavier meson exchanges. The calculated differential cross sections for elastic and charge exchange scattering are shown in Figure 2. They agree with experimental data (1). The forward differential cross section is not sensitive to the precise shape of the annihilation potential. The annihilation potential for the fits of Figure 2 (*left*) was generated from a quark-gluon model and a quark confinement potential (36, 59). A Gaussian annihilation

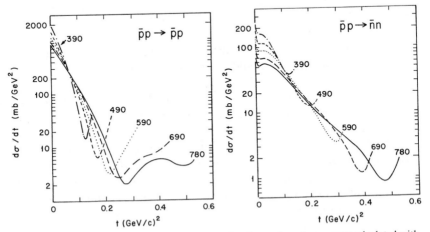

Figure 2   (*Left*) Elastic $\bar{p}p$ differential cross section for various $\bar{p}$ momenta calculated with the Bryan-Phillips potential and the imaginary part constructed from a quark model (36, 59). (*Right*) Charge exchange reaction with the same meson exchange potential but with a black sphere to simulate annihilation (11).

potential (or Woods-Saxon) works equally well and one finds an annihilation radius of 0.7 fm with a depth of ~ 1 GeV (60).

At small angles, the interference of the nuclear and Coulomb amplitudes yields the real part of the (spin-averaged) forward elastic scattering amplitude $f(0)$. The ρ parameter [$\rho = \mathrm{Re}\, f(0)/\mathrm{Im}\, f(0)$] was recently measured at LEAR below 300 MeV/c (61–63) and was discussed by Walcher (1). At zero momentum, ρ is related to the shift and broadening of the p̄p atomic 1S states, which are also measured at LEAR (64). (To be precise, ρ is the ratio of the real to imaginary parts of the hadronic scattering length, while x-ray experiments measure the Coulomb-modified scattering length. The correction is, however, much smaller than the present experimental error.) Using an optical potential, Batty (64) finds from the LEAR x-ray data $\rho = -1.08 \pm 0.09$. Hence ρ must be negative at very low momentum, although scattering data show it to rise toward positive values with decreasing momentum for momenta below 300 MeV/c, which means that ρ must increase very rapidly with increasing momentum very close to threshold. In Kroll & Schweiger's dispersion calculation of the forward p̄p amplitude these measurements of the ρ parameter generate a structure in the imaginary part of the p̄p amplitude ~ 20 MeV below threshold (65). These amplitudes below threshold (55, 65) have been used in a calculation by Fasano & Locher (66), who evaluate the annihilation reaction $\bar{p} + d \rightarrow N + n\pi$, which is sensitive to the imaginary part of the p̄p amplitude below threshold. They conclude that only a weak structure below threshold is compatible with p̄d data.

A fit to low energy cross section and x-ray data, using a coupled-channel effective range expansion of the spin-averaged N̄N S and P waves, shows that the averaged P wave is repulsive and turns attractive below 300 MeV/c, while the averaged S wave remains repulsive (67). Hence the turnover of the ρ parameter at low energy is attributed to this behavior of the P wave.

The p̄p analyzing power $P(\theta)$ has been measured between 439 and 1550 MeV/c (57, 58, 68). In one experimental approach, both proton and antiproton are detected and most of the background from interactions with heavier nuclei in the polarized target is eliminated by angular correlations (57, 68). At low momentum, however, absorption of the proton or antiproton in the target prevents measurements at small and large scattering angles. In another approach, the angle and the energy of one scattered particle (proton or antiproton) are measured in a magnetic spectrometer (58). This method allows coverage of the full angular range. The results from both experiments agree. Figure 3a shows $P(\theta)$ at 697 MeV/c compared with model predictions. The agreement is poor but the general trend (dip-bump structures) is reproduced. At lower energies, the $P(\theta)$ predictions in the forward direction are in better agreement with experi-

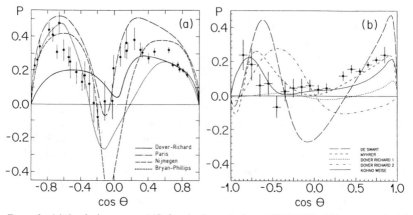

Figure 3 (*a*) Analyzing power $P(\theta)$ for elastic scattering at 697 MeV/c (58) compared with theoretical predictions (9, 10, 34, 69). (*b*) $P(\theta)$ for charge exchange at 656 MeV/c (70) compared with theoretical predictions (10, 11, 60; R. G. E. Timmermans et al, private communication from J. J. de Swart).

ments (71). [Note that the potential of the Nijmegen group (69) has an unphysical long-range phenomenological real part (12, 39).] One main difference between the various theoretical approaches is the short-range assumption for both the real and the imaginary parts of the potential. Hence polarization data, in contrast to cross-section data, are very sensitive to the short-range parametrization (12) and can provide new tests of the various $\bar{N}N$ models.

The measurement of other spin observables (72) is difficult because there are no polarized antiproton beams or suitable analyzers of antiproton polarization. The analyzing power on carbon is very small at low $\bar{p}$ momentum (73). On the other hand, the polarization transfer from the polarized target proton to the recoil proton can only be measured for energetic protons escaping from the target and for which the carbon analyzing power is large. These protons are associated with backward scattered $\bar{p}$, for which the elastic cross section is small. So far we have only a few data points with limited statistics for the depolarization parameter D between (the high momenta) 988 and 1259 MeV/c (74), and the points disagree with theoretical predictions of the low energy models.

2.2.3 CHARGE EXCHANGE The charge exchange $\bar{p}p \to \bar{n}n$ differential cross section is measured between 180 and 600 MeV/c (53, 70, 75). This cross section is strongly forward peaked, with a shoulder at small momentum transfer, a shape that agrees with OBEP models supplemented by strong annihilation, as discussed by Phillips (28) (see Figure 2*b*). Below

300 MeV/c, earlier data on the total charge exchange cross section (76) showed a much faster increase with energy just above threshold than data from LEAR (53). Also the various data disagree in the experimentally difficult range of extreme forward and backward directions.

Birsa et al measured $d\sigma/d\Omega$ and $P(\theta)$ for this reaction with a polarized penthanol target between 600 and 1300 MeV/c (70). The antineutron is detected by its annihilation in an iron/streamer-tubes calorimeter and the neutron in a scintillation counter array. The target polarization reached a record value of 96%. Figure 3b shows the analyzing power at 656 MeV/c together with theoretical predictions.

2.2.4 HYPERON-ANTIHYPERON PRODUCTION   The production of $\bar{\Lambda}\Lambda$ pairs in $\bar{p}p \rightarrow \bar{\Lambda}\Lambda$ has been studied at LEAR with high statistics close to the $\bar{\Lambda}\Lambda$ threshold (1435 MeV/c) (77, 78). The polarization of the Λ ($\bar{\Lambda}$) is derived from the known self-analyzing power in Λ ($\bar{\Lambda}$) decay into $p\pi^-$ ($\bar{p}\pi^+$). In addition, the spin correlations of the hyperon pairs have been determined at 1546 and 1695 MeV/c (79). The cross section for this reaction rises very quickly from threshold to ~ 100 μb at ~ 1650 MeV/c. A trigger was therefore required to collect large statistical samples. This experiment takes advantage of the narrow LEAR $\bar{p}$ beam: the small proton target was surrounded by veto counters to trigger on the emerging hyperon pairs, which have a short decay length.

The main motivation was a study of the $\bar{s}s$ production dynamics, since in a naive constituent quark model, the spin of the s ($\bar{s}$) quark is carried by the Λ ($\bar{\Lambda}$). Again, the S waves are strongly absorbed as a consequence of initial (and final) state interactions, which means that the $L = 0$ amplitude is almost purely imaginary and close to its unitarity limit. As before, we expect the P wave to be important because of the very attractive $\bar{N}N$ potential. A fit to the cross section $\sigma = b_0\varepsilon + b_1\varepsilon^3$ (where $\varepsilon = s^{1/2} - 2m_\Lambda$ and s is the square of the center-of-mass energy) near threshold indeed yields $b_0 = 1.51$ μb/MeV$^{1/2}$ and $b_1 = 0.26$ μb/MeV$^{3/2}$—a dominant S-wave part with an important P-wave part. This is also directly observed from the forward peak in the angular distribution (Figure 4) even at 0.8 MeV above threshold (78). The spin correlation data are consistent with the $\bar{\Lambda}\Lambda$ pair being produced in a pure spin triplet state (79). This is surprising, since in general multiple gluon exchanges and K exchange models (81) predict the spin triplet to dominate, although the spin singlet does not vanish. The spin singlet contribution only vanishes for single-gluon exchange, a questionable approximation at these low energies.

Despite the $\bar{p}p$ high energy, meson exchange potentials have been used here with some success to calculate $d\sigma/d\Omega$ and $P(\theta)$ in a DWBA formulation (81, 82). The main reason for this success is that, close to thresh-

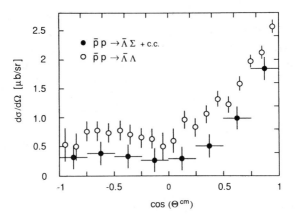

Figure 4 Differential cross section for $\bar{p}p \to \bar{\Lambda}\Sigma^0$ and $\Lambda\bar{\Lambda}$, each 15 MeV above their respective thresholds (80).

old, only S and P waves contribute. Since S waves for $\bar{p}p \to \bar{\Lambda}\Lambda$ are almost completely absorbed and the P waves are important, this reaction is sensitive to peripheral meson exchange forces (K exchange) (81). The long-range parts of K and K* t-channel exchanges (and even a naive timelike gluon exchange) give a reasonable fit to the data. However, according to Kohno & Weise (81), close to threshold, $P(\theta)$ is better predicted for only K or a timelike gluon exchange. It should be mentioned that the differential cross section of this peripheral reaction has also been described reasonably well at higher energies ($p_{\text{lab}} = 3$ to $6\ \text{GeV}/c$) with just K and K* exchanges, including a Regge-like form factor (83).

The roles of the different reaction mechanisms are being investigated by the same group by studying $\bar{\Lambda}\Sigma^0$ and $\bar{\Sigma}\Sigma$ since in these reactions K exchange is suppressed as a result of the coupling constant inequality $g_{K N \Sigma} \ll g_{K N \Lambda}$. Data for the $\bar{\Sigma}\Sigma$ reaction are not available yet. The angular distributions of the $\bar{\Lambda}$ for $\bar{p}p \to \bar{\Lambda}\Sigma^0$ and $\Lambda\bar{\Lambda}$, both 15 MeV above their respective thresholds, are very similar (Figure 4). A strong P wave is observed from the forward peak and the ratio of cross sections $\sigma(\bar{\Lambda}\Sigma^0)/\sigma(\Lambda\bar{\Lambda}) = 0.29 \pm 0.02$ agrees well with the prediction from the gluon model of Kohno & Weise (81). Polarization data for the Σ^0 are not yet of sufficient statistical quality to allow a meaningful comparison with theoretical models (80).

2.3 *Antiprotonic Atoms*

In protonium, the strong interaction models discussed above predict the measured 1S energy-level shift with respect to the expected QED value and the total 1S level width. The predicted level shift is typically $-0.8\ \text{keV}$

(that is, toward less binding) and the predicted width typically 1.0 keV (64), in good agreement with the experimental averages of -0.72 ± 0.04 and 1.11 ± 0.97 keV respectively (1, 64). So far, there is no measurement of the $^1S_0$ and $^3S_1$ level splitting. The predicted averaged 2P level width in protonium (84, 85) is in reasonable agreement with data. Batty reviewed the $\bar{p}p$ and $\bar{p}d$ atomic cascade measurements (64), and Reifenröther & Klempt presented recently refined cascade calculations (86). All models, which include the neutron-proton mass difference, find that the $\bar{p}p$ atomic wave function is strongly modified at distances $r < 5$ fm. The WKB calculation of Pilkuhn & Kaufmann (84) shows that the OPE potential introduces $\bar{n}n$ components into the atomic wave function, $\psi = a|\bar{p}p\rangle + b|\bar{n}n\rangle$, where $a^2 + b^2 = 1$. All calculations agree that $b \ll a$, but the calculation of Carbonell et al shows that the $|\bar{n}n\rangle$ component is large for $r < 2$ fm, where annihilation takes place (85). This is confirmed by the Aarhus group (87) and by estimates based on scattering lengths made by Jaenicke et al (88). Some aspects of Carbonell's work are discussed in Section 3.4, since modifications of the atomic wave function by strong interaction in the annihilation region might generate observable dynamical selection rules in some annihilation channels. This was remarked some time ago by Pilkuhn & Kaufmann (84).

The P-wave energy-level shifts (ΔE) and widths in $\bar{p}$ ^{3}He and $\bar{p}$ ^{4}He atoms have been measured (88a); if ΔE in $\bar{p}$ ^{3}He and the widths of the two isotopes are reproduced in calculations, then ΔE in $\bar{p}$ ^{4}He is about twice that expected from optical potential calculations (89). Only upper limits for the K x rays in $\bar{p}d$ are available (64), and there are conflicting measurement of L x-ray yields (86).

3. ANNIHILATION REACTIONS

There are two main motivations to study low energy annihilation:

1. Since the process involves the annihilation of quark pairs and the emission of gluons (g), $\bar{p}p$ is an excellent source for the production of exotic hadrons like glueballs (gg, ggg, . . .), hybrids ($q\bar{q}g$), and multi-quark mesons (for example, $\bar{q}^2q^2$). Recent results regarding these exotic mesons are reviewed in Section 4.

2. The annihilation mechanism itself is not understood. This is nontrivial since annihilation takes place in the nonperturbative regimes of QCD; we must resort to models. Nucleon-antinucleon annihilation at low energy generates an energy density of 1–2 GeV/fm^3, which hadronizes into several mesons. A theoretical effort is now under way to express annihilation into two or three mesons in terms of quark rearrangements

and $\bar{q}q$ annihilations (for a review, see 90) or through the excitation of intermediate meson resonances (32) (see Sections 3.3 and 3.4).

Since annihilation models are currently inspired by data, we first discuss the experimental situation and then review the models that explain some of the relations between the measured branching ratios or cross sections.

3.1 $\bar{p}p$ Annihilation at Rest

Earlier data stem from bubble chamber exposures taken in the 1960s at Brookhaven National Laboratory (BNL), Argonne National Laboratory, and CERN (91). In liquid hydrogen, the atomic $\bar{p}p$ S wave dominates annihilation at rest as a result of the Day-Snow-Sucher mechanism (92): following $\bar{p}$ capture in a high n orbital ($n \sim 30$) of the $\bar{p}p$ atom, the $\bar{p}$ is rapidly transferred by collisions with neighboring H_2 molecules to the nS level, where the system annihilates because there is no centrifugal barrier. Hence annihilation with the initial angular momentum $L = 0$ (S-wave annihilation) dominates in liquid hydrogen.

The largest data sample, collected by the CERN–Collège de France collaboration, consisted of 80,000 fully reconstructed pionic events and 20,000 fully reconstructed events with at least one $K_S (\to \pi^+ \pi^-)$. The largest fraction (60%) of all annihilations involving two or more π^0 or η could not be studied in bubble chamber experiments and the early kaonic data suffer from poor statistics. An experimental program has been initiated at LEAR to study annihilation into multineutral and kaonic channels. The Crystal Barrel and Obelix detectors have been commissioned and the first branching ratios from the Crystal Barrel will be published soon.

Recently, two-body branching ratios for $\pi^0 M$ and ηM, where $M = \pi^0$, η, η', ϕ, ω, ρ^0, have been determined in liquid hydrogen at KEK (93) and at LEAR (94, 95) by measuring the inclusive η or π^0 momentum spectrum in, for example, $\bar{p}p \to \pi^0(\eta) + $ anything. These branching ratios are displayed in Table 1, together with earlier measurements for two-body annihilation. The energies and angles of γ pairs were measured by a segmented γ detector. The momentum distribution of the pairs, consistent with π^0 or η decay, shows peaks corresponding to the two-body final states $\pi^0 M$ or ηM. However, since reflections distort the inclusive spectrum in an unpredictable way and the background is often very large, the branching ratios for broad mesons M or for weak annihilation channels cannot be reliably extracted from inclusive spectra. Consequently, exclusive measurements in which all particles are detected must be performed. Strictly speaking, branching ratios can be given only for channels involving narrow

Table 1 Branching ratios BR_{liq} for $\bar{p}p$ annihilation at rest in liquid[a]

Channel	BR_{liq}	Ref.	Channel	BR_{liq}	Ref.
$\gamma\gamma$	$< 1.7 \times 10^{-6}$	[94]	$\eta\phi$	$< 2.8 \times 10^{-3}$	[93]
e^+e^-	$3.2 \pm 0.9 \times 10^{-7}$	[96]	$\eta'\rho$	$1.29 \pm 0.81 \times 10^{-3}$	[100]
$\pi^0\gamma$	$1.74 \pm 0.22 \times 10^{-5}$	[94]	$\rho^0\rho^0$	$1.2 \pm 1.2 \times 10^{-3}$	[91]
$\pi^0\pi^0$	$2.06 \pm 0.14 \times 10^{-4}$	[94]	$\rho\omega$	$2.26 \pm 0.23 \times 10^{-2}$	[103]
	$4.8 \pm 1.0 \times 10^{-4}$	[97]		$3.9 \pm 0.6 \times 10^{-2}$	[104]
	$1.4 \pm 0.3 \times 10^{-4}$	[98]	$\omega\omega$	$1.4 \pm 0.6 \times 10^{-2}$	[105]
	$2.5 \pm 0.3 \times 10^{-4}$	[99]	$\omega\phi$	$6.3 \pm 2.3 \times 10^{-4}$	[106]
$\pi^0\rho^0$	$1.72 \pm 0.27 \times 10^{-2}$	[91][b]	$\omega f_2(1270)$	$3.26 \pm 0.33 \times 10^{-2}$	[103]
	$1.6 \pm 0.1 \times 10^{-2}$	[93]	$\rho f_2(1270)$	$1.57 \pm 0.34 \times 10^{-2}$	[91]
$\pi^0\omega$	$5.2 \pm 0.5 \times 10^{-3}$	[93]	$\pi^\pm\rho^\mp$	$3.44 \pm 0.54 \times 10^{-2}$	[91][b]
$\pi^0 f_2(1270)$	$4.1 \pm 1.2 \times 10^{-3}$	[91]		$3.0 \pm 0.3 \times 10^{-2}$	[104]
$\pi^0\phi$	$3.0 \pm 1.5 \times 10^{-4}$	[93]	$\pi^+\pi^-$	$3.33 \pm 0.17 \times 10^{-3}$	[91]
$\pi^0\eta$	$3.9 \pm 1.0 \times 10^{-4}$	[93][b]	K^+K^-	$10.1 \pm 0.5 \times 10^{-4}$	[91]
	$1.33 \pm 0.27 \times 10^{-4}$	[95]	$K^0\overline{K^0}$	$7.6 \pm 0.4 \times 10^{-4}$	[91]
$\pi^0\eta'$	$5.0 \pm 1.9 \times 10^{-4}$	[93]	$\pi^\pm a_0^\mp(980)$	$6.9 \pm 1.2 \times 10^{-3}$	[104]
$\eta\eta$	$8.1 \pm 3.1 \times 10^{-5}$	[95]	$\pi^\pm b_1^\mp(1235)$	$7.9 \pm 1.1 \times 10^{-3}$	[103]
	$1.6 \pm 0.8 \times 10^{-3}$	[93]		$1.96 \pm 0.27 \times 10^{-2}$	[104]
$\eta\eta'$	$< 1.8 \times 10^{-4}$	[95]	$\pi^\pm a_2^\mp(1320)$	$2.83 \pm 0.32 \times 10^{-2}$	[104]
$\eta\rho$	$6.5 \pm 1.4 \times 10^{-3}$	[100]		$4.74 \pm 0.61 \times 10^{-2}$	[91][b]
	$5.3 \pm 1.4 \times 10^{-3}$	[95]		$2.16 \pm 0.45 \times 10^{-2}$	[107, 108][b]
	$9.6 \pm 1.6 \times 10^{-3}$	[93]	$K^0\overline{K^{0*}}$	$1.57 \pm 0.11 \times 10^{-3}$	[91][b]
	$5.0 \pm 1.4 \times 10^{-3}$	[101]	$K^\pm K^{\mp*}$	$1.0 \pm 0.1 \times 10^{-3}$	[91][b]
	$2.2 \pm 1.7 \times 10^{-3}$	[102]		$1.42 \pm 0.14 \times 10^{-3}$	[104]
$\eta\omega$	$1.0 \pm 0.1 \times 10^{-2}$	[95]	$K^{0*}\overline{K^{0*}}$	$3.22 \pm 0.67 \times 10^{-3}$	[109]
	$4.6 \pm 1.4 \times 10^{-3}$	[93]	$K^{\pm*}K^{\mp*}$	$1.54 \pm 0.54 \times 10^{-3}$	[109][b]

[a] The pure S wave BR_S and pure P wave BR_P are related to BR_{liq} by the relation $BR_{liq} = BR_S(1-f_P) + BR_P f_P$, where $f_p = 8.6 \pm 1.1\%$ is the fraction of annihilation from atomic P states in liquid hydrogen (110). The branching ratios of the total annihilation rate include the experimentally unobserved decay modes.
[b] Average of two or more measurements.

mesons (π, K, η, η', and possibly ω and ϕ); the branching ratios for broad mesons are not defined because other channels leading to the same final state interfere, even in exclusive measurements. As seen in Table 1, the experimental situation is not satisfactory; many branching ratios are either poorly measured or in conflict with one another. To test the theoretical ideas on annihilation and improve our understanding of these processes, exclusive measurements of few-meson final states are clearly important.

Annihilation from atomic P states has been studied at LEAR by the Asterix collaboration using hydrogen gas at normal temperature and pressure (NTP, meaning here 20°C and 1 atm). In gaseous hydrogen, the molecular collision rate is reduced because of the lower density, the electromagnetic cascade can develop down to the 2P levels, and hence annihilation from atomic P states competes with annihilation from S states. In gaseous hydrogen at NTP, annihilation from all atomic P states occurs with a probability of $52.8 \pm 4.9\%$ (111). Experimental data on S- and P-state contributions were compared with a cascade model calculation by Reifenröther & Klempt (86). Annihilation from D states is expected to be negligible in hydrogen gas at NTP (112). The yield of L x rays (transitions to the 2P levels) is $13 \pm 2\%$ while the yield of K x rays (transitions to the 1S ground state) is $0.65 \pm 0.32\%$ (113). Once in the 2P levels, the $\bar{p}p$ atom annihilates with a probability of $98 \pm 1\%$ and therefore annihilation dominates over the K_{α} transition (113).

The Asterix group has developed a new technique of preparing the initial state in the atomic 2P states by triggering on the L x-ray transitions to the 2P states (114). A hardware trigger on the initial x-ray candidate typically yields 60–70% P-wave annihilation. The off-line analysis, requiring coincidence with L x rays, then leads to typically 90% P-wave annihilation. The residual 10% S-wave annihilation in the signal is due to background contamination from bremsstrahlung x rays under the L x-ray series (115). The bremsstrahlung is produced by the sudden acceleration of charge in the $\bar{p}p$ annihilation final state. The true P-wave contribution, when applying the coincidence with L x rays, depends on the annihilation channel. The exact P-wave fraction in the x-ray coincidence sample depends on the ratio of annihilation branching ratios between S and P states, since bremsstrahlung stems from the annihilation final states not necessarily associated with the emission of an L x ray.

The 105-MeV/c antiprotons from LEAR stopped in a NTP gaseous target, which was surrounded by a drift chamber where the x rays of the atomic cascade were converted. Charged particles were reconstructed in a solenoidal magnetic spectrometer and the γ angles were determined by conversion in lead sheets, albeit with a modest efficiency of 25%. Details on the apparatus are described by Ahmad et al (116). By measuring branching ratios in gas (with x-ray enhancing trigger and in off-line coincidence with L x rays) and in liquid, one can determine the branching ratios from pure initial S and P states. The statistics collected by Asterix in gaseous hydrogen exceed the bubble chamber sample by some two orders of magnitude.

3.1.1 ANNIHILATION INTO $\pi^+\pi^-$, K^+K^-, AND $K^0\overline{K^0}$ Figure 5a shows the

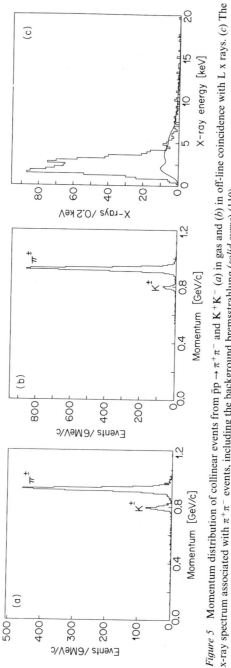

Figure 5 Momentum distribution of collinear events from $\bar{p}p \rightarrow \pi^+\pi^-$ and K^+K^- (*a*) in gas and (*b*) in off-line coincidence with L x rays. (*c*) The x-ray spectrum associated with $\pi^+\pi^-$ events, including the background bremsstrahlung (*solid curve*) (110).

momentum distribution of collinear two-prong events in gaseous hydrogen at NTP (110). The peaks are due to the annihilation channels $\bar{p}p \to K^+K^-$ and $\pi^+\pi^-$. Figure 5b shows the momentum distribution with off-line coincident L x rays. The x-ray spectrum associated with $\pi^+\pi^-$ events is shown in Figure 5c. The peak between 1.3 and 4 keV is due to the (unresolved) L x-ray series, the peak around 3.0 keV is due to argon fluorescence in the drift chamber.

A comparison of Figures 5a and b shows that the K/π ratio decreases with increasing atomic P-wave contribution. The branching ratios for these channels in liquid are given in Table 1, and the pure branching ratios for S and P waves are shown in Table 2. The S-wave annihilation rate in liquid

Table 2 The branching ratios of the total annihilation rate for $\bar{p}p$ S- and P-wave annihilation into the channels measured by the Asterix experiment[a]

	S			P		
K^+K^-	1.08	± 0.05	10^{-3}	2.87	± 0.51	10^{-4}
$K^0\overline{K^0}$	8.3	± 0.5	10^{-4}	8.8	± 2.3	10^{-5}
$\pi^+\pi^-$	3.19	± 0.20	10^{-3}	4.81	± 0.49	10^{-3}
$\rho^\pm\pi^\mp$	3.21	± 0.42	10^{-2}	1.50	± 0.20	10^{-2}
$\rho^0\pi^0$	1.56	± 0.21	10^{-2}	0.40	± 0.09	10^{-2}
$f_2(1270)\pi^0$	3.9	± 1.1	10^{-3}	1.83	± 0.23	10^{-2}
$\pi^+\pi^-\pi^0$	6.6	± 0.8	10^{-2}	4.5	± 0.6	10^{-2}
$\eta\pi^+\pi^-$	1.37	± 0.15	10^{-2}	3.35	± 0.84	10^{-3}
$\eta'\pi^+\pi^-$	3.46	± 0.67	10^{-3}	0.61	± 0.33	10^{-3}
$a_2^\pm(1320)\pi^\mp$	2.69	± 0.60	10^{-2}	9.03	± 4.76	10^{-3}
$\eta\rho$	3.29	± 0.90	10^{-3}	9.4	± 5.3	10^{-4}
$f_2(1270)\eta$	1.5	± 1.5	10^{-4}	1.1	± 0.5	10^{-3}
$\eta'\rho$	1.81	± 0.44	10^{-3}	~ 3		10^{-4}
$\pi^+\pi^-\omega$	6.55	± 0.68	10^{-2}	7.05	± 1.05	10^{-2}
$\rho\omega$	1.91	± 0.37	10^{-2}	6.38	± 1.28	10^{-2}
$\pi^\pm b_1^\mp(1235)$	0.83	± 0.12	10^{-2}	0.67	± 0.18	10^{-2}
$\phi\pi^0$	4.0	± 0.8	10^{-4}	≤ 3		10^{-5}
$\phi\pi^+\pi^-$	4.7	± 1.1	10^{-4}	6.6	± 1.5	10^{-4}
$\phi\rho$	3.4	± 1.0	10^{-4}	3.7	± 0.9	10^{-4}
$\phi\omega$	5.3	± 2.2	10^{-4}	2.9	± 1.4	10^{-4}
$\phi\eta$	3.0	± 3.9	10^{-5}	4.2	± 2.0	10^{-5}

[a] The branching ratios include the unobserved decay modes and the $b_1(1235)$ is assumed to decay exclusively into $\omega\pi$.

hydrogen has been determined by comparing the rates for $\pi^+\pi^-$ from P wave and for $\pi^0\pi^0$ from a liquid target (94). The former, from the P states, should be twice the latter. In fact, neutral pseudoscalar pairs cannot be produced from S waves. This explains the very small branching ratios in liquid. The fraction of P-wave annihilation in liquid into all channels is found to be $8.6 \pm 1.1\%$ (110) when the most precise $\pi^0\pi^0$ datum is used (94). This is the first determination of this important number. As seen in Table 2, the rates strongly depend on the initial angular momentum, with K^+K^- being suppressed by a factor of four when switching from S to P waves, the physics of which is discussed in Section 3.4.

The channel $K^0\overline{K}^0$ appears as K_SK_S from initial P states and as K_SK_L from initial S states. The former has not been observed in earlier bubble chamber experiments. Asterix observed the channels $\bar{p}p \to K_SK_S$ ($K_S \to \pi^+\pi^-$) and K_SK_L ($K_S \to \pi^+\pi^-$, K_L undetected) in hydrogen gas (111). The fraction of P-wave annihilation to all channels in gas was determined (a) from the $\pi^+\pi^-$ and K^+K^- branching ratios in gas, liquid, and in off-line coincidence with L x rays, and (b) from a comparison of K_SK_L in liquid and gas. The average fraction is $52.8 \pm 4.9\%$ (111), from which one derives the branching ratios for K_SK_S ($K^0\overline{K}^0$ from P waves) given in Table 2. The channel $K^0\overline{K}^0$ is hence suppressed by a factor of 10 when switching from S to P waves.

Since K^+K^- and $K^0\overline{K}^0$ are mixtures of $I = 1$ and $I = 0$ and since the two branching ratios are nearly equal for S states, one deduces that one isospin amplitude dominates, unless the admixtures of $\bar{n}n$ and $\bar{p}p$ are comparable (84, 85, 87, 88), in which case both isospin amplitudes could contribute as discussed by Jaenicke et al (88) (see also Sections 2.3 and 3.4). If one assumes that the $\bar{n}n$ atomic component for this reaction is small and that $\bar{p}n$ annihilates from an S state in deuterium, then one can use the rate for the pure $I = 1$ $\bar{p}n \to K^0K^-$, which is $(1.5 \pm 0.2) \times 10^{-3}$ (117), to find that $I = 1$ $\bar{p}p$ dominates $I = 0$ by a factor of three for annihilation into a $K\bar{K}$ pair (111). However, P waves could contribute significantly in $\bar{p}p$ and $\bar{p}n$ annihilation in deuterium (86), in which case the above argument would not hold. For P states, the much smaller rate for $K^0\overline{K}^0$ indicates that both isospins contribute from initial P states, contrary to theoretical expectations (84) (see Section 3.4).

3.1.2   ANNIHILATION INTO $\pi^+\pi^-\pi^0$ This annihilation channel was studied earlier in bubble chambers, where S-wave annihilation dominates (118). The salient feature is a dominating $\rho\pi$ channel produced mainly from the ($I = 0$) $^3S_1$ initial state and strongly suppressed from the ($I = 1$) $^1S_0$ state. This is sometimes referred to as the $\rho\pi$ puzzle (Section 3.4).

Asterix has analyzed the $\pi^+\pi^-\pi^0$ channel in gaseous hydrogen and in off-line coincidence with L x rays (meaning $91.8 \pm 1.0\%$ P wave) (119). Including data in liquid (118) one finds the pure $\bar{p}p$ S- and P-wave branching ratios given in Table 2. The Dalitz plot and the projections are shown in Figure 6. One notices a strong production of $f_2(1270)$ and a peak at 1565 MeV ascribed to a new resonance, $f_2(1565)$, discussed in Section 4.2.2. The state $f_2(1270)$ is only weakly produced in liquid and $f_2(1565)$ is not observed (118). A close examination of Figure 6 shows that $f_2(1270)$ and $f_2(1565)$ productions increase with increasing P-wave probability (Figure 6c vs Figure 6d), while the ratio of ρ^0 to $\rho^{\pm}$ decreases. This indicates that the $I = 1$ contribution (from which state the reaction $\bar{p}p \rightarrow \rho^0 \pi^0$ is forbidden) increases with increasing P wave, provided that the relative ratio of $\bar{n}n$ to $\bar{p}p$ remains very small and roughly the same for this reaction in both atomic S and P states.

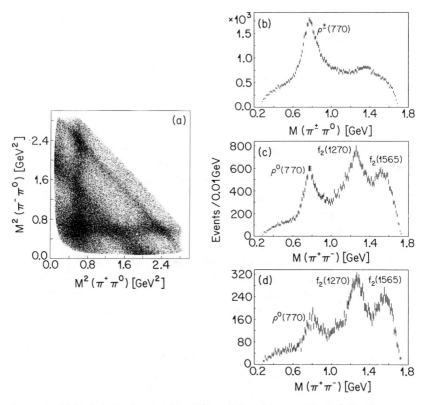

Figure 6 (a) Dalitz plot for the channel $\bar{p}p \rightarrow \pi^+\pi^-\pi^0$ in gas; (b) $\pi^{\pm}\pi^0$ invariant mass distribution; (c) $\pi^+\pi^-$ invariant mass distribution; (d) $\pi^+\pi^-$ invariant mass distribution requiring off-line coincidences with L x rays (119).

Table 3 Two-body and direct 3π contributions to $\pi^+\pi^-\pi^0$ for $\bar{p}p$ annihilation from S and P waves as a percentage of the total annihilation rate (119)

	$^3S_1$	$^1S_0$	$^1P_1$	$^3P_1$	$^3P_2$
$\rho^{\pm}\pi^{\mp}$	3.12 ± 0.42	0.09 ± 0.04	0.80 ± 0.17	0.67 ± 0.11	0.03 ± 0.01
$\rho^0\pi^0$	1.56 ± 0.21		0.40 ± 0.09		
$f_2(1270)\pi^0$		0.22 ± 0.06		0.95 ± 0.13	0.09 ± 0.03
$f_2(1565)\pi^0$				0.24 ± 0.04	0.14 ± 0.03
Direct $3\pi^a$		1.61 ± 0.03		1.18 ± 0.05	

^{a} The direct 3π contributions refer to the total S- or total P-wave rates.

A Dalitz plot analysis in terms of pure S waves has been performed by subtracting the Dalitz plot corresponding to Figure 6d (off-line coincidence with L x rays) with a floating normalization from the Dalitz plot of Figure 6a (annihilation in gas), thereby subtracting the P-wave contribution, and fitting all contributing amplitudes from the $^3S_1$ and $^1S_0$ initial states. The fit agrees perfectly with the results from the bubble chamber experiment (118), which indicates that corrections for detector acceptance have been applied properly. An analysis of the P-wave Dalitz plot has then been performed. Table 3 summarizes the branching ratios for all contributions to $\pi^+\pi^-\pi^0$. Note that the values in Table 3 are calculated from the amplitudes by adding all amplitudes incoherently, thereby neglecting interference effects.

3.1.3 ANNIHILATION INTO $\eta\pi^+\pi^-$, $\eta'\pi^+\pi^-$, AND $2\pi^+2\pi^-\pi^0$ Asterix has studied the channel $\bar{p}p \to \eta\pi^+\pi^-$ where (a) $\eta \to \pi^+\pi^-\gamma$ or (b) $\eta \to \pi^+\pi^-\pi^0$ and the channel $\bar{p}p \to \eta'\pi^+\pi^-$ where (c) $\eta' \to \pi^+\pi^-\gamma$ or (d) $\eta' \to \pi^+\pi^-\eta$. For reactions (b) and (d) the π^0 and η were kinematically reconstructed, while for reactions (a) and (c) the photon was detected by conversion in the lead sheets. These four-prong events were collected with the x-ray enhancing trigger (meaning 61% P wave) and in off-line L x-ray coincidence (meaning 86% P wave) (120). Figure 7a shows the $\pi^+\pi^-\pi^0$ invariant mass distribution for the final-state $2\pi^+2\pi^-\pi^0$ and 61% P wave. Figure 7b shows the $\pi^+\pi^-\gamma$ spectrum in the four-prong channel for 61% P wave. The background from $2\pi^+2\pi^-\pi^0$ with one γ escaping detection has been subtracted. An upper limit of 3×10^{-3} for $\omega \to \pi^+\pi^-\gamma$ is obtained by comparing ω production in $2\pi^+2\pi^-\pi^0$ and $2\pi^+2\pi^-\gamma$. This is an order-of-magnitude improvement on the upper limit for this decay. Figure 7c shows the $\eta\pi^+\pi^-$ spectrum in the η' region.

The branching ratios for the two reactions $\eta\pi^+\pi^-$ and $\eta'\pi^+\pi^-$ have been determined using the known η and η' decay branching ratios and averaging the results from reactions (a) and (b) with (c) and (d), respectively. These

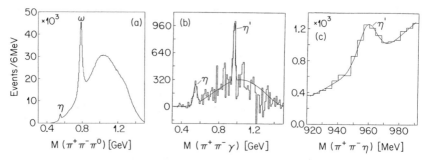

Figure 7 Invariant mass distritubtions for (*a*) $\pi^+\pi^-\pi^0$, (*b*) $\pi^+\pi^-\gamma$, and (*c*) $\pi^+\pi^-\eta$ for four-prong annihilation events (120).

branching ratios have also been measured in liquid (100, 101). The results for pure S and P waves are given in Table 2. Both branching ratios decrease with increasing P wave, typically by a factor of five when switching from S to P waves.

A fraction of $51 \pm 10\%$ of the $\eta'\pi^+\pi^-$ final state proceeds through the $\eta'\rho$ final state in liquid (100). This agrees with Asterix data for S states (although no Dalitz plot analysis has been performed because of the low statistics) and also for P states within the large statistical errors. A Dalitz plot analysis has been applied to the final-state $\eta\pi^+\pi^-$ with the decay $\eta \rightarrow \pi^+\pi^-\pi^0$ (reaction *b*). Initial S- and P-wave contributions were derived by subtracting the 86% P-wave Dalitz plot from the 61% P-wave Dalitz plot, as described in the previous section for the three-pion final state. The branching ratios for the intermediate states $a_2^\pm(1320)\pi^\mp$, $f_2(1270)\eta$, and $\eta\rho$ (corrected for the unobserved decay modes) are given in Table 2. One notes, as for the three-pion final state, a large enhancement of $f_2(1270)$ production when switching from S to P waves. A 2σ effect of $(1.3 \pm 0.7) \times 10^{-4}$ for the annihilation channel $\bar{p}p \rightarrow a_0^\pm(980)\pi^\mp$, with $a_0^\pm(980) \rightarrow \eta\pi^\pm$, is also observed.

The ratio of $\eta\rho$ to $\eta'\rho$ equals 2.5 ± 0.7 and thus is compatible with a larger $s\bar{s}$ component in the η', a component that does not contribute strongly to annihilation because of the Okubo-Zweig-Iizuka (OZI) rule. Ignoring phase-space differences, one naively expects a ratio of two, if one assumes a pseudoscalar mixing angle of $-20°$ and uses the standard SU(3) flavor meson wave functions of the simple constituent quark model (120). However, care should be taken when using these simple arguments: the ratio of $\pi^0\eta$ to $\pi^0\eta'$ is about one according to the branching ratios in Table 1, but is expected to be two. These ratios should be measured more precisely to test this simple model.

The Asterix collaboration has also derived branching ratios for the channels $\pi^+\pi^-\omega$, $\rho\omega$, and $b_1(1235)^\pm\pi^\mp$ for both S and P states (121), starting from the same four-prong data samples as in the previous section, but selecting events associated with ω production (see Figure 7a). The results are given in Table 2. The channel $\rho\omega$ proceeds from the spin-singlet S state or from the three spin-triplet P states. This may explain the strong enhancement of $\rho\omega$ production from P states.

3.1.4 ϕ PRODUCTION IN $\bar{p}p$ ANNIHILATION AT REST AND THE OZI RULE   It has been suggested that an enhanced production of ϕ mesons could be related to an excess of $s\bar{s}$ pairs in the nucleon (122) or to flavor mixing (123–125) with a $\bar{q}^2 q^2$ state below threshold containing an $\bar{s}s$ pair. Such a state C(1480) ($\rightarrow \phi\pi^0$) was recently reported at Serpukhov (126). If the nucleon and antinucleon do not contain any strange quarks, ϕ production should be suppressed in $\bar{p}p$ annihilation by the OZI rule. A naive estimate of the ratio R of the ϕ to ω rates is given by the deviation from ideal mixing in the vector meson nonet. Neglecting phase-space factors, one expects a value

$$R = \mathrm{BR}(\phi)/\mathrm{BR}(\omega) = \tan^2(\Phi - \Phi_i) = 4.2 \times 10^{-3}, \qquad 4.$$

where Φ_i is the ideal mixing angle of $35.3°$ and $\Phi = 39°$, when the quadratic mass formula is used. (An even smaller ratio occurs with the angle of $\Phi = 36°$ from the linear mass formula.)

The Asterix collaboration has analyzed the final states $\phi\pi^0$, $\phi\pi^+\pi^-$, $\phi\rho^0$, $\phi\eta$, and $\phi\omega$, where the ϕ is observed in its K^+K^- decay mode and the ω and η in their $\pi^+\pi^-\pi^0$ decay modes, from initial states with 61% P wave and with 86% P wave for four-prong events (59 and 93% P wave for two-prong events, respectively) (127). The charged kaons were identified by ionization sampling. The channel $\pi^0\phi$ and its $\pi^0\omega$ counterpart have been observed in liquid hydrogen from a study of the inclusive π^0 spectrum (93). Together with the Asterix data, one obtains the pure S- and P-wave branching ratios for $\phi\pi^0$ shown in Table 2. Ignoring phase-space differences, one finds $R = (7.7 \pm 1.7) \times 10^{-2}$ from the $\bar{N}N$ ${}^3S_1$ wave, an order of magnitude larger than the naive estimate of Equation 4.

For $\phi\pi^+\pi^-$ and $\phi\rho$ ($\rho \rightarrow \pi^+\pi^-$) the contributions from S and P waves are fitted for both the 61% and the 86% P-wave data samples (127). The branching ratios for pure S and P waves are given in Table 2. The S-wave branching ratio for $\phi\pi^+\pi^-$ agrees with an earlier measurement in liquid hydrogen (103). From the values shown in Table 2, one then finds for $\phi\pi^+\pi^-$ and $\omega\pi^+\pi^-$ $R = (7.1 \pm 1.8) \times 10^{-3}$ for S waves and $R = (9.4 \pm 2.5) \times 10^{-3}$ for P waves. Similarly for $\phi\rho$ and $\omega\rho$ one gets $R = (1.8 \pm 0.6) \times 10^{-2}$ for ${}^1S_0$ and $R = (5.8 \pm 1.8) \times 10^{-3}$ for P waves.

The branching ratios for $\phi\omega$ for pure S and P waves can again be calculated using the branching ratios in liquid (106) (Table 1). Taking the branching ratio for $\omega\omega$ in liquid (105) one finds $R = (3.8 \pm 2.2) \times 10^{-2}$ for $^1S_0$. For $\phi\eta$ one obtains together with the branching ratio for $\omega\eta$ in liquid $R = (2.9 \pm 3.8) \times 10^{-3}$. A test of OZI violation for $\phi\eta$ and $\phi\omega$ from P waves is not possible since the branching ratios for $\omega\eta$ and $\omega\omega$ are not known from P waves.

To discuss a possible OZI violation, one should correct for phase-space differences, which are largest when ϕ is slow. However, the phase-space correction is unclear. For large momenta p in the final state, the correction is simply p. For low momenta the phase-space factor following Blatt and Weisskopf is proportional to $p[p^2/(p^2+\mu^2)]^l$, where μ is the inverse range of the interaction (including any finite size of the particles) and l is the relative two-meson angular momentum. This certainly is the case for $\phi\rho$ ($\lambda = 4.5$ fm $\gg 0.2$ fm) and hence the ratio $\phi\rho/\omega\rho$ would increase by a factor of two. On the other hand, Vandermeulen's phase-space factor does not modify the ratio R significantly (123) (see also Section 3.4). Given the large experimental uncertainties, the size of the OZI violations in some branching ratios is sensitive to the theoretical prescription for the phase-space contribution used in each case.

The evidence for OZI rule violation is hence weak, except for the channel $\phi\pi^0$ compared to $\omega\pi^0$, for which the phase-space factors are similar (and possibly for $\phi\rho$ from the $^1S_0$). A strong violation has also been observed in liquid deuterium: from the average of three measurements for $\bar{p}n \to \pi^-\phi$ (108, 128, 129) and of three measurements for $\bar{p}n \to \pi^-\omega$ (130–132), one finds a very large value, $R = 0.14 \pm 0.02$. It should also be noted that, at higher energies, ϕ production seems enhanced in both pp and $\bar{p}p$ interactions (133).

3.1.5   EXPERIMENTAL STATUS OF NARROW $\bar{N}N$ BOUND STATES   As discussed in Section 2.1.3, baryonium states could arise from the strongly attractive meson exchange potential. Narrow baryonia have been sought by searching for narrow lines in the γ, π^0, and $\pi^\pm$ inclusive spectra of $\bar{p}p$ annihilation at rest. The earlier evidence for narrow states associated with mono-chromatic γ emission (134) has neither been confirmed at KEK (135) nor at LEAR (136, 137). The 95% confidence level upper limit for the production of states narrower than 25 MeV is 8×10^{-5} of all annihilations for a mass of 1100 MeV, rising to 5×10^{-4} for a mass of 1780 MeV (for a review, see 18, 138). For monochromatic $\pi^\pm$ emission, the experimental limit is typically 4×10^{-4} (139) and for π^0 monochromatic emission, typically 2×10^{-3} (99).

Asterix has also looked for states associated with the emission of a $\pi^\pm$

from initial atomic P states, since narrow states were predicted to have a high angular momentum and might hence be enhanced from P states. No narrow state was found in the mass range 1100 to 1670 MeV with a 95% confidence level upper limit of 7×10^{-4} (138, 140).

The experimental evidence pertains to the existence of narrow states ($\Gamma \leq 25$ MeV). Broad states cannot be observed from inclusive spectra and are therefore not excluded. In Sections 4.2.2 to 4.2.4 on spectroscopy, we discuss the evidence for the broad (~ 170 MeV) $f_2(1565)$ proposed to be a deeply bound 2^{++} ($I = 0$) baryonium (141). Finally, we point out that narrow states very close to the $\bar{N}N$ threshold, which could be produced by the emission of soft photons (≤ 100 MeV), are not excluded.

3.2 $\bar{p}p$ Annihilation in Flight

Little progress has been made in the recent years in studying $\bar{p}p$ annihilation into exclusive final states above the $\bar{N}N$ threshold. The angular distribution for the two-body final states $\pi^-\pi^+$ and K^-K^+ has been studied by KEK between 360 and 760 MeV/c (142). This experiment did not determine the charges of the emitted mesons and hence could not distinguish between forward and backward scattering. The angular distributions for $\pi^-\pi^+$ and K^-K^+ show a peak for $|\cos\theta|$ close to unity.

An earlier experiment at KEK, which measured the full (unfolded) angular distribution, reported a strong forward peak for $\pi^-\pi^+$ and strong peaks in both forward and backward directions for K^-K^+ (143). The backward peak in K^-K^+ was also observed earlier at 790 MeV/c (144). However, the energy dependence of the integrated cross sections for the new KEK experiment (142) is at variance with the old KEK experiment (143) and in particular does not confirm the strong enhancement seen in the K^-K^+ channel around 500 MeV/c. Hence the existence of a backward peak in K^-K^+ at very low energy needs to be clarified. Data for these reactions have been collected at LEAR between 360 and 1550 MeV/c in connection with a measurement of the analyzing power (145), but results are not available yet. A second experiment at LEAR has measured the angular distributions below 300 MeV/c (146). The $\pi^-\pi^+$ angular distribution is very strongly asymmetric, even at very low momentum (225 MeV/c), rising quickly in the forward hemisphere, while K^-K^+ remains flat. However, the angular coverage of this second experiment does not extend to small enough forward nor large enough backward scattering angles to clarify the earlier KEK experimental results. A backward K^-K^+ peak cannot be reproduced by any quark model (60) nor by a simple baryon exchange model (147–149) but a backward peak is found in the coupled-channel calculation of Liu & Tabakin (39). However, this latter work found some deep minima not observed in the $\pi^-\pi^+$ differential cross section. This backward peak could signal the formation of an s-channel resonance or, as has also been speculated, it could be due to the exchange of a strangeness $+1$ exotic baryon in the t channel, enhanced by the admixture of strangeness in the proton (122).

The integrated cross section for $\pi^-\pi^+$ rises very quickly with decreasing $\bar{p}$ momentum, while K^-K^+ seems to be constant from about 800 down to 200 MeV/c (142, 146, 150), a trend that is reproduced by quark models (149, 151).

Data for the analyzing power in $\pi^-\pi^+$ and K^-K^+ are now available from LEAR between 360 and 1550 MeV/c (145). The analyzing power is very large for broad regions of the angular range, even reaching the maximum value of one. This latter means that the π^- or K^- are always scattered to the left of the beam for a target proton 100% polarized along the normal to the scattering plane. This remarkable behavior is not understood (37).

The annihilation into an e^-e^+ pair has also been studied at LEAR between 416 and 888 MeV/c in order to determine the form factor of the proton in the timelike region (152). The isotropic angular distribution at low momentum is consistent with the electric and magnetic form factors being equal. More surprisingly, the dependence of the form factor on the incident $\bar{p}$ energy is steeper than predicted by the vector dominance model and furthermore seems to oscillate. This latter behavior leads to speculations about the existence of $\bar{N}N$ resonances or bound states close to $\bar{N}N$ threshold (153). However, these states should also be seen in the $\bar{p}d \rightarrow p5\pi$ data, since these latter reactions probe the $\bar{N}N$ amplitude above and below threshold, as discussd by Fasano & Locher (66).

3.3   Models for the Global Annihilation Process

On the average, $\bar{N}N$ annihilates into five pions at low $\bar{p}$ energies, although $\bar{p}p$ at rest can in principle decay into $13\pi^0$. For a given number of pions, the charge is distributed statistically according to the model of Pais (30) discussed in Section 2.1.2. This statistical notion of the bulk properties of annihilation is corroborated by the hot gas model, which describes the energy spectrum of charged pions emitted in the annihilation reaction $\bar{p}p \rightarrow$ hot gas $\rightarrow \pi +$ anything (29). It is assumed that $\bar{N}N$ annihilates into fragments that are in thermal (or nearly thermal) equilibrium and that the pions evaporated from this hot gas have the same energy distribution as the fragments. Following Kimura & Saito (29) and others, the cross section is

$$d^3\sigma/dk^3 \propto k^2 \exp(-\omega/T),   \qquad 5.$$

where $\omega = (m_\pi^2 + k^2)^{1/2}$ is the pion energy and the factor k^2 is required by

Adler's soft pion consistency condition. The temperature T is determined from the measured cross section and found to be ~ 100 MeV (29). This type of model should be developed further for $\bar{p}p$ annihilation, since the arguments are analogous to the ones used in relativistic heavy-ion physics to search for the quark-gluon plasma. In fact, $\bar{p}p$ annihilation is a small-scale laboratory for concepts used in describing the quark-gluon plasma.

Another successful model, which describes the bulk properties of annihilation, is the threshold dominance model (154), discussed by Vandermeulen (32). He assumes that $\bar{N}N$ annihilation proceeds via two-meson intermediate doorway states and that the intermediate two-meson thresholds closest to the actual $\bar{N}N$ total energy dominate (see Equation 6 below). The reaction is $\bar{N}N \rightarrow a + b \rightarrow$ final-state pions (and kaons). Here a and b are all possible pairs of intermediate S- and P-wave $\bar{q}q$ mesons that decay (with known branching ratios) into the specific final state under consideration. Vandermeulen further assumes that the different intermediate two-meson channels leading to the same final state add incoherently. The branching ratio for the production of a pair of nonstrange mesons a and b is parametrized for $E_{c.m.} > m_a + m_b$ as

$$BR = pC_{ab} \exp\left\{ -A[E_{c.m.}^2 - (m_a + m_b)^2]^{1/2} \right\}, \qquad 6.$$

which when multiplied by the measured total annihilation cross section gives the cross section for each annihilation channel. The factor p is the two-meson cms momentum and the coefficient C_{ab} is the average weight of the spin and isospin factors for the reaction. The threshold parameter $A = 1.2$ GeV^{-1} is determined from a fit to the total cross section for $\bar{p}p \rightarrow \pi^- \pi^+$ as a function of energy. This model reproduces the increasing value of the average multiplicity $\langle n \rangle$ with increasing $\bar{p}$ energy (32). Its success is impressive. It also reproduces the various $\bar{p}p$ cross sections into the different charge combinations of two and up to eight pions in the final state, from threshold and up to $p_{lab} \simeq 3.5$ GeV/c (32). Again, as in the model of Pais, it is the statistical distribution of the charges for each value of n that is partly responsible for this success.

Equation 6 is multiplied by a factor 0.15 to describe the suppression of strange final states ($\bar{K}K$ plus pions). This factor is determined from the total amount of K production and is sufficient to reproduce correctly the cross sections for the final states $\bar{K}K$ with one and up to five pions. Only the experimental cross section for $\bar{p}p \rightarrow K^- K^+$ turns out to be larger than what this simple model predicts. A suppression of strange final states is expected, according to Dosch & Gromes (155), who studied $\bar{q}q$ creation in a background chromoelectric field similar to Schwinger's $e^+ e^-$ creation in an electric field. As pointed out by Dover & Fishbane (123), the suppression of strange annihilation channels does not permit a large $\bar{s}s$ admixture in the nucleon (or antinucleon), in contrast to the claim of Ellis et al (122). Further efforts along Vandermeulen's line of thought, such as the work of Mundigl and collaborators (149), is called for, as we discuss in the next section in connection with specific meson final states. However, an understanding of the crucial two-meson threshold dominance mechanism in this model is necessary.

These three models, especially the last one, give clues as to what is required for describing annihilation through coupled-channel calculations (CCC) in baryon exchange models (39, 69). Accordingly, the lightest mesons are not too important in CCC to account for the major part of the annihilation. They are suppressed by meson-baryon vertex factors or by the threshold factor of Equation 6. Instead, heavier doorway pairs of mesons should be included with increasing $\bar{N}N$ energy. Threshold dominance arises naturally in the baryon exchange model, where the dominant pairs of meson channels are the ones with minimum energy and momentum transfer at the two meson-baryon vertices. The baryon exchange models (37, 39, 40, 47, 147) should therefore be developed further to enhance our understanding of Vandermeulen's results.

3.4   Annihilation to Specific Meson States

For annihilation in flight, many initial $\bar{p}p$ waves are involved and the assumption of statistical distribution of quantum numbers should be reasonable. A doorway model with two-meson intermediate states in the s channel, similar to the threshold dominance model (32), is used in several DWBA calculations (37, 47, 147). In a CCC, Liu & Tabakin introduce four effective (fictitious) channels (in addition to the $\pi\pi$ and $\bar{K}K$ channels) to simulate annihilation, and they calculate the differential cross sections for $\bar{p}p \to \pi^-\pi^+$ and K^-K^+, as well as the differential cross sections for elastic and charge exchange scattering (39). As mentioned in Section 3.2, Liu & Tabakin and others (60, 147, 151) have problems reproducing the measured differential cross sections of Tanimori et al (143).

At rest, annihilation occurs from well-defined $\bar{N}N$ states (S and P waves) and Vandermeulen's model has very limited success (32). One main experimental result is that the branching ratio for $\bar{p}p \to \pi^+\pi^-$ is roughly the same from atomic S and P states, whereas the reaction $\bar{p}p \to K^+K^-$ is suppressed from initial P states relative to S states by a factor of about four (110) (see Table 2). The Aarhus group presented a very thorough model-independent analysis of $\bar{p}p$ annihilating into two pseudoscalar mesons (156). When applying this analysis to $\bar{p}p \to K^+K^-$, using the transition matrix $\bar{N}N \to \bar{K}K$ from the quark model of Kohno & Weise (KW) (60), they derive a branching ratio from atomic S states of $\sim 2 \times 10^{-3}$, twice the measured value, but find a branching ratio of

2.2×10^{-4} from P states, in rough agreement with data; hence $\bar{K}K$ is suppressed by one order of magnitude when going from atomic S to P states. For the $\pi^+\pi^-$ final state the Aarhus group, using the quark model of KW, finds that their calculated branching ratio from the atomic S state agrees with LEAR data, but that from the spin-averaged atomic P state is too large by a factor of 40. However, the helicity amplitudes for $\bar{N}N \rightarrow \pi\pi$ scattering calculated by Martin & Morgan (157) (who used crossing symmetry and dispersion relations) fit the $\pi^+\pi^-$ branching ratio from both S and P states within a factor of two (156). The Aarhus group points out that the $\pi^+\pi^-$ rate from the atomic $^3P_0$ state is too large in the KW model and causes the large discrepancy with data. This work is recommended reading for researchers in this field (87, 156). Further model-dependent discussions regarding the rate of $\bar{K}K$ from P and S states are given by Furui et al (158). They use the $^3P_0$ quark annihilation model, which has some success describing meson decays, and explain the rate by arguing that one particular quark diagram dominates the annihilation process.

There are two major weak points in these discussions. The branching ratio for annihilation from a particular atomic state is strongly affected both by the atomic state wave function, which is distorted by strong interaction for $\bar{N}N$ distances less than 2 fm, and by the effective operator describing the transition of $\bar{N}N$ to mesons. Carbonell et al, in a highly recommended paper, used the atomic wave function (distorted by a MEP model) with effective transition operators from a quark model to calculate branching ratios into various two-meson final states (85). However, they have only limited success when comparing with experimental branching ratios. In the spirit of the works of Povh & Walcher (159) and Shibata (46), they plot the calculated annihilation densities as a function of $\bar{N}N$ separation distance for the various initial atomic states. They and other groups find that the spatial region of maximal annihilation density varies with the atomic state and extends from 0.5 to 1.2 fm (47, 160). The precise region depends on the MEP model.

The long-range MEP has coherent tensor forces from the different meson exchanges, as discussed by Dover & Richard (161). For coupled partial waves like $^3S_1$ and $^3D_1$, the overall D-wave probability in the atomic S state is tiny, but for $r < 1$ fm the S and D waves are equally important in some annihilation channels (see Figure 1e of Ref. 85). The importance of the D wave was exploited by Maruyama, who used a particular MEP model and a quark model transition matrix to explain the $\rho\pi$ puzzle (162) (Section 3.1.2). However, Mundigl et al, using a MEP model different from Maryuama and postulating an effective $\bar{N}N$ to two-meson transition potential, did not resolve the $\rho\pi$ puzzle (160). One weak point of these calculations is the arbitrary short-range parametrizations of the MEP for

$\bar{N}N$ distances shorter than 1 fm (12). The strength of the tensor potential at short distances is very uncertain and can strongly affect the calculated branching ratio, therefore easily changing the conclusions (12, 40, 160). For the same reason, the strong tensor force used by Carbonell et al has been criticized by the Aarhus group in their model-independent helicity amplitude analysis of protonium annihilation into two mesons (87). The $\rho\pi$ puzzle remains unsolved.

The short-distance uncertainty of the MEPs also affects the isospin mixing in the short-distance atomic wave function. This mixing is very sensitive to which MEP (Bonn or Paris) model is employed (163). The initial-state isospin mixing strongly affects the calculated neutral and charged $\bar{K}K$ and $\overline{K^*K}$ branching ratios (see also 158, 160). Because of the short-distance theoretical uncertainty ($r < 1$ fm), an analysis to determine the isospin mixture of the various $\bar{N}N$ initial states, like the one of Klempt (164), should be pursued.

A surprising result is found by Mundigl et al when they reproduce the shape of the $\pi^+\pi^-$ mass spectrum in the $\pi^+\pi^-\pi^0$ final state from atomic P states, using Vandermeulen's model, and also find the relative rate of the ρ to $f_2(1270)$ peaks (165). No interference or rescattering of the final mesons are included in this calculation. However, the experimental results in Table 3 for P-wave annihilation into $\pi^+\pi^-\pi^0$ indicate that 27% of the rate proceeds through the direct emission of three pions. The two-meson doorway assumption ignores this direct three-pion contribution and therefore cannot reproduce the overall measured rate correctly.

As discussed, the initial-state interaction and the effective transition matrices for $\bar{N}N$ into mesons are both very important to calculate the meson branching ratios from $\bar{p}p$ at rest. Since the overlap of the quark cores allows annihilation to take place, it is natural to resort to quark models to calculate these transition matrices. These quark model calculations were reviewed by Green & Niskanen (90). They employ many parameters and various arguments to neglect one type of quark diagram and not others. (For recent quark model calculations, see 163, 166–170.)

At this time, it is not clear what we have learned from these quark model calculations or which measured branching ratios are sensitive to explicit quark dynamics. Part of the problem is that there still are conflicting measurements of branching ratios; furthermore we do not yet have an agreed set of reasonable model approximations for the quark-gluon dynamics in the nonperturbative region of QCD. Regarding the quark diagram calculations, a word of caution comes from the work of Green et al (171), reiterated by Carbonell et al (85). The size of the final mesons can influence markedly the annihilation rate, whereas the size of the nucleon just determines the spatial interaction volume. In fact, some quark diagrams do not contribute if the final two mesons are pointlike. To investigate the various effective transition operators, both effective meson-baryon models and quark models should be used. One hopes a consensus will emerge as to which approximations are viable and whether or not we are seeing signs of the underlying quark dynamics.

A minimalist approach based on the quark line rule (QLR) is taken by Genz et al (166) and Hartmann et al (167), who argue that only the flavor flow in the different quark diagrams is relevant (the effective transition operators being unity and the initial-state interactions being neglected). An example of a QLR argument is the OZI rule, which is applied to ϕ-meson final states in Section 3.1.4. Another example of the QLR is the following: the QLR does not allow (a) the final-state configuration $(\bar{u}u)(\bar{d}d)$ if two $\bar{q}q$ pairs annihilate and a new pair is created (annihilation or planar graph) or (b) the configuration $(\bar{d}d)(\bar{d}d)$ if only one pair annihilates while the other two pairs are reshuffled (rearrangement graph) (166). The ratios of branching ratios will depend on whether annihilation or rearrangement dominates the annihilation process. For instance, the branching ratios for $\rho^0\rho^0$ and $\omega\omega$ are equal if the annihilation graph dominates. This does not appear to be true experimentally (Table 1). To really test the QLR in $\bar{N}N$ annihilation, the branching ratios for the neutral two- and three-meson final states have to be measured with the Crystal Barrel detector at LEAR.

One topic warrants further study: Is the argument of Richards and coworkers (85, 172) correct that the various annihilation reactions take place at different $\bar{N}N$ distances? These authors make the analogy with muonium (μ^+e^-)-antimuonium (μ^-e^+) annihilation. In this pure QED process, the complete annihilation reaction, with only photons in the final state, takes place at very short distances on the atomic scale. On the other hand, the rearrangement process giving $\mu^+\mu^- + e^+e^-$ in the final state is governed by the spatial overlap of the initial- and final-state wave functions. Richard et al apply these arguments to $\bar{N}N$ annihilations (85, 172) and state that the reaction $\bar{N}N \rightarrow \phi\phi$, in which all initial $\bar{q}q$ pairs are annihilated, presumably occurs at a much shorter $\bar{N}N$ separation than, for example, the reaction $\bar{N}N \rightarrow 3\pi$, which can be generated by rearrangement of the three initial $\bar{q}q$ pairs (no $\bar{q}q$ annihilations) (172). The latter reaction may occur when the N and $\bar{N}$ quark cores overlap.

In summary, we only have a rudimentary model-dependent understanding of the measured two-meson annihilation branching ratios. One reason is the large inherent short-distance uncertainties of the meson exchange potential. Another open question is which effective transition operators should be used to mimic the annihilation dynamics. With the forthcoming results from LEAR, we are at least able to resolve a few well-defined questions: (a) Is $\rho\rho$ really so weak compared to $\rho\omega$ or $\omega\omega$? (b) What is the origin of the $\rho\pi$ puzzle? Do other final states exhibit similar behavior? (c) What is behind the suppression or enhancement of various annihilation channels, like $\bar{K}K$, when switching from atomic S to atomic P states? (d) Why is the ratio $BR(\phi\pi)/BR(\omega\pi)$ so large compared to the ratio $BR(\phi\pi\pi)/BR(\omega\pi\pi)$? Is this due to an exotic meson $(\bar{q}^2q^2) = C(1480)$? Can we use the observed global suppression of strangeness in the final states to argue against a large $\bar{s}s$ component in the nucleon? As always in strong interaction physics, progress is slow, but with the new data we are increasing our understanding of some aspects of the annihilation reactions.

4. MESON SPECTROSCOPY

4.1 Overview

In the past, $\bar{p}p$ annihilation has been a rather successful tool to investigate the spectrum of light quark mesons. The E meson, now $\eta(1440)$, and the $K_1(1270)$ were discovered in $\bar{p}p$ annihilation at rest (173, 174), while the ω and $f_1(1285)$ were first observed in annihilation in flight (175, 176). These experiments were performed in the 1960s in bubble chambers.

At LEAR, a large statistical sample of $\bar{p}p$ annihilation at rest in hydrogen gas has been collected by the Asterix collaboration. These studies mainly pertain to annihilation from atomic P states. Quantum number conservation restricts the possible final-state configurations, and hence the rates for the production of meson resonances are different from S and P states. For example, the production of two identical neutral pseudoscalar mesons is forbidden from S states but allowed from P states, and is thus suppressed in liquid hydrogen. Furthermore, the production of neutral $J^{PC} = 1^{++}$ mesons (for example, $\bar{p}p \rightarrow \pi^0 f_1$) is forbidden from S states for the same reason, while $\pi\pi f_1$ is phase-space suppressed from S states. Hence the production of $f_1(1285)$ and $f_1(1420)$ is suppressed in bubble chamber experiments (see also Section 4.2.1). Annihilation from P states might therefore reveal new states. Asterix has already found, apart from a strong signal for $f_2(1270)$ production from P states, a new state $f_2(1565)$ in the three-pion final state (discussed below).

Antiproton annihilation at rest in gas can be performed only at LEAR because only the cooled low energy beams have the necessary small-range straggling and narrow momentum spread. Furthermore, with the advent of new technologies like CsI scintillators read out by photodiodes in strong magnetic fields, large solid-angle and modular γ detectors can be built. The Crystal Barrel collaboration at LEAR is investigating low energy annihilation into final states involving several neutral particles (π^0, η, η',

ω). The detection of the 2γ decay modes (or 3γ for the ω) not only provides strong constraints in the event reconstruction (through kinematical fitting), but also avoids the serious combinatorial background of final-states charged pions for the η and ω decay mode $\pi^+\pi^-\pi^0$. Furthermore, the amplitude analysis for final states involving neutral pion pairs is generally simpler than for those involving $\pi^+\pi^-$ pairs, since the strongly produced ρ^0 meson does not decay into $\pi^0\pi^0$, and since C-parity conservation reduces the number of contributing $\bar{p}p$ initial states.

The main motivation for light quark spectroscopy is to search for mesons that are not made of $\bar{q}q$ pairs (glueballs, hybrid mesons, multiquark states, two-meson molecules, or $\bar{N}N$ states). (For recent reviews, see 16, 17.) Since $\bar{p}p$ annihilation is a good source of gluons, one expects to produce some of these states in addition to the standard quark model $\bar{q}q$ states. Therefore both initial and final states must be carefully selected to enhance the signals. At rest, only a few initial states can contribute, depending on the final state (0^{-+}, 1^{--} for initial S states and 0^{++}, 1^{++}, 2^{++}, and 1^{+-} for initial P states). However, phase space limits the accessible mass range to a maximum of ~ 1650 MeV. Higher masses can be reached by $\bar{p}p$ annihilation in flight, but then many partial waves contribute in the initial state.

Although we do not have any straightforward recipe to identify exotic mesons, guidelines are well established. For example, for gluonic hadrons the relevant decay modes are those involving strange quarks ($\eta\eta$, $\eta\eta'$, and $\bar{K}K$) since gluons are flavor blind. Also, decay channels with quantum numbers 0^{--}, 0^{+-}, 1^{-+}, 2^{+-}, etc, which do not couple to $\bar{q}q$, are useful to identify exotic particles unambiguously, for instance the decay modes $\omega\pi$ and $\omega\eta$, which can couple to 0^{--}, or $\eta\pi$ and $\eta\eta'$, which can couple to 1^{-+}. A 1^{-+} state at 1405 MeV, decaying to $\eta\pi$, was recently reported by the GAMS collaboration at CERN (177).

Finally, all the $\bar{q}q$ states need to be identified. In the mass range accessible to annihilation at rest, essentially only S- and P-wave $\bar{q}q$ mesons need to be considered. These are the nonets 0^{-+}, 1^{--}, 1^{+-}, 0^{++}, 1^{++}, 2^{++} and their radial excitations. In quark model spectroscopy, a principal problem is the mass assignment for the radially excited S-state $\bar{q}q$ mesons. We know that the 2S $\bar{q}q$ state can be lower in mass than the lightest $\bar{q}q$ P state if the (u,d,s) quarks experience an attractive Yukawa-like force (178). An example of an unusual mass ordering is the $J^P = 1/2^+$ Roper resonance N*(1440), which is lighter than the lightest negative-parity state N(1520) with $J^P = 3/2^-$. This unusual mass ordering cannot be explained with power law quark potentials (179). Baryons are not mesons; nevertheless one might have several radially excited $\bar{q}q$ mesons in the 1–2 GeV mass range.

The scalar meson nonet is not well established because the $a_0(980)$ and $f_0(975)$ are believed to be $\bar{K}K$ molecules (180). Instead, the scalar isovector $\bar{q}q$ meson of the 0^{++} nonet could be the $a_0(1320)$ decaying to $\eta\pi$, reported by the GAMS collaboration (181), while one of the isoscalar mesons could be the (nearly pure) $\bar{s}s$ meson $f_0(1525) \to \bar{K}K$ reported by LASS (182). In addition, there is the excess candidate $f_0(1590) \to \eta\eta$, $\eta\eta'$, $4\pi^0$ (183; for review, see 17). However, these states need confirmation. In the axial vector meson nonet, the missing 1^{+-} isoscalar is reported by LASS at 1380 MeV, decaying to $K\bar{K}^*$ (184). The 1^{++} states are discussed by Burnett & Sharpe (17). There are two candidates for the $\bar{s}s$ 1^{++} states, $f_1(1420)$ and $f_1(1510)$. The latter fits better in the nonet, assuming ideal mixing. The former might be an exotic state (see 17). The excited η and f_2 mesons are discussed below.

Above ~ 1650 MeV the experimental situation is very confused. All nonets are incomplete (with the possible exception of 3^{--}) although many candidates exist.

4.2 Recent Results in $\bar{p}p$ Annihilation

4.2.1 THE E MESON   The E(1420) was first discovered in $\bar{p}p$ annihilation at rest in liquid into $(K^{\pm}K_s\pi^{\mp})\pi^+\pi^-$. Its quantum numbers had been determined to be 0^{-+} (173). However, this assignment has been the subject of a long controversy. Several experiments (inelastic πp, pp central collisions, and $\gamma\gamma$ collision) report either a 0^{-+} (185, 186) or a 1^{++} state [$f_1(1420)$] (187–189) at the same mass. [Following an early observation of a 1^{++} state at 1420 MeV (187), the E meson was renamed $f_1(1420)$ by the Particle Data Group. As we argue below, the renaming is not justified.] Furthermore, a glueball candidate around 1440 MeV, $\eta(1440)$ (formerly ι), is observed in radiative J/ψ decay. The question naturally arises as to whether the states observed in $\bar{p}p$ annihilation and in J/ψ decay are identical. According to a recent analysis by the Mark III collaboration, the $\eta(1440)$ is made of three states, $f_1(1443)$, $\eta(1416)$, and $\eta(1490)$ (190).

The Asterix collaboration confirmed the original 0^{-+} assignment for the E meson, but with a hydrogen gas target (191). Figure 8a shows the $K^{\pm}K^0\pi^{\mp}$ invariant mass distribution. The charged $K^{\pm}$ was detected by ionization sampling while the neutral $K^0(\bar{K}^0)$ here was a K_L escaping detection or a $K_s \to \pi^0\pi^0$. Since there are two possible combinations for the $\pi^{\mp}$, the wrong-charge invariant mass distribution $K^{\mp}K^0\pi^{\mp}$ (one entry per event) was subtracted from the $K^{\pm}K^0\pi^{\mp}$ (two entries per event). The peak with mass 1413 ± 8 MeV and width 62 ± 16 MeV is fully consistent with the original E(1420) (173). In addition, a 3σ peak is seen, consistent with the well-known $f_1(1285)$, which was not observed earlier in liquid (173). A Dalitz plot analysis of the E meson was attempted but one could

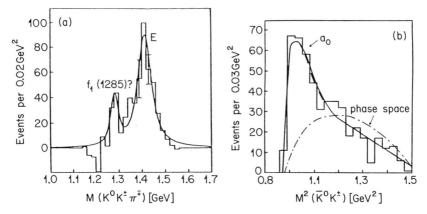

Figure 8 (*a*) Distribution of $K^0K^-\pi^+$ invariant mass ($+$c.c.) in $\bar{p}p \to K^0K^-\pi^+\pi^+\pi^-$ ($+$c.c.). (*b*) $K^0\overline{K}^-$ invariant mass ($+$c.c.) in the E meson region (191, 192).

not distinguish between 0^{-+} and 1^{++} (191). The dominant decay mode for the E meson is $a_0(980)\pi$ [$a_0(980) \to \overline{K}K$], at variance with the $K\overline{K}^*$ decay mode of the $f_1(1420)$ (Figure 8*b*) and in agreement with the decay mode of $\eta(1440)$ found in radiative J/ψ decay.

The 0^{-+} assignment can nevertheless be established from the E meson production rate, as a 1^{++} state cannot be produced with a pair of pions from $\bar{p}p$ S states: because of the limited phase space, a 0^{++} di-pion recoils against the E meson with zero relative angular momentum. Hence the 0^{-+} E meson is produced from the $^1S_0$ $\bar{p}p$ state. A 1^{++} meson [like $f_1(1285)$] would be produced mainly from the $^3P_1$ state and hence should not be seen in liquid, as confirmed by the earlier data (173). If this phase-space argument holds, then the 0^{-+} E meson should be produced from P states while $f_1(1285)$ should be prominent. The branching ratio for $\bar{p}p \to (E \to K^\mp K^0\pi^\pm)\pi^+\pi^-$ is $(7.1 \pm 0.4) \times 10^{-4}$ in liquid (173). In the gas target (Figure 8), the fraction of P wave is 61% and the branching ratio is $(3.0 \pm 0.9) \times 10^{-4}$ (191). With these two measurements one extrapolates to 100% P wave and finds that the branching ratio is consistent with zero, i.e. the E meson is 0^{-+}.

A similar suppression also operates for the pseudoscalars in the channels $\bar{p}p \to \eta\pi^+\pi^-$ and $\eta'\pi^+\pi^-$ (see Section 3.1.3 and Table 2). The phase-space suppression is, of course, less dramatic than for $E\pi\pi$ since η and η' are lighter than the E meson.

4.2.2   THE $f_2(1565)$ IN ITS $\pi^+\pi^-$ DECAY MODE   As discussed in Section 3.1.2, the Asterix collaboration has presented new data (119) for $\bar{p}p$ annihilation into $\pi^+\pi^-\pi^0$ in gaseous hydrogen (53% P wave) and in off-line coincidence with L x rays (92% P wave) (see Figure 6). Apart from ρ^0 and $f_2(1270)$, a new resonance $f_2(1565)$ (called AX by the collaboration) is observed to decay into $\pi^+\pi^-$ from P states.

Since $f_2(1565) \rightarrow \pi^+\pi^-$ and no charged partner is observed in $\pi^{\pm}\pi^0$, the quantum numbers are $J^{PC}(I^G) = 0^{++}$ or 2^{++} (0^+). Higher even spins are not expected at this low mass. Without $f_2(1565)$ the fit is poor. A better fit is obtained with a 2^{++} (or a 0^{++}) state with mass 1565 ± 20 MeV and width of 170 ± 40 MeV. The analysis is complicated by the occurrence of $\rho^+\rho^-$ interference under the $f_2(1565)$, which could shift the mass and change the width of this meson. Since 2^{++} is preferred by the fit, but 0^{++} is not excluded, a phase-shift analysis was performed. The relativistic Breit-Wigner amplitude for $f_2(1565)$ was parametrized by the amplitude ae^{ib}, which was allowed to interfere with the other amplitudes describing the Dalitz plot. The 2^{++} phase advances through $90°$ at the correct $\pi^+\pi^-$ mass and it is concluded that a new isoscalar 2^{++} state has been observed (119). The branching ratio for $\bar{p}p \rightarrow f_2(1565)\pi^0$ with $f_2(1565) \rightarrow \pi^+\pi^-$ is $(3.7 \pm 0.6) \times 10^{-3}$, and $f_2(1565)$ is produced mainly from the isovector 1^{++} and 2^{++} $\bar{p}p$ states while $f_2(1270)$ is produced dominantly from 1^{++} (see Table 3).

4.2.3   THE $f_2(1565)$ IN ITS $\rho^0\rho^0$ DECAY MODE   A state (called ζ) with mass 1477 and width 116 MeV was reported from bubble chamber exposures in the reaction $\bar{p}n \rightarrow (\zeta \rightarrow 2\pi^+2\pi^-)\pi^-$, where the momentum of the spectator proton was less than 200 MeV/c (193). The ζ peak is best seen by subtracting the wrong-charge invariant mass $(3\pi^-\pi^+)\pi^+$ distribution (two combinations per event) from the $(2\pi^+2\pi^-)\pi^-$ distribution (three combinations per event) as shown in Figure 9. An amplitude analysis of the five-pion system led to the spin-parity 2^{++} and $\zeta \rightarrow \rho^0\rho^0$. The branching ratio for production and decay into $\rho^0\rho^0$ is $3.7 \pm 0.3\%$. The isospin is presumably zero since $I = 1$ does not couple to $\rho^0\rho^0$. For $I = 2$, the decay rate into $\rho^+\rho^-$ can be related to $\rho^0\rho^0$ by Clebsch-Gordan coefficients. With $I = 2$ the rate for production and decay into $\rho^0\rho^0 + \rho^+\rho^-$ would be 50% of all $\bar{p}n$ annihilations, an unreasonably high branching ratio.

The Asterix collaboration has confirmed the existence of an enhancement compatible with $\rho^0\rho^0$ decay at the mass 1504 MeV with a width of 206 MeV for the same reaction in gaseous deuterium with five times more events (121, 194). No spin-parity analysis has been performed. These data are also shown in Figure 9. However, the mass and width of the enhancement depend on the spectator momentum, as demonstrated by the Asterix data (Figures 9b,c). Higher spectator momenta are associated with a lower resonance mass (Figure 9c). The contribution from pion rescattering (inset of Figure 9b) has been calculated (195). The scattering

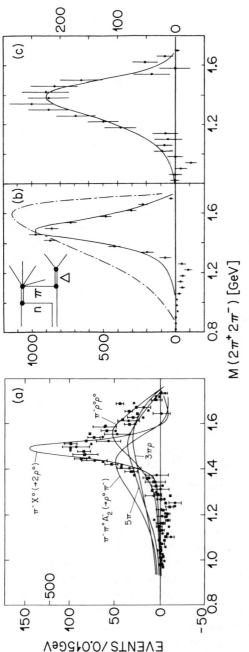
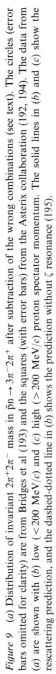

Figure 9 (*a*) Distribution of invariant $2\pi^{+}2\pi^{-}$ mass in $\bar{p}n \rightarrow 3\pi^{-}2\pi^{+}$ after subtraction of the wrong combinations (see text). The circles (error bars omitted for clarity) are from Bridges et al (193) and the squares (with error bars) from the Asterix collaboration (192, 194). The data from (*a*) are shown with (*b*) low (<200 MeV/*c*) and (*c*) high (>200 MeV/*c*) proton spectator momentum. The solid lines in (*b*) and (*c*) show the rescattering prediction, and the dashed-dotted line in (*b*) shows the prediction without ζ resonance (195).

of one of the ρ decay pions leads to a lower $\rho\rho$ invariant mass and boosts the "spectator" proton to higher momenta. The prediction is in excellent agreement with data.

Since $f_2(1565)$ and ζ have the same quantum numbers, both are produced from $I = 1$ $\bar{p}N$ states, and their masses and widths are comparable, one concludes that the states are probably identical. As discussed in the next section, the mass of $f_2(1565)$ may actually be close to 1515 MeV as observed in its $\pi^0\pi^0$ decay mode. Notice that $f_2(1565)$ is produced from P states only, which implies that P wave in $\bar{p}n$ annihilation in deuterium should be strong, even for low spectator momenta, or alternatively, much stronger in the $I = 1$ than in the $I = 0$ states. This is supported by the large $f_2(1270)$ production rate in $\bar{p}n \rightarrow \pi^+\pi^-\pi^-$ (130) and in the ($I = 1$) $3\pi^0$ final state (next section). The rate R for $\pi^-\pi^0$ on the neutron should be twice the rate for $\pi^+\pi^-$ on the proton if S wave dominates annihilation into $\pi\pi$ [after correcting for the ratio of $\bar{p}p$ to $\bar{p}n$ annihilation in deuterium: 1.33 ± 0.07 (196)]. A recent measurement in a liquid deuterium target at LEAR indeed finds $R = 2.07 \pm 0.05$ (197), which is at variance with an earlier measurement that led to $R = 0.68 \pm 0.07$ (198), and which therefore disputed the S-wave dominance. However, these measurements (197, 198) apply to the fraction of P and S waves for annihilation into 2π, not necessarily into 3π or 5π.

The need for $\rho\rho$ resonances in the five-pion channels is not new. A 2^{++} (0^+) $\rho^0\rho^0$ resonance was introduced earlier to obtain satisfactory fits to $2\pi^+2\pi^-\pi^0$ in $\bar{p}p$ annihilation at rest (199).

4.2.4   THE $f_2(1565)$ IN ITS $\pi^0\pi^0$ DECAY MODE   The Crystal Barrel collaboration at LEAR has presented new data on the annihilation channel $\bar{p}p \rightarrow \pi^0\pi^0\pi^0$ at rest in liquid hydrogen (200). The Crystal Barrel detects charged particles over a solid angle of 95% $\times$ 4π and photons with 100% efficiency over 97% $\times$ 4π. The 200-MeV/c antiprotons enter a solenoid magnet and annihilate in a liquid hydrogen target. Two cylindrical proportional wire chambers provide the trigger for the final-state charged multiplicity (here zero prong). Photons are detected in a barrel-shaped assembly of 1380 CsI(Tl) crystals with photodiode readout. Details on the apparatus can be found in the paper by Aker et al (201).

The $3\pi^0$ Dalitz plot and its projection are shown in Figure 10. One observes three bands corresponding to $f_2(1270) \rightarrow \pi^0\pi^0$ interfering at the edge of the Dalitz plot. The additional three bands correspond to a state around 1515 MeV decaying into $\pi^0\pi^0$. A Dalitz plot analysis has been performed including the contributing 0^{-+} (S-wave) and the 1^{++} and 2^{++} (P-wave) $\bar{p}p$ initial states. The energy-dependent amplitudes of Au et al (202) were introduced into the fit to describe the 0^{++} $\pi^0\pi^0$ partial wave.

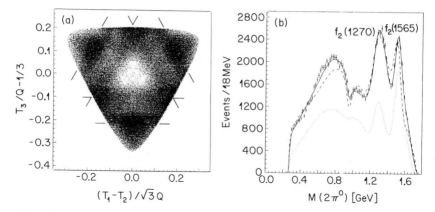

Figure 10 (*a*) Dalitz plot for the reaction $\bar{p}p \to 3\pi^0$. The marks indicate the $f_2(1565)$ bands (corners of the plot) and the $f_2(1270)$ bands (interfering at the edge of the plot). (*b*) Distribution of $\pi^0\pi^0$ invariant mass. The dotted line shows the total $\bar{p}p$ P-wave contribution, the dashed line shows the fit for a 2^{++} state at 1515 MeV, and the full curve includes the tail of $f_2(1810)$ (200).

A 0^{++} resonance at 1515 MeV does not describe the data (200). The dashed line in Figure 10*b* shows the fit for a 2^{++} resonance; the full curve shows the same fit but includes the broad $f_2(1810)$ observed earlier in the partial wave analysis of K^-K^+ and $\pi^-\pi^+$ (203). The various contributions are given in Table 4.

One concludes that a tensor meson with mass 1515 ± 10 MeV and width 120 ± 10 MeV is observed. This state is probably identical to $f_2(1565)$, although the mass and width are slightly lower. However, in contrast to $\pi^+\pi^-\pi^0$, the difficulty of the interfering $\rho^+\rho^-$ bands below the $f_2(1565)$ is avoided. Therefore, the mass and width obtained in this experiment are more reliable. The state at 1515 MeV is produced from $^1S_0$, $^3P_1$, and $^3P_2$ with comparable rates (Table 4). The sizeable contribution from $^1S_0$ is not in contradiction with Asterix data since the branching ratio for $\bar{p}p \to 3\pi^0$

Table 4 Contribution (in %) to $\bar{p}p$ annihilation into $3\pi^0$ in liquid hydrogen (200)[a]

	$^1S_0$	$^3P_1$	$^3P_2$
$\pi^0\pi^0$-S wave	28.3 ± 2.5		
$f_2(1270)\pi^0$	2.9 ± 1.1	19.3 ± 1.1	
$f_2(1565)\pi^0$	9.5 ± 0.7	8.0 ± 0.5	8.8 ± 1.1
$f_2(1810)\pi^0$			23.2 ± 2.5

[a] The branching ratio for $\bar{p}p \to 3\pi^0$ is $0.76 \pm 0.23\%$ (204).

in liquid hydrogen is only $0.76 \pm 0.23\%$ (204), compared to $6.6 \pm 0.8\%$ for $\pi^+\pi^-\pi^0$ in liquid (119). Hence the S-wave contribution to $f_2(1565)$ computed from the $3\pi^0$ data is expected to be tiny in the charged channel and further hidden under the $\rho^+\rho^-$ interference. In addition, both $f_2(1270)$ and $f_2(1565)$ are strongly produced from P states, although liquid hydrogen is used in the Crystal Barrel. This is also not a contradiction since the $3\pi^0$ final state is exclusively $I = 1$ while $\pi^+\pi^-\pi^0$ is dominantly $I = 0$ (119). One therefore expects that the $f_2(1565)$ signal will be hidden below the strong $I = 0$ background in liquid hydrogen.

Enhancements around 1500–1600 MeV in $\pi^+\pi^-$ and $\pi^0\pi^0$ have been reported earlier in $\bar{p}n \rightarrow \pi^+\pi^-\pi^-$ (129) and $\bar{p}p \rightarrow 3\pi^0$ (129, 204) albeit with limited statistics. The meson $f_2(1565)$ is not $f_2'(1525)$ since the production branching ratio for $\bar{p}p \rightarrow f_2(1565)\pi^0$ at rest would be $\sim 50\%$, $f_2'(1525)$ decaying mainly to $\bar{K}K$. In fact, no signal is observed in $\bar{K}K$ (129). The 2^{++} $\bar{q}q$ ground-state mesons are known and the first radial excitation is predicted around 1820 MeV (205), although the mass of a radial excitation is very model dependent and difficult to calculate (Section 4.1). The state $f_2(1515)$ is a candidate for the first radial excitation. This state could be, if not a radial excitation, a glueball, a hybrid, a $\bar{q}^2q^2$ state (206), or an $\bar{N}N$ molecule (141). Furthermore, $I = 1$ or $I = 2$ states are not observed by Asterix in the $\pi^\pm\pi^0$ invariant mass (Figure 6).

A natural parity band of $I = 0$ $\bar{N}N$ molecular states is predicted by potential models (161). If $f_2(1565)$ is the 2^{++} member of this band, then 1^{--} and 0^{++} states are expected around 1250 and 1100 MeV respectively (141). Candidates for these states have been reported. A scalar state decaying into two pions is suggested at this mass in the final states $\omega\pi^+\pi^-$ (103), $\rho^0\pi^+\pi^-$ (207), and in $\bar{p}n - \rho^-\pi^+\pi^-$ (208). A vector meson has also been reported at 1250 MeV in $\bar{p}p \rightarrow (e^+e^-)X$ (209), although with very limited statistics. For further theoretical considerations, see the reviews (16, 17).

5. ANTIPROTON-NUCLEUS INTERACTIONS

The antiproton interacts only with the nuclear surface at a maximum of 10% of the central nuclear density. The interior of the nucleus is black to antiprotons, which means that the short-range (heavy-meson) exchanges in the $\bar{N}N$ potential will be partially absorbed because of the nuclear density, and the longer-range (OPE) will be the dominant exchange. Therefore, in inelastic $(\bar{p}, \bar{n})$ nuclear reactions, the pionlike states $(0^-, 1^+, 2^-,$ etc) should be preferentially excited, as opposed to the many more states in (n, p) reactions. This is one of the useful aspects of antiproton-nucleus reaction studies. So far this type of experiment has not been performed.

The experimental data in antiproton-nucleus interactions have been reviewed by Guaraldo (210).

The antiproton-nucleus interaction is successfully described by the intranuclear cascade model (INC); see the review by Cugnon & Vandermeulen (211). The momentum distribution of pions following $\bar{p}$ annihilation on a nucleus shows two apparent temperature scales, $T \sim 100$ MeV and $T \sim 50$ MeV (211, 212). The first temperature is associated with $\bar{N}N$ annihilation (Section 3.3), the other is associated with pion rescattering from nucleons, which also leads to a proton momentum spectrum with a high energy tail. These distributions are reproduced in the cascade calculations (211). Furthermore, these calculations find that, after a few energetic protons remove some 200 MeV, the nucleus loses excitation energy by evaporating a few nucleons or fragmenting into many pieces. For a more detailed discussion, see the review by Cugnon & Vandermeulen (211).

It has been speculated that the production of strangeness (for example, the production of kaons and hyperons) could be enhanced in $\bar{p}$-nucleus annihilation either by the formation of quark-gluon plasma (213) or by annihilation on a cluster of several nucleons. The fraction of strangeness production measured in the ITEP xenon bubble chamber is $6.2 \pm 0.9\%$ for antiprotons at rest and $6.2 \pm 0.8\%$ in the 0.4–0.9 GeV/c range (214). The value obtained in an INC calculation is 6.25% at 0.65 GeV/c (215). Rescattering of annihilation mesons, mainly strangeness exchange ($\bar{K}N \rightarrow \Lambda\pi$), accounts for the observed Λ production on several nuclei (215, 216). Hence no unusual yield is observed in strangeness production for low energy $\bar{p}$-nucleus annihilation.

The so-called Pontecorvo reaction $\bar{p}d \rightarrow \pi^- p$ occurs with a rate $\sim 2 \times 10^{-5}$ (196, 217, 218). It is not clear which process (rescattering or multiquark clusters) dominates this interaction since the interpretation is model dependent. There are two calculations of the rescattering contributions for this reaction, one finding a rate of the right order of magnitude but with large theoretical uncertainties (219), the other finding too small a rate (220). The relative rates for other Pontecorvo reactions (such as $\bar{p}d \rightarrow \pi^- p$, $\Sigma^- K^+$, $K^0\Lambda$) are less model dependent, but unfortunately no data exist except for $\pi^- p$.

6. OUTLOOK

A wealth of interesting new data has emerged from the LEAR facility. Some of the low energy scattering data can be readily understood by effective meson-baryon models. Other measurements might require a treatment of QCD in the nonperturbative regime. The scattering and annihilation data are still quite fragmentary. However, the first round of LEAR experiments provides guidelines as to which phenomena should be studied next. For example, the peculiar oscillatory behavior of the ρ parameter below 200 MeV/c should be studied at LEAR with a gaseous hydrogen target.

We have shown that polarization measurements are sensitive to the behavior of the interaction at short distances. Apart from the analyzing power data at a few energies, we do not have precise measurements of other spin observables. These experiments require intense polarized antiproton beams and good analyzers of antiproton polarization. However, a few observables such as spin correlations can possibly be measured with an internal polarized antiproton beam and a polarized hydrogen gas jet (221). Polarization effects should be investigated in the charge exchange reaction and in (the pure $I = 1$) elastic $\bar{n}p$ scattering for which only measurements of the total cross section and of integrated annihilation cross sections are available (222, 223).

The charge exchange reaction is unique since it is the only NN or $\bar{N}$N reaction where we can measure the *difference* of the two large isospin amplitudes in order to probe the isovector part of the meson exchange forces. This reaction is a critical test of the models used to explain the data. For example, for the charge exchange reaction, a very large transfer of longitudinal polarization from the proton to the antineutron is predicted in the forward direction (224), a feature that could be used to generate polarized antineutron beams. Present experimental results at the lower energies show that our model understanding is reasonable, and further specific polarization measurements could be very beneficial to our understanding of these reactions and the intermediate-range $\bar{N}$N (and possibly NN) forces.

For $\bar{p}p$ and $\bar{n}p$ annihilation, the Crystal Barrel and Obelix experiments will provide the large data samples of neutral and kaonic final states required to study both the annihilation dynamics and the production of new and possibly exotic mesons. The study of specific two-body annihilation reactions in flight has been limited so far. Two-body differential cross sections as a function of momentum are only available for $\pi^-\pi^+$, K^-K^+ (142, 144–146, 150), and $\pi^0\pi^0$, $\pi^0\eta$ (225) above 1 GeV/c. A partial wave amplitude analysis of these reactions has been performed and shows resonance behavior in nearly all partial waves (226). These resonances cannot be observed in the total annihilation cross section because of the dominating nonresonant background. Both reactions $\bar{p}p \rightarrow K^-K^+$ and $\pi^-\pi^+$ possess the dramatic property of a maximum left-right asymmetry in a wide angular range. Other two-body final states ($\pi^0\omega$, $\omega\omega$, $\eta\omega$, K_SK_S, K_SK_L, etc), possibly with a polarized target, should be systematically studied to clarify the dynamics involved in these reactions. In addition, the glueball-sensitive channels like $\phi\phi$ (row being studied by the Jetset collaboration) are accessible at LEAR, but only in a very narrow energy window limited by threshold (2.04 GeV) and the maximum available center-of-mass energy of 2.4 GeV.

At LEAR, the accessible mass range for the production of meson resonances, associated with the emission of one or two pions, is limited by phase space. Hence glueballs or hybrids might not be observed in production at LEAR, if their masses lie above 2 GeV or if they are broad and lie in the mass range 1.7 to 2 GeV. The formation and production of exotic light quark mesons might be investigated at the SuperLEAR facility that is currently being evaluated, or with antiprotons at the KAON factory.

Charmonium states ($\bar{c}c$) or charmed hybrids ($\bar{c}cg$) (227) might be studied in formation at the SuperLEAR facility. In parallel, antiprotons interacting with nuclei could be used to investigate the formation and interaction of J/ψ in nuclear matter (important in relativistic heavy-ion searches for the quark-gluon plasma), in charmonium-nucleus bound states, and in charmed hypernuclei, as well as to test the hypothesis of color transparency (228, 229).

ACKNOWLEDGMENT

We thank Prof. J. Vandermeulen for useful discussions related to the antiproton-nucleus interaction. This work was supported in part by a grant from the National Science Foundation (FM).

Literature Cited

1. Walcher, Th., *Annu. Rev. Nucl. Part. Sci.* 38: 67–95 (1988)
2. Kerbikov, B. O., Kondratyuk, L. A., Sapozhnikov, M. G., *Sov. Phys. Usp.* 32: 739 (1989)
3. Sedlak, J., Simak, V., *Sov. J. Part. Nucl.* 19: 191 (1988)
4. DeGrand, T., et al., *Phys. Rev.* D12: 2060 (1975)
5. Close, F. E., *An Introduction to Quarks and Partons.* New York: Academic (1979)
6. Isgur, N., Karl, G., *Phys. Rev.* D18: 4187 (1978); D19: 2653 (1979)
7. Thomas, A. W., *Adv. Nucl. Phys.* 13: 1 (1983)
8. Myhrer, F., In *Quarks and Nuclei, Int. Rev. Nucl. Phys.*, ed. W. Weise. Singapore: World Scientific (1984), 1: 325–407
9. Bryan, R. A., Phillips, R. J. N., *Nucl. Phys.* B5: 201 (1968)
10. Dover, C. B., Richard, J. M., *Phys. Rev.* C21: 1466 (1980)
11. Dalkarov, O. D., Myhrer, F., *Nuovo Cimento* 40A: 152 (1977)
12. Myhrer, F., *Nucl. Phys.* A508: 513c (1990)
13. Serber, R., *Phys. Rev. Lett.* 10: 357 (1963)
14. Brückner, W., et al., *Phys. Lett.* B166: 113 (1986); CERN-PPE preprint 91-41 (1991)
15. Anderson, B., et al., *Phys. Rep.* 97: 33 (1983)
16. Close, F. E., *Rep. Prog. Phys.* 51: 833 (1988)
17. Burnett, T. H., Sharpe, S. R., *Annu. Rev. Nucl. Part. Sci.* 40: 327–55 (1990)
18. Amsler, C., *Adv. Nucl. Phys.* 18: 183 (1987)
19. Myhrer, F., Wroldsen, J., *Rev. Mod. Phys.* 60: 629 (1988)

20. Vinh Mau, R., et al., *Phys. Lett.* B44: 1 (1973)
21. Ericson, T. E. O., Rosa-Clot, M., *Annu. Rev. Nucl. Part. Sci.* 35: 271–94 (1985)
22. Lacombe, M., et al., *Phys. Rev.* D12: 1495 (1975)
23. Brown, G. E., Jackson, A. D., *Nucleon-Nucleon Interaction.* Amsterdam: North-Holland (1976)
24. Machleidt, R., Holinde, K., Elster, Ch., *Phys. Rep.* 149: 1 (1987)
25. Hamilton, J., Oades, G. C., *Nucl. Phys.* A424: 447 (1984)
26. Nagels, M. M., Rijken, T. A., de Swart, J. J., *Phys. Rev.* D12: 744 (1975); D15: 2547 (1977); D17: 768 (1978)
27. Oka, M., Yazaki, K., See Ref. 8, 1: 489–568
28. Phillips, R. J. N., *Rev. Mod. Phys.* 39: 681 (1967)
29. Kimura, M., Saito, S., *Nucl. Phys.* B178: 477 (1981); Kim, J. H., Toki, H., *Prog. Theor. Phys.* 78: 616 (1987); Gregory, P., et al., *Nucl. Phys.* B102: 189 (1976)
30. Pais, A., *Ann. Phys. NY* 9: 548 (1960)
31. Fett, E., et al., *Nucl. Phys.* B130: 1 (1977)
32. Vandermeulen, J., *Z. Phys.* C37: 563 (1988)
33. Martin, A., *Phys. Rev.* 124: 614 (1961)
34. Côté, J., et al., *Phys. Rev. Lett.* 48: 1319 (1982)
35. Hippchen, T., Holinde, K., Plessas, W., *Phys. Rev.* C39: 761 (1989)
36. Tegen, R., Mizutani, T., Myhrer, F., *Phys. Rev.* D32: 1672 (1985)
37. Moussallam, B., *Nucl. Phys.* A407: 413 (1983); A429: 429 (1984)
38. Hippchen, T., et al., *Nucl. Phys. B. Proc. Suppl.* 8: 116 (1989); Hippchen, T., et al., Jülich preprint KFA-IKP(TH)-1991-3 (1991); *Phys. Rev. C.* In press (1991)
39. Liu, G. Q., Tabakin, F., *Phys. Rev.* C41: 665 (1990)
40. Holinde, K., et al., In *First Biennial Conf. on Low Energy Antiproton Physics*, ed. P. Carlson, et al. Singapore: World Scientific (1991), p. 92
41. Dalkarov, O. D., Mandelzweig, V. B., Shapiro, I. S., *JETP Lett.* 10: 257 (1969); *Nucl. Phys.* B21: 88 (1970)
42. Montanet, L., Rossi, G. C., Veneziano, G., *Phys. Rep.* 63: 149 (1980)
43. Myhrer, F., *AIP Conf. Proc.* 41: 357 (1978)
44. Myhrer, F., Gersten, A., *Nuovo Cimento* 37A: 21 (1977)
45. Myhrer, F., Thomas, A. W., *Phys. Lett.* B64: 59 (1976)
46. Shibata, T.-A., *Phys. Lett.* B189: 232 (1987)
47. Haidenbauer, J., et al., *Z. Phys.* A334: 467 (1989); *Nucl. Phys.* A508: 329c (1990)
48. Shapiro, I. S., *Phys. Rep.* 35: 129 (1978)
49. Shapiro, I. S., *Nucl. Phys. B. Proc. Suppl.* 8: 100 (1989)
50. Clough, A. S., et al., *Phys. Lett.* B146: 299 (1984)
51. Bugg, D. V., et al., *Phys. Lett.* B194: 563 (1987)
52. Brückner, W., et al., *Z. Phys.* A335: 217 (1990)
53. Brückner, W., et al., *Phys. Lett.* B169: 302 (1985)
54. Agnello, M., et al., *Phys. Lett.* B256: 349 (1991)
55. Kroll, P., Schweiger, W., *Nucl. Phys.* A503: 865 (1989)
56. Brückner, W., et al., *Phys. Lett.* B197: 463 (1987)
57. Kunne, R., et al., *Nucl. Phys.* B323: 1 (1989)
58. Bertini, R., et al. *Phys. Lett.* B228: 531 (1989)
59. Tegen, R., Myhrer, F., Mizutani, T., *Phys. Lett.* B182: 6 (1986)
60. Kohno, M., Weise, W., *Nucl. Phys.* A454: 429 (1986); *Phys. Lett.* B152: 303 (1985)
61. Linsen, L., et al., *Nucl. Phys.* A469: 726 (1987)
62. Brückner, W., et al., *Phys. Lett.* B158: 180 (1985)
63. Schiavon, P., et al., *Nucl. Phys.* A505: 595 (1989)
64. Batty, C. J., *Rep. Prog. Phys.* 52: 1165 (1989)
65. Kroll, P., Schweiger, W., *Nucl. Phys. B. Proc. Suppl.* 8: 121 (1989)
66. Fasano, C. G., Locher, M. P., *Z. Phys.* A336: 469 (1990)
67. Mahalanabis, J., Pirner, H. J., Shibata, T.-A., *Nucl. Phys.* A 485: 546 (1988)
68. Kunne, R., et al., *Phys. Lett.* B206: 557 (1988)
69. Timmers, P. H., van der Sanden, W. A., de Swart, J. J., *Phys. Rev.* D29: 1928 (1984)
70. Birsa, R., et al., *Phys. Lett.* B246: 267 (1990)
71. Perrot-Kunne, F., See Ref. 40, p. 251
72. Bradamante, F., See Ref. 40, p. 219
73. Martin, A., et al., *Nucl. Phys.* A487: 563 (1988)
74. Kunne, R. A., et al., See Ref. 40, p. 241
75. Nakamura, K., et al., *Phys. Rev. Lett.* 53: 885 (1985)
76. Hamilton, R. P., et al., *Phys. Rev. Lett.* 44: 1179 (1980)
77. Barnes, P. D., et al., *Phys. Lett.* B189: 249 (1987)

78. Barnes, P. D., et al., *Phys. Lett.* B229: 432 (1989)
79. Barnes, P. D., et al., *Nucl. Phys.* A526: 575 (1991)
80. Barnes, P. D., et al., *Phys. Lett.* B246: 273 (1990)
81. Kohno, M., Weise, W., *Nucl. Phys.* A479: 433c (1988); *Phys. Lett.* B179: 15 (1986); B206: 584 (1988)
82. Timmermans, R. G. E., Rijken, T. A., de Swart, J. J., *Nucl. Phys.* A479: 383c (1988)
83. Høgaasen, H., Høgaasen, J., *Nuovo Cimento* 40: 560 (1965)
84. Kaufmann, W. B., Pilkuhn, H., *Phys. Rev.* C17: 215 (1978)
85. Carbonell, J., Ihle, G., Richard, J. M., *Z. Phys.* A334: 329 (1989)
86. Reifenröther, G., Klempt, E., *Phys. Lett.* B245: 129 (1990)
87. Mandrup, L., et al., *Nucl. Phys.* A512: 591 (1990)
88. Jaenicke, S., Kerbikov, B., Pirner, H. J., *Z. Phys.* A339: 297 (1991)
88a. Schneider, M., et al., *Z. Phys.* A338: 217 (1991)
89. Batty, C. J., *Nucl. Phys.* A508: 89c (1990)
90. Green, A. M., Niskanen, J. A., *Prog. Part. Nucl. Phys.* 18: 93 (1987)
91. Armenteros, R., French, B., In *High Energy Physics*, ed. E. H. S. Burhop. London: Academic (1969), 4: 237
92. Day, T. B., Snow, G. A., Sucher, J., *Phys. Rev.* 3: 864 (1960)
93. Chiba, M., et al., *Phys. Rev.* D38: 2021 (1988); D39: 3227 (1989)
94. Adiels, L., et al., *Z. Phys.* C35: 15 (1987)
95. Adiels, L., et al., *Z. Phys.* C42: 49 (1989)
96. Bassompierre, G., et al., *Phys. Lett.* B64: 475 (1976)
97. Devons, S., et al., *Phys. Rev. Lett.* 27: 1614 (1971)
98. Bassompierre, G., et al., In *4th Eur. Antiproton Symp.*, ed. A. Fridman. Paris: CNRS (1978), 1: 139
99. Chiba, M., et al., *Phys. Lett.* B202: 447 (1988)
100. Foster, M., et al., *Nucl. Phys.* B8: 174 (1968)
101. Espigat, P., et al., *Nucl. Phys.* B36: 93 (1972)
102. Baltay, C., et al., *Phys. Rev.* 145: 1103 (1966)
103. Bizzarri, R., et al., *Nucl. Phys.* B14: 169 (1969)
104. Smith, G. A., In *The Elementary Structure of Matter*, ed. J. M. Richard, et al. Berlin: Springer (1988), p. 197
105. Bloch, M., Fontaine, G., Lillestøl, E., *Nucl. Phys.* B23: 221 (1970)
106. Bizzarri, R., et al., *Nucl. Phys.* B27: 140 (1971)
107. Conforto, B., et al., *Nucl. Phys.* B3: 469 (1967)
108. Bettini, A., et al., *Nuovo Cimento* 63A: 1199 (1969)
109. Barash, N., et al., *Phys. Rev.* 145: 1095 (1966)
110. Doser, M., et al., *Nucl. Phys.* A486: 493 (1988)
111. Doser, M., et al., *Phys. Lett.* B215: 792 (1988)
112. Richard, J. M., Sainio, M. E., *Phys. Lett.* B110: 349 (1982)
113. Ahmad, S., et al., *Phys. Lett.* B157: 333 (1985)
114. Ziegler, M., et al., *Phys. Lett.* B206: 151 (1988)
115. Schaefer, U., et al., *Nucl. Phys.* A495: 451 (1989)
116. Ahmad, S., et al., *Nucl. Instrum. Methods* A286: 76 (1990)
117. Bettini, A., et al., *Nuovo Cimento* A62: 1038 (1969)
118. Foster, M., et al., *Nucl. Phys.* B6: 107 (1968)
119. May, B., et al., *Z. Phys.* C46: 191; C46: 203 (1990); *Phys. Lett.* B225: 450 (1989)
120. Weidenauer, P., et al., *Z. Phys.* C47: 353 (1990)
121. Weidenauer, P., et al., Personal communication (1991)
122. Ellis, J., Gabathuler, E., Karliner, M., *Phys. Lett.* B217: 173 (1989)
123. Dover, C. B., Fishbane, P. M., *Phys. Rev. Lett.* 62: 2917 (1989)
124. Close, F. E., Lipkin, H. J., *Phys. Rev. Lett.* 41: 1263 (1978); *Phys. Lett.* B196: 245 (1987)
125. Achasov, N. N., *JETP Lett.* 43: 526 (1986)
126. Bityukov, S. I., et al., *Phys. Lett.* B188: 383 (1987)
127. Reifenröther, G., et al., *Phys. Lett. B.* In press (1991)
128. Bizzarri, R., et al., *Nuomo Cimento* 20A: 393 (1974)
129. Gray, L., et al., *Phys. Rev.* D27: 307 (1983)
130. Bridges, D., et al., *Phys. Rev. Lett.* 56: 215 (1986)
131. Bettini, A., et al., *Nuovo Cimento* 67A: 642 (1967)
132. Bizzarri, R., et al., In *Physics at LEAR with Low-Energy Cooled Antiprotons*, ed. U. Gastaldi, R. Klapisch. New York: Plenum (1984), p. 193
133. Cooper, A. M., et al., *Nucl. Phys.* B146: 1 (1978)
134. Richter, B., et al., *Phys. Lett.* B126: 284 (1983)

135. Chiba, M., et al., *Phys. Lett.* B177: 217 (1986)
136. Adiels, L., et al., *Phys. Lett.* B182: 405 (1986)
137. Angelopoulos, A., et al., *Phys. Lett.* B178: 441 (1986)
138. Tauscher, L., In *Antiproton 86*, ed. S. Charalambous et al. Singapore: World Scientific (1987), p. 247
139. Angelopoulos, A., et al., *Phys. Lett.* B159: 210 (1985)
140. Ahmad, S., et al., *Phys. Lett.* B152: 135 (1985)
141. Dover, C. B., Gutsche, T., Faessler, A., *Phys. Rev.* C43: 379 (1991)
142. Tanimori, T., et al., *Phys. Rev.* D41: 744 (1990)
143. Tanimori, T., et al., *Phys. Rev. Lett.* 55: 1835 (1985)
144. Eisenhandler, E., et al., *Nucl. Phys.* B96: 109 (1975)
145. Birsa, R., et al., *Nucl. Phys. B. Proc. Suppl.* 8: 141 (1989)
146. Kochowski, C., et al., See Ref. 40, p. 173
147. Mull, V., et al., Jülich preprint KFA-IKP(TH)-1991-4 (1991); *Phys. Rev. C.* In press (1991)
148. Moussallam, B., In *Physics with Antiprotons at LEAR in the ACOL Era*, ed. U. Gastaldi et al. Singapore: Ed. Frontières (1985), p. 203
149. Mundigl, S., Vicente Vacas, M., Weise, W., *Nucl. Phys.* A523: 499 (1991)
150. Bardin, G., et al., *Phys. Lett.* B192: 471 (1987)
151. Maruyama, M., Ueda, T., *Prog. Theor. Phys.* 78: 841 (1987)
152. Dalpiaz, P., et al., See Ref. 40, p. 346
153. Dalkarov, O. D., Protasov, K. V., *Nucl. Phys.* A504: 845 (1989)
154. Chew, G. F., In *Proc. 14th Conf. on Physics*, Univ. Brussel. London: Wiley (1968), pp. 72–74
155. Dosch, H. G., Gromes, D., *Phys. Rev.* D33: 1378 (1986); *Z. Phys.* C34: 139 (1987)
156. Oades, G. C., et al., *Nucl. Phys.* A464: 538 (1987)
157. Martin, B. R., Morgan, D., See Ref. 98, 2: 101
158. Furui, S., et al., *Nucl. Phys.* A516: 643 (1990)
159. Povh, B., Walcher, Th., *Comments Nucl. Part. Phys.* 16: 85 (1986)
160. Mundigl, S., Vicente Vacas, M., Weise, W., *Z. Phys.* A338: 103 (1991)
161. Dover, C. B., Richard, J. M., *Ann. Phys. NY* 121: 47, 70 (1979)
162. Maruyama, M., et al., *Phys. Lett.* B215: 223 (1988)
163. Furui, S., *Nucl. Phys. B. Proc. Suppl.* 8: 231 (1989)
164. Klempt, E., *Phys. Lett.* B244: 122 (1990)
165. Mundigl, S., Vicente Vacas, M., Weise, W., *Nucl. Phys. B. Proc. Suppl.* 8. 228 (1989)
166. Genz, H., Martinis, M., Tatur, S., Z. *Phys.* A335: 87 (1990)
167. Hartmann, U., Klempt, E., Körner, J., *Z. Phys.* A331: 217 (1988)
168. Gutsche, T., Maruyama, M., Faessler, A., *Nucl. Phys.* A472: 643 (1987); A503: 737 (1989)
169. Kalashnikova, Yu. S., Yurov, V. P., *Phys. Lett.* B231: 341 (1989)
170. Henley, E. M., Oka, T., Vergados, J., *Phys. Lett.* B166: 274 (1986)
171. Green, A. M., Niskanen, J. A., Richard, J. M., *Phys. Lett.* B121: 101 (1983)
172. Richard, J. M., *Nucl. Phys. B. Proc. Suppl.* 8: 128 (1989)
173. Baillon, P., et al., *Nuovo Cimento* A50: 393 (1967)
174. Armenteros, R., et al., *Phys. Lett.* 9: 207 (1964)
175. Maglic, B. C., et al., *Phys. Rev. Lett.* 7: 178 (1961)
176. d'Andlau, C., et al., *Phys. Lett.* 17: 347 (1965)
177. Alde, D., et al., *Phys. Lett.* B205: 397 (1988)
178. Baumgartner, B., Grosse, H., Martin, A., *Nucl. Phys.* B254: 528 (1985); Martin, A., Stubbe, J., *Europhys. Lett.* 14: 287 (1991)
179. Høgaasen, H., Richard, J. M., *Phys. Lett.* B124: 520 (1983)
180. Weinstein, J., Isgur, N., *Phys. Rev. Lett.* 48: 659 (1982); *Phys. Rev.* D27: 588 (1983)
181. Boutemeur, M., Poulet, M., In *Hadron '89*, ed. F. Binon et al. Gif-sur-Yvette: Ed. Frontières (1989), p. 119
182. Aston, D., et al., *Nucl. Phys.* B301: 525 (1988)
183. Alde, D., et al., *Phys. Lett.* B201: 160 (1988); B198: 286 (1987)
184. Aston, D., et al., *Phys. Lett.* B201: 573 (1988)
185. Chung, S. U., et al., *Phys. Rev. Lett.* 55: 779 (1985)
186. Ando, A., et al., *Phys. Rev. Lett.* 57: 1296 (1986)
187. Dionisi, C., et al., *Nucl. Phys.* B169: 1 (1980)
188. Armstrong, T. A., et al., *Phys. Lett.* B221: 216 (1989)
189. Aihara, H., et al., *Phys. Lett.* B209: 107 (1988)
190. Bai, Z., et al., *Phys. Rev. Lett.* 65: 2507 (1990)

191. Duch, K. D., et al., *Z. Phys.* C45: 223 (1989)
192. Amsler, C., *Nucl. Phys.* A508: 501c (1990)
193. Bridges, D., et al., *Phys. Rev. Lett.* 57: 1534 (1986); 56: 211 (1986)
194. Ahmad, S., et al., In *Physics at LEAR with Low Energy Antiprotons*, Nucl. Sci. Res. Conf. Ser., ed. C. Amsler et al. Chur: Harwood Academic (1988), p. 447
195. Kolybasov, V. M., Shapiro, I. S., Sokolskikh, Y. N., *Phys. Lett.* B222: 135 (1989)
196. Riedlberger, J., et al., *Phys. Rev.* C40: 2717 (1989)
197. Angelopoulos, A., et al., *Phys. Lett.* B212: 129 (1988)
198. Gray, L., et al., *Phys. Rev. Lett.* 30: 1091 (1973)
199. Defoix, C., Espigat, P., CERN Yellow Rep. 74-18: 28 (1974)
200. Aker, E., et al., *Phys. Lett.* B260: 249 (1991)
201. Aker, E., et al., Submitted to *Nucl. Instrum. Methods* (1991)
202. Au, K. L., Morgan, D., Pennington, M. R., *Phys. Rev.* D35: 1633 (1987)
203. Longacre, R. S., et al., *Phys. Lett.* B177: 223 (1986)
204. Devons, S., et al., *Phys. Lett.* B47: 271 (1973)
205. Godfrey, S., Isgur, N., *Phys. Rev.* D32: 189 (1985)
206. Jaffe, R. L., *Phys. Rev.* D15: 267 (1977)
207. Diaz, J., et al., *Nucl. Phys.* B16: 239 (1970)
208. Daftari, I., et al., *Phys. Rev. Lett.* 58: 859 (1987)
209. Bassompierre, G., et al., *Phys. Lett.* B65: 397 (1976)
210. Guaraldo, C., *Nuovo Cimento* A102: 1137 (1989)
211. Cugnon, J., Vandermeulen, J., *Ann. Phys. Fr.* 14: 49 (1989)
212. Dover, C. B., In *1st Workshop on Intense Hadron Facilities and Antiproton Physics*, ed. T. Bressani et al. Bologna: SIF (1989), p. 55
213. Rafelski, J., *Phys. Lett.* B91: 281 (1980)
214. Dolgolenko, A., et al., See Ref. 40, p. 205
215. Cugnon, J., Deneye, P., Vandermeulen, J., *Phys. Rev.* C41: 1701 (1990); *Nucl. Phys.* A517: 533 (1990); Vandermeulen, J., Private communication
216. Kharzeev, D. E., Sapozhnikov, M. G., See Ref. 40, p. 59
217. Smith, G. A., See Ref. 104, p. 219
218. Bizzarri, R., et al., *Lett. Nuovo Cimento* 2: 431 (1969)
219. Oset, E., Hernandez, E., See Ref. 194, p. 753
220. Kondratyuk, L. A., Sapozhnikov, M. G., *Phys. Lett.* B220: 333 (1989)
221. Haeberli, W., See Ref. 194, p. 195
222. Armstrong, T., et al., *Phys. Rev.* D36: 659 (1987)
223. Mutchler, G. S., et al., *Phys. Rev.* D38: 742 (1988)
224. Dover, C. B., Richard, J. M., *Phys. Rev.* C25: 1952 (1982)
225. Dulude, R. S., et al., *Phys. Lett.* B79: 329 (1978)
226. Martin, A. D., Pennington, M. R., *Nucl. Phys.* B169: 216 (1980)
227. Hasenfratz, P., et al., *Phys. Lett.* B95: 299 (1980)
228. Brodsky, S. J., See Ref. 40, p. 15
229. Brodsky, S. J., Mueller, A. H., *Phys. Lett.* B206: 685 (1988)

Annu. Rev. Nucl. Part. Sci. 1991. 41: 269–320
Copyright © 1991 by Annual Reviews Inc. All rights reserved

SEARCHES FOR NEW MACROSCOPIC FORCES

E. G. Adelberger and B. R. Heckel

Physics Department, University of Washington, Seattle, Washington 98195

C. W. Stubbs

Center for Particle Astrophysics, University of California, Berkeley, California 94720

W. F. Rogers

Department of Physics and Astronomy, SUNY Geneseo, Geneseo, New York 14454

KEY WORDS: gravitation, equivalence principle, inverse-square law

CONTENTS

269

0163–8998/91/1201–0269$02.00

1. INTRODUCTION

The search for new interactions and their associated bosons is a central theme in modern physics. Such searches are particularly topical today, when there is a consensus that the Standard Model cannot be complete, and that new interactions and their mediating particles are required. However, because of the wide variety of theoretical ideas for overcoming the perceived difficulties of the Standard Model, there is no consensus on the expected properties of mediating particles. One normally supposes that the bosons mediating these new interactions have not been observed because they are too massive to have been produced by existing accelerators. Therefore particle searches are usually conducted at the highest energy accelerators. However, it has recently been recognized that there is another quite different and largely unexplored window on possible "new physics," namely the search for new bosons with masses so low that they would generate a macroscopic force. This review covers recent theoretical and experimental work that addresses this possibility.

There has been considerable experimental progress in the last four years. In 1986, experimental hints of apparent violations of both the inverse-square law and the universality of free fall were presented as evidence for the existence of a "fifth" force. A series of experiments undertaken specifically to test this possibility have now produced results. In Section 2 we summarize theoretical speculations about new macroscopic interactions. Experimental signatures of such interactions are discussed in Section 3. Sections 4 and 5 review the techniques and results of inverse-square law and universality of free-fall experiments, respectively. Searches for new macroscopic spin-dependent interactions are presented in Section 6. In Section 7 we summarize the body of experimental results, extract some of the implications of the data, and discuss prospects for future improvements. Our review covers the literature through December 1990.

2. THEORETICAL SPECULATIONS

2.1 *General Considerations*

Our discussion concentrates on Yukawa interactions mediated by scalar or vector bosons. These produce a potential between two unpolarized point test bodies of the form

$$V_{12}(r) = \mp \frac{g_5^2}{4\pi}[(q_5)_1(q_5)_2]\frac{\exp(-r/\lambda)}{r}, \qquad 1.$$

where the $-$ and $+$ signs refer to scalar or vector interactions respectively, g_5 is the coupling constant, q_5 is the dimensionless scalar or vector test body charge, and $\lambda = \hbar/(m_b c)$ with m_b the mass of the mediating boson. Note that a boson with mass $m_b c^2 = 1$ μeV would generate a Yukawa force with a range of ≈ 20 cm. Of course, $g_5^2/(4\pi)$ must be very small, or the Yukawa force would not be feeble enough to have escaped detection. Thus, feeble macroscopic forces would differ fundamentally from the known vector interactions (such as the electroweak) where feebleness at low energies is due not to a small coupling constant but rather to the large mass of the mediating boson that enters into the propagator. Were such a vector interaction to be confirmed it would be the first evidence for a new hierarchy of gauge coupling constants. Were such a scalar interaction to be discovered it would be the first confirmed example of a fundamental scalar force.

One could also imagine macroscopic forces generated by the exchange of pseudoscalar or pseudovector particles. Such forces are necessarily spin dependent and do not produce a net force between unpolarized test bodies. For example, the exchange of a pseudoscalar particle between two fermions generates a potential with the asymptotic form (1)

$$V_{12}(\boldsymbol{\sigma}_1, \boldsymbol{\sigma}_2, \mathbf{r}) = \frac{\hbar^2}{m_1 m_2 c^2}\frac{(g_P)_1(g_P)_2}{16\pi}$$

$$\times \left[(\boldsymbol{\sigma}_1 \cdot \hat{r})(\boldsymbol{\sigma}_2 \cdot \hat{r})\left(\frac{1}{\lambda^2} + \frac{3}{\lambda r} + \frac{3}{r^2}\right) - (\boldsymbol{\sigma}_1 \cdot \boldsymbol{\sigma}_2)\left(\frac{1}{\lambda r} + \frac{1}{r^2}\right) - \right]\frac{\exp(-r/\lambda)}{r}, \qquad 2.$$

where the g_P's are pseudoscalar coupling constants, the σ's are Pauli matrices describing the spins, and the m's are masses of the two interacting fermions.

Moody & Wilczek proposed a third class of macroscopic forces ("monopole-dipole" interactions) that could be generated by the exchange, for example, of a particle that has a *CP*-violating pseudoscalar-scalar admixture (1). Such an interaction would violate time-reversal and parity invariance and have the form

$$V_{12}(\boldsymbol{\sigma}_1, \mathbf{r}) = \frac{\hbar}{m_1 c}\frac{(g_P)_1(g_S)_2}{8\pi}\left[(\boldsymbol{\sigma}_1 \cdot \hat{r})\left(\frac{1}{\lambda} + \frac{1}{r}\right)\right]\frac{\exp(-r/\lambda)}{r}, \qquad 3.$$

where g_S and g_P are scalar and pseudoscalar coupling constants. Note that the existence of an interaction of this form necessarily implies the existence (at some level) of the interactions given in Equations 1 and 2.

2.2 *Models*

Theoretical work featuring new Yukawa interactions generally falls into one of three categories:

1. phenomenological attempts to account for discrepancies in the experimental results,
2. demonstrations that a feeble macroscopic interaction is in accord with certain extensions of the Standard Model, or
3. examination of astrophysical or cosmological consequences of such an interaction.

The phenomenological models were motivated by attempts to fit the apparently discrepant early experimental results by adjusting the coupling strengths, ranges, and charges of one or more forces. While much was learned from this exercise, we argue below that the existing data do not provide convincing evidence for any anomalous effects. Under these circumstances the purely ad hoc phenomenological models are mainly of historical interest and are not reviewed here. The reader is referred to specialized conferences for further information (2).

It is now recognized that a wide variety of theoretical scenarios naturally accommodate a new long-range interaction. The connection between local gauge transformations, conservation laws, and possible new macroscopic interactions was explored in 1955 by Lee & Yang (3). They pointed out that the observed conservation of baryon number might arise from a local gauge invariance, and appealed to tests of the weak equivalence principle to test for a long-range field associated with this symmetry. They concluded that the null result reported by von Eötvös et al (4) indicated that any infinite-range force arising from a vector boson coupled to baryon number had to be extremely weak, less than 10^{-5} of gravity. This connection between apparent modifications to Newtonian gravity and new particle physics remains valid today.

Fujii (5, 6) was among the first to claim that a breaking of scale invariance could bring about an intermediate-range interaction, and predicted a large effect that was soon ruled out experimentally. Recent work by Peccei, Solà & Wetterich has carried this idea much further, showing how the spontaneous breaking of dilatation symmetry could give rise to a scalar "cosmon" (pseudo-dilaton) that could mediate an intermediate-range interaction (7). If the cosmon's couplings to ordinary matter are suppressed by a mass scale M and if the cosmon originates from unification with gravity, then it is natural to take $M = M_{\text{Planck}}$, and its couplings are then of gravitational strength. Furthermore, the cosmon can acquire a mass as a result of anomalies in the broken symmetry. Peccei, Solà & Wetterich estimated a cosmon mass of order $\Lambda_{QCD}^2/M_{Planck}$, corresponding to an interaction range of roughly 10 km. The cosmon couples to a scalar charge that is predominantly proportional to mass, with small corrections proportional to baryon number B and $N - Z$. A particularly interesting feature of their model was the claim that cosmons would dynamically drive the cosmological constant to zero, although this has been questioned (8).

Fayet has shown that supersymmetric theories can accommodate an intermediate-range interaction (9). He treated the general case in which the force is mediated by a gauge boson associated with the spontaneous breaking of an extra U(1), potentially of supersymmetric origin. This particle could interact (via both vector and axial-vector couplings) with a charge that is (for ordinary neutral matter) a linear combination of baryon and lepton number. As there is only one small parameter in the theory, the strength and range of the new interaction are related by $g_5^2 \propto 1/\lambda^2$.

It has been argued that attempts to build quantum theories of gravity naturally lead to additional forces, mediated by spin-0 and spin-1 partners of the conventional spin-2 graviton (10–12). If these partners have sufficiently low masses, the forces they generate will have ranges long enough to produce interesting macroscopic effects. The exchange of even-spin bosons (such as the ordinary graviton or its proposed spin-0 partner) between unpolarized particles of the same kind generates attractive forces, while the exchange of odd-spin bosons (such as the photon or the proposed spin-1 graviton) leads to forces that are repulsive. Thus in the everyday world consisting of ordinary matter, the attractive gravi-scalar and repulsive gravi-vector interactions could potentially cancel and may therefore have escaped detection. An imperfect cancellation could appear as a small modification to Newtonian gravitation. For antimatter the situation would be strikingly different. As the particle-antiparticle forces generated by scalar and vector interactions are both attractive, an antiparticle in the gravitational field of the earth would experience gravi-scalar and gravi-vector interactions of the same sign, and thus could fall with an acceleration greater than g.

Higher dimensional gravitational theories can also account for feeble intermediate-range interactions. A five-dimensional toy model explored by Bars & Visser results in both scalar and vector interactions, each with ranges of about 200 m (13). The authors argue that a variant of their approach should apply to more realistic 10-dimensional theories as well.

Four-dimensional string theories almost inevitably predict dilatons that are scalar partners of the graviton and are expected to couple to matter with roughly gravitational strength. Strong loop corrections cause these couplings to violate the weak equivalence principle (14) so that current experiments set stringent limits on the allowable mass of the dilaton. In addition, moduli fields are scalars that arise in the matter sector of superstring theories, and if the supersymmetry-breaking mechanism does not endow these with a mass greater than the μeV scale, then the current generation of experiments can rule out certain candidate string vacuua (15). Although superstring theories readily accommodate long-range *scalar* interactions, the unambiguous detection of a long-range *vector* interaction with a strength many orders of magnitude below that of the photon would deal a severe blow to attempts at unified theories that postulate a single gauge coupling. This led D. Gross to remark that "a fifth force that is attractive, would be very attractive; whereas a fifth force that is repulsive is very repulsive" (16).

Anselm & Uraltsev (17) and Chang, Mohapatra & Nussinov (18) discussed long-range interactions that would be generated by the exchange of Goldstone bosons arising from the breaking of a global symmetry. The γ_5 couplings of these pseudoscalar particles would produce a spin-spin interaction of the form given in Equation 2. Chang et al noted that the QCD θ anomaly would induce scalar couplings of their Goldstone bosons, and that these would lead to spin-independent interactions of the form given in Equation 1, as well as to a time-reversal-violating interaction of the form given in Equation 3.

Hill & Ross (19) extended these ideas, investigating how both broken global family symmetries and supersymmetry breaking could keep the mass of the resulting pseudo-Golstone boson—termed a "schizon" because of its *CP*-violating pseudoscalar-scalar character—small. They show in some detail how schizons could generate measurable effects in gravitational redshift experiments, and in free-fall and inverse-square law tests. Schizons could also play an important cosmological role, as they may allow for phase transitions to occur after matter-radiation decoupling (20).

The theoretical speculations summarized above provide one of the rationales to search for new macroscopic forces. Another, possibly more basic, motivation is simply the desire to explore an unknown regime. Clearly such searches address a wide range of physics beyond the Standard Model, and complement the ongoing (and so far unsuccessful) accelerator-based efforts to determine the limits of the validity of the Standard Model.

3. EXPERIMENTAL SIGNATURES

Any experimental search for extremely feeble new forces confronts a severe problem. How does one avoid false detection due to subtle backgrounds from the known and very much stronger interactions? Naturally, one must take great care to ensure that the test bodies do not "feel" forces from small magnetic impurities, thermal gradients, etc. In principle almost all such perturbing influences can be eliminated by careful shielding, although it may be quite difficult to do so in practice. Gravity, however, cannot be shielded so one must distinguish the feeble macroscopic interactions of scalar or vector bosons from the unavoidable gravitational background. Suppose a point test body, t, is placed in the field of a massive source, s. The interaction potential given in Equation 1 will cause the test body to accelerate with

$$|a| = \frac{g_5^2}{4\pi}\left(\frac{q_5}{m}\right)_t\left(\frac{q_5}{M}\right)_s \int\left(\frac{1}{r^2} + \frac{1}{\lambda r}\right)\exp\left(-r/\lambda\right)\rho(r)\,\mathrm{d}^3r, \qquad 4.$$

where m and M are the test body and source masses, and ρ is the mass density of the source. There are two ways in which this acceleration may be distinguished from a gravitational acceleration:

1. It violates the gravitational inverse-square law $(1/r^2)$ unless λ is infinite.
2. It violates the universality of free fall unless the "charge"-to-mass ratio q_5/m is identical for all materials.

These two tests are partly complementary. The first is sensitive to any interaction mediated by a boson with nonvanishing mass; while the second is sensitive to any interaction whose "charge" is not exactly proportional to mass, even if the exchanged boson is massless.

3.1 Violation of the Inverse-Square Law

Any experimental test of the $1/r^2$ law is only sensitive to Yukawa interactions with λ's within a certain range. This window is set by the scale of the probe. Suppose one compares the orbits of two different satellites about the Earth. The scale length is a distance between the radii of the two orbits. A force with a range much smaller than this scale length will have a negligible effect on the orbits, while one with a range much greater than the scale length will be indistinguishable from a $1/r^2$ interaction and will effectively renormalize gravity. Thus very different techniques are needed to survey the region of λ's between the millimeter scale and the astronomical scale. Unfortunately, inverse-square law tests have little sensitivity for interactions with λ much greater than 1 AU, as the conditions required for precise tests of orbital dynamics are not yet achievable.

3.2 Violation of the Universality of Free Fall

That any vector interaction and essentially all scalar interactions generate apparent violations of the weak equivalence principle follows from the behavior of vector and scalar charges under C, the particle-antiparticle transformation. A vector charge of a particle is the negative of the charge of its antiparticle, while the scalar charges of a particle and its antiparticle are identical. By imagining that binding energy induces fermion-anti-fermion loops, one can see that binding energy does not affect the vector charge of a composite object but it will, in general, affect the scalar charge. Because mass is affected by binding energy, the vector charge-to-mass ratio of composite objects cannot be independent of the material but must vary from one material to another. It is convenient to express the charge-to-mass ratio in dimensionless units as (q/μ) with $\mu = m/u$ where, throughout this paper, u denotes the atomic mass unit (amu).

The most general form of q_V for stable matter is

$$q_V = N_p q_p + N_n q_n + N_e q_e, \qquad\qquad 5.$$

where q_p, q_n, and q_e are the vector charges of protons, neutrons, and electrons; N_p, N_n, and N_e are the numbers of protons, neutrons, and electrons contained in the test body. For electrically neutral bodies $N_e = N_p$ so that q_V generally depends on two parameters, which can be chosen to be $(q_e + q_p)$ and q_n, or alternatively the baryon and lepton numbers B and L. We often use the parameterization

$$q_V = B \cos\theta_5 + L \sin\theta_5, \qquad\qquad 6.$$

where θ_5 is an arbitrary mixing angle.

The variation in q_V/μ from one material to another may easily be computed once the form of the vector charge is specified, because all of the "real physics" is contained in the mass, which can readily be measured. The amount of this variation depends upon the choice of the vector charge, q_V. If $q_V = B$, where B is baryon number, then the variation in q_V/μ is quite small, having a maximum value of 0.9% for ^{1}H/^{56}Fe test body pairs. For practical reasons neither of these materials is suitable for a test body, so that the actual variation one can obtain is considerably smaller. Essentially any choice for q_V other than $q_V = B$ will yield a stronger composition dependence. For example, if $q_V = B - L$ (a quantity conserved even in Grand Unified Theories), q_V/μ varies by 15% between ^{27}Al and ^{197}Au.

Scalar charges of test bodies cannot be specified as easily as vector charges because scalar charges are typically not conserved. Not only is a scalar charge of a composite object affected by binding energy as shown above, it is also affected by the motion of the constituents. Although a scalar charge density is Lorentz invariant, the volume experiences a Lorentz contraction so that the scalar charge contains a factor of $1/\gamma = \sqrt{1 - v^2/c^2}$. (For vector charges the dilation of the time component of the 4-current density is compensated by the contraction of the volume.) Thus, although the scalar couplings may be simple in terms of the fundamental fields (for example, the expectation value of the number density of quarks plus antiquarks), the scalar charge of an atom reflects its complex structure and may depend in a complicated fashion on the atomic and nuclear binding energies. In fact it is very difficult to arrange a scalar interaction that does not, at some level, have a composition dependence.

Tests of the universality of free fall are inherently precise because they can be conducted as differential null experiments. If all extraneous influences are rendered negligible and the gravitational field is sufficiently uniform, any difference in the acceleration of two dissimilar test bodies would constitute evidence for new physics. Furthermore, unlike the $1/r^2$ tests, a given experiment is sensitive to a broad region of ranges λ. For example, tests that compare the acceleration of test bodies toward the Earth have constant sensitivity for ranges $\lambda \geq R_{\oplus}$, where $R_{\oplus}$ is the radius of the Earth. Therefore differential acceleration measurements do probe the region $\lambda > 1$ AU where $1/r^2$ tests yield little information.

3.3 Spin Dependence

Searches for spin-dependent interactions of the forms given in Equations 2 and 3 do not have a gravitational background as Newtonian gravity is spin independent. But they face a much more severe problem from magnetic effects because they require polarized test bodies (and, for spin-spin interactions, polarized sources as well). The sensitivities of macroscopic experiments are further reduced because, in practice, only a small fraction of the test body constituents have their spins aligned. As a result, useful constraints are also provided by atomic experiments where the polarization can be complete and the electromagnetic effects may be reliably computed.

3.4 Experimental Evidence for New Macroscopic Yukawa Interactions?

Recent experimental and theoretical activity regarding feeble macroscopic interactions was largely stimulated by Fischbach et al (21) when in 1986 they reanalyzed the classic von Eötvös equivalence principle data (4) and discovered a striking correlation between the differential acceleration of test body pairs and their $\Delta(B/\mu)$ values (see Figure 1). Fischbach et al cited this as evidence for a fifth force—a Yukawa interaction coupled to B. They argued that, if $g_5^2/(4\pi\hbar c) \approx 4 \times 10^{-41}$ and $10 \leq \lambda \leq 1000$ m, their interpretation of the von Eötvös anomaly could also account for the claim of Stacey et al, based on measurements of the acceleration of gravity g in mines; that there exists a repulsive Yukawa component of gravity (22). A large number of experiments were launched to test this conjecture, several of which (including both $1/r^2$ and equivalence principle tests) supported

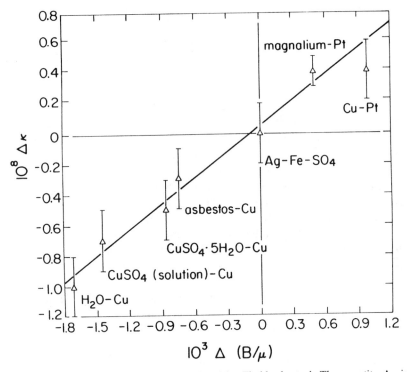

Figure 1 The von Eötvös data as reanalyzed by Fischbach et al. The quantity $\Delta\kappa$ is proportional to $\delta a_\perp$, the differential horizontal acceleration of a test body pair, via the relation $\Delta\kappa = \Delta a_\perp/g \sin \delta$ (our notation is defined in Section 5.2.2). The striking correlation between the observed horizontal acceleration of test bodies and their baryon content was taken as evidence for a "fifth force" coupled to B.

the fifth force claims. But most experiments found no evidence for any new interactions. This experimental evidence is reviewed in the next three sections.

4. TESTS OF THE GRAVITATIONAL INVERSE-SQUARE LAW

Some twenty years ago, it was noted that a lack of experimental data on the strength of gravity over distance scales larger than a meter permitted the existence of new Yukawa interactions with macroscopic ranges (5, 6, 23, 24). This observation led to a series of experiments, continuing to this day, designed to detect a new force through an apparent breakdown of the inverse-square law of gravity.

An interaction with finite range, λ, acting between two point objects in addition to the normal gravitational interaction produces a net force given by

$$
\mathbf{F} = \frac{m_1 m_2 \hat{r}}{r^2} G_N [1 + \alpha(1 + r/\lambda)e^{-r/\lambda}] = \frac{m_1 m_2 \hat{r}}{r^2} G(r), \qquad 7.
$$

where G_N is the Newtonian gravitational constant. By comparison with Equation 1, we see

$$
\alpha = \pm \left(\frac{q_5}{\mu}\right)_1 \left(\frac{q_5}{\mu}\right)_2 \frac{g_5^2}{4\pi u^2 G_N}. \qquad 8.
$$

If the charge-to-mass ratios can be considered as approximately material independent, then the Yukawa term produces an effective gravitational constant, $G(r)$, that varies with distance. Were it possible to measure $G(r)$ at two distances, a and b, that are very small ($a \ll \lambda$) and very large ($b \gg \lambda$) respectively, then a difference between $G(a) \approx G(0) = (1+\alpha)G_N$ and $G(b) \approx G(\infty) = G_N$ would provide unambiguous evidence for a finite-range force. For $\lambda \geq 1$ m, laboratory Cavendish experiments measure $G(0)$ with good precision (25). However, uncertainties in the masses or mass distributions in Equation 7 along with the inherent weakness of the interaction make it essentially impossible to measure $G(\infty)$ with enough accuracy to detect a value of $|\alpha| \ll 1$. It has been argued that considerations of stellar structure (26–29) and the density of the Earth's core (22) allow a difference between $G(0)$ and $G(\infty)$ as large as 10% and 15% respectively.

Because an absolute measurement of $G(\infty)$ is difficult, tests of the gravitational inverse-square law generally vary the distance between a given pair of test bodies and compare the measured $\mathbf{F}(r)$ in Equation 7 with that expected from Newtonian gravity alone. Different experimental techniques are required for different scales of λ because any single experiment can only vary r over a limited range. We discuss below three types of experiments that are sensitive to different ranges of λ. Several other discussions of tests of the gravitational inverse-square law may be found in the literature (22, 30–32). Although it is customary to quote experimental limits on α vs λ, it is really the product αG_N that is constrained by the data. This distinction is only important as $\lambda \to \infty$.

4.1 Laboratory Tests

4.1.1 CAVENDISH EXPERIMENTS
Variants of the torsion-balance method used by Michell & Cavendish (33) in 1798 are commonly referred to as

Cavendish experiments. Figure 2 shows the basic principle of the Cavendish balance and some of the geometries that have been employed. The need to isolate the test body, m, from external disturbances and the rapid decrease of the gravitational force with increasing separation have limited Cavendish balance gravitational measurements to mass separations of a few millimeters up to 10 meters.

A Cavendish balance measurement of G demands careful calibration of the response to external torques and precise metrology of the masses and mass separations. For tests of the $1/r^2$ law, however, only the *variation* of G with r is important, and the need for absolute measurements can be reduced by clever experimental design. For example, if a single attracting source, M, is moved over a known distance, a test of the $1/r^2$ law will be insensitive to uncertainties in the mass of M. By using a source in the shape of a ring (34) or a cylinder with the appropriate aspect ratio (35), the dependence upon the positional uncertainty of the source mass can be

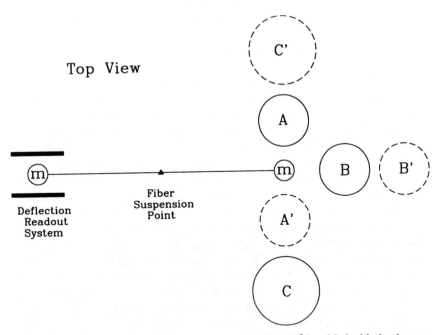

Figure 2 Basic principles of Cavendish balance tests of the $1/r^2$ law. Method I: the change in the equilibrium position of the pendulum is measured as a source mass is moved from A to A'. Method II: the change in the torsional oscillation period is measured as the source mass is moved from position B to B'. Method III: two opposing sources of different mass are positioned so that a $1/r^2$ interaction would produce no net force on m. One searches for a deflection of m as the two source masses are moved from A and C to A' and C'.

reduced. Similarly, if two opposing sources are employed (with different masses and distances from the test body) such that their net gravitational force on the test body vanishes (a null experiment), the dependence of the result on the calibration of the balance is reduced. Combinations of all of these techniques have been used in experiments to test the $1/r^2$ law.

In 1974, Long noted that the value of G obtained from Cavendish experiments with different mass separations allowed (and was even best fitted by) a substantial deviation from the $1/r^2$ law (36). A subsequent measurement by Long compared the force between a spherical test body and two rings, 29.9 and 4.48 cm from the test body, and found that $G(29.9$ cm) exceeded $G(4.48$ cm) by $0.37 \pm 0.07\%$ (34). This result motivated a number of new experiments to test the $1/r^2$ law, the results of which were consistent with Newtonian gravity. Figure 3 shows the limits on α vs λ obtained from Cavendish experiments.

Several interesting experimental geometries and techniques have been used in the Cavendish tests of the $1/r^2$ law. For a $1/r^2$ force law, a test body inside a spherical shell or a cylindrical shell of infinite length feels no net force from the shell as the shell is moved, while a changing force would be felt for a Yukawa law potential. The Irvine group suspended a test body inside a long, hollow, vertical cylinder and observed no force on the test body as the cylinder was translated horizontally (37, 38). They obtained an upper limit on $|\alpha|$ of $\approx 10^{-4}$ over a range from 2 to 5 cm. Liu et al reported a less precise result from a similar experiment with a test body suspended inside a hollow horizontal cylinder (39). The results of Long and of the Irvine group disagree unless, as noted by Long (40), the deviation from the $1/r^2$ law comes from a gravitational vacuum polarization effect [also discussed by Gibbons & Whiting (41)] that would be suppressed in experiments in which the source produces a uniform gravitational field.

Experiments in which the gravitational field from the source was not uniform were carried out by several groups. Panov & Frontov (42) placed spherical source masses 0.4, 3, and 10 m from the test mass, as illustrated by Method I in Figure 2. They achieved modest sensitivity out to a range of 10 m. Chen et al (35) and Hoskins et al (38) achieved the highest sensitivity (see Figure 3) by using opposing cylindrical sources separated from the test body by 5 and 9 cm, 12 and 19.5 cm, or 5 and 105 cm, as illustrated by Method III in Figure 2. Their results were inconsistent with those of Long. Milyukov (43) achieved a similar sensitivity by measuring the resonant frequency of the torsion balance as a cylindrical source was moved between 11 and 21 cm from the test body, as illustrated in Method II of Figure 2. Mitrofanov & Ponomareva (44) were able to test the $1/r^2$ law at a shorter length scale, 3.8 to 6.5 mm, by a dynamic method. A 0.7-g

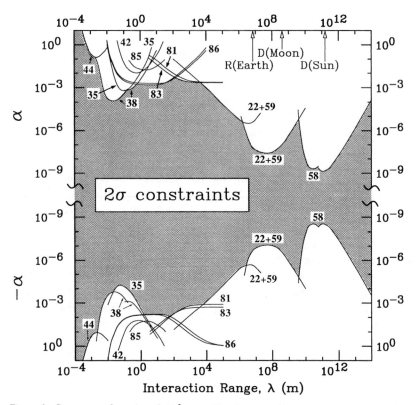

Figure 3 Summary of results of $1/r^2$ tests. Numbers on the constraint curves refer to references in the text. For example, Reference 58 establishes with 95% confidence that for $\lambda \sim 10^{11}$ m, α lies within the region from $\sim -3 \times 10^{-9}$ to $\sim 1 \times 10^{-9}$. Only the most precise results are shown in this plot.

sphere was oscillated toward and away from the test body at the resonant frequency of the torsion balance; the resonant enhancement of the torsion oscillation was used to amplify the weak signal from the small source.

Dynamic tests of the $1/r^2$ law at laboratory ranges have also been developed by Hirakawa and his colleagues in Tokyo (45–48), and by Paik et al (49). In the Tokyo experiments, a mechanical oscillation in an antenna is resonantly driven by a rotating quadrupolar source mass. By moving the source rotor relative to the antenna and measuring the resulting change in the antenna's oscillation amplitude, the experimenters placed an upper limit on $|\alpha|$ of $\approx 10^{-2}$ over a range from 7 mm to 10 m. Paik et al used a superconducting gravity gradiometer to measure the Laplacian of the gravitational potential of a 1600-kg lead pendulum at an average distance of 2.3 m from the gradiometer. They found that $\alpha = 0.24 \pm 0.36$ at a range of 1 m (49). Additional discussions of laboratory length scale tests of the $1/r^2$ law can be found in the literature (30, 31, 35, 50).

It has been argued (51, 52) that measurements of the Casimir force (Van der Waals interaction) between a small plate and a plano-convex lens (53, 54) separated by less than 10^{-6} m may be used to test the $1/r^2$ law at even shorter length scales, i.e. $\lambda < 1$ mm. In these experiments, the change in the force between the plate and lens as their separation was varied from submicrometer distances out to 1 μm was found to agree with the $1/r^2$ Casimir force law expected for this geometry (55). This agreement was used to deduce that a Yukawa force could only contribute at a level of 10^{-7} dyne, $\approx 10\%$ of the Casimir force at 1 μm (51, 52). However, these limits on α are overly optimistic. For $\lambda \gg 10^{-6}$ m, a Yukawa force, the normal gravitational force, and spurious electrostatic forces from surface charges all vary more slowly with distance than the Casimir force and are difficult to disentangle. An absolute force measurement at these distances and a thorough understanding of surface effects are required to constrain α. For $\lambda \approx 10^{-6}$ m, where a Yukawa force could alter the slope of the measured Casimir force versus distance in these experiments, only weak limits ($|\alpha|$ of order 10^{+9}) may be deduced. Tests of the $1/r^2$ law of Newtonian gravity for ranges less than 10^{-3} m remain a challenge for experimenters.

4.2 Astronomical Tests

Observations of the orbits of natural and artificial satellites about the Earth and of the planets around the Sun set strong limits on possible deviations from the $1/r^2$ law for ranges between 10^3 and 10^{14} m. The most stringent astronomical constraints are obtained either (a) by observing a single planet and seeking any anomalous precession of its orbit (after general relativistic corrections), or (b) by comparing the behavior of objects orbiting at different distances around a common attractor.

The astronomical limits deduced by Mikkelsen & Newman (26) were reassessed by De Rújula (56) and Stacey et al (22) after Fischbach et al proposed a fifth force (21). Talmadge et al recently improved these limits by including new solar system data (57, 58). There is no evidence for a new interaction in any of the astronomical data.

The anomalous perihelion precession per orbit, $\delta\phi$, due to the influence of a Yukawa potential of the form given in Equation 7 is, for $|\alpha| \ll 1$ (56, 57),

$$\delta\phi = \pi\alpha\left(\frac{a}{\lambda}\right)^2 e^{-a/\lambda}[1 + O(\alpha, \varepsilon^2)], \qquad 9.$$

where a is the mean value of the semimajor axis of the orbit, and ε is the eccentricity of the orbit. The discrepancies between the measured and calculated values of $\delta\phi$ for Mercury and Mars are -0.8 ± 2.1 and -1.3 ± 1.8 nrad per orbit, respectively (58). The resulting limits on α vs λ are shown in Figure 3.

The second technique uses independent measurements of the period, T, and the semimajor axis, a, of a planet's orbit to determine an effective value of $GM_\odot$ (we consistently use the subscripts $\odot$ and $\oplus$ to refer to the Sun and the Earth respectively) via Kepler's third law:

$$a^3 = G(a)M_\odot \left(\frac{T}{2\pi}\right)^2, \qquad\qquad 10.$$

where $G(a)$ is the gravitational constant at a mean distance a from the Sun. A variation in $G(a)$ for different planets could be attributed to a new finite-range interaction. Accurate period data are known for all of the planets. Radar ranging and data from the *Mercury, Mariner, Viking,* and *Voyager* spacecraft provide measurements of the semimajor axes of Mercury, Venus, Mars, and Jupiter. Defining $G(a) = G(a_\oplus)(1+\epsilon)$, the values of ϵ from the orbits of Mercury, Venus, Mars, and Jupiter are found to be, in units of 10^{-10}, 13 ± 17, -18 ± 12, -0.3 ± 0.9, and 67 ± 133, respectively (57). The corresponding limits on α vs λ are slightly less stringent than those obtained from the perihelion precessions.

The values of $GM_\oplus$ inferred from the orbits of the Moon and artificial satellites, and from measurements at the surface of the Earth, constrain possible departures from the $1/r^2$ law at smaller length scales than the solar-based analyses (22, 41, 59). The most useful artificial satellite is the *LAGEOS* orbiter with an average elevation of 5900 km. The value of $GM_\oplus$ obtained from the *LAGEOS* orbital parameters (60) agrees with the determination from lunar ranging (61) at the level of 2.5 parts in 10^8. Furthermore, the average value of g at the Earth's surface inferred from the *LAGEOS* data agrees at the parts per million level with the actual average of surface g measurements (22, 59). The resulting constraints on α vs λ are shown in Figure 3. Similar analyses of GM using spacecraft flybys of Mercury and Venus and lunar surface gravity data set less stringent limits on deviations from the $1/r^2$ law (26).

We note that the inverse-square law is not experimentally well established on the galactic scale. In fact the "missing mass" problem (62) has been considered by some (63, 64) as evidence for a breakdown of Newtonian gravity on the cosmic scale, as an alternative to the dark matter hypothesis. However, these models for a departure from the $1/r^2$ law are inconsistent with laboratory tests of the universality of free fall (65).

4.3   Geophysical Tests

Geophysical tests of the inverse-square law typically use a gravimeter (essentially a mass on a spring) to measure at different locations the attraction, $g_z(x, y, z)$, between the gravimeter proof-mass and the Earth. Comparing these measurements to the prediction of Newtonian gravity provides a test for new forces. Such measurements, made over a vertical span of several hundred meters (up a tower or down a borehole), probe Yukawa forces with ranges of $10 < \lambda < 1000$ m, where neither the Cavendish nor astronomical tests of the $1/r^2$ law are very sensitive.

Newtonian gravity predicts that at a depth z below the surface of a nonrotating, spherical, homogenous Earth

$$g(z) = \gamma z - 4\pi G_N \rho z, \qquad\qquad 11.$$

where γ is the free air gradient (which gives the expected variation if all the Earth's mass were at its center), ρ is the Earth's density, and the second term accounts for the shell of mass above the gravimeter. The first term in Equation 11 is slightly larger than the second, so that g actually increases as one penetrates the Earth's surface. For the real Earth, corrections must be made for tidal effects, the Earth's rotation, the ellipsoidal deformation of the Earth, local topography, and density variations within the Earth (66). The latter two corrections prove to be most troublesome. At points a distance z above the surface, the same considerations apply, except the second term in Equation 11 is absent (if the atmosphere is ignored).

A Yukawa force, as given in Equation 7, will produce a discrepancy $\Delta g(z) = g(z)_{\text{meas.}} - g(z)_{\text{Newt.}}$ between the measured and predicted results (assuming the Newtonian prediction is reliable). For $\lambda \ll R_\oplus$, the acceleration in a borehole satisfies

$$\Delta[g(z) - g(0)] = \frac{2\pi G_N \rho \alpha}{1 + \alpha}[2z - \lambda(1 - e^{-z/\lambda})], \qquad\qquad 12.$$

where here ρ is the average local density (22, 66). The term linear in z in Equation 12 reflects the influence of the Yukawa force on the second term in Equation 11 and makes it possible for borehole measurements to set a limit on α alone. For tower experiments, Equation 12 still applies but without the term linear in z; therefore tower measurements are mainly sensitive to the product $\alpha\lambda$. Any curvature in $\Delta[g(z) - g(0)]$ can be used to constrain λ.

La Coste Romberg model D or G gravimeters with tungsten proof-masses are commonly used for the $g(z)$ measurements and provide a sensitivity of ≈ 1 μGal (1 Gal = 1 cm/s^2). Marson & Faller give a detailed description of the operation and performance of these relative gravimeters that must be calibrated periodically at reference stations where absolute gravity measurements have been made (67). The gravimeter output typically drifts in time, and a complete sequence of measurements must contain enough redundancy to allow a subtraction of the monotonic drift component. It is the consensus of experimenters using these devices that the readings are reliable and are not the limiting factor in tests of the $1/r^2$ law.

The most serious limitation in geophysical tests of the $1/r^2$ law is the need for an accurate understanding of the Earth's gravity field at the experimental site. Variations in the subsurface and surface structure affect both terms in Equation 11 (69, 72–74). Auxillary measurements of ρ and surface gravity are made to ensure that the Newtonian gravity model incorporates higher order contributions. Even so, the task of predicting $g(z)$ values within the Earth ("downward continuation") is, in principle, ill defined because the surface gravity data alone do not uniquely determine the distribution of masses below the lowest depth sampled (75). Attempts to solve Poisson's equation within the Earth to match the surface gravity data (68, 69) are also limited by noise in the surface data, the effects of which are amplified going down. Extrapolating $g(z)$ above the surface ("upward continuation") is less problematic. As the contribution from the atmosphere is negligible, a complete map of surface gravity data would determine the gravity field in the source-free region above the surface. However, problems associated with a discrete set of noisy data points remain (70, 71). In all $g(z)$ measurements, the burden is on the experimenter to argue or prove that the Newtonian gravity model used has the required accuracy. There is at present no evidence from geophysical experiments for a violation of the $1/r^2$ law.

4.3.1 BOREHOLE MEASUREMENTS   Airy was the first to obtain G from measurements of $g(z)$ in boreholes (76). Stacey and collaborators exploited the Airy method in a series of measurements in mineshafts at Mt. Isa and Hilton in Queensland, Australia (22, 66). The Hilton data were considered more reliable because the 1000 m deep shaft was several hundred meters away from the nearest ore deposits. Density data were obtained for 2300 boreholes drilled within 1 km of the Hilton site to establish a local three-dimensional density structure. The Hilton and Mt. Isa data yielded a value of G 0.7% larger than the laboratory value for a range less than ≈ 1 km, with a significance of 10 standard deviations. (The larger value of G was evidence for a repulsive Yukawa interaction because a force with $\lambda \ll 1000$ m would average to zero in the borehole, but would affect the laboratory measurements.) Hsui analyzed measurements from a borehole in Michigan and found a deviation from predicted gravity in agreement with the

Queensland result, but with experimental uncertainties too large to be conclusive (77). Following the recent observation by Bartlett & Tew (73) that effects from the local terrain not included in the gravity model could explain the anomalous gravity signal at Hilton, the Queensland group reported a revised analysis that yielded no evidence for a violation of the $1/r^2$ law at Hilton (78).

To reduce the uncertainties about the homogeneity of the local mass through which a borehole passes, Ander et al lowered a gravimeter down a 2 km deep borehole in the Greenland icecap (79). Surface gravity measurements and a radar echo survey of the underlying ice-rock interface helped constrain the Newtonian gravity prediction. Although the measured $g(z)$ deviated from the Newtonian prediction, it was found that an appropriate distribution of density contrasts in the underlying rock could explain the data. The likelihood that such a geological structure exists in the rock below the Greenland icecap is an open question, but must be compared to the likelihood of a defect in Newtonian gravity.

Thomas & Vogel analyzed data from boreholes at the Nevada Test Site, comparing $g(z)$ data from a borehole at the center of a cluster of five boreholes to a Newtonian gravity model (72). The model used density measurements from all five boreholes to generate a lateral mass profile. The matter beneath the holes was assumed to occur in homogeneous ellipsoidal layers. They observed a 4% deviation from Newtonian gravity, too large to be consistent with the original result of the Queensland group, and concluded that unmodeled underlying mass anomalies accounted for their large discrepancy and must plague all borehole measurements at worrisome levels.

4.3.2 TOWER MEASUREMENTS Eckhardt and collaborators reported the first results from tower measurements of $g(z)$ (80). They selected a television tower 600 m high in Garner, North Carolina, for its mechanical stability and for its location in a well-mapped, featureless local terrain. Surface gravity measurements were made at 77 sites within 5 km of the tower, and gravity data for 1784 sites between 5 and 220 km were used to derive a Newtonian gravity model. A deviation from Newtonian gravity of -500 ± 35 μGal was observed at the top of the tower.

Bartlett & Tew (74) then pointed out that local topography and a vertical bias in the sample of surface gravity stations could explain much of the discrepancy observed by Eckhardt et al. Meanwhile Thomas et al used a 465-m tower at the Nevada Test Site to measure $g(z)$ (82). Their Newtonian gravity model was generated using surface gravity measurements made at 281 locations within 2.6 km of the tower and gravity survey data for the region within 300 km of the tower. Thomas et al found a maximum difference between their measured and predicted $g(z)$ values of -60 ± 95 μGal at the top of the tower. Jekeli, Eckhardt & Romaides subsequently supplemented their original surface gravity map with additional points to reduce any biases, and included detailed topographic data (81). Their new gravity model agreed well with their tower data, which allowed them to set a limit of $|\alpha| < 10^{-3}$ for $\lambda > 100$ m, as shown in Figure 3. Most recently, Speake et al measured $g(z)$ up a 300 m tower at Erie, Colorado, which stands on nearly flat terrain (83). Local surface gravity measurements at 265 stations within 8 km of the tower were supplemented by gravity library data covering a $4° \times 4°$ area. No deviation from Newtonian gravity was found at the level of 10 μGal. The constraint on α vs λ from this experiment is shown in Figure 3.

4.3.3   MOVING SOURCE MEASUREMENTS   By leaving a gravimeter fixed and measuring the effect of a large moving mass, one removes the need for a detailed characterization of the Earth's gravity field and eliminates the major uncertainty in $g(z)$ experiments (although one does need to understand the elasticity of the supporting terrain). Yu et al used an exploration gravimeter to map the change in local field as a large oil tank was emptied and filled (84). This early experiment measured $G(a)$ to an accuracy of 10% for a around 10 m. Moore et al operated a beam balance atop a hydroelectric reservoir (85). The balance compared the weights of two 10-kg masses, one above the water level and the other inside an evacuated cylinder at various depths below the surface; the mean effective distance between the gravitating masses was 22 m. Results were taken as the water level changed by up to 10 m in a day; the measurement uncertainty was limited by the stability of the platform that held the apparatus. The experimental result was consistent with the laboratory value of G within the quoted 1σ error of 0.9%. Most recently, Muller et al measured G with gravimeters above and below the water line of an artificial lake that has an asphalt bottom to reduce water leakage (86). The daily variation in water level was between 5 and 22 m. Muller et al found values for G that agreed with the laboratory value to within 0.4% ($2-3\sigma$) at effective ranges of 39 and 68 m.

5.   TESTS OF THE UNIVERSALITY OF FREE FALL

Searches for differences in the gravitational accelerations of various test bodies have been highly developed as precise tests of the Weak Equivalence Principle (WEP), a violation of which would appear as an infinite-range modification of conventional gravity. In our context, the WEP experiments are used to search for ultrafeeble composition-dependent interactions with a range, λ, that may have any value. From Equation 1, we see that the signal in any given experiment (that is, the differential acceleration toward the source) is

$$\Delta \mathbf{a} = \tilde{\alpha} G \Delta \left(\frac{q}{\mu}\right)_{\mathrm{d}} \left(\frac{q}{\mu}\right)_{\mathrm{s}} \mathbf{I}(\lambda),$$

13.

where $\tilde{\alpha}$ is a dimensionless parameter customarily used to analyze experimental results

$$\tilde{\alpha} = \pm \frac{g_5^2}{4\pi u^2 G},$$

14.

$\Delta(q/\mu)_{\mathrm{d}}$ is the difference in charge-to-mass ratios of the two test materials in the "detector," $(q/\mu)_{\mathrm{s}}$ refers to the "source" material, and

$$\mathbf{I}(\lambda) = \mathbf{V} \int \frac{\exp(-r/\lambda)}{r} \rho(\mathbf{r}) \, \mathrm{d}^3 r$$

15.

points toward the source (ρ is the source density). Note that $\tilde{\alpha}$ differs from the α defined in Equation 8. Equations 6 and 13 show that an unbiased search for a new vector interaction requires at least two different pairs of detector materials and at least two source materials, as any single choice of source or detector materials may produce a misleading result because of a very low value of $\Delta(q/\mu)_{\mathrm{d}}$ or q_{s}. Similar arguments apply to searches for scalar interactions as well.

5.1 *Direct Free-Fall Experiments*

5.1.1 EXPERIMENTS WITH MACROSCOPIC TEST BODIES   The main advantage of testing for a composition-dependent vertical acceleration is that the vertical component of $\mathbf{I}(\lambda)$ can be reliably calculated for all $\lambda \gg R_{\mathrm{a}}$ (where R_{a} is the scale of the apparatus) and thus avoids a difficulty that arises in interpreting the complementary horizontal tests discussed below. For the artificial case of a spherical Earth of uniform density, we obtain

$$I_z(\lambda, r) = 3M_\oplus \left(\frac{\lambda}{R_\oplus}\right)^3 \left(\frac{R_\oplus}{\lambda} \cosh \frac{R_\oplus}{\lambda} - \sinh \frac{R_\oplus}{\lambda}\right) \left(\frac{1}{r^2} + \frac{1}{r\lambda}\right) e^{-r/\lambda},$$

16.

where r is the distance from the test body to the center of the sphere. More realistic layered models of the Earth may be treated with Equation 16 using superposition (87).

Vertical free-fall experiments set the most stringent limits on any violation of the WEP for λ between 20 and 1000 km. The main difficulty with this approach is that one must detect a small difference between the behavior of two objects that are each accelerating at ≈ 980 Gal. Potential sources of systematic error in the vertical free-fall tests include gravitational gradients, electromagnetic forces, artifacts from rotation of the reflectors, recoil motion imparted to the platform upon release of the test objects, the mutual attraction of the objects, misalignment between the interferometer legs and local vertical, frequency instability in the laser, and differential viscous drag from residual gases. Unless the two objects travel along exactly the same path, a spatial variation in g can impart a difference in acceleration at the level of interest. This can be overcome either by exchanging the positions of the objects used, or by arranging for them to fall along the same trajectory. Minimizing differences between the gravitational multipole moments of the two test objects suppresses effects of gradients in the local gravitational field. The optical center of the falling object must coincide with its center of mass to reduce the effect of reflector rotation to a tolerable level. False signals arising from reproducible and correlated recoil motion of the platform or from rotation of the test objects tend to be the limiting factors in these experiments.

The first modern version of the Galileo experiment was reported by Niebauer, McHugh & Faller (88). They attached corner reflectors to Cu and U test bodies and used them to define the two arms of a Michelson interferometer. The bodies fell along two parallel, but horizontally separated, evacuated paths. Drag from residual gases was minimized by small enclosures that were controlled by servomechanisms to track the falling bodies. One of the bodies was intentionally released 25 ms before the other. This initial velocity difference introduced a constant fringe shift rate of roughly 800 Hz. Any differential vertical acceleration would alter this beat frequency. The device was automated so that drops could be performed at 15 s intervals, without breaking vacuum. Niebauer et al made measurements with the test bodies exchanged between the two descent paths, and alternated which of the bodies was released first. This allowed the effects of horizontal and vertical gravity gradients to be distinguished from any composition-dependent acceleration difference, which was established to be $\Delta a(\text{Cu} - \text{U}) = 0.13 \pm 0.50$ μGal.

Kuroda & Mio introduced a refinement in their free-fall comparisons of Be, C, Al, and Cu (89, 90). The centers of mass of their test bodies coincided and the two nested objects fell along the same trajectory, as shown in Figure 4. This eliminated problems with gravity gradients, but the close proximity of the test objects required that their mutual gravitational interaction be accounted for, and that any electrostatic interaction between the two be minimized. Their work set limits of $\Delta a(\text{Al} - \text{Be}) = 0.43 \pm 1.23$, $\Delta a(\text{Al} - \text{C}) = -0.18 \pm 1.38$, and $\Delta a(\text{Al} - \text{Cu}) = -0.13 \pm 0.78$ μGal, where

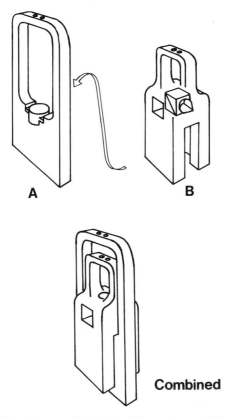

Figure 4 Nested falling objects used in the free-fall experiments of Kuroda & Mio. A laser beam entered from above through one of the small holes in test bodies A and B. The beam was split inside B, and reflected from corner reflectors in A and B. The recombined beam exited through the other holes. The interference beats revealed any relative motion of A and B.

the errors are statistical. Systematic errors were estimated to be well below 1 μGal.

5.1.2 EXPERIMENTS WITH ELEMENTARY PARTICLES   A direct comparison of the free-fall properties of matter and antimatter requires measurements on elementary particles. Such experiments are technically demanding, particularly if they involve electrically charged particles. For example, a single proton 5 m above an electron produces an electric field large enough to cancel the gravitational force on the electron. Witteborn & Fairbank, intending ultimately to measure the "gravitational" acceleration of positrons, used a time-of-flight technique to measure the acceleration of electrons and found that $a < 0.09g$ (91). Positron results, which are considerably more difficult to obtain, were not reported. Witteborn & Fairbank's result was originally attributed to the effect of gravity on the electrons in the conductors that shielded the falling electron (92). These conduction electrons "sagged" in the gravitational field until they built up an electric field $\mathbf{E} = mg/e$ that just cancelled the gravitational force on the free electrons. Subsequently, it was pointed out that the electric field induced by the gravitational compression of the lattice was $\sim 10^3$ times greater than the field from electron "sag" and had the opposite sign (93). This, of course, was in strong contradiction to Witteborn & Fairbank's data. There is still no satisfactory explanation for these results.

As problems with stray electric fields are smaller on particles with lower charge-to-mass ratios, subsequent attention has been focused on neutrons and antihydrogen, which are uncharged, and on antiprotons for which the electrical effects are ≈ 2000 times smaller. The free-fall acceleration of neutrons has been determined to be $(1.00011 \pm 0.00017)g$ by comparing the neutron scattering length measured in a high-precision total cross-section measurement to that obtained with a gravity refractometer (94). Unfortunately the inadequacy of antineutron sources precludes a particle-antiparticle comparison that capitalizes on the relatively good precision obtained in the neutron experiment. Attempts to measure the acceleration of antiprotons (95) (with an anticipated precision of $0.01g$) and neutral antihydrogen (96) are currently under consideration.

5.2   Torsion-Balance Experiments with Terrestrial and Solar Sources

The most sensitive searches for intermediate-range composition-dependent forces have used torsion balances to compare the horizontal accelerations of two test bodies. The underlying principle is shown in Figure 5. The torsion-balance experiences a torque about the vertical axis if and only if the net forces on the test bodies do not lie in the same plane.

The torque, τ, on the pendulum is proportional to the amount of "active" mass, and the angular deflection signal $\theta = \tau/\kappa$ is greatest when κ, the torsional constant of the fiber, is small. The largest mass, M, that can be supported by the fiber, scales as d^2, where d is the fiber diameter, while κ scales as d^4. Therefore the signal from an applied torque is proportional to $1/d^2$, or $1/M$.

Two methods are used to determine torques acting on the pendulum: measuring its equilibrium angle θ or its period P. Because θ and P of the pendulum are not known a priori, the torque must be monitored as the relative orientation of the pendulum and source is varied. A composition-

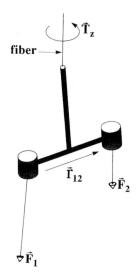

Figure 5 Principle of the torsion balance. The total force on a test body has contributions from gravity, centrifugal force, and possibly from a new macroscopic interaction. The pendulum will experience a torque T_z if and only if $\mathbf{r}_{12}$, $\mathbf{F}_1$, and $\mathbf{F}_2$ are not coplanar. Therefore, a torsion balance placed in a uniform gravitational field has $T_z = 0$ even if the test bodies have unequal masses.

dependent interaction between the pendulum and its surroundings would produce a torque that depends sinusoidally on orientation.

The θ and P techniques, which are sensitive to fluctuations in the phase and amplitude respectively of the torsional oscillation, promise comparable performance against the fundamental limit set by the Brownian motion of this oscillation (97). A thermally driven oscillator in equilibrium with an environment at temperature T has a root-mean-square (rms) amplitude of $\theta_{\mathrm{rms}} = \sqrt{kT/\kappa}$. As the oscillation is coherent over many cycles, only the fluctuations in this amplitude produce noise in the experiment. When averaging the equilibrium angle over one torsional cycle the mean square uncertainty in θ is given (97) by

$$\delta\theta^2 = 3kTb/(2\pi\kappa\omega_0 I), \qquad\qquad 17.$$

where I is the moment of inertia of the pendulum, ω_0 is its natural frequency, and b is the damping coefficient given by $Q = \omega_0 I/b$ and $1/Q$ is the fractional energy loss per radian of oscillation. The dependence of $\delta\theta^2$ on M is obtained by noting that $\kappa \propto M^2$, $\omega_0 I \propto \sqrt{\kappa I} \propto M^{3/2}$, so that $\sqrt{\delta\theta^2} \propto M^{-7/4}$. If Brownian fluctuations limit device performance, the system signal-to-noise ratio scales as $M^{3/4}$.

For a pendulum limited by thermal noise, the optimum design has a large active mass and a large moment arm, with minimum damping. Because the performance of torsion balances is typically limited by processes other than thermal noise, a practical optimized design is quite different. For example, if the noise is dominated by the torque readout system, it is advantageous to maximize the signal associated with torques applied to the pendulum; this drives the design toward small masses and thin fibers, limited by attainable machining tolerances and ease of handling.

The main issue in torsion-balance experiments is not sensitivity, but rather control of systematic errors. In general, the more symmetrical the system the less susceptible it is to sources of systematic error.

Eliminating electromagnetic couplings is relatively straightforward, although care is required to avoid magnetic impurities in the pendulum. Most contemporary devices employ multilayer magnetic shielding and Helmholtz coils to minimize magnetic fields and their gradients in the vicinity of the pendulum.

Gravitational couplings are more troublesome, particularly in experiments undertaken on sloping terrain, where gravitational gradients can be large. The gravitational torque exerted on the pendulum by its surroundings is given by

$$\tau_{\text{grav}}(\phi) = -4\pi i G \sum_{l=0}^{\infty} \frac{1}{2l+1} \sum_{m=-l}^{+l} m\bar{q}_{lm}Q_{lm}e^{-im\phi}, \qquad 18.$$

where

$$\bar{q}_{lm} = \int \rho_{\text{det}}(\mathbf{r})r^l Y_{lm}^*(\hat{\mathbf{r}})\,\mathrm{d}^3r \qquad 19.$$

characterizes the pendulum mass distribution and is evaluated in a body-fixed frame centered on the pendulum center of mass. The source moments,

$$Q_{lm} = \int \rho_{\text{source}}(\mathbf{r})r^{-(l+1)}Y_{lm}(\hat{\mathbf{r}})\,\mathrm{d}^3r, \qquad 20.$$

are evaluated in the laboratory frame. The $m = \pm 1$ components are particularly important because they have the same ϕ dependence as the torque from a non-Newtonian interaction. The leading-order gravitational $|m| = 1$ torque arises in the q_{21} order; q_{11} vanishes because the pendulum center of mass must hang directly under the support of a perfectly flexible fiber. Most experiments employ pendula with minimized $m = \pm 1$ moments and also reduce the ambient $Q_{l,m=1}$ moments with compensating mass distributions.

Temperature variations can produce systematic errors by a wide variety of mechanisms. These are typically minimized by controlling the thermal environment of the apparatus and by maximizing the symmetry of the experiment. Thermal systematic errors are difficult to assess, as it is not always possible to determine which element of the apparatus is sensitive, or whether the sensitivity is to temperature changes or to thermal gradients.

Mechanical disturbances of the suspension fiber can easily produce systematic errors. For example, a tilt of the upper fiber attachment correlated with the orientation of the pendulum can produce rotations of the torsion fiber that mimic an external torque. Seismic activity also tends to excite the angular mode of torsion pendula; although the seismic noise is peaked at frequencies considerably higher than the typical torsion period, nonlinearities in the fiber can convert vertical seismic noise into the torsional mode.

5.2.1 SOLAR SOURCE EXPERIMENTS The torsion-balance WEP experiments of Roll, Krotkov & Dicke (98) and of Braginsky & Panov (99) used the Sun as a source. The Earth's rotation smoothly and continuously reoriented the balances relative to the source, while the balances remained stationary in the laboratory frame of reference. The source strength can be approximated by computing the component of the ordinary gravitational acceleration normal to the torsion fiber. For a composition dipole oriented along the north-south axis, the solar source produces a maximum gravitational acceleration of $g_\perp = g_\odot = 0.593$ Gal.

The signature of WEP violation would be a torque with a 24-hr period. This is not a particularly convenient frequency, because mechanical $1/f$ noise is large and some sources of systematic error (e.g. thermal variations) have the same frequency as the signal. Because these two elegant experiments were thoroughly reviewed by Everitt (100), we do not discuss them in detail. Roll et al and Braginsky & Panov reported null results of $\Delta a/g_\odot(\text{Al}-\text{Au}) = (-1.3 \pm 1.5) \times 10^{-11}$ and $\Delta a/g_\odot(\text{Al}-\text{Pt}) = (0.3 \pm 0.45) \times 10^{-12}$ respectively (we quote 1σ errors). The charge-to-mass ratios of Au and Pt are expected to be so similar that these two experiments tested essentially the same physics. While both experiments probed for WEP violation at levels far more sensitive than those of von Eötvös, an intermediate-range interaction with $\lambda \ll 1$ AU would have escaped detection.

5.2.2 TERRESTRIAL SOURCE EXPERIMENTS To obtain sensitivity to interactions with $\lambda \ll 1$ AU, one needs terrestrial or laboratory sources. The strength $I_\perp(\lambda)$ of the terrestrial source can be readily calculated in two regimes: $\lambda < R_<$ and $\lambda > R_>$, where $R_<$ is roughly 10 km and $R_>$ is roughly 1000 km. These bounds are determined by geophysical considerations. For $\lambda > R_\oplus$, a Yukawa force would be aligned with the gravitational force. The Earth's rotation causes a misalignment between the torsion fiber and these forces by an angle

$$\delta = \frac{\omega^2 R_\oplus}{2g_\oplus} \sin 2\Theta, \qquad\qquad 21.$$

where Θ is the latitude of the experiment. Terrestrial source experiments are maximally sensitive to long-range interactions at a latitude of $45°$, where $\delta = 1.7$ mrad, giving a value of $g_\perp = g_\oplus \sin \delta$ roughly three times larger than for the solar source test. The situation is more complicated for $\lambda \ll R_\oplus$. For example, on the open sea $\delta = 0$ because a short-range Yukawa force will point along the suspension fiber. This condition will also occur at a level site on land because the Earth, on the average, has nearly the shape of a fluid in equilibrium under the actions of gravitational and centrifugal forces. To obtain sensitivity to short-range interactions, one must work on sloping terrain (at the base of an effectively infinite vertical cliff $\delta = 45°$). For $\lambda < R_<$, $I_\perp(\lambda)$ can be computed by integrals over the local topography.

Three groups have reported results from contemporary terrestrial source torsion-balance experiments. One of these obtained a nonzero result (101); while the other two saw no anomalous effects (102–105).

The "Eöt-Wash" experiment (106), shown in Figure 6, uses a highly symmetrical torsion pendulum in an evacuated apparatus surrounded by magnetic shielding and an isothermal jacket. The instrument sits on a rotating turntable that continuously varies the angle between the torsion balance and the source, as in the solar source tests, but with a 2-hr rather than 24-hr period. The instrument is operated at a hillside location on the University of Washington campus. The angular deflection of the torsion pendulum is monitored and any component sinusoidal in the angle of the composition dipole in the laboratory frame of reference is treated as a signal.

A primary objective in the design of this experiment was minimizing systematic errors by making the test bodies essentially indistinguishable: they are cylinders with identical external dimensions, and are coated with

---

Figure 6 The pendulum and innermost magnetic shield of the torsion balance used by the Eöt-Wash group. The instrument was operated on a hillside and compared the acceleration of different test bodies in the field of the Earth. The horizontal tray held two pairs of cylindrical test bodies made of different materials and arranged as a composition dipole. The two masses at the ends of the vertical axle minimized errors from gravity gradients by making the mass quadrupole moment of the entire pendulum vanish. A light beam was directed onto one of four identical mirrors on the pendulum. The angle of the returning beam monitored the torsional oscillation of the pendulum. The pendulum was suspended inside a thermally shielded vacuum can that rotated about the vertical axis with a period of ≈ 2 hr. A torque on the pendulum that varied sinusoidally with the can rotation period constituted the signature of a new macroscopic force.

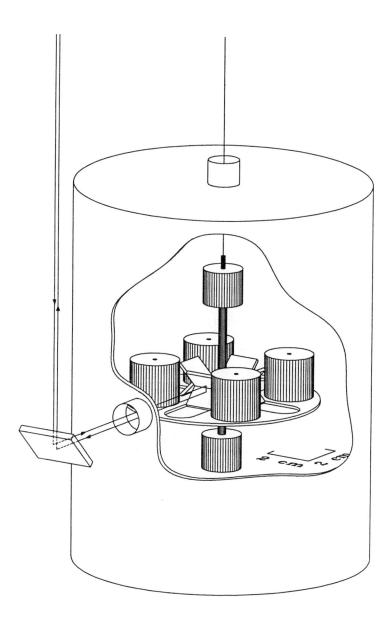

an evaporated gold layer to match their surface properties. Differences in the densities of the materials being tested were accommodated by machining the denser objects with internal voids. The entire torsion pendulum is designed to have vanishing $m = 1$ gravitational moments up to $l = 5$, apart from machining imperfections. The test bodies could be easily and reproducibly rearranged on the pendulum tray, allowing, for example, both composition dipole and quadrupole configurations. This was very useful in diagnosing and eliminating various systematic errors. Gradients in the local gravitational field were suppressed by a factor of 145 by an appropriately shaped distribution of Pb. The compensation mass was tailored to cancel the dominant ambient gradient term without introducing higher-order $m = 1$ terms at any significant level. This cancellation, together with the small gravitational moments of the pendulum, reduced gradient effects to a negligible level. Null results were obtained with two different test body pairs: $\Delta a_\perp(\text{Be} - \text{Al}) = (-2.1 \pm 2.1) \times 10^{-11}$ Gal and $\Delta a_\perp(\text{Be} - \text{Cu}) = (0.8 \pm 1.7) \times 10^{-11}$ Gal. These null results represent the most stringent upper limits to date on the existence of a new composition-dependent interaction over much of interesting region of parameter space. The performance of the device was limited by seismic activity and by small residual temperature fluctuations.

Boynton et al developed an apparatus for operation at remote sites (101). The first version of their experiment compared the accelerations of Be and Al at the base of a 330 m high cliff. They observed a positive effect that was consistent with a new Yukawa interaction at the 3.5σ level. The torsion pendulum was a ring made up of half-rings of two different materials. The torsional period was timed following successive discrete rotations of the entire apparatus that excited $\approx 30°$ of torsional amplitude, and data were taken with two orientations of the pendulum within the apparatus. A sinusoidal dependence of the torsional period on the orientation of the pendulum in the laboratory frame constituted the signal of interest. Gravitational gradients at the site used by Boynton et al are significant. An appropriate distribution of Pb suppressed these gradients by a factor of seven. Residual gravitational couplings were subtracted in the final stage of data analysis, using data taken with and without the gradient compensator in place, and with an intentionally tilted ring. The initial result indicated an apparent composition-dependent differential acceleration of $\Delta a_\perp(\text{Be} - \text{Al}) = (2.5 \pm 0.8) \times 10^{-9}$ Gal. A subsequent comparison of Cu and CH_2 at the same site showed no anomalous effects (107); the experimenters intend to repeat their Be-Al comparison with an improved apparatus in the near future.

Fitch et al (105) compared Cu and CH_2 at a site that sits atop a large tear fault with a significant density discontinuity, which enhances any signal from a Yukawa force with $\lambda \sim 500$ m. This experiment used closed loop systems to control both the torsion pendulum's angular position and the overall tilt of the apparatus. A radiofrequency capacitance bridge sensed the pendulum position and a resulting direct-current error signal was applied to the capacitor plates to provide a restoring torque. The device was operated in air, and the servo signal required to maintain a constant angular position of the pendulum was monitored as a function of the orientation of the apparatus. The null result of this experiment, $\Delta a_\perp = (3.0 \pm 4.9) \times 10^{-9}$ Gal, is particularly significant when compared to the positive effect observed in Thieberger's comparison of Cu and H_2O, discussed below. The materials used by Fitch et al differ only in substituting C for the O, and both experiments were performed using terrestrial (rock) sources.

5.3   Experiments with Moveable Sources

Instead of using local topography as the source for a composition-dependence experiment, one can use a moveable source of known dimensions and composition and extract the signal correlated with source position. There are three main advantages to such local-source experiments:

1. They can probe shorter ranges than is possible with terrestrial sources because a stationary apparatus can have very high sensitivity and the source can be very close to the detector.
2. They can determine g_5^2 independent of λ, if the source is within a distance λ of the detector.
3. They can test for a coupling to a charge (such as $N-Z$) for which terrestrial material is essentially neutral by using a source with a large neutron or proton excess (such as Pb or H_2O).

Local-source experiments have some disadvantages: the amount of source material is limited, and there is an increased danger of introducing unwanted gravitational gradient couplings as the source is necessarily close to the detector. Laboratory sources are consequently designed to have minimal Q_{lm} moments for $l > 1$. Maintaining mechanical and thermal stability while making a major perturbation in the immediate vicinity of the apparatus is another concern.

The five moving-source experiments discussed below were largely designed to test for a coupling with $q_5 \propto (N-Z)$ that at one stage was a candidate for reconciling the apparently discrepant terrestrial source results (101, 103). These experiments all produced null results.

Speake & Quinn performed an experiment with a beam balance that could compare forces with an accuracy of 10^{-11} Newton (108). The weights of 2.3-kg C, Cu, and Pb test masses were compared as 1782-kg sources of

Pb and brass were alternately placed under the balance. The experimenters looked for any change in the relative weights of the test bodies that could be attributed to an anomalous interaction between them and the sources. None was found.

Bennett conducted a torsion-balance experiment adjacent to navigational locks on the Snake River (109). The angular position of a Cu/Pb torsion pendulum was monitored as 1.7×10^8 kg of water rose and fell in the lock. No statistically significant correlations were observed between the torque and the water level. This experiment was unique in its use of a proton-rich substance as the source material.

A moving-source experiment was performed with an earlier version of the rotating Eöt-Wash apparatus described above, using a 1295-kg Pb source instead of the hillside (110). The source was placed on alternate sides of the apparatus, and the signal of interest was any change in the Fourier coefficients of the deflection that tracked the source position. No significant signal was observed. This was the first experiment with enough sensitivity to bring into question the proposed reconciliation of the composition-dependent results in terms of a coupling to $N - Z$.

Cowsik et al obtained high sensitivity by resonantly driving a torsion pendulum with an external source moved at an appropriate frequency (111, 112). This essentially allowed for a Q-fold enhancement of their sensitivity over the intrinsic signal-to-noise ratio of the readout system. This resonant suppression of the readout noise allowed the apparatus to be larger than the other torsion-balance experiments described in this review. The pendulum, consisting of two 700-g semicircular rings of Cu and Pb, was operated at 10^{-8} Torr to attain a Q for the torsional mode that was high enough to give several hundred hours of signal integration. The Pb and brass source material with a total mass of 1280 kg was configured as a composition dipole, and was designed to minimize any gravitational torques on the system. With the source and pendulum composition dipoles at right angles, the orientation of the source dipole was reversed every torsional half-period, which drove the system close to resonance. The equilibrium angle of the pendulum was measured and the rate of growth of torsional amplitude constituted the signal, which was extracted from the data by a least-squares fit to the angle time series. The experiment produced a null result. A powerful test for systematic uncertainties from positioning errors was performed by driving the system with the source and detector composition dipoles aligned; this eliminated any composition-dependent torque. Tests were also made with the source material configured so that its composition dipole moment vanished.

Nelson et al (113) conducted an experiment at the University of California, Irvine, in which a torsion balance with Pb and Cu test bodies was driven by placing a 320-kg Pb ring on alternate sides of the apparatus, as shown in Figure 7. The equilibrium angle of the system was measured optically under successive discrete rotations of the source mass. The Irvine data analysis procedure included the generation of a random offset in the data in a self-consistent fashion, and only after all systematic tests and data reduction were completed did the experimenters remove their self-imposed "blindfold" and compare their final result with the value of the offset. This procedure was designed to overcome the natural tendency to cease searching for sources of systematic errors when the experimental result agrees with the experimenter's prejudice. Nelson et al obtained a null result: $\Delta a_\perp(\mathrm{Cu-Pb}) = (1.7 \pm 1.9) \times 10^{-11}$ Gal.

5.4 *Floating Ball Experiments*

A neutrally buoyant object floating in a fluid can also be used to test for possible composition-dependent interactions. Any difference in the horizontal acceleration experienced by the float and the fluid it displaces will cause the float to drift sideways. As the forces being sought are exceedingly small, the float will soon reach terminal velocity in the fluid, presumably in accord with Stokes' Law. Potential systematic errors in floating-ball experiments in addition to the gravitational, thermal, and mechanical effects that also afflict torsion balances include

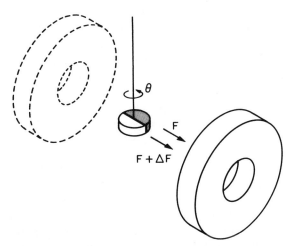

Figure 7   Rotating-source torsion balance used by Nelson et al. The experimenters searched for a change in θ as the Pb source ring was alternated between the solid and dashed positions. The shaded and unshaded halves of the pendulum were composed of Cu and Pb, respectively. An autocollimator (not shown) monitored the angular deflection of the pendulum.

1. convection cells in the fluid driving motion of the float,
2. displacement of the center of mass of the float from that of the displaced water,
3. surface tension effects if any part of the float breaks the fluid surface, and
4. possible departures from Stokes' Law.

Thieberger obtained the first result from the contemporary generation of WEP tests and reported an apparent 3σ WEP violation consistent with the initial "fifth force" hypothesis (114). His experiment was conducted at the edge of a large cliff overlooking the Hudson River. Thieberger's device contained a Cu ball floating in water. Vertical equilibrium was maintained by a vertical post on the float that extended above the waterline. The position of the float was monitored by a video camera, and the Cu was seen to drift toward the cliff edge at 4.7 mm hr^{-1}. The thermal convection problem was minimized by operating with the water at 4°C, where its density is a maximum and convection does not occur. Thieberger also took data with the apparatus tilted and with its temperature perturbed, and found no significant difference in the motion of the float.

A subsequent, more refined, floating-ball experiment by Bizzeti et al, using the apparatus shown in Figure 8, yielded a null result; the drift velocity of a nylon sphere floating in a stratified solution of KBr was found to be less than 10 μm hr^{-1} (115). The density gradient in the dissolved KBr ensured vertical equilibrium of the float, which could now be a spherical object with no protrusion above the surface. The experimenters made a number of tests to determine the magnitude of the systematic effects discussed above, including a calibration of Stokes' Law in which they magnetically drove the float through the water using the difference in magnetic susceptibility between the nylon and the fluid. The experiment was conducted on the slope of a large mountain, which provided a source of considerable strength for λ up to 15 km.

In light of these null results, Thieberger thinks it likely that his positive effect was due to some, as yet unidentified, spurious effect (116).

6.   SEARCHES FOR NEW SPIN-DEPENDENT INTERACTIONS

The main experimental challenge in searches for anomalous spin-dependent interactions is avoiding spurious effects from the magnetic fields that inevitably accompany spin polarization. This is done in two ways: by exploiting the properties of superconductors as highly efficient shields, or by arranging for a cancellation of electronic spin magnetism by electronic orbital magnetism, while retaining a net spin polarization.

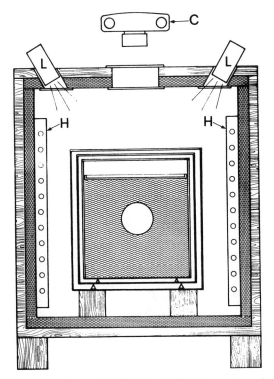

Figure 8 Floating-ball apparatus used by Bizzeti et al. The transverse motion of a nylon ball floating in a stratified saline solution was monitored by flash lamps L and camera C. A refrigerating fluid circulated through heat exchangers H. This instrument was operated on a sloping mountainside.

The most recent research for new macroscopic spin-spin interactions of the form given in Equation 2 was reported by Ritter et al (117). Their torsion-balance apparatus is shown in Figure 9. A special compound, Dy_6Fe_{23}, was used for both the source and test masses. In this compound, the magnetic moments of the Fe and Dy cancel to within 2% at room temperature but, because the Fe magnetization is entirely from spin while that in Dy is 50% orbital, a net spin of 0.4 ± 0.1 polarized electrons per atom remains. The spin orientation of the sources was periodically reversed, and the experimenters sought any change in the pendulum's period that was correlated with this spin orientation. They found that any anomalous spin-spin interaction was $(1.6 \pm 6.9) \times 10^{-12}$ of the normal magnetic interaction. For $\lambda > 1$ m, this corresponds to a 2σ constraint on the coefficient in Equation 2 of $(g_P)_e^2/(4\pi\hbar c) = (01. \pm 1.0) \times 10^{-13}$. This

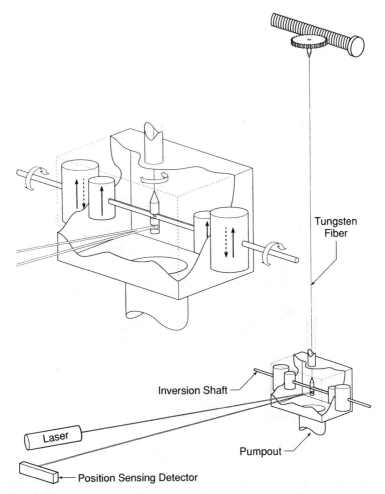

Figure 9   Torsion balance used by Ritter et al to search for a spin-spin interaction. Both the test bodies and the sources were made from magnetically compensated materials (Dy_6Fe_{23}) that had net electron spin polarization; the arrows on the test bodies and the sources denote the electron polarizations. The horizontal rods were used to rotate periodically the source polarizations. A correlated torque about the fiber would provide evidence for a new spin-spin force.

error is roughly an order of magnitude smaller than that of a previous experiment by Graham & Newman, who used a low-temperature torsion balance with split toroids and superconducting shields (118).

Vorobyov & Gitarts obtained a very precise limit on spin-spin interactions by surrounding a SQUID-coupled ferromagnetic sample with concentric superconducting shields, and driving the system with an external rotating magnetic field (119). The SQUID (superconducting quantum interference device) output was monitored for any component synchronous with the drive field that would have indicated an anomalous spin-spin force penetrating the magnetic shields. None was seen, at a level of 5×10^{-14} times the normal magnetic spin-spin interaction, although Ritter et al question the reliability of the calibration of this experiment (117).

Limits on the spin-spin couplings $(g_P)_e^2/(4\pi\hbar c) \leq 1.3 \times 10^{-9}$ (2σ) for $\lambda \geq 2 \times 10^{-12}$ m, and $(g_P)_p^2/(4\pi\hbar c) \leq 4.8 \times 10^{-5}$ (2σ) for $\lambda \geq 7.4 \times 10^{-11}$ m may be obtained from the agreement between QED calculations and precise measurements of the electron $g-2$ factor (120–122) and the splitting between the *ortho* and *para* states of the H_2 molecule (123).

Only a few searches for $\boldsymbol{\sigma} \cdot \hat{\boldsymbol{r}}$ interactions have been reported. Hsieh et al used a pan balance to compare the weights of a Dy_6Fe_{23} test body with 0.1 polarized electrons per atom (including the spacers and shielding material) in the spin-up and spin-down configurations (124). No difference was observed at the 7.8×10^{-7} level, and the experimenters concluded that any spin-up/spin-down difference in the electron free-fall rate is less than $0.01g$. To evaluate these experiments one needs the source integral

$$\mathbf{J}(\lambda) = \mathbf{V} \int \left(\frac{1}{r^2} + \frac{1}{r\lambda} \right) e^{-r/\lambda} \rho(\mathbf{r}) \, d^3r. \qquad 22.$$

For a spherical Earth of uniform density, the integral is

$$J_z(\lambda, r) = 3M_\oplus \left(\frac{\lambda}{R_\oplus} \right)^3 \left[\frac{R_\oplus}{\lambda} \cosh \frac{R_\oplus}{\lambda} - \sinh \frac{R_\oplus}{\lambda} \right] \left[\frac{2}{z^3} + \frac{2}{r^2\lambda} + \frac{1}{r\lambda^2} \right] e^{-r/\lambda}, \qquad 23.$$

where r is the distance from particle 1 to the center of the sphere. For $\lambda \gg 1$ m, Hsieh et al's result corresponds to a 2σ limit on the coefficient in Equation 3 of $(g_P)_e(g_S)_N/(4\pi\hbar c) \leq 5.3 \times 10^{-24}$, with $(g_S)_N$ assumed to couple to nucleons.

Leitner & Okuba have discussed limits on $\boldsymbol{\sigma} \cdot \hat{\boldsymbol{r}}$ interactions from measurements of the hyperfine splittings of atomic transitions (125). Wineland & Ramsey (126) reported a result for deuterium ($\Delta\nu \leq 10^{-4}$ Hz) corresponding to a 2σ limit of $(g_P)_D(g_S)_N/(4\pi\hbar c) < 2.3 \times 10^{-34}$ for an interaction with $\lambda \gg R_\oplus$. It has also been reported that any spin-up/spin-down difference in the gravitational acceleration of neutrons is "less than a few percent of g" (127).

7. SUMMARY AND CONCLUSIONS

7.1   Summary of Experimental Results

Considerable experimental progress has occurred in the four years since 1986 when Fischbach et al proposed a "fifth force." New experimental techniques have been introduced, and sensitivities have increased dramatically. The situation regarding inverse-square law tests is now clear. No violations are observed in astronomical or laboratory experiments (Figure 3). The claims of $1/r^2$ violations in geophysical tests that probed $g(z)$ in boreholes and on towers have all been retracted, and replaced by improved upper limits on any violation of the gravitational Gauss' law. The earlier erroneous claims are now understood to have been due to inadequate accounting for the local terrain.

The experimental situation in tests of the universality of free fall, summarized in Table 1, has also been greatly improved. With two exceptions, experiments show no evidence for a new macroscopic interaction. The most sensitive (in terms of differential acceleration resolution) results in each category—von Eötvös experiments with Earth and laboratory sources, Galileo experiments, and floating-ball experiments—give null results. Although reports of positive effects by Thieberger and by Boynton et al have not been retracted, these authors themselves do not claim evidence for new physics. However, because no two experiments are alike, there is always the possibility that the positive effects occurred because some special feature of the detectors or sources used by Thieberger or Boynton et al allowed them to see new physics that was not detected in other experiments. So one must consider possible differences in elemental composition of the detectors and/or sources, and differences in the spatial structure of the sources. A single Yukawa vector interaction coupled to an arbitrary linear combination of B and L cannot reconcile all the experimental results, as illustrated in Figures 10–12. It has been suggested that the results could be explained in terms of multiple Yukawa interactions or by interactions that do not have a Yukawa form (2). We prefer to adopt "Occam's razor" and, until the positive results have been reproduced, assume that they are due to as yet unidentified systematic errors in very difficult experiments. This point is made indirectly in Figure 13, where we compare the intriguing correlation Fischbach et al uncovered in the von Eötvös data to recent results. It is probably significant that the only modern experiments to report positive effects were conducted at remote sites where experimental conditions are less favorable than in laboratories.

7.2   Some Implications of the Experimental Results

7.2.1   CONSTRAINTS ON ULTRA-LOW-MASS BOSONS   Figure 14 shows constraints inferred from the experimental results discussed above on hypo-

Table 1 Modern tests of the universality of free fall

Ref.	detector $\Delta(B/\mu)$ (10^{-4})	$\Delta(L/\mu)$ (10^{-2})	source[a] (B/μ)	(L/μ)	$\bar\alpha\,\Delta(q/\mu)_{\bar\mu}(q/\mu)_s$ [b] $\lambda = 1$ m	$\lambda = 30$ m	$\lambda = 1000$ m	$\lambda = \infty$
88	7.0	5.6	1.00	0.50		$(0.4 \pm 1.4)10^{-4}$	$(1.1 \pm 4.3)10^{-6}$	$(1.3 \pm 5.1)10^{-10}$
90	17.6	3.3	1.00	0.50		$(1.2 \pm 3.5)10^{-4}$	$(0.4 \pm 1.1)10^{-5}$	$(0.4 \pm 1.3)10^{-9}$
90	4.3	-2.6	1.00	0.50		$(0.4 \pm 2.2)10^{-4}$	$(1.1 \pm 6.6)10^{-6}$	$(1.3 \pm 8.0)10^{-10}$
90	6.4	-1.6	1.00	0.50		$(-0.5 \pm 3.9)10^{-4}$	$(-0.2 \pm 1.2)10^{-5}$	$(-0.2 \pm 1.4)10^{-9}$
98	5.14	8.073	0.994	0.859				$(-1.3 \pm 1.5)10^{-11}$
99	5.01	8.198	0.994	0.859				$(3.0 \pm 4.5)10^{-13}$
101	20.36	3.797	1.0006	0.496	$(0.6 \pm 1.4)10^{-6}$	$(1.5 \pm 0.4)10^{-6}$	$(1.2 \pm 0.3)10^{-7}$	d
104	24.69	-1.253	1.0005	0.501		$(1.4 \pm 2.9)10^{-8}$	$(1.4 \pm 4.2)10^{-9}$	$(-0.2 \pm 1.0)10^{-11}$
104	20.36	3.797	1.0005	0.501		$(-2.1 \pm 3.6)10^{-8}$	$(-5.1 \pm 5.1)10^{-9}$	$(-0.5 \pm 1.3)10^{-11}$
105	22.35	-11.40	1.00	0.50	$(-0.9 \pm 1.7)10^{-6}$	$(-3.0 \pm 5.2)10^{-6}$	$(-1.1 \pm 2.0)10^{-7}$	$(1.8 \pm 12.9)10^{-9}$
108	10.01	6.073	0.00098[c]	0.060[c]	$(-7.2 \pm 7.6)10^{-4}$	$(-7.2 \pm 7.6)10^{-4}$	$(-7.2 \pm 7.6)10^{-4}$	$(-7.2 \pm 7.6)10^{-4}$
108	1.19	-10.39	0.00098[c]	0.060[c]	$(-3.2 \pm 3.7)10^{-4}$	$(-3.2 \pm 3.7)10^{-4}$	$(-3.2 \pm 3.7)10^{-4}$	$(-3.2 \pm 3.7)10^{-4}$
109	10.01	6.073	0.9994	0.555		$(-0.9 \pm 1.9)10^{-5}$	$(-0.7 \pm 1.4)10^{-5}$	$(-0.7 \pm 1.4)10^{-5}$
110	20.36	3.797	1.0001	0.396	$(0.5 \pm 4.1)10^{-6}$	$(0.4 \pm 3.9)10^{-6}$	$(0.4 \pm 3.9)10^{-6}$	$(0.4 \pm 3.9)10^{-6}$
112	10.01	6.073	0.00098[c]	0.060[c]		$(-1.4 \pm 0.9)10^{-6}$	$(-1.4 \pm 0.9)10^{-6}$	$(-1.4 \pm 0.9)10^{-6}$
113	9.52	5.31	1.0002	0.407	$(1.1 \pm 1.2)10^{-6}$	$(1.1 \pm 1.2)10^{-6}$	$(1.1 \pm 1.2)10^{-6}$	$(1.1 \pm 1.2)10^{-6}$
114	17.04	-9.87	1.0006	0.494		$(-6.8 \pm 2.2)10^{-5}$	$(-5.1 \pm 1.7)10^{-6}$	d
115	3.73	-0.652	1.0006	0.498			$(0.0 \pm 1.1)10^{-7}$	$(0.0 \pm 1.4)10^{-9}$

[a] Charges of terrestrial sources depend slightly on the value of λ, but this has little practical significance.
[b] Uncertainties are $\pm 1\sigma$ errors.
[c] Because of the geometry of this source these are $\Delta(B/\mu)$ and $\Delta(L/\mu)$ values.
[d] A positive effect was observed in a direction not consistent with $\lambda = \infty$.

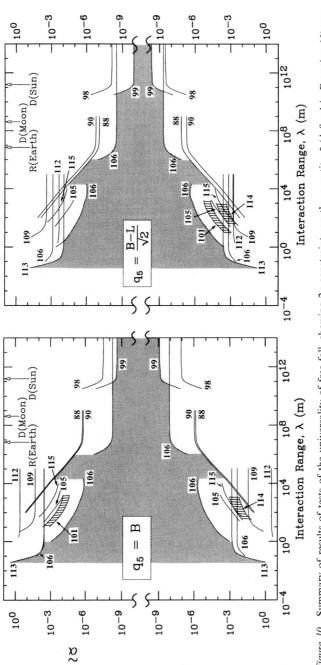

Figure 10 Summary of results of tests of the universality of free fall, showing 2σ constraints on the quantity $\tilde{\alpha}$ (defined in Equation 13) as a function of range, for two possible values for the charge. (*Left*) $q = B$. (*Right*) $q = (B-L)/\sqrt{2}$. Numbers on the constraints refer to citations in the text. Note the discrepancy between the positive effects reported by Thieberger (114) and Boynton et al (101) and null results observed in other experiments. The inner shaded region is allowed at 95% confidence.

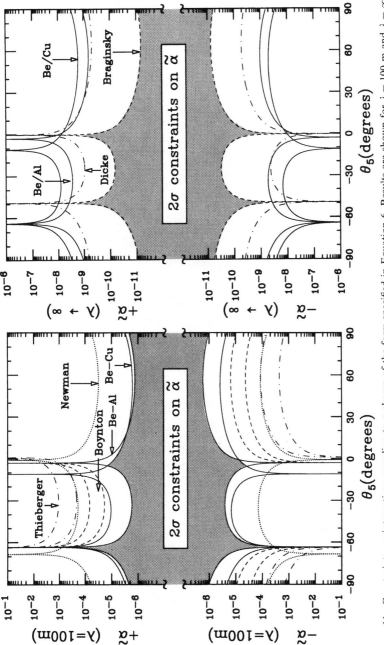

Figure 11 Constraints on interactions coupling to a charge of the form specified in Equation 6. Results are shown for $\lambda = 100$ m and $\lambda = \infty$. Names on the constraints refer to citations in the text. The poles near $\theta_5 = -63°$ occur because the strength of terrestrial sources vanishes when $q \sim N - Z$. The region around these poles is shown in expanded form in Figure 12.

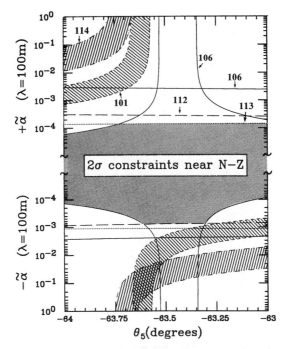

Figure 12 Constraints on interactions coupled to a charge approximately equal to $N-Z$. Data include results from both moving and terrestrial source experiments.

thetical vector interactions with charges of $q_5 = B$ and $q_5 = (B-L)/\sqrt{2}$. It is interesting to note that for ranges between 1 m and 1 AU the constraints from WEP experiments are typically better than those from the $1/r^2$ tests, and for $\lambda > 1$ AU the only significant constraints come from the WEP experiments. Thus our only solid experimental evidence for the correctness of classical gravitation at scales larger than 1 AU rests on data that test only for composition dependence and not for a low-mass scalar that couples exactly to mass.

Figure 15 shows constraints on hypothetical L and B vector bosons that couple only to leptons or quarks respectively. This figure includes results from searches for macroscopic forces along with those deduced from other data by Hawkins & Perl (120) and by Nelson & Tetradis (128). The laboratory constraints on $(g_P)_e^2$, $(g_P)_p^2$, and $(g_P)_e(g_S)_N$ as a function of λ are summarized in Figure 16. Considerably more stringent bounds on $(g_P)^2$ have been inferred from astrophysical considerations such as the cooling rates of red giant stars, but are not discussed here as they lie outside the scope of this article and were recently reviewed elsewhere

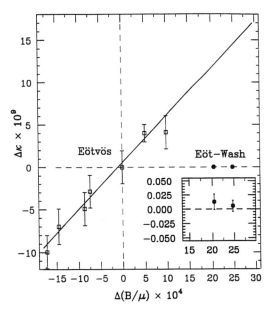

Figure 13 Updated version of Figure 1 showing the reanalyzed von Eötvös data along with the two recent Eöt-Wash points.

(129, 130). Similar astrophysical constraints on $(g_S)^2$ and on $(g_V)^2$, the corresponding vector coupling, are not competitive with results from laboratory experiments, as the coherence of spin-independent interactions allows laboratory experiments to reach very high sensitivity.

7.2.2 THE STRONG EQUIVALENCE PRINCIPLE   Laboratory equivalence principle experiments employ test bodies whose masses are too small to have a measurable contribution from gravitational binding energy

$$\frac{(BE)_{grav}}{Mc^2} \sim \frac{GM}{Rc^2} \sim 10^{-27}. \qquad 24.$$

Thus they cannot test whether gravitational binding energy obeys the equivalence principle, as predicted by general relativity but not by many alternative theories of gravity. However, the strong principle (which includes gravitational binding energy) can be tested by comparing the acceleration of the Earth and Moon toward the Sun because gravitational binding energy makes a significant contribution ($\approx 5 \times 10^{-10}$) to the Earth's mass. Any relative Earth-Moon acceleration can be detected because it will distort the Moon's orbit around the Earth in a characteristic way (131). The best analysis (132) of the lunar laser ranging data yields

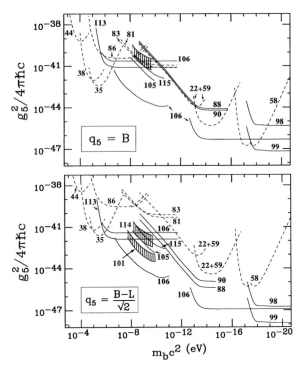

Figure 14   The 2σ constraints on ultra-low-mass vector bosons coupled to B or to $(B-L)/\sqrt{2}$. Constraints denoted by solid lines are obtained from tests of the universality of free fall, dashed constraints are from tests of the inverse-square law.

3 ± 4 cm for the amplitude of the main distortion term; this corresponds to a differential acceleration of $(1.5 \pm 2.0) \times 10^{-12} \, g_{\odot}$.

Nordtvedt pointed out that this test is accurate only to the extent that one can be sure that no interfering force, such as a very feeble composition-dependent interaction, fortuitously cancels a gravitational binding energy anomaly (133). It has recently been argued (134) that the 2σ limits on such "fifth force" accelerations from existing WEP experiments (99, 104), $\Delta a(\text{Earth-Moon})/g_{\odot} \leq 1 \times 10^{-12}$, are sufficiently tight that one can take full advantage of the recent lunar ranging analysis. Thus the gravitational acceleration of gravitational binding is rigorously tested at the 1% level (95% confidence level).

7.2.3   THE GRAVITATIONAL PROPERTIES OF ANTIMATTER   The "quantum gravity" models mentioned in Section 2.2 that postulate low-mass, spin-0

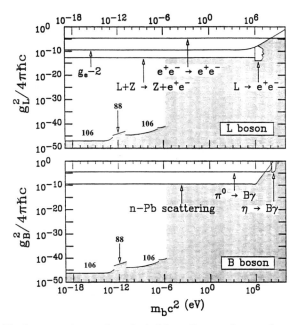

Figure 15   The 2σ constraints on hypothetical L or B vector bosons that couple only to leptons or to quarks respectively, based on results reviewed in this paper plus other data analyzed elsewhere (120, 128).

(gravi-scalar) and spin-1 (gravi-vector) partners of the conventional spin-2 graviton predict unusual gravitational properties of antimatter due to the gravi-vector interaction. It is not easy to test this prediction directly, as was discussed in Section 5.1.2. On the other hand, WEP experiments with ordinary matter have strong implications for the gravitational acceleration of antimatter.

It has been suggested that the gravi-vector and gravi-scalar forces could nearly cancel and thereby escape detection in WEP experiments. Any such cancellation requires that the two force vectors be parallel. But the λ dependence of the direction of $I_\perp(\lambda)$ sets stringent constraints on the strength and range of the hypothesized forces (135). In addition, it was recently shown that WEP experiments with ordinary matter probe the predicted gravi-vector acceleration of antimatter in precisely the same way as do direct experiments of antiparticles in free fall. Consider, for example, free-fall experiments with protons, electrons, neutrons, and their anti-particles in the presence of ordinary gravitational, electric, and gravi-vector fields **g**, **E**, and **V**, respectively. These particles experience forces

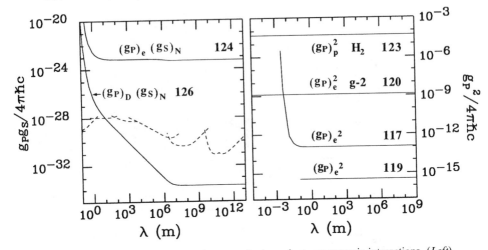

Figure 16 Laboratory 2σ constraints on spin-dependent macroscopic interactions. (*Left*) $\sigma \cdot \hat{r}$ interactions. For simplicity, we assume the experiments were conducted 1.5 m above a uniformly dense Earth. The different dependences on λ for the results from Wineland & Ramsey (126) and from Hsieh et al (124) occur because one experiment measured a potential and the other a force. The dashed curve shows constraints implied by Vorobyov & Gitarts' limit on $|(g_P)_e|$ and the limits on $|(g_S)_N|$ from the $1/r^2$ data summarized in Figure 3. (*Right*) $\sigma \cdot \sigma$ interactions.

$$F_p = M_p a_p = M_p g + q_p V + eE \qquad F_{\bar{p}} = M_p a_{\bar{p}} = M_p g - q_p V - eE$$

$$F_{e^-} = M_e a_e = M_e g + q_e V - eE \qquad F_{e^+} = M_e a_{e^+} = M_e g - q_e V + eE$$

$$F_n = M_n a_n = M_n g + q_n V \qquad F_{\bar{n}} = M_n a_{\bar{n}} = M_n g - q_n V, \qquad 25.$$

where q_p, q_e, and q_n are the gravi-vector charges of the proton, electron, and neutron, respectively. One can obtain $q_n V$ simply by comparing a_n and $a_{\bar{n}}$. A complication arises for charged particles because E can be known with sufficient accuracy only by measuring the acceleration of a test charge. To eliminate the dependence on E and solve for qV, one must measure the acceleration of two different particles. This dependence cannot be removed by comparing acceleration of a particle and its antiparticle, so one must compare the forces, say, on an electron and an antiproton. As a result, one cannot determine q_p and q_e separately but only the combination $(q_p + q_e)V$. But the quantities $q_n V$ and $(q_p + q_e)V$ are precisely the same ones that can be extracted from WEP experiments with electrically neutral test bodies composed of ordinary matter.

If a gravi-scalar field, S, is present along with the gravi-vector field, it will affect both WEP experiments and experiments with charged antimatter.

However, the WEP experiments provide null tests for anomalous gravitational forces while measurements of the "free fall" acceleration of charged particles do not. For example, if one compares the forces on an antiproton and an electron to eliminate electrical effects, a substantial signal will be observed even if $V = 0$ and $S = 0$:

$$\mathbf{F}_{\bar{p}} - \mathbf{F}_{e^-} = (M_p - M_e)g + (r_p - r_e)\mathbf{S} - (q_p + q_e)\mathbf{V}, \qquad 26.$$

where r is the gravi-scalar charge. On the other hand, the Cu/Be acceleration difference,

$$\mathbf{a}_{Cu} - \mathbf{a}_{Be} = \left[\left(\frac{r}{\mu}\right)_{Cu} - \left(\frac{r}{\mu}\right)_{Be}\right]\mathbf{S} + \left[\left(\frac{q}{\mu}\right)_{Cu} - \left(\frac{q}{\mu}\right)_{Be}\right]\mathbf{V}, \qquad 27.$$

vanishes in the absence of new physics.

Could the gravi-scalar and gravi-vector interactions have fortuitously cancelled in all WEP experiments? Adelberger et al argue that this is precluded by the inherently different response of scalar and vector charges to binding energy (135). Because the effects of scalar and vector forces cannot cancel within current experimental uncertainties for *all* the different pairs of materials used in recent WEP experiments, the data with ordinary matter place stringent constraints (typically less than $10^{-6}g$) on the predicted gravi-vector acceleration of antiprotons, antineutrons, or antihydrogen, as shown in Figure 17.

7.3   *Prospects for Future Improvements*

7.3.1   OVERCOMING LIMITATIONS OF CONVENTIONAL EXPERIMENTS   Inverse-square law tests are least sensitive in the "geophysical window," and we expect it will be difficult to improve these significantly because the main limitations are not in the gravimeters but in uncertainties in the density distribution of the Earth.

The main practical limitations of WEP tests are in the torsion fiber suspension systems. For example, the fiber is the transducer by which seismic vibrations are converted into a rotation of the pendulum. Direct free-fall tests are also limited by vibration of the apparatus. Although considerable work has been done on vibration isolation systems (for example 136, 137), it has generally not been applied to WEP experiments. A completely independent approach, replacing the torsion fiber by a fluid suspension system, is also being explored (138, 139). In principle, a fluid suspension allows one to separate the functions of supporting the detector and of providing a restoring torque on the detector; the vertical support is provided by the fluid, while the restoring torque is provided by electrostatic

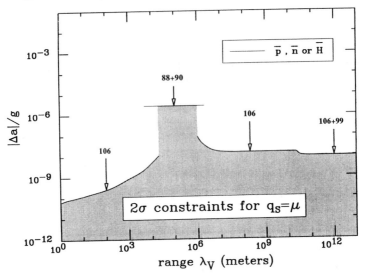

Figure 17 The 2σ constraints from WEP experiments on the gravi-vector acceleration of antimatter predicted by "quantum gravity" models that predict spin-0 and spin-1 partners of the conventional spin-2 graviton (from 135). These particular constraints assume the existence of a gravi-scalar interaction coupling exactly to mass.

forces. However, problems with systematic errors currently prevent these devices from competing successfully with torsion balances.

7.3.2 REDSHIFT EXPERIMENTS Will (140) presented a clear discussion of redshift experiments and pointed out that they probe the behavior of the clocks in the potential of the Earth, and not the free-fall characteristics of the photon. Such experiments played an important role in classical experimental gravitation, but they have not yet provided significant constraints on new Yukawa interactions although both the composition dependence and the finite-range aspects of the new interactions could produce departures from the classical redshift expectations (141). There are two main reasons for this state of affairs: the expected effects are very small, as emphasized by Hill & Ross (19) and by Hughes (141), and it is difficult to achieve a significant improvement over the existing limits (142, 143).

7.3.3 SPACE-BORNE EXPERIMENTS Spacecraft offer some opportunities for making large improvements in experimental sensitivity. We mention two such possibilities.

Paik and coworkers have done extensive analysis and development of superconducting three-axis gradiometers (144). Such devices would permit direct searches for violation of the gravitational Gauss' law by checking if $\nabla^2 V = 0$, and they would find numerous applications in other areas as well. In essence, one measures the accelerations of test bodies on the six faces of a cube and combines the result to obtain a signal proportional to $\nabla^2 V$. Maintaining calibrations and orthogonality of the six sensors at the requisite level of precision is a challenging technical problem.

Worden has proposed using nested test bodies in a drag-free satellite to test the universality of free fall (145). In principle it should be possible to measure any differential acceleration of two nested cylinders made of different materials with very high precision and thereby greatly improve the sensitivity of WEP tests for $\lambda \geq h$, where h is the altitude of the satellite.

ACKNOWLEDGMENTS

We are grateful to numerous colleagues with whom we have shared stimulating conversations concerning the work reviewed in this paper. We particularly acknowledge Ephraim Fischbach who directly or indirectly motivated much of the work covered in this review, and we thank him, D. F. Bartlett, and P. E. Boynton for their helpful comments on the manuscript. We are grateful to Y. Su for help with the illustrations. This work was supported in part by the NSF, under grants PHY-9104541 and ADT-8809616.

Literature Cited

1. Moody, J. E., Wilczek, F., *Phys. Rev.* D30: 130 (1984)
2. Fischbach, E., In *Tests of Fundamental Laws in Physics*, ed. O. Fackler, J. Tran Thanh Van. Gif-sur-Yvette: Editions Frontières (1989), p. 445
3. Lee, T. D., Yang, C. N., *Phys. Rev.* 98: 1501 (1955)
4. von Eötvös, R., Pekár, D., Fekete, E., *Ann. Phys.* 68: 11 (1922)
5. Fujii, Y., *Nature* 234: 5 (1971)
6. Fujii, Y., *Phys. Rev.* D9: 874 (1972)
7. Peccei, R. D., Solà, J., Wetterich, C., *Phys. Lett.* B195: 183 (1987)
8. Ellis, J., Tsamis, N. C., Voloshin, M., *Phys. Lett.* B194: 291 (1987); Linde, A. *Phys. Lett.* B201: 437 (1988)
9. Fayet, P., *Phys. Lett.* B171: 261 (1986); B172: 363 (1986)
10. Scherk, J., In *Unification of Fundamental Particle Interactions*, ed. S. Ferrara, S. Ellis, P. Nieuwenhuizen, Ettore Majorana Int. Sci. Ser. 7: 381 (1980); *Phys. Lett.* B85: 265 (1979)
11. Macrae, K. I., Riegert, R. J., *Nucl. Phys.* B224: 513 (1984)
12. Goldman, T., Hughes, R. J., Nieto, M. M., *Phys. Lett.* B171: 217 (1986)
13. Bars, I., Visser, M., *Gen. Rel. Gravit.* 19: 219 (1987)
14. Taylor, T. R., Veneziano, G., *Phys. Lett.* B213: 450 (1988)
15. Cvetic, M., *Phys. Lett.* B229: 41 (1989)
16. Gross, D., In *Proc. 24th Conf. on High Energy Phys.*, ed. R. Kotthaus, J. H. Kuhn. Berlin: Springer-Verlag (1989), p. 310
17. Anselm, A. A., Uraltsev, N. G., *Phys. Lett.* B116: 161 (1982)
18. Chang, D., Mohapatra, R. N., Nussinov, S., *Phys. Rev. Lett.* 55: 2835 (1985)
19. Hill, C. T., Ross, G. G., *Nucl. Phys.* B311: 253 (1988)
20. Hill, C. T., Schramm, D., Fry, J. N., *Comments Nucl. Part. Phys.* 19: 25 (1989)
21. Fischbach, E., et al., *Phys. Rev. Lett.* 56: 3 (1986)

22. Stacey, F. D., et al., *Rev. Mod. Phys.* 59: 157 (1987)
23. Wagoner, R. V., *Phys. Rev.* D1: 3209 (1970)
24. O'Hanlon, J., *Phys. Rev. Lett.* 29: 137 (1972)
25. Luther, G. G., Towler, W. R., *Phys. Rev. Lett.* 48: 121 (1982)
26. Mikkelsen, D. R., Newman, M. J., *Phys. Rev.* D16: 919 (1977)
27. Sugimoto, D., *Prog. Theor. Phys.* 48: 699 (1972)
28. Blinnikov, S. I., *Astrophys. Space Sci.* 59: 13 (1978)
29. Hut, P., *Phys. Lett.* B99: 174 (1981)
30. Newman, R. D., In *Proc. 3rd Marcel Grossmann Meet. on General Relativity*, ed. Hu Ning. Amsterdam: Science Press and North-Holland (1983), p. 755
31. Fischbach, E., et al., *Ann. Phys.* 182: 1 (1988)
32. Fischbach, E., Talmadge, C., In *5th Force—Neutrino Physics*, ed. O. Fackler, J. Tran Thanh Van. Gif-sur-Yvette: Editions Frontières (1988), p. 369
33. Cavendish, H., *Philos. Trans. R. Soc. London* 88: 469 (1798)
34. Long, D. R., *Nature* 260: 417 (1976)
35. Chen, Y. T., Cook, A. H., Metherell, A. J. F., *Proc. R. Soc. London A* 394: 47 (1984)
36. Long, D., *Phys. Rev.* D9: 850 (1974)
37. Spero, R., et al., *Phys. Rev. Lett.* 44: 1645 (1980)
38. Hoskins, J. K., et al., *Phys. Rev.* D32: 3084 (1985)
39. Liu, H., Pinghua, Z., Rongxian, Q., See Ref. 30, p. 1501
40. Long, D. R., *Nuovo Cimento* B55: 252 (1980)
41. Gibbons, G. W., Whiting, B. F., *Nature* 291: 636 (1981)
42. Panov, V. I., Frontov, V. N., *Sov. Phys. JETP* 50: 852 (1979)
43. Milyukov, V. K., *Sov. Phys. JETP* 61: 187 (1985)
44. Mitrofanov, V. P., Ponomareva, O. I., *Sov. Phys. JETP* 67: 1963 (1988)
45. Hirakawa, H., Tsubono, K., Oide, K., *Nature* 283: 184 (1980)
46. Ogawa, Y., Tsubono, K., Hirakawa, H., *Phys. Rev.* D26: 729 (1982)
47. Kuroda, K., Hirakawa, H., *Phys. Rev.* D32: 342 (1985)
48. Mio, N., Tsubono, K., Hirakawa, H., *Phys. Rev.* D36: 2321 (1987)
49. Chan, H. A., Moody, M. V., Paik, H. J., *Phys. Rev. Lett.* 49: 1748 (1982)
50. Gillies, G. T., *Metrologia Suppl.* 24: 1 (1987)
51. Kuz'min, V. A., Tkachev, I. I., Shaposhnikov, M. E., *JETP Lett.* 36: 59 (1982)
52. Mostepanenko, V. M., Sokolov, I. Yu., *Phys. Lett.* A132: 313 (1988)
53. Derjaguin, B. V., Abrikosova, J. J., Lifshitz, E. M., *Q. Rev. London* 10: 295 (1956)
54. Hunklinger, S., Geisselmann, H., Arnold, W., *Rev. Sci. Instrum.* 43: 584 (1972)
55. Lifshitz, E. M., *Sov. Phys. JETP* 2: 73 (1956)
56. De Rújula, A., *Phys. Lett.* B180: 213 (1986)
57. Talmadge, C., et al., *Phys. Rev. Lett.* 61: 1159 (1988)
58. Talmadge, C., Fischbach, E., See Ref. 32, p. 413
59. Rapp, R. H., *Geophys. Res. Lett.* 14: 730 (1987)
60. Smith, D. E., et al., *J. Geophys. Res.* 90: 9221 (1985)
61. Williams, J., Dickey, J., *Lunar Laser Ranging: Geophysical Parameters, Ephemerides, and Modeling*, Proc. 5th Annu. NASA Geodynamics Program Conf. and Crustal Dynamics Project Rev., Washington, DC: NASA (1983)
62. Trimble, V., *Annu. Rev. Astron. Astrophys.* 25: 425 (1987)
63. Saunders, R. H., *Astron. Astrophys.* 136: L21 (1984); 154: 135 (1986)
64. Visser, M., *Gen. Rel. Gravit.* 20: 77 (1988)
65. Saunders, R. H., *Mon. Not. R. Astron. Soc.* 223: 539 (1986)
66. Holding, S. C., Stacey, F. D., Tuck, G. J., *Phys. Rev.* D33: 3487 (1986)
67. Marson, I., Faller, J. E. *J. Phys.* E(GB)19: 22 (1986)
68. Stacey, F. D., Tuck, G. J., Moore, G. I., *J. Geophys. Res.* 93: 10575 (1989)
69. Thomas, J., et al., In *Proc. 5th Marcel Grossmann Meet. on General Relativity*, ed. D. Blair, M. J. Buckingham. Singapore: World Scientific (1989), p. 1573
70. Romaides, A. J., et al., *J. Geophys. Res.* 94(B2): 1563 (1989)
71. Thomas, J., *Phys. Rev.* D40: 1735 (1989)
72. Thomas, J., Vogel, P., *Phys. Rev. Lett.* 65: 1173 (1990)
73. Bartlett, D. F., Tew, W. L., *Phys. Rev.* D40: 763 (1989); *J. Geophys. Res.* 95: 17363 (1990)
74. Bartlett, D. F., Tew, W. L., *Phys. Rev. Lett.* 63: 1531 (1989)
75. Parker, R. L., *Geophysics* 39: 644 (1974)
76. Airy, G. B., *Philos. Trans. R. Soc. London* 146: 297, 343 (1856)
77. Hsui, A. T., *Science* 237: 881 (1987)
78. Tuck, G. J., Contributed talk at the 12th Int. Conf. on General Relativity and Gravitation, Boulder, Colorado (1989) (unpublished)

79. Ander, M. E., et al., *Phys. Rev. Lett.* 62: 985 (1989)
80. Eckhardt, D. H., et al., *Phys. Rev. Lett.* 60: 2567 (1988); *J. Geophys. Res.* 94: 1563 (1989)
81. Jekeli, C., Eckhardt, D. H., Romaides, A. J., *Phys. Rev. Lett.* 64: 1204 (1990)
82. Thomas, J., et al., *Phys. Rev. Lett.* 63: 1902 (1989)
83. Speake, C. C., et al., *Phys. Rev. Lett.* 65: 1967 (1990)
84. Yu, H. T., *Phys. Rev.* D20: 1813 (1979)
85. Moore, G. I., et al., *Phys. Rev.* D38: 1023 (1988)
86. Muller, G., et al., *Phys. Rev. Lett.* 63: 2621 (1989)
87. Stacey, F. D., Tuck, G. J., Moore, G. I., *Phys. Rev.* D36: 2374 (1987)
88. Niebauer, T. M., McHugh, M. P., Faller, J. E., *Phys. Rev. Lett.* 59: 609 (1987)
89. Kuroda, K., Mio, N., *Phys. Rev. Lett.* 62: 1941 (1989)
90. Kuroda, K., Mio, N., *Phys. Rev.* D42: 3903 (1990)
91. Witteborn, F. C., Fairbank, W. M., *Phys. Rev. Lett.* 19: 1049 (1967); *Nature* 220: 436 (1968)
92. Schiff, L. I., Barnhill, M. V., *Phys. Rev.* 151: 1067 (1966)
93. Dressler, A. J., et al., *Phys. Rev. Lett.* 168: 737 (1968)
94. Schmeidmayer, J., *Nucl. Instrum. Methods* A284: 59 (1989)
95. Jarmine, N., *Nucl. Instrum. Methods* B24/25: 437 (1987)
96. Gabrielse, G., *Hyperfine Interactions* 44: 349 (1988)
97. Boynton, P., See Ref. 32, p. 431
98. Roll, P. G., Krotkov, R., Dicke, R. H., *Ann. Phys.* 26: 442 (1964)
99. Braginsky, V. B., Panov, V. I., *Zh. Eksp. Teor. Fiz.* 61: 873 (1971) [transl. *Sov. Phys. JETP* 34: 463 (1972)]
100. Everitt, C. F. W., In *Proc. 1st Marcel Grossmann Conf. on General Relativity*, ed. R. Ruffini. Amsterdam: North-Holland (1975), p. 545
101. Boynton, P. E., et al., *Phys. Rev. Lett.* 59: 1385 (1987)
102. Stubbs, C. W., et al., *Phys. Rev. Lett.* 58: 1070 (1987)
103. Adelberger, E. G., et al., *Phys. Rev. Lett.* 59: 849 (1987); erratum 59: 1790 (1987)
104. Heckel, B. R., et al., *Phys. Rev. Lett.* 63: 2705 (1989)
105. Fitch, V. L., Isaila, M. V., Palmer, M. A., *Phys. Rev. Lett.* 60: 1801 (1988)
106. Adelberger, E. G., et al., *Phys. Rev.* D42: 3267 (1990)
107. Boynton, P., Aronson, S., In *New and Exotic Phenomena '90*, ed. O. Fackler, J. Tran Thanh Van. Gif-sur-Yvette: Editions Frontières (1990) p. 207
108. Speake, C. C., Quinn, T. L., *Phys. Rev. Lett.* 61: 1340 (1988)
109. Bennett, W. R. Jr., *Phys. Rev. Lett.* 62: 365 (1989)
110. Stubbs, C. W., et al., *Phys. Rev. Lett.* 62: 609 (1989)
111. Cowsik, R., et al., *Phys. Rev. Lett.* 61: 2179 (1988)
112. Cowsik, R., et al., *Phys. Rev. Lett.* 64: 336 (1990)
113. Nelson, P. G., Graham, D. M., Newman, R. D., *Phys. Rev.* D42: 963 (1990)
114. Thieberger, P., *Phys. Rev. Lett.* 58: 1066 (1987)
115. Bizzeti, P. G., et al., *Phys. Rev. Lett.* 62: 2901 (1989)
116. Thieberger, P., *Phys. Rev. Lett.* 62: 2333 (1989)
117. Ritter, R. C., et al., *Phys. Rev.* D42: (1990)
118. Graham, D. M., PhD thesis, Univ. Calif., Irvine (1987)
119. Vorobyov, P. V., Gitarts, Ya. I., *Phys. Lett.* B208: 146 (1988)
120. Hawkins, C. A., Perl, M. L., *Phys. Rev.* D40: 823 (1989)
121. Van Dyck, R. S., In *Quantum Electrodynamics*, Adv. Ser. on Directions in High Energy Phys. 7, ed. T. Kinoshita. Singapore: World Scientific (1990), p. 322
122. Kinoshita, T., See Ref. 121, p. 218
123. Ramsey, N. F., *Physica* 96A: 285 (1979)
124. Hsieh, C.-H., et al., *Mod. Phys. Lett.* A4: 1597 (1989)
125. Leitner, I., Okuba, S., *Phys. Rev.* 136B: 1542 (1964)
126. Wineland, D. J., Ramsey, N. F., *Phys. Rev.* A5: 821 (1972)
127. Dabbs, J. W. T., et al., *Phys. Rev.* 139: B756 (1965)
128. Nelson, A. E., Tetradis, N., *Phys. Lett.* B221: 80 (1989)
129. Turner, M. S., *Phys. Rep.* 197: 67 (1990)
130. Raffelt, G. G., *Phys. Rep.* 198: 1 (1990)
131. Nordtvedt, K., *Phys. Rev.* 170: 1186 (1968)
132. Dickey, J. O., Newhall, X. X., Williams, J. G., *Adv. Space Res.* 9: 975 (1989)
133. Nordtvedt, K., *Phys. Rev.* D37: 1070 (1988)
134. Adelberger, E. G., et al., *Nature* 347: 261 (1990)
135. Adelberger, E. G., et al., *Phys. Rev. Lett.* 66: 850 (1991)
136. Rinker, R. L., PhD thesis, Univ. Colorado (1983)

137. Saulson, P. R., *Rev. Sci. Instrum.* 55: 1315 (1984)
138. McHugh, M. P., See Ref. 107, p. 233
139. Bartlett, D. F., Tew, W. L., *Bull. Am. Phys. Soc.* 35: 1073 (1990)
140. Will, C. M., *Phys. Rev.* D10: 2330 (1974)
141. Hughes, R. J., *Phys. Rev.* D41: 2367 (1990)
142. Pound, R. V., Snyder, J. L., *Phys. Rev.* B140: 788 (1965)

143. Vessot, R. F. C., et al., *Phys. Rev. Lett.* 45: 2081 (1980)
144. Chan, H. A., Paik, H. J., *Phys. Rev.* D35: 3551 (1987); Chan, H. A., Moody, M. V., Paik, H. J., *Phys. Rev.* D35: 3572 (1987)
145. Worden, P. W., In *Near Zero: New Frontiers of Physics*, ed. J. F. Fairbank, C. W. F. Everitt, P. Michelson. New York: Freeman (1988), p. 766

Annu. Rev. Nucl. Part. Sci. 1991. 41: 321–55

SUPERDEFORMED NUCLEI[1]

Robert V. F. Janssens and Teng Lek Khoo

Argonne National Laboratory, Argonne, Illinois 60439

KEY WORDS: superdeformation, high spin states, collective nuclear models

CONTENTS

1.   INTRODUCTION

The ever increasing accuracy with which nuclear structure studies can be performed experimentally and theoretically has yielded a wealth of fascinating new results. It is by now well known that throughout the

[1] The US Government has the right to retain a nonexclusive royalty-free license in and to any copyright covering this paper.

periodic table nuclei can adopt a rich variety of shapes, particularly when rotated. The shape changes result from the interplay between macroscopic (liquid-drop) and microscopic (shell-correction) contributions to the total energy of the nucleus. The shell correction is a quantal effect arising from the occupation of nonuniformly distributed energy levels. Rotation affects the nuclear shape by modifying both the liquid-drop moment of inertia and the nucleonic occupation of specific shape-driving orbitals. Many interesting phenomena have been reported in rapidly rotating nuclei; examples are the alignment of particle spins along the rotation axis (backbending), the transition from prolate collective to oblate aligned-particle structures (band termination), the possible occurrence of triaxial shapes, and the onset of octupole instabilities. The most spectacular experimental result to date is the discovery of superdeformation, in which nuclei are trapped in a metastable potential minimum associated with very elongated ellipsoidal shapes corresponding to an axis ratio of roughly 2:1.

Superdeformation was first proposed some twenty years ago to explain the fission isomers observed in some actinide nuclei (1, 2). It was later realized that superdeformed shapes can occur at high angular momentum in lighter nuclei (3–7). The interest in the mechanisms responsible for these exotic shapes has increased enormously with the discovery of a superdeformed band of nineteen discrete lines in ^{152}Dy (8). At about the same time, evidence for highly deformed nuclei (axis ratio 3:2) was also reported near ^{132}Ce (9). Striking properties emerged from the first experiments, such as the essentially constant energy spacing between transitions ("picket-fence" spectra), the unexpectedly strong population of superdeformed bands at high spins, and the apparent lack of a link between the superdeformed states and the yrast levels.

These findings were reviewed by Nolan & Twin (10). The present article follows upon their work and discusses the wealth of information that has since become available. This includes the discovery of a new "island" of superdeformation near $A = 190$, the detailed spectroscopy of "ground" and excited bands in the superdeformed well near $A = 150$ and $A = 190$, the surprising occurrence of superdeformed bands with identical transition energies in nuclei differing by one or two mass units, and the improved understanding of mechanisms responsible for the feeding into and the decay out of the superdeformed states.

2. A NEW REGION OF SUPERDEFORMATION

2.1 *Calculations of Superdeformed Shapes*

As alluded to above, the single-particle energy spectrum plays an essential role in determining the nuclear shape. The stability of spherical nuclei at closed shells is related to large gaps in the energy level spectrum of the various orbitals. Shell gaps also occur in the single-particle spectrum for specific deformations. In an axially symmetric harmonic oscillator (11), these shell gaps occur when the lengths of the principal axes form integer ratios and, in particular, when the ratio is 2:1:1, which corresponds to a quadrupole deformation β_2 (12) of 0.65. These shell corrections, superimposed on a smooth liquid-drop contribution, can generate local minima in the potential energy surface. The cranked-Strutinsky method is often used to calculate potential energy surfaces. In this approach, the total energy of a nucleus is calculated as a sum of liquid-drop and shell-correction terms. The liquid-drop energy is a sum of Coulomb, surface, and rotational energies. All of these terms are calculated as a function of nuclear deformation. In the heaviest nuclei, strong Coulomb forces favor large deformations, balancing the surface energy (which favors compact shapes). Because of this balancing, shell corrections at large deformations give rise to pronounced potential minima that are responsible for the fission isomers (1). At high spin in the $A = 150$ region, it is the rotational energy term that balances out the surface term at large deformation and makes superdeformation possible. Several cranked-Strutinsky calculations, using either an anharmonic oscillator potential (5, 13) or a Woods-Saxon potential (6, 7), have been quite successful in predicting the existence of an "island" of superdeformation for $Z \approx 64$ and $N \approx 86$ as well as the occurrence of "nearly" superdeformed nuclei ($\beta_2 = 0.4$) for $Z \approx 58$ and $N \approx 74$. The calculations indicate that pronounced secondary minima are obtained only when both proton and neutron shell corrections are favorable and this leads to the occurrence of these "islands" in the periodic table.

Superdeformed minima in a great number of nuclei with $Z \geq 80$ were originally predicted by Tsang & Nilsson (3) at zero spin and later confirmed in several other calculations. From general expectations (e.g. 5), these minima survive and come closer to the yrast line with increasing spin. More recent calculations (13–15) suggest that the superdeformed minimum becomes yrast at spins in excess of $30\hbar$ in nuclei with $Z \approx 80$, $N \approx 112$. Calculated potential energies (15) as a function of quadrupole deformation are presented in Figure 1 for different spin values in the nucleus ^{191}Hg. From the figure, it is clear that a deep minimum exists at very large deformation ($\beta_2 \approx 0.5$, axis ratio $\approx 1.65:1$) and this minimum is calculated to persist down to the lowest spins, even though the well depth diminishes. Here, strong Coulomb and rotational effects both play a significant role. The occurrence of a superdeformed minimum at zero spin in the $A = 190$ region has also been predicted in calculations using static, self-consistent Hartree-Fock and Hartree-Fock-Bogoliubov calculations (16, 17).

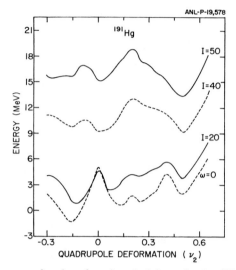

Figure 1   Total energy as a function of quadrupole deformation for different spin values in ^{191}Hg [from (15); v_2 is related to β_2, $v_2 = 0.5$ when $\beta_2 = 0.55$].

2.2   *The First Superdeformed Band in the* A = *190 Region*

Experimental evidence for a new region of superdeformation near $A = 190$ first became available in 1989. As can be seen in Figure 2, a weak rotational band of 12 transitions was observed in the nucleus ^{191}Hg (18). The measured properties of this band identify it as a superdeformed band: (*a*) the average energy spacing is small (37 keV) and corresponds to an average dynamic moment of inertia $J^{(2)}$ of $110\hbar^2$ MeV^{-1}, which agrees well with expectations based on cranked-Strutinsky calculations (15); and (*b*) the measured average quadrupole moment of 18 ± 3 eb implies a large quadrupole deformation ($\beta_2 \approx 0.5$). The band also exhibits other characteristics that are similar to those noted for superdeformed bands in the $A = 150$ region; (*a*) the small total intensity with which the band is fed (the flow through the band represents $\sim 2\%$ of the ^{191}Hg intensity); (*b*) the intensity pattern (which shows the γ-ray intensity in the band decreases gradually with increasing γ-ray energy, while at the bottom of the cascade it remains constant over the last 3–4 transitions); and (*c*) the fact that the transitions linking the superdeformed band to the yrast states could not be observed.

Several important questions were raised by the original experimental result. First, $J^{(2)}$ was found to increase steadily with the rotational frequency $\hbar\omega$ (defined as $E_\gamma/2$). Mean-field calculations that attempt to reproduce variations in $J^{(2)}$ suggest that such a rise may be attributed to three major factors, contributing either separately or cooperatively: (*a*) shape changes as a function of $\hbar\omega$ (e.g. centrifugal stretching), (*b*) changes in

ANL-P-19,423

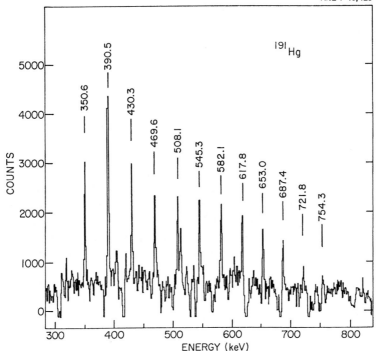

Figure 2 The γ-ray spectrum of the first superdeformed band observed in the $A = 190$ region (18). The 390-keV transition is an unresolved doublet consisting of a transition in the superdeformed band and the $17/2^+$-$13/2^+$ ground-state transition in ^{191}Hg.

pairing at large deformations (19), and (*c*) occupation of specific high-$\mathcal{N}$ orbitals (19, 20) (i.e. high-$\mathcal{N}$ orbitals from two major shells higher, which plunge down as a function of deformation and approach the Fermi surface at very large deformations, in this case $i_{13/2}$ protons and $j_{15/2}$ neutrons). Second, this superdeformed band was found to decay only to the $17/2^+$ yrast state of ^{191}Hg. Thus one could neither obtain a firm indication of the spins of the superdeformed states nor assess whether the link between superdeformed states and the yrast levels is statistical in nature, as in the $A = 150$ region (10), or whether it occurs only through a few specific transitions. Finally, questions concerning the existence of other superdeformed nuclei [as predicted by theory (15)] and the limits in N and Z of the superdeformed region also needed attention.

2.3 *Superdeformation in* ^{192}Hg

An impressive number of new results in the $A = 190$ region have become available recently. However, before presenting these we first summarize the present experimental situation concerning ^{192}Hg, the nucleus regarded as the analogue of ^{152}Dy for this region in the sense that shell gaps are calculated (15, 21) to occur at large deformation for both $Z = 80$ and $N = 112$ (see below). The gamma spectrum of the superdeformed band in ^{192}Hg, measured with the ^{160}Gd(^{36}S,4n) reaction at 162 MeV (22), is presented in Figure 3a. [The 16 transitions have also been observed in an independent experiment (23).] The total flow through the band represents 1.9% of all transitions in ^{192}Hg. From Figure 3a it is clear that the band feeds the known levels up to 8^+ in the positive-parity yrast sequence and up to 9^- in the negative-parity band. Transitions linking the super-deformed band with known yrast levels could not be found. It is likely that many different decay paths share the intensity and that the link is statistical in nature. This assumption is supported by the observation that the feeding into the yrast states is spread over several states belonging to bands of opposite parity with rather different intrinsic structure (24). Thus, the mechanism of deexcitation out of the superdeformed bands in this region appears to be similar to that discussed for the $A = 150$ region (10).

As with ^{191}Hg, the transition energies in the superdeformed band of ^{192}Hg extend to much lower energy than in superdeformed bands of the $A = 150$ region (the lowest transition energy in ^{152}Dy is 602 keV, as opposed to 257 keV in ^{192}Hg). If transition energies (and rotational frequencies) can be related to spin, this result indicates that superdeformation persists to lower spin in the new region. The data on ^{192}Hg allow for the verification of this assertion. The spin of the lowest level in the super-deformed band was estimated to be $10\hbar$ from the average entry spin ($8\hbar$) into the yrast states (22, 23) and from the assumption of a $\Delta I = 2\hbar$ angular momentum removal by the transitions linking the superdeformed states and the yrast line. The same spin value is also obtained from a procedure in which $J^{(2)}$ is fit by a power series expansion in ω^2, which is then integrated to give the spin (see Section 4.2) (23). With these assumed spin values, a static moment of inertia $J^{(1)}$ can be derived: the latter is presented as a function of $\hbar\omega$ together with the values of $J^{(2)}$ in Figure 3b. It is striking that (a) there is a large monotonic increase (40%) in $J^{(2)}$ with $\hbar\omega$; (b) the $J^{(2)}$ values for ^{192}Hg are similar to, but always higher than, those for ^{191}Hg at the same frequency; and (c) $J^{(2)}$ is significantly larger than $J^{(1)}$ for all values of $\hbar\omega$.

Crucial information regarding the properties of the ^{192}Hg super-deformed band were provided by detailed measurements of lifetimes with the Doppler-shift attenuation method (DSAM) (25). In contrast with previous measurements for superdeformed states, where only fractional Doppler shifts $F(\tau)$ were reported (10, 18), Moore et al analyzed detailed

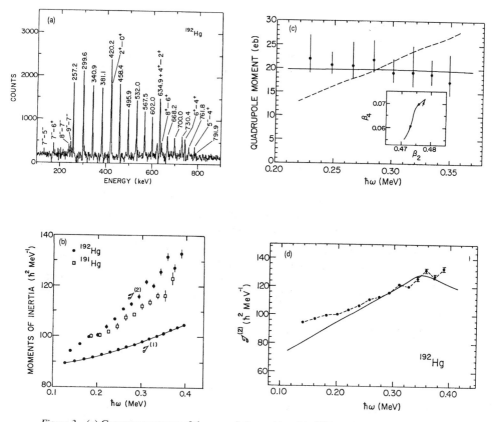

Figure 3 (*a*) Gamma spectrum of the superdeformed band in ^{192}Hg obtained from the sum of selected coincidence spectra (22). The energies of the superdeformed transitions are indicated and transitions between yrast states are also given.

(*b*) Comparison between the dynamic moments of inertia $J^{(2)}$ for the superdeformed bands in ^{191,192}Hg. The static moment of inertia $J^{(1)}$ is also given for ^{192}Hg assuming spin values for superdeformed states discussed in the text.

(*c*) Comparison between the measured and calculated transition quadrupole moments Q_t in the superdeformed band of ^{192}Hg (25). The dashed line represents a calculation assuming that the rise in $J^{(2)}$ is due to centrifugal stretching, the solid line is the result of a cranked shell model calculation discussed in the text, and the inset shows the calculated change in the deformation parameters over the frequency range of interest (the arrows point toward increasing frequency).

(*d*) Comparison between the measured and calculated $J^{(2)}$ values for the superdeformed band in ^{192}Hg (25).

line shapes for individual transitions between superdeformed states. Such an analysis allows one to determine the variation of transition quadrupole moment (Q_t) as a function of $\hbar\omega$, as opposed to previous studies in which Q_t was assumed to be constant for all states in the band. The measured lifetimes τ were transformed into transition quadrupole moments $Q_t(I) = (1.22\langle I020|I-20\rangle^2 \tau E_\gamma^5)^{-1/2}$ assuming the spin values given above. The Q_t values are displayed as a function of $\hbar\omega$ in Figure 3c. As can be seen, the quadrupole moment Q_t, and hence the deformation, remain essentially constant $(Q_t \approx 20 \pm 2$ eb) over the entire frequency range. This result rules out centrifugal stretching as an explanation for the rise in $J^{(2)}$: this is illustrated by the dashed line in Figure 3c, where the values of Q_t have been derived assuming that the change in $J^{(2)}$ is entirely due to a variation in deformation.

Bengtsson et al have shown that the occupation of specific high-$\mathcal{N}$ intruder orbitals plays an important role in understanding the variations of $J^{(2)}$ with mass number and rotational frequency in the $A = 150$ region (20). This effect alone cannot account for the variation in $J^{(2)}$ in ^{191,192}Hg: mean-field calculations without pairing, such as those by Chasman (15) or Åberg (13), give proton and neutron contributions to $J^{(2)}$ that remain essentially constant with $\hbar\omega$. This finding emphasizes the need to examine the effects of pairing (changes in pairing at large deformations) as first pointed out by Ye et al (22). This is best done in the framework of cranked deformed shell model calculations that include the effects of static and dynamic pairing, using either the Woods-Saxon or the modified-oscillator approach. The basis for these calculations is discussed in detail elsewhere (26, 27) as are the first applications to the Hg nuclei (21, 25, 28, 29).

The neutron and proton Woods-Saxon routhians for large deformation are presented in Figure 4 (21). Notice that the large $Z = 80$ shell gap remains at all rotational frequencies. In the neutron system there are two single-particle gaps at $N = 112$ and $N = 116$, separated by two high-K levels [512]5/2 and [624]9/2. Because of these shell gaps, $^{192}_{80}$Hg$_{112}$ can be regarded as a doubly magic nucleus in the superdeformed well. For this nucleus the relevant high-$\mathcal{N}$ intruder orbitals occupied in the super-deformed configuration are four $(i_{13/2})$ protons and four $(j_{15/2})$ neutrons. Adopting the nomenclature of Bengtsson et al (20), we can label the superdeformed configuration in ^{192}Hg as $(\pi 6^4 \nu 7^4)$. Figure 3d compares the calculated dynamic moment of inertia with the data. In the calculation, pairing correlations were treated self-consistently by means of the particle number projection procedure (26), but the neutron pairing interaction strength was reduced. The rise in the calculated $J^{(2)}$ can be ascribed to the combined gradual alignment of a pair of $\mathcal{N} = 6$ $(i_{13/2})$ protons and of a pair of $\mathcal{N} = 7$ $(j_{15/2})$ neutrons within the frequency range under con-

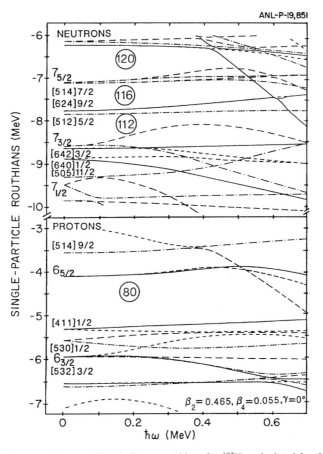

Figure 4 Neutron and proton Woods-Saxon routhians for ^{192}Hg calculated for the deformation parameters β_2, β_4, and γ indicated. The orbitals are labeled by their asymptotic Nilsson quantum numbers $[\mathcal{N} n_3 \Lambda]\Omega$ or, in the case of the high-$\mathcal{N}$ intruder orbitals, by $\mathcal{N}_\Omega$. The conventions for labeling the orbitals are from Riley et al (21).

sideration. The data are reproduced rather well. The evolution of the nuclear shape with $\hbar\omega$ was also calculated. The inset in Figure 3*c* illustrates that within the frequency range of interest the predicted changes in the β_2 and β_4 deformation parameters are very small. The resulting Q_t values agree well with the measured values, as is shown by the solid line in Figure 3*c*. The success of the calculations in reproducing all aspects of the data allows one to propose that quasi-particle alignments and the resulting changes in pairing are major contributors to the rise in $J^{(2)}$ in ^{192}Hg and, probably, in other nuclei in this region. Further evidence for this conclusion, as well as for the power of calculations, is discussed below.

2.4 *Neutron and Proton Excitations in the Superdeformed Minimum*

Figure 5 presents the dynamic moments of inertia of all the superdeformed bands observed so far in the $A = 190$ region: bands have been identified in all Hg isotopes with $A = 189$–194 (18, 21–23, 28, 30–33), in ^{193,194}Tl (34, 35), and in ^{194,196}Pb (36, 37). Furthermore, in many nuclei several superdeformed bands have been reported. Thus, just as in the $A = 150$ region discussed below, it has been possible to perform detailed spectroscopy in the superdeformed well. Most of the data can be understood in the framework of the cranked shell model calculations with pairing (introduced in Section 2.3), and specific configurations have been proposed for many of the bands of Figure 5 (21, 29). Neutron configurations involving the intruder orbital $7_{3/2}$ and/or the [642]3/2, [512]5/2, and [624]9/2 levels (Figure 4) have been proposed for the superdeformed bands in the Hg isotopes (the proton configuration is always $\pi 6^4$). In ^{194}Hg, for example, the superdeformed yrast band still contains the $\pi 6^4 \nu 7^4$ configuration, but the lowest neutron excitations are predicted to involve promotion from the [512]5/2 to the [624]9/2 levels. This leads to two pairs of strongly coupled bands with negative parity showing no signature splitting. Three superdeformed bands have been observed in this nucleus (21, 33). Their properties match the expectations: the two excited bands are interpreted as one signature partner pair, since the γ-ray energies in one of the bands are observed to lie midway between those of the other within 1 keV over the entire frequency range, and both bands are of similar intensity. Equally successful comparisons between the data and the calculations can be made for the bands of the other Hg isotopes. The only notable exception is ^{193}Hg, where some of the observed features, such as the irregularities in the evolution of $J^{(2)}$ with $\hbar\omega$ for two of the bands (Figure 5), cannot be readily understood within the framework of the calculations (32). The suggestion has been made that octupole effects must be taken into account. This is discussed in Section 5.

The cranked shell model calculations with pairing appear to be equally successful in describing proton excitations. Considering the proton routhians of Figure 4, the [411]1/2, [530]1/2, and [532]3/2 levels are important for superdeformed nuclei with $Z \leq 80$, while the third $i_{13/2}$ ($6_{5/2}$ in Figure 4) intruder orbital and the [514]9/2 state should characterize superdeformed bands in the Tl and Pb nuclei. No superdeformed bands have yet been observed in Au or Pt nuclei, which may underline the importance of the $Z = 80$ shell gap. On the other hand, superdeformed bands have been

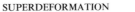

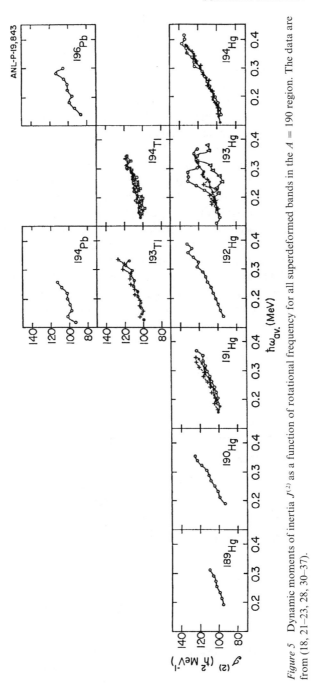

Figure 5   Dynamic moments of inertia $J^{(2)}$ as a function of rotational frequency for all superdeformed bands in the $A = 190$ region. The data are from (18, 21–23, 28, 30–37).

seen in several Pb and Tl nuclei (34–37). In ^{193}Tl, for example, the yrast superdeformed configuration is labeled as $\pi6^5$ in the calculations, and it exhibits some signature splitting at $\hbar\omega \geq 0.2$ MeV (Figure 4). Thus one would expect to observe two signature partner bands, and indeed they were seen experimentally (34).

A direct comparison of the $J^{(2)}$ values for ^{193}Tl with those observed in ^{192}Hg and in the first ^{191}Hg superdeformed band (which, as ^{193}Tl, is thought to contain a single nucleon in a high-$\mathcal{N}$ intruder orbital) is also particularly revealing (Figure 6). The occupation of the third $i_{13/2}$ orbital in ^{193}Tl increases the value of $J^{(2)}$ with respect to ^{192}Hg at the lowest frequencies. In both ^{191}Hg and ^{193}Tl, $J^{(2)}$ is essentially constant at the lowest frequencies before it exhibits the characteristic rise described above. This feature is also present in the calculations and is proposed to be a signature for the occupation of these high-j intruder orbitals by a single nucleon (34). In the ^{191}Hg band, the alignment of the $j_{15/2}$ neutron is blocked, and the rise in $J^{(2)}$ is attributed to the alignment of the $i_{13/2}$ protons; in ^{193}Tl the opposite situation occurs. The fact that the rise in $J^{(2)}$ with $\hbar\omega$ is very similar in both cases implies that neutron and proton alignments make contributions of comparable magnitude. Furthermore, the rise of $J^{(2)}$ in ^{193}Tl starts at lower frequency that that in ^{191}Hg. This suggests that the neutrons align at somewhat lower frequency than the protons, in agreement with the calculations.

From the discussion above, it can be concluded that a good description

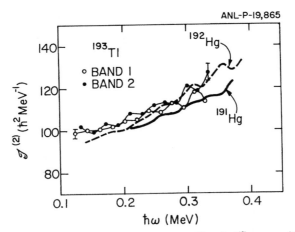

Figure 6 Comparison between the dynamic moments of inertia $J^{(2)}$ measured in the super-deformed bands of ^{193}Tl, ^{192}Hg, and ^{191}Hg (yrast superdeformed band). The data are from (22, 28, 34).

of the superdeformed bands within the framework of cranked shell model calculations with pairing can be achieved. Two points need to be emphasized, however. First, the transition energies in some of the superdeformed bands are surprisingly close to those of bands in neighboring nuclei and several bands can be related to ^{192}Hg (40). This property is discussed in Section 4. Second, the inclusion of pairing is crucial for reproducing the data and, in particular, the smooth increase of $J^{(2)}$ with $\hbar\omega$. In the proton system, pairing is reduced by the presence of the $Z = 80$ shell closure. As was shown by Riley et al (21) and by Carpenter et al (28), the calculations require that the neutron pairing be reduced as well if one wants to reproduce the similarities in the behavior of $J^{(2)}$ with $\hbar\omega$ observed in *all* nuclei in this region.

Reduced pairing is to be expected on the basis of general arguments (38). Pairing is sensitive to the overlap between orbitals of interest. At the very large deformations being considered here, states originating from different shells approach the Fermi level, and these states will only be very weakly coupled through the pairing interaction. Moreover, the coupling between the all-important unique-parity levels (that is, the various components of the high-$\mathcal{N}$ intruders) is also severely reduced because of their sizable energy splitting at large deformation. First attempts to provide a quantitative estimate of the reduction in pairing have recently been made (38, 39).

3. SUPERDEFORMATION IN THE $A = 150$ REGION: "IDENTICAL" BANDS

3.1 High-$\mathcal{N}$ Orbital Assignments in the Superdeformed Minimum

Since the review by Nolan & Twin (10) discussing the two first superdeformed bands of the $A = 150$ region (^{152}Dy and ^{149}Gd), the "island" of superdeformed nuclei in this mass region has expanded considerably. Superdeformed bands have now been identified in all Gd isotopes with $A = 146$–150 (41–46), in ^{150,151}Tb (44, 45, 47), and in $^{151-153}$Dy (8, 48, 49). Furthermore, in many cases several superdeformed bands have been seen in the same nucleus: a summary of the available data on the $J^{(2)}$ moments of inertia is presented in Figure 7. Preliminary reports of similar band structures in ^{145}Gd and ^{142}Eu have also become available (50, 51). As in the $A = 190$ region, differences in the variations of $J^{(2)}$ with $\hbar\omega$ from nucleus to nucleus have been attributed to the occupation of specific high-$\mathcal{N}$ intruder orbitals. In the $A = 150$ region, $^{152}_{66}$Dy$_{86}$ can be described as the "doubly magic" nucleus: all available calculations indicate the presence

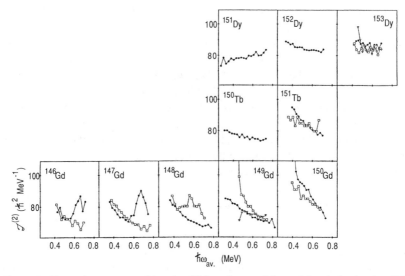

Figure 7 Dynamic moments of inertia $J^{(2)}$ for all superdeformed bands in the $A = 150$ region. The data are from (8, 10, 41–49).

of very large shell gaps at $Z = 66$ and $N = 86$ for a quadrupole deformation $\beta_2 \approx 0.6$.

The occupation of specific high-$\mathcal{N}$ intruder orbitals can have dramatic effects: in the Dy isotopes for example, $J^{(2)}$ rises smoothly over the entire frequency range in ^{151}Dy (Figure 7), while a smooth decrease is seen in ^{152}Dy and an essentially constant value of $J^{(2)}$ is observed for the yrast superdeformed band in ^{153}Dy (it is assumed that the most intensely populated band is the "ground" band in the second well). For these isotopes, the proton contribution has been assigned as $\pi 6^4$ (i.e. $i_{13/2}$) and the observed differences have been attributed to changes in the occupation of the $j_{15/2}$ neutron orbitals ($\nu 7^1$, $\nu 7^2$, and $\nu 7^3$ for $^{151-153}$Dy, respectively), which result in contributions of varying magnitude to $J^{(2)}$ (20). While the calculations of Bengtsson et al (20) were performed without pairing and at a fixed deformation, more recent Woods-Saxon cranked shell model calculations take into account small variations of the shape with spin as well as changes in deformation from one nucleus to another (19, 26). These calculations also treat pairing correlations self-consistently (see also 52).

The best indication for the importance of including pairing correlations and shape effects in this mass region comes from the data on the Gd isotopes. In the yrast superdeformed band of ^{150}Gd, for example, $J^{(2)}$ not only decreases with $\hbar\omega$, but also falls off dramatically at the lowest frequencies (Figure 7). In the calculations, this band is assigned a $\pi 6^2 \nu 7^2$ configuration. This is in agreement with the assignment made by Bengtsson et al (20), but the alignment of a pair of $j_{15/2}$ neutrons at $\hbar\omega \approx 0.4$ MeV (i.e. a band crossing) has to be invoked to account for the sharp drop in $J^{(2)}$. The observation that the deexcitation out of the superdeformed band in ^{150}Gd is extremely abrupt, with essentially all the intensity being lost over a single transition, has been interpreted as additional evidence for this band crossing, which in turn requires the presence of the pairing correlations (45). Marked irregularities in the behavior of $J^{(2)}$ can also be seen in the yrast superdeformed bands of ^{146,147}Gd as well as in an excited band in ^{148}Gd (Figure 7). At present, there is some argument regarding the exact orbitals involved in these crossings as well as concerning the role of pairing and/or octupole correlations (26, 41, 42, 46). The general conclusion, however, remains that in all yrast superdeformed bands near $A = 150$ the variations of $J^{(2)}$ with $\hbar\omega$ reflect the major role played by the few nucleons in high-$\mathcal{N}$ orbitals, and the adopted configurations are $\pi 6^{2,3,4}$ for Gd, Tb, and Dy respectively; $\nu 7^0$ or $\nu 7^1$ for $N = 82$–85; $\nu 7^2$ for $N = 86$, and $\nu 7^3$ for $N = 87$. The number of occupied intruder orbitals varies in some of the excited bands.

3.2   Superdeformed Bands with Identical Energies

The discovery of multiple superdeformed bands within a single nucleus has made it possible to investigate the microscopic structure of both the ground and excited states in the second well. However, a greater impetus for detailed studies of excited bands has been the unexpected discovery that several pairs of related bands have almost identical transition energies. The first reported cases consisted of the pairs (^{151}Tb*, ^{152}Dy) and (^{150}Gd*, ^{151}Tb)—the asterisk denotes an excited superdeformed band—where transition energies in the pair were found to be equal to within 1–3 keV over a span of 14 transitions (53). Later, another similar pair (^{149}Gd*, ^{150}Tb) was found by Haas et al (44) (they used the notation ^{149}Gd** because they suggested that the second excited superdeformed band in ^{149}Gd is involved). This is illustrated in Figure 8, where the difference between "identical" transition energies ΔE_γ is plotted versus the transition energy. On average, the deviation is less than 1 keV for the first pair and only slightly larger for the other two. This implies that transition energies are equal to better than 3 parts in 1000. This is a rather surprising equality and is quite unprecedented in nuclear physics! The γ-ray energies should scale with the moment of inertia J, which is proportional to $A^{5/3}$ ($J \approx MR^2$); hence, the energies of adjacent mass nuclei would differ by ~ 14 keV. Furthermore, the spins of corresponding transitions in each pair necessarily differ by $1/2\hbar$, leading to differences in E_γ of ~ 13 keV.

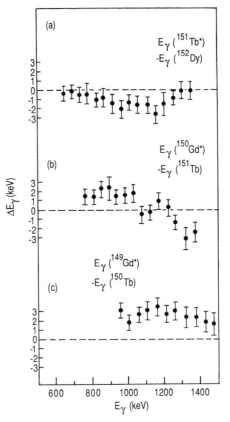

Figure 8 Differences in the γ-ray energies ΔE_γ between the superdeformed bands in (*a*) (^{151}Tb*, ^{152}Dy), (*b*) (^{150}Gd*, ^{151}Tb), and (*c*) (^{149}Gd*, ^{150}Tb). The data are from (44, 53).

For ($A-1$*, A) pairs, with A even, these differences would reinforce each other.

A word of caution is in order since it is assumed that each pair of transitions being compared has the appropriate spins ($I+1/2$, I), but the spins of the superdeformed bands have not been measured. Since the spacing between consecutive transitions in each band is ~50 keV, the maximum difference in energy in a pair of bands is ~25 keV if no spin correlation is involved. However, with three cases and not just an isolated one, it is unlikely that these degeneracies are accidental. Furthermore, in all three cases the excited ($A-1$)* superdeformed band is proposed to be characterized by a hole in the same specific orbital (see below).

A related case of identical transition energies occurs in ^{153}Dy—the first case in which excited superdeformed bands were reported (49). Here, two excited bands have been interpreted as signature partners, and the averages of the transition energies in the partners reproduce the γ-ray energies in ^{152}Dy within 1–3 keV. Finally, the two superdeformed bands of ^{147}Gd have been related to the yrast superdeformed bands of ^{146}Gd and ^{148}Gd respectively, although the average ΔE_γ values are somewhat larger in this case (~ 5 keV) (42). In ^{147}Gd, another relation applies as well: one of the bands has γ-ray energies following closely (1–4 keV) the average of two successive transition energies in ^{148}Gd, while the other band shows the same property when compared with the superdeformed band of ^{146}Gd.

3.3   *Strong Coupling and Identical Bands*

The first attempt at an explanation of this surprising phenomenon was presented by Nazarewicz et al (38, 54). The interpretation is done within the framework of the strong coupling limit of the particle-rotor model, in which one or more particles are coupled to a rotating deformed core and follow the rotation adiabatically. We note that for rotors the γ-ray energy for an $I \to I-2$ transition is given by

$$E_\gamma = \frac{\hbar^2}{2J}(4I-2). \qquad\qquad 1.$$

If the Coriolis force causes alignment i of the particle along the rotation vector **R**, giving $R = I - i$, then

$$E_\gamma = \frac{\hbar^2}{2J}[4(I-i)-2]. \qquad\qquad 2.$$

Equation 2 shows that odd and even nuclei can have identical transition energies if $i = 1/2$, when $J^{\text{odd}} = J^{\text{even}}$. In the strong coupling limit, no alignment is present, i.e. $i = 0$, and the transition energies in an odd nucleus, relative to those in an even-even core, obey simple relations, which are shown in Figure 9 (11). Here the moments of inertia for all cases are assumed to be identical. When $K \neq 1/2$, one can see (Figure 9a–c) that

$$1/2[E_\gamma(R+1/2)+E_\gamma(R-1/2)]_{\text{odd}} = E_\gamma(R)_{\text{even}}.$$

This strong coupling relation provides a straightforward explanation for the (^{153}Dy*, ^{152}Dy) pair. The two excited bands in ^{153}Dy have been interpreted as a ^{152}Dy $\otimes$ $\nu[514]9/2$ structure with no signature splitting (26, 49)—the relevant single-particle levels for protons and neutrons at large deformation are presented in Figure 3 of the article by Twin (55).

For a $K = 1/2$ band, the transition energies in the odd nucleus are affected by the decoupling parameter, a, and obey the relation

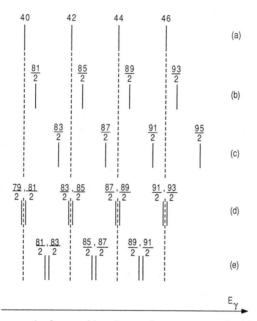

Figure 9 The γ-ray energies for transitions in a perfect rotor, assuming identical moments of inertia in all cases: (*a*) even core; (*b*) and (*c*) odd-even nucleus in the strong coupling limit; (*d*) and (*e*) odd nucleus with $K = 1/2$, $a = +1$, and $a = -1$, respectively. Note that (*d*) and (*e*) also apply for an odd nucleus with alignment $\pm 1/2$, i.e. $I = R \pm 1/2$.

$$E_\gamma = \frac{\hbar^2}{2J}[4I - 2 + 2(-1)^{I+1/2}a\delta_{K,1/2}].$$

The transition energies for $a = \pm 1$ are indicated in Figure 9*d, e*. Transitions from both signatures form degenerate doublets, with the $a = 1$ case giving energies identical to those of the core, while the $a = -1$ case has energies midway between those of adjacent transitions in the core. The alignment in a $K = 1/2$ band is given by $i = (-)^{I-1/2} \cdot a/2$.

The three pairs $(A - 1^*, A)$ that have identical energies can be interpreted as having $a = 1$. In each case, the $(A-1)^*$ configuration is $A \otimes \pi([301]1/2)^{-1}$. If the decoupling parameter a is calculated from the Nilsson wave function of the [301]1/2 orbital, a value of 0.85 is obtained, very close to but not exactly unity (54). On the other hand, for a $[\mathcal{N}n_3 \Lambda]\Omega$ orbital, the decoupling parameter can be calculated from the asymptotic quantum numbers, by

$$a = (-)^{\mathcal{N}}\delta_{\Lambda,0},$$

which would result in $a = 0$ for the [301]1/2 orbital. However, if one employed a different coupling scheme in terms of pseudo-spin (see below), the appropriate quantum numbers are [200]1/2, and

$$a = (-)^{\tilde{\mathscr{N}}} \delta_{\tilde{\Lambda},0} = 1,$$

providing a natural explanation for the identical energies of the three pairs. A case where $a = -1$ has not been seen so far, although it is predicted within the pseudo-spin scheme to apply to an excited superdeformed band in ^{151}Dy (38, 54).

It must be stressed again that the explanation of identical energies presented here follows only if the moments of inertia for odd and even nuclei are identical. $A^{5/3}$ scaling would give differences of $\sim 1\%$, as mentioned above. Thus, it still is a puzzle why the moments of inertia J for adjacent nuclei appear to be constant to within $\sim 0.2\%$.

3.4 Pseudo-spin in Rotating Deformed Nuclei

A simplified analysis of nucleonic motion and of the effect of deformation can be made within the framework of the pseudo-spin scheme (56–59). In heavy nuclei, this formalism is based on the fact that a major shell, labeled by the total quantum number $\mathscr{W}$, consists of the members of an oscillator shell modified by the removal of the state with largest j ($= \mathscr{N} + 1/2$) and the addition of an intruder orbital with opposite parity and $j = \mathscr{N} + 3/2$. The remaining normal-parity orbitals form close-lying doublets with quantum numbers l_1, $j_1 = l_1 + 1/2$ and $l_2 = l_1 + 2$, $j_2 = l_2 - 1/2 = j_1 + 1$. Examples are the $(d_{5/2}, g_{7/2})$ and $(f_{7/2}, h_{9/2})$ doublets. The doublets can be relabeled in terms of pseudo-quantum numbers $\tilde{l} = l_1 + 1$, $\tilde{\mathscr{N}} = \mathscr{N} - 1$. Thus, the doublets $(d_{5/2}, g_{7/2})$ and $(f_{7/2}, h_{9/2})$ become $\tilde{f}$ and $\tilde{g}$, respectively. A deformed potential preserves this degeneracy: in a Nilsson diagram, close-lying nearly parallel orbits are observed as function of deformation [see, for example, Figure 1 in the article by Bohr et al (59)]. The pair of orbits

$$[\mathscr{N} n_3 \Lambda, \Omega = \Lambda + 1/2] \quad \text{and} \quad [\mathscr{N} n_3 \Lambda + 2, \Omega = \Lambda + 3/2]$$

can be relabeled with pseudo-quantum numbers

$$[\tilde{\mathscr{N}} = \mathscr{N} - 1, n_3, \tilde{\Lambda} = \Lambda + 1, \Omega = \tilde{\Lambda} \pm 1/2].$$

The doublet can be viewed as pseudo-spin-orbit partners. The important feature is that the $\tilde{l} \cdot \tilde{s}$ interaction is substantially smaller than the normal $l \cdot s$ coupling. Since the $\tilde{l} \cdot \tilde{s}$ coupling is weak, the Coriolis force readily causes $\tilde{s}$ to align along the rotation vector, i.e. $\tilde{s}_1 = \pm 1/2$. One sees, therefore, that pseudo-spin can lead to $i = 1/2$ in Equation 2 and account for identical transition energies in odd and even nuclei by offsetting the intrinsic spin difference of $1/2\hbar$ (60).

The case for $K = 1/2$ bands has been discussed above, where it was pointed out that the pseudo-spin formalism naturally provides the correct decoupling parameter for explaining bands with identical transition energies. For orbits with $K \neq 1/2$, the alignment in an odd nucleus relative to an even core will, in general, have nonzero contributions from alignment $\tilde{l}_1$ of the pseudo-orbital angular momentum (59). (Only if $\tilde{l}_1 = 0$ will one obtain the simple spectrum given in Figure 9d, e.) Thus, it will not be easy to find experimental evidence for pseudo-spin alignment by comparing data in odd and even nuclei for orbits with $K \neq 1/2$. However, Hamamoto has suggested that evidence for pseudo-spin alignment may be found in an alignment difference of $1\hbar$ for pairs of orbits that constitute doublets with identical $\tilde{l}_1$ and $\tilde{s}_1 = \pm 1/2$ (I. Hamamoto, private communication) [see also Figures 3 and 4 in the paper by Bohr et al (59)]. Examples of such doublets have so far not been identified. Detection of those doublets will probably require detection of more (weak) excited superdeformed bands, as well as firm spin assignments.

3.5   Identical Moments of Inertia

The identical bands discussed above require moments of inertia J in different nuclei to be equal to a remarkable degree. Since J depends on several factors (mass, deformation, polarization effects, alignment, and pairing) the equality in J is very striking and leads one to wonder about the possibility of a fundamental explanation. The standard models are unable to reproduce transition energies with an accuracy of ~ 1 keV. We note that only a few identical bands are observed among the many superdeformed bands in this region and that three of these pairs of bands involve a common orbital. This suggests that the phenomenon is associated with only a few specific orbitals. Indeed, it is recognized that occupation of different high-$\mathcal{N}$ intruder orbits will not result in identical bands because the orbital energies tend to have steep slopes as a function of both deformation and rotational frequency, i.e. strong polarization and alignment effects will result. Ragnarsson used a simple harmonic oscillator model to show that for particles or holes in certain orbitals there can be cancellation among the different terms contributing to changes in J (62). He has been able to reproduce the "identical" energies observed in (^{152}Dy, ^{153}Dy*), where the orbital occupied in ^{153}Dy* is either $\nu[402]5/2$ or $\nu[514]9/2$. These orbitals slope upward with increasing β_2—so-called oblate orbitals—so that a particle here would tend to decrease the deformation of the $(A+1)$ nucleus, compensating the increase in J due to the larger mass. Agreement for other cases calculated by Ragnarsson, including that involving the $\pi[301]1/2$ hole (responsible for the three pairs of bands with identical energies), is not as satisfactory, which suggests that other effects, such as changes in pairing, should be included. A further requirement for the orbital is that the alignment be small, a criterion satisfied by the "oblate" orbitals.

4.   IDENTICAL BANDS IN THE $A = 190$ REGION

4.1   *Identical Transition Energies and Relation to* ^{192}Hg

There are even more examples of identical superdeformed bands in the $A = 190$ region than in the $A = 150$ region (63, 64). However, two features distinguish the bands in this region from those near $A = 150$: (*a*) many of the bands occur in pairs separated by two mass units, and (*b*) a large number of bands can be related to the superdeformed band in ^{192}Hg, which appears to serve as a doubly magic core. These features are illustrated in Figure 10*a*, which shows the degree of similarity between transition ener-

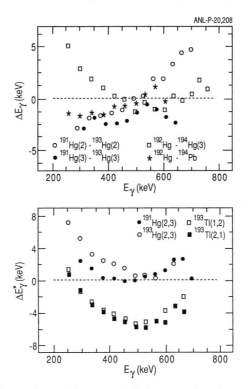

Figure 10   (*a*) Difference in transition energy ΔE_γ, between selected superdeformed bands in the $A = 190$ region.

(*b*) Energy difference ΔE_γ^* (see text) referenced to ^{192}Hg for strongly coupled bands in ^{191,193}Hg and ^{193}Tl. For both (*a*) and (*b*), the data are from (21–23, 28, 34); the figure is adapted from (64).

gies in the pairs [^{191}Hg(2), ^{193}Hg(2)], [^{191}Hg(3), ^{193}Hg(3)], [^{192}Hg, ^{194}Hg(3)], and [^{192}Hg, ^{194}Pb]. (The numbers in parentheses correspond to the labels given to different bands in the original publications; numbers larger than 1 designate excited bands.) Figure 10b relates transition energies of ^{192}Hg to those of bands in adjacent nuclei ^{191,193}Hg and ^{193}Tl through the relationship

$$\Delta E_\gamma^* = 1/2 \, [E_\gamma^f(R \pm 1/2) + E_\gamma^u(R \mp 1/2)] - E_\gamma^{core}(R),$$

where R is the core angular momentum, and u and f designate transitions from favored and unfavored states in signature partner bands of the odd-even nucleus. As discussed by Satula et al (64) and in Section 2 above, the proposed configurations for these superdeformed bands are characterized by $K \neq 1/2$. As a result both features in Figure 10 can be understood in a straightforward way in the strong coupling scheme, which gives $\Delta E_\gamma^* = 0$. However, there is again the requirement of equal moments of inertia in all nuclei.

Another way of relating the energies of different bands to those of a reference, ^{192}Hg, has been proposed by Stephens et al (40, 63), using the quantity

$$\Delta i = 2 \frac{\Delta E_\gamma}{\Delta E_\gamma^{ref}},$$

where $\Delta E_\gamma = E_\gamma - E_\gamma^{ref}$ is obtained by subtracting the transition energy E_γ in a band of interest from the closest transition energy in ^{192}Hg (E_γ^{ref}), and ΔE_γ^{ref} is calculated as the energy difference between the two closest transitions in the reference. Stephens et al called Δi incremental alignment but it is not necessarily related to any physical alignment. In the strong coupling limit, $\Delta i = \pm 1/2$ for an odd nucleus referred to an even core, as can readily be seen in Figure 9a–c. For even nuclei, bands with the same or different signatures have $\Delta i = 0$ or 1, respectively. Plots of Δi are shown in Figure 11, which illustrates that the superdeformed bands of the $A = 190$ region can be classified in two families. When additional particles (holes) with respect to ^{192}Hg occupy lower-K members of the high-$\mathcal{N}$ intruder orbitals, values of Δi scatter significantly (Figure 11a). This is not surprising since particles in these intruder orbitals tend to both increase deformation and align with rotation, thus making the strong coupling scheme inappropriate. In contrast, when the orbitals involved do not show much variation with $\hbar\omega$ (see Figure 4), that is, when there is little alignment, Δi values fall close to the limits of 0, $\pm 1/2$, ± 1, which are expected in the strong coupling limit (Figure 11b).

Stephens et al suggested that many of the cases that exhibit $\Delta i = 0$,

ANL-P-20,219

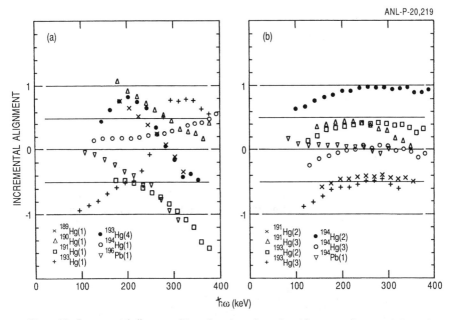

Figure 11   Incremental alignment Δi as a function of rotational frequency for superdeformed bands in $^{189-194}$Hg (21–23, 28–33) and ^{194,196}Pb (36, 37), with ^{192}Hg as a reference. As discussed in the text, the data are divided into two groups in order to illustrate the existence of two families; in (*b*) Δi clusters around integer and half-integer values. The data on superdeformed bands in the Tl isotopes have been omitted for clarity; these are discussed by Azaiez et al (35).

$\pm 1/2$, or ± 1 also have integer alignment differences with respect to the core, which they refer to as quantized alignment (40, 63). While the strong coupling limit corresponds to a zero alignment difference, Stephens et al instead suggest alignments (with respect to ^{192}Hg) of $1\hbar$ in, for example, ^{191}Hg(2,3) and ^{194}Hg(2,3), which are then attributed to pseudo-spin alignment (40). An alignment of $1\hbar$ in ^{191}Hg, instead of the expected $1/2\hbar$, is not easy to understand. However, a word of caution is in order. Whereas extraction of Δi does not require knowledge of spin, determination of alignment does. So far the spins of superdeformed band members have not been assigned using the conventional, tested techniques of γ-ray spectroscopy. Methods to infer the spin have been suggested (23, 65, 66), and are discussed below. At present, it is not established whether these methods give the exact spin or have uncertainties of at least $1\hbar$ (67, 68). Thus, until spins can be assigned firmly, one cannot be certain whether the alignments have the putative value of $1\hbar$ (40) or simply have the value $0\hbar$ expected from strong coupling.

4.2   *Spin Assignments*

Several related procedures for assigning spins to superdeformed states in the $A = 190$ region have been proposed recently (23, 65, 66). These procedures all start from a Harris expansion (69) of the moment of inertia

$$J^{(2)} = J_0 + J_1\omega_x^2 + \cdots$$

and the relation

$$\frac{dI_x}{d\omega_x} = J^{(2)}, \qquad\qquad 3.$$

where I_x is the projection of spin along the rotation vector, and $\omega_x = \Delta E_\gamma / \Delta I_x$ is the corresponding rotational frequency. Integrating Equation 3 gives

$$I_x = J_0\omega_x + \frac{J_1}{3}\omega_x^3 + i_0, \qquad\qquad 4.$$

where i_0 is the constant of integration. The parameters J_0 and J_1, obtained by fitting $J^{(2)}$ as a function of ω_x, are used to determine I_x with Equation 4. The spin I is then derived from the relation

$$I_x = \sqrt{I(I+1) - K^2},$$

where K is the projection of angular momentum along the symmetry axis. It has been shown that reasonable choices of K do not sensitively affect the derived spin (70). Two assumptions need to be valid for the derived spin to be correct. First, it must be assumed that the fit of $J^{(2)}$ vs ω_x can be correctly extrapolated to zero frequency. [A minimal requirement for this assumption to be correct is that $J^{(2)}$ varies smoothly and gradually in the low frequency domain.] Second, the constant of integration i_0, which corresponds to the alignment at zero frequency must be known. Becker et al (65) and Draper et al (66) assumed that $i_0 = 0$, since this results in calculated values of I that are within $0.1\hbar$ of the integer or half-integer value appropriate for an even or odd nucleus. However, it has not been unambiguously demonstrated that both assumptions are valid. In fact, pairing may well change as $\hbar\omega$ approaches zero, which would lead to uncertainty in the extrapolation of $J^{(2)}$ and result in an "effective" alignment (i.e. an effect that mimics alignment). The extent to which these effects are subsumed into the parameters J_0 and J_1 is not clear. Other attempts at spin determination have concluded that spin uncertainties of at least $1\hbar$ exist (67, 68). Thus, there is a clear need to determine the actual spins of superdeformed band members experimentally in order to resolve these uncertainties.

4.3   *Strong Coupling Limit and Identical Moments of Inertia*

The discussion above suggested that the identical bands can be understood in terms of the strong coupling limit of the particle-rotor model, with the additional requirement that the moments of inertia be identical. The following questions must then be raised: why does this strong coupling limit work so well and why are the moments of inertia so close? These are interesting questions for theory to address quantitatively, but qualitative comments can be made at this point. As discussed in Section 3.3, the strong coupling limit applies when the alignment is zero. (Assuming equal J values for the bands, the identical energies suggest $i \le 0.1\hbar$.) At the large deformations discussed here, alignment effects are expected to be small: the large value of the moments of inertia and the increased separation between Nilsson levels from a given shell both reduce the Coriolis coupling responsible for the alignment. In addition, the reduced pairing discussed in Section 2 also leads to less alignment. These qualitative arguments, however, do not guarantee that $i \le 0.1\hbar$—as implied by the equal transition energies—and, indeed, cranked shell model calculations suggest that alignments of the order of ~ 0.5–$1\hbar$ are obtained (70).

It has been emphasized above that identical transition energies require equal moments of inertia in the bands. This feature is unexpected since the masses involved differ by as much as two units. In the $A = 190$ region, scaling by $J \approx A^{5/3}$ results in a difference of ~ 5 keV per mass unit. Furthermore, bands are compared where the number of quasiparticles associated with the superdeformed configurations differ by one or two. These bands can be expected to have different moments of inertia J because blocking may be present, which would reduce pairing. A major question remains whether the equality of the moments of inertia reflects some hitherto undiscovered symmetry.

Another interesting observation can be added. When $J^{(2)}$ values for all the known superdeformed bands in the $A = 190$ region are compared (see Figure 5), they are all found to cluster within $\lesssim 10\%$ (with the exception of four cases). This is indeed a surprise since bands that differ by 0, 1, or 2 quasiparticle excitations are included. For rotational bands at normal deformation, $J^{(2)}$ values increase by $\sim 15\%$ per quasiparticle because of reduced pairing due to blocking. One wonders if the reduced pairing in superdeformed bands, which is partially responsible for smaller alignment, is also responsible for the noted clustering in $J^{(2)}$ values.

5. OTHER IMPORTANT EFFECTS AT LARGE DEFORMATION

Some pairs of orbitals such as the $i_{13/2}$-$f_{7/2}$ and the $j_{15/2}$-$g_{9/2}$ orbitals that are responsible for strong octupole correlations in light actinide nuclei (71) also appear close to the Fermi level in superdeformed configurations around ^{152}Dy and ^{192}Hg. Several recent calculations indicate that for many superdeformed nuclei the minima in the total energy surfaces exhibit considerable octupole softness, which is expected to persist even at the highest spins (32, 72–75). As a result, collective octupole vibrational excitations can be mixed with low-lying one- and two-quasiparticle states and the excitation pattern near the superdeformed yrast line can be different from that expected for axial symmetry. The calculations suggest that the first excited state in the doubly magic superdeformed nuclei ^{152}Dy and ^{192}Hg should be of collective octupole character and an analogy can be drawn with the well-known collective 3^- state in the doubly magic spherical nucleus ^{208}Pb. Octupole correlations are also expected to reduce single-particle alignments, increase band interactions, and modify the de-excitation pattern of the superdeformed states because of enhanced $B(E1)$ rates (71).

Most of the anticipated effects still remain to be observed experimentally. However, first evidence for strong mixing of quasiparticle excitations with octupole vibrations may have been seen in ^{193}Hg (32). In this nucleus, two of the four superdeformed bands are characterized by moments of inertia $J^{(2)}$ strikingly different from those of all other superdeformed bands in this region (see Figure 5), i.e. one shows a strong upbend at the frequency where the other shows a strong downbend. Furthermore, there is indirect experimental evidence for enhanced E1 transitions linking one of the irregular bands with the yrast superdeformed band. These observations, together with the reduced alignments observed and the strong interaction between the crossing bands, have been interpreted as evidence for strong octupole correlations (32). In the mass $A = 150$ region, one of the proposed explanations for the behavior of one of the superdeformed bands in ^{147}Gd invokes collective octupole excitations as well (42, 76), but other interpretations are also possible (46).

The residual n-p interaction may also play a role in superdeformed nuclei (77). Such an interaction is expected to be strong when protons and neutrons occupy rotationally aligned high-$\mathcal{N}$ intruder orbitals with large spatial overlap, such as the $\pi i_{13/2}$ and $\nu j_{15/2}$ orbitals involved in the superdeformed bands of the $A = 150$ and 190 regions. Such an interaction is expected to lower the energy of superdeformed bands based on these specific intruder configurations with respect to other excitations in the superdeformed well (77). Firm experimental evidence for the importance of this interaction is currently lacking even though it has been invoked to explain the strong feeding of a superdeformed band in ^{142}Eu (51) (interaction between $\mathcal{N} = 6$ odd proton and odd neutron) and the smooth rise of $J^{(2)}$ with $\hbar\omega$ in the lighter Hg isotopes (interaction between four $\mathcal{N} = 6$ protons and four $\mathcal{N} = 7$ neutrons) (30).

6.   FEEDING AND DECAY OF SUPERDEFORMED BANDS

Superdeformed bands are populated to much higher spins than states with smaller deformation (referred to as "normal" states hereafter) and have about an order of magnitude larger intensity than might be expected from the intensity pattern of the normal states (10, 78). These features make it interesting to try to understand the population mechanism. Important elements affecting the feeding are the level densities of both normal and superdeformed states, the mixing between the two classes of states at moderate excitation energy, and the electromagnetic decay rates in the normal and superdeformed minima (78, 79), as well as the barrier separating the two minima (80). Thus, investigations of the feeding mechanism also yield information on these aspects, providing perhaps the main incentive for such studies. Furthermore, knowledge of this mechanism allows selection of optimal conditions (in terms of reaction and bombarding energy) for spectroscopic studies of superdeformation.

Insight on the feeding mechanism may be obtained from data for both superdeformed and normal states on (a) population intensities as a function of spin and beam energy for one or several projectile-target combinations; (b) distributions in spin and energy at entry into the final nucleus; (c) shapes of the quasicontinuum and statistical spectra associated with the feeding of the two classes of states; and (d) the properties of ridge structures in E_γ-E_γ correlation matrices. Unfortunately, data on this problem have not kept pace with the steady flow of information on the many superdeformed bands discussed above. Only (a) and (b) are discussed here. [Moore et al (81) discuss (c); Nolan & Twin (10) and Schiffer et al (79) discuss (d).]

6.1   Intensities of Superdeformed Bands

Figure 12 presents, as a function of the rotational frequency $\hbar\omega$, the relative intensities of transitions in ^{149}Gd and ^{192}Hg, which are representative of superdeformed nuclei in their respective regions. In the $A = 150$ region, the number of coincident γ rays is generally larger and the transition

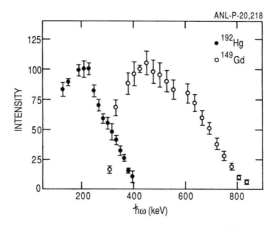

Figure 12   Relative intensities as a function of frequency $\hbar\omega$ for ^{192}Hg (22) and ^{149}Gd (44), normalized to 100 at their respective maxima.

energies are almost twice as large. The larger transition energy is principally due to the larger spins of the emitting states and, to a lesser extent, to the smaller moments of inertia. The first and last transitions in the super-deformed bands in ^{149}Gd and ^{192}Hg correspond to estimated spins of ($63/2$, $131/2\hbar$) and (10, $42\hbar$), respectively (22, 23, 44). In both cases, the population extends to much higher spin than in the normal states. Figure 12 also shows another similarity between bands in the two regions: as the spin decreases, the intensity gradually increases, flattens out at the maximum, and then drops rapidly within a couple of transitions as the bands decay to the lower-lying normal states. The superdeformed bands are weakly populated in all cases, with maximum intensities reaching about 1 and 2% in the $A = 150$ and $A = 190$ regions, respectively.

6.2 Entry Distributions

The entry distribution, representing the population distribution in spin and excitation energy after neutron emission, can be measured for super-deformed and normal states with 4π detector calorimetric arrays. Together with Compton-suppressed Ge detectors, these arrays constitute the detection systems used in all superdeformation studies. Data on complete distributions have not been published, but information exists on the average entry points, that is, the centroids of the distributions. The latter have been derived from data on the sum-energy (the total γ-ray energy emitted) and fold (average number of detectors firing) measured in coincidence with discrete normal or superdeformed transitions. The entry points for normal and superdeformed states in ^{192}Hg, measured at several bombarding energies with the ^{160}Gd(^{36}S,4n) reaction, are shown in Figure 13*a* (80). The entry spins for the superdeformed states are higher than those for the normal states, and the initial population leading to superdeformed states is colder, i.e. the entry points are lower in excitation energy than those of normal states at the same spin. Similar data on entry points for the $A = 150$ region have not been published, but the sum-energy and fold for normal and superdeformed states in ^{149}Gd have been measured (Figure 13*b*) (82). Because spin scales approximately linearly with fold, feeding of superdeformed bands also originates from higher spin and lower energy, as for ^{192}Hg. Thus, feeding of superdeformed bands in both $A = 150$ and $A = 190$ regions share common characteristics.

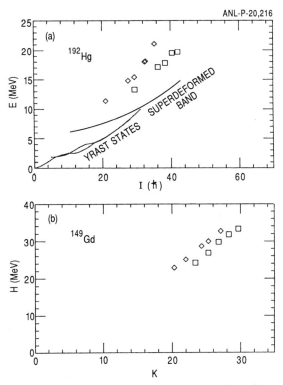

Figure 13   (*a*) Entry points for normal (*diamonds*) and superdeformed (*squares*) states in ^{192}Hg measured in the ^{160}Gd(^{36}S,4n) reaction at several bombarding energies (80). The known yrast states are indicated; the energy of the superdeformed band is not known and is arbitrarily chosen.

(*b*) Average sum-energy H and fold K for normal and superdeformed states in ^{149}Gd measured at a number of projectile energies in the ^{124}Sn(^{30}Si,5n) reaction (82).

Since the entry spins for ^{149}Gd must necessarily be larger than the average spin where the discrete superdeformed band is fed ($\sim 111/2\hbar$), the spins of the states from which the superdeformed bands originate are larger in ^{149}Gd than in ^{192}Hg. On the other hand, the maximum angular momenta brought into the compound nuclei are comparable in the two cases. The lower entry spin in ^{192}Hg is attributed to the depletion of the highest partial waves by fission, which predominates over evaporation residue formation beyond spin $40\hbar$ (80). (Thus, superdeformed bands have been identified in this nucleus near to the limit of fission instability.) It is striking that the feeding of the superdeformed bands originates from the highest spins in the evaporation residues. This occurs because the level densities of superdeformed states exceed those of the normal states only for the highest spins.

Schiffer and Herskind, in an extensive series of calculations, have successfully described many of the observed feeding features in ^{152}Dy and have provided insight into the population mechanism (78, 79). The calculations start by describing the first phase of the decay of the compound nucleus (neutron evaporation and fission) to obtain an entry distribution in spin and energy from which γ decay occurs. The decay is then followed through Monte Carlo simulations and consists of a competition at each step between electromagnetic decay (E1 statistical γ decay, which cools the nucleus, and collective E2 γ decay, which removes two units of angular momentum) in the normal and superdeformed wells, and hopping between the normal and superdeformed wells. As the energy increases, the level density for the superdeformed states increases less rapidly than for the normal states. This results in a limited region in excitation energy in which the density of superdeformed states is larger than that of normal states. This region corresponds to the highest partial waves and is located roughly beyond the point where the superdeformed band crosses the normal states and becomes yrast (Figure 13). Schiffer and Herskind suggest that the entry states for the superdeformed band originate from this region and that this feature accounts for the "colder" feeding of the superdeformed bands. Although the barrier between the superdeformed and normal states was included in the calculations and governed the tunneling between the two wells, its role was not emphasized. The barrier may, in fact, play the pivotal role since mainly entry states within the superdeformed well (states that are below the barrier) are likely to be trapped within the well (80). If this is indeed the case, then the entry energy is closely related to the barrier energy. As a result, a measurement of the entry points could provide a determination of the barrier and also of the superdeformed well depth, although the latter requires knowledge of the excitation energy of the band. These quantities are directly related to the shell correction responsible for the superdeformed pocket, and thus they are interesting to compare with theoretical values.

6.3   Deexcitation of Superdeformed Bands

The sudden decrease of superdeformed transition intensities at low frequencies (Figure 12) indicates that, after a long cascade of consecutive intraband transitions, a rapid decay toward the lower-lying normal states takes place. So far, none of the decay paths linking the superdeformed bands with the normal states has been identified in the $A = 150$ or 190 regions, which suggests that the decay is fragmented into many pathways, each too weak to define, and is probably statistical in nature (78). It has, therefore, not been possible to define the spins and excitation energies of superdeformed band levels with standard spectroscopic techniques. This task remains the major challenge in the study of superdeformation.

The transition from a superdeformed to a normal shape represents a large amplitude motion involving a drastic structural rearrangement. It is perhaps not surprising that the decay is fragmented into many intermediate excited states involving, in some cases, a time delay before the yrast line is reached (21, 83).

The deexcitation of superdeformed bands represents an interesting problem addressed by a number of recent calculations (78, 84, 85). As γ decay occurs within a superdeformed band, a curious situation arises where the excitation energy of the superdeformed states with respect to the yrast levels increases (see Figure 13a). At the point of decay, the superdeformed levels have an estimated energy of 3–6 MeV above yrast. Hence, a cold state, isolated within its own potential well, is immersed in a hot sea of normal states with very high density. The decay then occurs through admixtures with this sea of states. Properties of the decay, e.g. the spin dependence of the out-of-band decay probability, can provide information on the mixing between the superdeformed and normal states and on the barrier separating them. The gross energy distribution of the decay γ rays may also provide additional information. (It may be easier to determine this distribution than to trace out the individual pathways.)

7.   SUMMARY AND OUTLOOK

Superdeformed nuclei have now been discovered in four distinct regions of the periodic table with masses around $A = 130$, 150, 190, and 240, and with respective axis ratios of 3:2, 1.9:1, 1.7:1, and 2:1. These coincide with the regions where theory, which incorporates a macroscopic liquid-drop term and a quantal shell-correction term, also predicts the occurrence of nuclei with very large deformation. This success represents a triumph for the Strutinsky method, and for mean-field theories in general, in describing the macroscopic and microscopic aspects of nuclear behavior.

In the $A = 150$ and 190 regions reviewed here, it has been possible to perform spectroscopic studies in the superdeformed secondary well, and both "ground" and excited bands have been observed. An unexpected discovery is that a number of the excited bands have energies identical to those of the lowest superdeformed bands in adjacent nuclei. The degeneracies of better than one part in 500 require that the moments of inertia be identical and also that the strong coupling limit applies, both to a remarkable degree. It is not clear whether this is a result of an accident or is a consequence of a symmetry that has yet to be identified. Perhaps models that exploit symmetries, such as the interacting boson model (86), may shed light on this question. One also wonders if the identical energies of bands in even and odd nuclei might be a manifestation of supersymmetry which covers both fermion and boson degrees of freedom (87). Pseudo-spin alignment partially accounts for identical energies in superdeformed bands of some adjacent even and odd nuclei, but it is not clear if it plays a significant role in other identical bands.

Although there has been much progress in research on super-deformation, many questions remain and, indeed, new questions have been raised by the new discoveries. Certainly, a major challenge is to determine the excitation energies and spins of superdeformed levels since there is not a single superdeformed band in the $A = 150$ and 190 regions for which these properties are known. The borders of the regions of super-deformation in the periodic table need to be established. The known excited superdeformed bands are believed to correspond to particle or quasiparticle excitations, but the collective modes associated with states of large deformation—for example the beta, gamma, or octupole vibrations—are yet to be found. If the deformation is stiff with respect to quadrupole distortion, the beta vibrations may lie at high excitation energies, but soft octupole modes may exist at lower energies (88), possibly giving rise to exotic bending modes. To shed more light on the cause of identical bands, it will probably be necessary to identify higher-lying particle states that can be used, for instance, to search for evidence of pseudo-spin orbit doublets and pseudo-spin alignment, as described in Section 3.4.

Investigations of the mixing between states in the normal and super-deformed well will give information on the barrier separating them and on the superdeformed well depth. Both quantities are important to study because they are direct manifestations of the shell corrections leading to the formation of the superdeformed pocket. Mixing in excited states can be investigated through the properties of ridges in the E_γ-E_γ matrices and, at higher energies, by studying the feeding mechanism of superdeformed bands. Finally, mixing between cold superdeformed states and excited normal ones can be studied through the decay properties involved in the deexcitation out of superdeformed bands.

At present, experiments are limited by the detection capabilities of current γ-ray detectors. The next generation of detector arrays currently under construction (GAMMASPHERE in the United States and EURO-GAM in Europe) will improve the detection sensitivity by about two orders of magnitude and provide answers to the questions raised above. Many fascinating discoveries about superdeformation—and about nuclear structure in general—lie ahead.

ACKNOWLEDGEMENTS

The data and ideas reviewed above are the result of dedicated work by many colleagues and friends. It is impossible to name them all here. We would like, however, to acknowledge the contributions of our collaborators at Argonne, Notre Dame, Purdue, I.N.E.L., Stockholm, and Warsaw. We also thank Mike Carpenter, Patricia Fernandez, and Frank Moore for carefully reading the manuscript. This work was supported by the Department of Energy, Nuclear Physics Division, under contract number W-31-109-ENG-38.

Literature Cited

1. Strutinsky, V. M., *Nucl. Phys.* A95: 420–42 (1967); A122: 1–33 (1968)
2. Polikanov, S. M., et al., *Sov. Phys. JETP* 15: 1016–21 (1962)
3. Tsang, C. F., Nilsson, S. G., *Nucl. Phys.* A140: 275–88 (1970)
4. Bengtsson, R., et al., *Phys. Lett.* B57: 301–5 (1975)
5. Ragnarsson, I., et al., *Nucl. Phys.* A347: 287–315 (1980)
6. Dudek, J., Nazarewicz, W., *Phys. Rev.* C31: 298–301 (1985)
7. Chasman R. R., *Phys. Lett.* B187: 219–23 (1987)
8. Twin, P. J., et al., *Phys. Rev. Lett.* 57: 811–14 (1985)
9. Kirwan, A. J., et al., *Phys. Rev. Lett.* 58: 467–70 (1987)
10. Nolan, P. J., Twin, P. J., *Annu. Rev. Nucl. Part. Sci.* 38: 533–62 (1988)
11. Bohr, A., Mottelson, B. R., *Nuclear Structure*, Vol. 2. Reading, Mass: Benjamin (1975)
12. Nilsson, S. G., et al., *Nucl. Phys.* A131: 1–66 (1969)
13. Åberg, S., *Phys. Scr.* 25: 23–27 (1982)
14. Bengtsson, T., Ragnarsson I., *Nucl. Phys.* A436: 14–82 (1985)
15. Chasman, R. R., *Phys. Lett.* B219: 227–31 (1989)
16. Girod M., et al., *Phys. Rev. Lett.* 62: 2452–56 (1989)
17. Bonche, P., et al., *Nucl. Phys.* A500: 308–22 (1989)
18. Moore, E. F., et al., *Phys. Rev. Lett.* 63: 360–63 (1989)
19. Nazarewicz, W., et al., *Phys. Lett.* B225: 208–14 (1989)
20. Bengtsson, T., et al., *Phys. Lett.* B208: 39–43 (1988)
21. Riley, M. A., et al., *Nucl. Phys.* A512: 178–88 (1990)
22. Ye, D., et al., *Phys. Rev.* C41: R13–16 (1990)
23. Becker, J. A., et al., *Phys. Rev.* C41: R9–12 (1990)

24. Hubel, H., et al., *Nucl. Phys.* A453: 316–48 (1986)
25. Moore, E. F., et al., *Phys. Rev. Lett.* 64: 3127–30 (1990)
26. Nazarewicz, W., et al., *Nucl. Phys.* A503: 285–330 (1989)
27. Bengtsson T., et al., *Nucl. Phys.* A496: 56–65 (1989)
28. Carpenter M. P., et al., *Phys. Lett.* B240:44–49 (1990)
29. Janssens, R. V. F., et al., *Nucl. Phys.* A520: 75c–90c (1990)
30. Drigert, M. W., et al., et al., *Nucl. Phys.* A530: 452–74 (1991)
31. Henry E. A., et al., *Z. Phys.* A335: 361–62 (1990)
32. Cullen, C. M., et al., *Phys. Rev. Lett.* 65: 1547–50 (1990)
33. Beausang, C. W., et al., *Z. Phys.* A335: 325–30 (1990)
34. Fernandez, P. B., et al. *Nucl. Phys.* A517: 386–98 (1990)
35. Azaiez, F., et al., *Z. Phys.* A336: 243–44 (1990); *Phys. Rev. Lett.* 66: 1030–33 (1991)
36. Brinkman, M. J. et al., *Z. Phys.* A336:115–16 (1990)
37. Theine, K., et al., *Z. Phys.* A336: 113–14 (1990)
38. Nazarewicz, W., In *Proc. XXV Zakopane Sch. on Physics*, to be published (1991)
39. Chasman, R. R., *Phys. Lett.* B242: 317–22 (1990)
40. Stephens, F. S., et al., *Phys. Rev. Lett.* 64:2623–26 (1990); 65: 301–4 (1990)
41. Hebbinghaus, G., et al., *Phys. Lett.* B240: 311–16 (1990)
42. Zuber, K., et al., *Nucl. Phys.* A520: 195c–200c (1990); *Phys. Lett.* B254: 308–14 (1991)
43. Deleplanque, M. A., et al., *Phys. Rev. Lett.* 60:1626–29 (1988)
44. Haas, B., et al., *Phys. Rev. Lett.* 60: 503–6 (1988); *Phys. Rev.* C42: R1817–21 (1990)
45. Fallon, P., et al., *Phys. Lett.* B218: 137–42 (1989); B257: 269–72 (1991)
46. Janzen, V. P., et al., In *Proc. Int. Conf. on High Spin Physics and Gamma-soft Nuclei.* Pittsburgh: World Scientific (1991), pp. 225–43
47. Deleplanque, M. A., et al., *Phys. Rev.* C39: 1651–54 (1989)
48. Rathke, G.-E., et al., *Phys. Lett.* B209: 177–81 (1988)
49. Johansson, J. K., et al., *Phys. Rev. Lett.* 63: 2200–3 (1989)
50. Lieder, R. M., et al., *Nucl. Phys.* A520: 59c–66c (1990)
51. Twin, P. J., *Nucl. Phys.* A520: 17c–33c (1990); Mullins S. M., et al., *Phys. Rev. Lett.* 66: 1677–80 (1991)
52. Shimizu, Y. R., et al., *Phys. Lett.* 198B: 33–38 (1989); *Nucl. Phys.* A509: 80–116 (1990)
53. Byrsky, T., et al., *Phys. Rev. Lett.* 64: 1650–53 (1990)
54. Nazarewicz, W., et al., *Phys. Rev. Lett.* 64: 1654–58 (1990)
55. Twin, P. J., *Nucl. Phys.* A522: 13c–30c (1991)
56. Hecht, K. T., Adler, A., *Nucl. Phys.* A137: 129–43 (1969)
57. Arima, A., et al., *Phys. Lett.* B30: 517–22 (1969)
58. Ratna-Raju R. D., et al., *Nucl. Phys.* A202: 433–66 (1973)
59. Bohr, A., et al., *Phys. Scr.* 26: 267–72 (1982)
60. Mottelson, B., *Nucl. Phys.* A520: 711c–22c (1990)
61. Deleted in proof
62. Ragnarsson, I., *Nucl. Phys.* A520: 67c–74c (1990)
63. Stephens, F. S., *Nucl. Phys.* A520: 91c–104c (1990)
64. Satula, W., et al., *Nucl. Phys.* In press (1991)
65. Becker, J. A., et al., *Nucl. Phys.* A520: 188c–94c (1990)
66. Draper, J. E., et al., *Phys. Rev.* C42: R1791–95 (1990)
67. Wu, C., et al., *Phys. Rev. Lett.* 66: 1377–78 (1991)
68. Wyss, R., Pilotte, S., *Phys. Rev. C.* In press (1991)
69. Harris, S. M., *Phys. Rev.* B509: 138–54 (1965)
70. Carpenter, M. P., et al., to be published (1991)
71. Nazarewicz, W., *Nucl. Phys.* A520: 333c–51c (1990) and references therein
72. Höller, J., Åberg, S., *Z. Phys.* A336: 363–64 (1990)
73. Åberg, S., *Nucl. Phys.* A520: 35c–57c (1990)
74. Dudek, J. et al., *Phys. Lett.* 248B: 235–42 (1990)
75. Bonche, P. et al., *Phys. Rev. Lett.* 66: 876–79 (1991)
76. Szymanski, Z., *Nucl. Phys.* A520: 1c–16c (1990)
77. Wyss, R., Johnson, A., see Ref. 46, pp. 123–32
78. Herskind, B., Schiffer, K., *Phys. Rev. Lett.* 59: 2416–19 (1987)
79. Schiffer, K., et al., *Z. Phys.* A332: 17–27 (1989)
80. Khoo, T. L., et al., *Nucl. Phys.* A520: 169c–77c (1990)
81. Moore, E. F., et al., In *Proc. Workshop on Nuclear Structure and Heavy-ion Reaction Dynamics, Inst. Phys. Conf. Ser.* 109: 171–78 (1991)
82. Taras, P. et al., *Phys. Rev. Lett.* 61:

1348–51 (1988); Haas, B., et al., *Phys. Lett.* B245: 13–16 (1990)

83. Carpenter, M. P., et al., *Nucl. Phys.* A520: 133c–37c (1990)

84. Vigezzi, E., et al., *Phys. Lett.* B249: 163–67 (1990)

85. Bonche, P., et al., *Nucl. Phys.* A519: 509–20 (1990)

86. Iachello, F., Arima, A., *The Interacting Boson Model.* Cambridge Univ. Press (1987); Iachello, F., *Nucl. Phys.* A522: 83c–98c (1991)

87. Iachello, F., van Isacker, P., *The Interacting Boson Fermion Model*, Cambridge Univ. Press (1990)

88. Mizutori, S., et al., to be published (1991)

Annu. Rev. Nucl. Part. Sci. 1991. 41: 357–88

STORAGE RINGS FOR NUCLEAR PHYSICS

R. E. Pollock

Department of Physics and IUCF, Indiana University, Bloomington, Indiana 47405

KEY WORDS: cooled particle beams, accelerator facilities

CONTENTS

INTRODUCTION

A basic tool for the study of nuclear forces and structures is the fast particle probe. While the first sources were naturally occurring radioisotopes, and while cosmic rays and astrophysical processes have continued to provide alternative paths to the understanding of nuclear phenomena, much of the progress in the field is linked to the man-made particle accelerator as the source of particle beams.

The classical cyclotron of Lawrence and his colleagues was not only an accelerator but an early example both of a particle trap and of a beam

357

0163–8998/91/1201–0357$02.00

storage device. Storage allowed the repetitive passage of beam particles through a radiofrequency structure giving a useful energy multiplication. Successive extensions of this concept led to one important type of beam source: the modern circular accelerator. Phase focussing added longitudinal stability, alternating gradients added robust transverse focussing, cyclic operation and momentum compaction reduced the bending field volume, and extraction and transport systems carried beams to the user.

The energy frontier, a shock wave pushed upward so rapidly by these technical advances, left in its wake a severe dichotomy between the closely related fields of particle and nuclear physics, from which we are only beginning to emerge. The many-body nature of a complex nucleus results in a rich spectrum of closely spaced energy levels. To be useful in a study of this system, particle beams must have qualities other than raw energy. Beam sources specialized for nuclear studies place emphasis on phase space density ("brightness"), variability of energy or species, and other properties that set them apart from the machines on the energy frontier. In consequence, nuclear accelerators have had a disparate technical evolution.

Within the past decade or so, a reconsideration of the role of beam storage has changed our perception of the ideal machine and has begun to bring together the once separate branches of the accelerator community. During the same period, some particle theorists have recognized the nucleus as a potentially useful laboratory for nonperturbative applied QCD and have interacted more strongly with their nuclear colleagues.

A beam storage device may be used as a particle beam processor. Acceleration is one among many useful processes. Time structure alteration and phase space compression through cooling are other possible beam manipulations. Storage buys processing time. For the price of a ring of magnets, one gains access to a variety of slow processes that would be ineffective in the brief flight of a beam from accelerator to target.

The reasons for beam extraction need also to be reconsidered. Historically, getting the experiment away from the cramped, high background environment of the inside of the traditional accelerator was a major improvement. Now, however, we can design storage rings with ample space for internal target experiments. A stored beam particle can pass through an internal target many times before being lost, so very thin targets can give useful interaction rates. This efficient use of the beam may actually reduce backgrounds relative to extracted beams.

The storage ring is unlikely to replace completely the conventional accelerator/beam-line/target facility as the primary tool for nuclear research. In particular, the very high production rate of secondary beams in meson factories cannot now be obtained via internal targets. However, there are some experiments that are only practicable or at least are better performed with a stored beam, so these facilities may be expected to play an important role in the future. The precise boundaries to the regimes in which one or the other technology is most appropriate are still being established.

In this review, the second section is a brief tutorial for the physicist who is not an accelerator expert, introducing terminology and concepts useful in understanding the basic properties of beam storage devices and the beam processing that they make possible. The third section deals with performance-limiting phenomena: stored beam intensity limits; beam-target interactions and their effects on target thickness and beam quality; and the luminosity domains at different energies and for various ion species for applicability of the stored beam technology. In the final section, examples are presented from among facilities in operation or under construction to show some of the diversity of this rapidly evolving field. A glossary of terms and abbreviations used in this review is included for the reader's convenience.

THE STORAGE RING AS BEAM PROCESSOR

The Container

To carry out an experiment with a storage ring may require of the experimenter a more intimate knowledge of the machine than may be necessary when an accelerated beam is delivered to a target at a remote location. Experience has shown that defining some of the machine terminology is a helpful prelude to communicating the ideas. This section is offered for that purpose and may be skipped by the expert reader.

SPATIAL CONFINEMENT   A ring is a topological trap: escape of particles from the open ends is circumvented by bending the particle path back onto itself. Usually the bending fields are produced by magnetic dipoles. The reference frame for describing particle motion is noninertial and relative to a nominal "design orbit." The "closed orbit" may deviate appreciably from the design orbit because of errors such as small misalignments. The azimuthal pattern of deviations is called the "closed orbit distortion." The ring design will typically include position monitors and dipole correction elements to measure and reduce these deviations.

Dipoles, quadrupoles, and possibly higher multipoles form the complement of "lattice elements," which constitute the confinement system of the ring. Transverse stability with respect to the closed orbit is provided by focussing elements, typically magnetic quadrupoles of alternating polarity. Not all arrangements of lattice elements provide stable confinement. There is typically a band structure to the allowed lens strengths arising from the

Glossary of terms and abbreviations used

Adone	A 1.5-GeV electron-positron collider (INFN Frascati, Italy)
AmPS	Amsterdam Pulse Stretcher
ASTRID	Århus Storage Ring in Denmark
B	Bottom (or beauty) quark
BNL	Brookhaven National Laboratory (New York)
CELSIUS	Cooler for Electron Storage in Uppsala, Sweden
CERN	Centre European de Recherche Nucleaire (Geneva)
CoSy	COoler-SYnchrotron
CRYEBIS	CRYogenic Electron Beam Ion Source
CRYSIS	CRYogenic Stockholm Ion Source
CRYRING	CRYsis synchrotron-storage RING
ESR	Experimental Storage Ring (Darmstadt)
FNAL	Fermi National Accelerator Laboratory (Chicago)
GSI	Gesellschaft für SchwerIonenforschung mbH (Darmstadt)
HISTRAP	Heavy Ion STorage Ring for Atomic Physics
INS	Institute for Nuclear Studies (Tokyo)
ISR	Intersecting Storage Rings (CERN)
IUCF	Indiana University Cyclotron Facility
JINR	Joint Institute for Nuclear Research (Dubna)
KFA	KernForschungsAnlage (now Forschungszentrum) (Jülich)
LAMPF	Los Alamos Meson Physics Facility (New Mexico)
LEAR	Low Energy Antiproton Ring (CERN)
MIMAS	Injector for (moon of) Saturne (Saclay, France)
MIT	Massachusetts Institute of Technology
MP	"Emperor" class tandem electrostatic accelerator
MPI	Max Planck Institute
MSI	Manne Siegbahn Institute (Stockholm)
NEP	New Electron Ring (Novosibirsk)
NIKHEF-K	Nationaal Instituut voor Kernfysica en Hoge Energie Fysica–sectie K(ernfysica) (Amsterdam)
NSF	National Science Foundation (a US funding agency)
ORNL	Oak Ridge National Laboratory (Tennessee)
PEGASYS	PEP GAs jet Spectrometer target sYStem
PEP	Positron Electron Proton collider (Stanford)
PSR	Proton Storage Ring (Los Alamos)
QCD	Quantum ChromoDynamics
RCNP	Research Center for Nuclear Physics (Osaka)
RFQ	RadioFrequency Quadrupole (a type of linear accelerator)
RHIC	Relativistic Heavy Ion Collider (at BNL)
SHR	South Hall Ring (at MIT)
SIS	Schwerionen Synchrotron (at GSI)
SLAC	Stanford Linear Accelerator Center
TARN	Test Accelerator Ring for NUMATRON (at INS Tokyo)
TSL	The Svedberg Laboratory (formerly GWI) (at Uppsala)
TSR	Test Storage Ring (at MPI Heidelberg)
VEPP	A collider for electron-positron beams
ZGS	Zero Gradient Synchrotron (formerly at Argonne Lab, Chicago)

periodicity imposed by orbit closure, in close analogy to the band structure of electrons in crystal lattices, from which the name derives. The confined particle executes periodic "betatron" motion in both transverse planes with respect to the closed orbit. The number of oscillations per revolution in each of the two transverse dimensions characterizes the strength of the transverse focussing. This pair of wave numbers is called the "tune" of the ring.

It is useful to think of the focussing elements in terms of a confining pseudopotential, with the closed orbit following the path of minimum transverse potential energy. At any ring azimuth, a particle passing by may have its transverse position and momentum in either dimension specified as a point in the two-dimensional phase space for that dimension. A particle on a closed orbit traces out a closed curve upon enough successive returns to this azimuth. The area within the closed curve is the "particle emittance" for the chosen transverse direction. For linear optics, the figure is an ellipse. By convention, the Hamiltonian conjugate (position, momentum) pair is modified by dividing the transverse momentum by the (much larger) longitudinal momentum to form an angle with respect to the design orbit; thus the emittance has dimensions of length, and is best stated with an explicit factor of π to avoid ambiguity in the specification of the ellipse area. The "beam emittance" is the appropriate moment of the distribution for an ensemble of particles in the beam.

Projections onto position space of the bounded transverse motion at all azimuthal positions trace out the beam "envelope." An envelope minimum ("waist") or maximum occurs where the beam ellipse is upright. The beam transverse dimension is given by the geometric mean of emittance divided by π and a scaling quantity called the beta function (1) that describes the envelope. The ring "acceptance" is the largest emittance for which the beam envelope avoids obstacles such as the vacuum chamber.

Confinement may exist for a band of longitudinal momenta extending to either side of the design momentum. The width of the band is the "momentum acceptance," commonly expressed as the fraction of the nominal momentum, in percent. In general a different momentum leads to a different closed orbit. The momentum dispersion $\eta_x = p(dx/dp)$ and angular dispersion $\eta_x' = p(d\theta/dp)$ (and corresponding quantities in the other transverse direction) are functions of azimuth that express the form of the deviation of the off-momentum closed orbit. If the longitudinal momentum of a particle is changed (by radiofrequency or by a collision) at an azimuth where there is dispersion, its closed orbit shifts position so the transverse emittance is also altered. This "dispersion-coupling" phenomenon is important to a number of the limiting processes discussed in following sections.

Higher momentum particles are less strongly focussed, so the tune decreases with increasing momentum. The slope is called the "chromaticity." Sextupole elements placed in regions of high dispersion are commonly used to reduce the chromaticity, which may increase the phase space volume available for storage. Nonlinear elements have undesirable side effects (2), however, which may reduce acceptance in the vicinity of "orbit resonances" where the tune is a ratio of small integers. A complete discussion of resonance phenomena is outside the scope of this section. The "dynamic aperture" (3) is the effective acceptance when nonlinearities are correctly treated. This quantity may have to be determined by particle tracking calculations as the nonlinearities and couplings can lead to chaotic motion.

SPIN CONFINEMENT   The storage ring allows manipulations of the spin of particles. Just as there is a closed orbit defining a stable transverse position at every azimuth, there is a stable direction for spin at each azimuth, given by an eigenvector of the spin precession matrix for one revolution. Often the dominant magnetic field seen by the beam is the dipole component perpendicular to the ring bend plane. If other fields may be neglected, the stable spin direction is parallel to the bend field and perpendicular to the ring plane at all azimuths. However, other fields can be added to alter the spin direction. An example would be a strong solenoid creating a $180°$ precession of spin about the beam axis. The effect of such an element is to change the stable spin direction so it lies in the ring plane and is parallel to the beam direction at the azimuth opposite to the solenoid.

Combinations of dipoles or dipoles plus solenoids can be devised (4) to give arbitrary precessions about any axis. The orbit path, if the spin precessor contains dipoles, may become rather contorted, which is why these devices are referred to as "snakes." Snakes are called type I, II, or III if the added precession is about the longitudinal, radial, or vertical axis, respectively. More than one snake may be used in a ring. For example, a combination of a type I snake with a type II snake opposite gives spin up for half the circumference and spin down for the other half.

Just as certain betatron frequencies excite orbit resonances and therefore lack stability, certain spin precession frequencies in the ring excite depolarization resonances. These can make the stable spin direction undefined at some energies and cause a loss of polarization or a spin flip during acceleration. A snake may be employed (5) to simplify the passage through such resonances.

Conservative Processes

Properties of the stored beam may be altered in a manner that largely conserves phase space volume. An example is "adiabatic capture" with a radiofrequency system (6), in which time structure is imposed on a previously coasting beam. Other processes such as beam cooling, in which phase space density is altered, are discussed in the next section. The distinction is not precise because the density-conserving condition may not be satisfied in all regions of the phase space, giving rise to some localized dilution. A more extreme dilution can occur in a bunched beam when the radiofrequency is switched off. The faster particles overtake the slower until time structure is lost. In principle this process is reversible but it is usually not practicable to recover the time structure (7), so a large net reduction of phase space density may occur.

LONGITUDINAL MANIPULATION   A ring may be used to alter the beam time structure. A stretcher ring increases the macroscopic duty factor, while a compressor ring reduces it. In a stretcher, the injection occurs in one to a few orbit periods, while extraction is stretched over many orbit periods. In a compressor, the injection covers many periods and the extraction takes place in a single turn or less. Either operation may be useful, depending on the application. The proton storage ring (PSR) at LAMPF (8) offers both modes. A stored beam of long lifetime always has nearly 100% macroscopic duty factor since the current contains only frequencies that are multiples of the orbit frequency, modulated by the decaying beam envelope. An extracted beam can have slow time structure superimposed by properties of the extraction process. In favorable cases, the macroscopic duty factor of an extracted beam may exceed 80%.

The processes of rebunching and debunching use a radiofrequency (rf) system to alter the width/height ratio of beam bunches in longitudinal phase space. The beam is said to be debunched if the time spread is increased and energy spread decreased, and rebunched if the opposite occurs. Debunching may occur adiabatically, by gradually decreasing the rf amplitude with the beam following the change in shape of the phase stable region, or suddenly, by allowing a one-quarter period of synchrotron oscillation of a beam bunch having an initial time duration much smaller than the rf period. Debunching may also be possible in a beam line if the transit time is long enough. However, the debunching used as part of rf stacking in the IUCF Cooler would require a beam line with an effective length of a few kilometers (9; X. Pei, PhD thesis, Indiana Univ., in preparation).

A ring may of course accommodate alteration of the beam energy: in a synchrotron, ramping magnets and rf acceleration with phase stability are employed.

TRANSVERSE MANIPULATION   The analog of rebunching or debunching is performed in the transverse phase space by focusing elements. A lens system can make the beam more parallel, reducing its transverse momentum spread. Since the aperture function governing beam size varies with position in the ring, the beam "temperature" tensor, defined as $\langle p^2 \rangle / 2m$ for each momentum component, is not independent of azimuthal position. The beam is hottest approaching a tight waist, and is coolest where it is largest, within the lenses.

The use of temperature to describe a system far from equilibrium is a source of confusion. The temperature at a waist can be modified by space charge forces. A dense beam (10) will "bounce" at the waist, with the transverse kinetic energy partially converted to electrostatic potential energy, and thus briefly reduce the temperature. On the other hand, particles of a dilute beam will follow undeflected trajectories, so the beam will not change temperature in passing through the waist. However, a section of the distribution will show a maximum temperature at the waist. To avoid ambiguity the concept of beam temperature ought to be replaced by a description of the transverse beam quality in terms of emittance, which is an approximate invariant during passage through the ring lattice. Beam "cooling" then refers to a reduction of phase space volume. In the longitudinal dimension, the same difficulty of terminology is seen for a beam performing coherent quadrupole synchrotron oscillations; such a beam exhibits a periodic modulation in the longitudinal temperature but at most a slow variation in longitudinal emittance.

COUPLING MANIPULATIONS   Skew quadrupoles (rotated about the beam axis) or solenoids in the lattice can introduce mixing between the two transverse dimensions, tending to equalize transverse emittances. Robinson (11) pointed out a use for this mixing in stabilizing an emittance growth in the radial direction.

Nonconservative Processes

A heating or cooling process increases or reduces phase space volume, respectively. Heating may result from interaction of the beam with either residual gas or with a target in the ring, or it may result from random fluctuations of ring electromagnetic fields. Normally, heating is an undesirable effect to be minimized. Cooling, on the other hand, is normally a desirable effect to be maximized. It may occur naturally or it may be deliberately imposed. Useful reviews of cooling phenomena are available (12, 13). An example of natural cooling is synchrotron radiation emission in combination with rf acceleration. Examples of imposed cooling are immersion in a cold bath (electron cooling); active feedback to reduce beam density fluctuations (stochastic cooling); or directed resonant photon absorption in combination with isotropic re-emission (laser cooling). Each process is discussed below.

Under suitable conditions, charge-changing injection may increase phase space density markedly, thus enhancing beam brightness (14) without cooling. Consider the time-reversed dilution process: a partial charge change followed by magnetic separation into two beams of lower phase space density, one still stored and the other proceeding backward through the injection channel.

Interaction of a beam with an inhomogeneous energy degrader (for example, a wedge of material) in a region with dispersion may cause a phase space contraction in a transverse dimension accompanied by a growth in the longitudinal dimension, or the converse. To cool longitudinally, one must make the degrader thicker on the high momentum side. To a first approximation, the six-dimensional volume is not reduced, but the aspect ratio of the temperature tensor is altered. The possibility of achieving net cooling in a thick degrader is discussed by Skrinsky & Parkhomchuk (12) and named by them "ionization cooling." A difficulty is that the cooling accompanying energy loss is normally weaker than the heating from beam-target interactions. The most favorable case is for energetic muon beams. The related process of positron brightness enhancement through focussing, stopping, and reacceleration (15) lies outside the scope of this review.

SYNCHROTRON RADIATION DAMPING   For relativistic beams with $E/mc^2 = \gamma \gg 1$, bremsstrahlung emission induced by lattice fields is directed forward within a narrow cone of half angle $1/\gamma$. If the betatron motion contains angles larger than this, the reactive kick accompanying photon emission opposes the motion, which leads to transverse cooling with equilibrium rms angle of about $1/\gamma$. More energetic particles emit radiation more often, giving longitudinal cooling. The longitudinal component of the momentum loss is made up by an rf system that maintains the average stored beam energy. The damping rates can be quite large, giving cooling times of a few milliseconds at 1 GeV. Fluctuations in the quantized emission determine an equilibrium momentum spread typically of order 10^{-3} for a 1-GeV electron ring (16). This cooling mechanism is exploited in filling of electron rings. Being rapid, it should be capable of maintaining thermal equilibrium of a stored electron beam in the presence of an internal target of order 10 μg cm^{-2} thickness.

ELECTRON COOLING   During the 1970s three test rings established the basic characteristics of electron cooling (17–19). The theory is nontrivial and by now fairly well developed (20–22). An excellent recent review by the late H. Poth is available (23).

Ions slow down in matter by transfer of energy to electrons through collisions with a wide range of impact parameters. The stopping force

(dE/dx) varies directly with electron density and inversely with the square of the velocity of the ion relative to the electrons. There is also a logarithmic factor with argument the ratio of maximum to minimum impact parameter. This "Coulomb logarithm" reduces the strength of the force in the "end-of-range" region where the relative velocity is small. Electron cooling is a similar stopping process in which the electrons bound to atoms of ordinary matter are replaced by electrons confined magnetically in a "one-component plasma." The weaker binding of these electrons permits energy transfer for collisions of larger impact parameter, so the end-of-range effect is deferred to much lower relative velocities, and a useful stopping force can be obtained despite the lower density of these electrons. By employing electrons in the form of a beam moving at the same velocity as the stored ions, the stopping (in the electron rest frame) occurs at a large (laboratory) velocity. The ion "stops" in the electron frame, or at least comes to approximate thermal equilibrium with the "cold" electrons.

Electron cooling is a form of immersion cooling in which thermal energy in a stored ion beam is transferred to a colder medium. Electrostatic acceleration produces a very low electron longitudinal temperature, in comparison to a transverse temperature that is typically the 0.1 eV of the hot cathode source of electrons. In the rest frame of the average electron, the ion beam should come to thermal equilibrium at some intermediate temperature well below 0.1 eV, much colder than the temperature of several eV of a typical ion source. Ion beam momentum spreads of 10^{-5} to 10^{-4} are typical, and emittances of 0.1π μm have been observed. This is superb beam quality in comparison to the output of most accelerators.

The time to reach equilibrium is typically of order one second for a cooling system occupying about 1% of the ring circumference, a time that shows why storage is essential. The maximum strength of the cooling drag forces for a $Q = 1$ ion is of order 0.1 eV per turn, or 0.1 MeV s^{-1}. The force is a highly nonlinear function (24) of relative velocity of ion to electron beam and is maximum for relative velocities of order 10^5 m s^{-1}. For ions of charge Qe and mass Am_0, the cooling force scales as Q^2 and the cooling rate as Q^2/A.

The technology of electron cooling as it now stands employs a magnetic field parallel to the electron beam to counteract space-charge repulsion within the intense beam. The thermionic cathode is held at a high negative static potential so the electrons accelerate to the desired velocity in the cooling region. Normally, the beam is decelerated again prior to collection to recover most of the beam power. The energy limit is set by electrostatic breakdown phenomena in the accelerating tube or in the cathode or collector regions. Careful design and the use of insulating gases such as SF$_6$ have allowed some cooling devices to operate near 300 keV. Figure 1

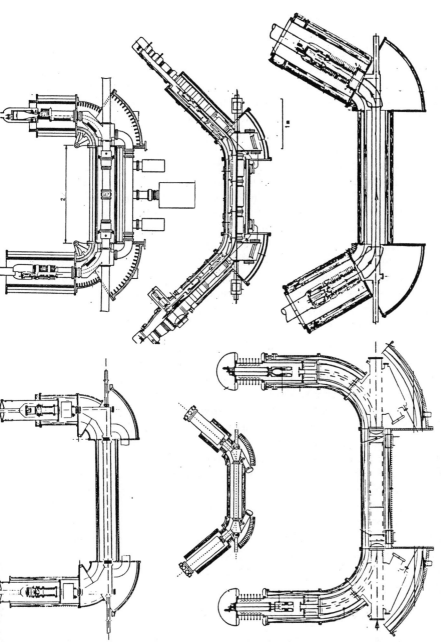

Figure 1 Geometries of six typical electron cooling devices, reduced to a common scale. From left to right; top to bottom the devices are for CELSIUS, CoSy; CRYRING, TARN-II; ESR and the IUCF Cooler. In each case, the stored ion beam merges with the electron beam in the central solenoid. The electron beam is diverted by the toroidal field segments from gun to collector. Major differences in the electrostatic configuration reflect the difficulties of obtaining up to 300 keV electrons in a confined space.

shows sections of some representative electron cooling systems. The rings to which they belong are described in a later section.

STOCHASTIC COOLING An excellent review on the process of stochastic cooling is available (25). If a sufficiently sensitive and fast detector could measure the displacement of a single ion from the closed orbit of a ring and apply an optimally amplified corrective kick at a suitable odd multiple of a betatron quarter-wavelength downstream, then the displacement error would be removed after a few revolutions. If two ions moved together, the detected signal would be twice as strong, but the same kick would be needed, so for stability the feedback gain has to be reduced as the number of stored ions seen by the pickup increases, and this makes the cooling time proportional to beam intensity. Even when many ions are observed together, the feedback, by reducing the error in centroid position, also gradually reduces the incoherent motion, and thereby produces beam cooling. The method has been highly successful when applied to accumulation of antiprotons. Cooling times of order 10^2 s are typical. The equilibrium temperature contains a contribution from amplifier noise that may be reduced by lowering the gain if the longer cooling time is acceptable. The technique is extremely flexible and may be applied to any ion species at any one energy, but is somewhat less conveniently applied than electron cooling to variable energy operation.

LASER COOLING Electrons attached to a stored ion may allow a strong interaction of the ion with a photon beam. Photons from a directed beam that are absorbed and isotropically re-emitted exert a net force on the ion. Narrow electronic energy levels give a cross section that is large and strongly dependent on frequency. If the frequency is tuned to one side of the resonance, the Doppler effect makes this force strongly dependent on the ion velocity, and therefore useful for longitudinal ion cooling. The cooling rate in favorable cases can be very high (26), and the limiting longitudinal temperature, determined by fluctuations in the re-emission, can be well below 10^{-4} eV. There are relatively few ion species that exhibit a strong resonance at a frequency accessible to existing lasers, so the technique for the moment has not become applicable to the ions of most general interest for nuclear research. Perhaps in the future laser cooling may be employed to cool a beam that is in thermal contact with the beam of interest. Such indirect immersion cooling has been given the name "sympathetic cooling."

PERFORMANCE BOUNDARIES

To use a storage ring for nuclear physics, a beam interacting with target material is obviously required. The target may be internal or external; it may be a solid, a gas, or even another beam. The product of beam current and target thickness, called the "luminosity," determines the event rate for a given cross section. The user needs to know how target thickness affects luminosity and beam quality. For internal targets, heating of the beam by the target affects the cooled beam equilibrium and may increase the beam emittance. By increasing the emittance, the target may also increase a brightness-dependent stability limit to the stored current. Target-induced beam losses, by reducing the beam lifetime, may reduce the duty factor and the average stored current. The problem is interesting in its complexity. In this section we introduce some of the known physical limitations. A more complete treatment must await a fuller confrontation of the concepts by experiment.

Stored Beam Phenomena

A stored beam is in a metastable equilibrium, far from thermodynamic equilibrium because of its coherent longitudinal momentum. Many destabilizing mechanisms lurk in the storage environment, waiting to assist the tendency toward energy equipartition. The beam stability decreases with increasing stored current or phase space density. In this section we discuss the upper limits to beam current and beam brightness.

ION-ION INTERACTIONS   Scattering of beam particles from one another ("intrabeam scatter") may cool, heat, or transfer thermal energy among dimensions, depending on the circumstances. In the absence of dispersion, a relaxation toward equipartition in the beam frame is expected. With dispersion, intrabeam scatter couples coherent longitudinal energy into incoherent thermal energy, resulting in beam heating or beam loss (27). The intrabeam scattering rate is brightness dependent and so may establish a lower bound for equilibrium momentum resolution and emittance (28) in the absence of a target in the presence of strong cooling. The heating by this process is greatest at a tight waist in a region of high dispersion. As discussed in the section on target heating below, just these envelope conditions may be experimentally desirable.

If the stored ions have internal degrees of freedom, for example from excitation of attached electrons or of vibrational and rotational degrees of freedom in molecular ions, intrabeam scatter may transfer incoherent kinetic energy to internal excitation where it may be radiated away. This process, related to the collisional cooling thought to be important in the formation of galaxies, may be useful for beam cooling (29), but it has not yet been demonstrated in a storage ring.

ION-BEAM INTERACTIONS   The electrostatic repulsion within a dense beam reduces the betatron frequency. If the tune is lowered enough by this "space charge tune shift" (30), an orbit resonance may be encountered, which may lead to beam loss. This interaction between an ion and the mean field limits the number of particles stored. The limit increases with kinetic energy at low energy and with total energy at high energy, where the effect is partly compensated by magnetic attraction and image currents. The current limit is reduced for a well-cooled beam. Observed maximum beam currents for electron-cooled beams are of order of 10 mA, while uncooled beams of large emittance may have currents of several amperes. If a radiofrequency system confines the stored beam within a bucket in the longitudinal phase space, the stored current may be further reduced in proportion to the microscopic duty factor. Cooling may lead to a quite small time spread, so the current limit of a cooled, bunched beam may be of order 0.1 mA.

For a sufficiently cold, dense beam, potential energy may exceed thermal kinetic energy, which leads to emittance growth during mismatched transport (31). The stable density profile of such a beam has been derived for special cases, for example, uniform focussing and axial symmetry (32). The beam has a uniform core, with a surface gradient determined by beam temperature. The beam diameter is determined by density and only indirectly related to emittance. In a ring, this limiting case corresponds to the tune being reduced nearly to zero. It is not yet clear whether this high density state can be formed in a storage ring. If it can, with further cooling the beam profile might begin to develop ordering (a phase transition from gas or liquid to solid or crystalline form) (33) of the type observed in laser-cooled ion traps (34).

BEAM-WALL INTERACTIONS   The electromagnetic field of a single stored particle induces image charges and image currents in the conducting walls of the vacuum container. The wall resistivity causes the image to lag behind the particle, which creates a retarding force resulting in slow deceleration. This effect was measured in the intersecting storage rings (ISR) (35), a decreasing fractional energy of about 10^{-14} per turn being attributed to single particle fields. A beam in the form of a uniform fluid should create a static field and no deceleration, while a particle bunch would generate a larger field giving a faster deceleration. Thus a uniform beam may become unstable when a beam density fluctuation is amplified by coupling some of the coherent energy of longitudinal motion into a growing density disturbance. Both longitudinal (36) and transverse (37) wall instabilities may occur. The growth rate depends on stored current, and may be damped by cooling at low intensities, yet be insufficiently damped above some instability threshold. High impedance structures in the ring such as rf cavities and diagnostic pickups may lower the instability thresholds appreciably. There are many known instabilities with colorful names and differing characteristics. Each ring will have a threshold for each instability. Special wall treatment may be employed to raise a particular threshold. Active feedback may be useful in raising the threshold for a given mode. Useful reviews (38) and treatments related to cooled beams (39) may be consulted for more information.

BEAM-BEAM INTERACTIONS   The electromagnetic interaction between beams in a ring or rings in which two beams overlap sets a limit to luminosity of such a collider. The electromagnetic interaction is dependent on the beam density profiles and the geometry in the overlap region and is highly nonlinear. From a theoretical standpoint, the disruption caused by the interaction and the onset of chaos in this coupled nonlinear system (40) makes the prediction of the maximum luminosity a nontrivial computational task. Empirical limits have been established for operating facilities. Active studies are under way of the limits for unequal beam energies that may be used in a future B factory.

Beam-Target Interactions

BEAMS AS TARGETS   An optional form of internal target for a stored beam is another beam. Relativistic colliders increase the center-of-mass energy in a collision. At nonrelativistic energies there is no economic incentive for constructing two rings, each of half the momentum. Moreover the energy dependence of the luminosity limit set by beam-beam interactions is unfavorable for colliders at low energy.

A beam target may be of interest for other reasons. The electron beam of a cooler has been used as a target (41) for atomic physics in radiative and dielectronic recombination studies. A backscattered laser beam (42) can serve as a hard photon source. Interactions of stored ions with atomic beams are possible. At a synchrotron light source at Brookhaven National Laboratory (BNL), a stored ion beam as a target for the external photon beam has been proposed (43).

It is possible to store two different ion species having similar magnetic rigidity simultaneously in a single ring. If the charge-to-mass ratios are nearly identical, both beams may be cooled by a single electron beam (44). In this case the relative velocity of the ions is too low to be useful for nuclear interaction studies. However, if the ring had space to allow cooling by two electron beams of different velocities, collisions with useful relative velocity would be possible.

Within the momentum acceptance of a ring, a single beam species might be split into two components of slightly different energy. If the lattice design contains a point with large angular dispersion, the beams will cross with relative transverse velocity. This option has been considered (45) in the lattice design of the ESR at GSI Darmstadt, so beams of bare nuclei or few-electron ions (which are stored at high laboratory energy to reduce losses through electron pickup from the residual gas) can collide with energies near the Coulomb barrier.

A high energy electron beam could be stored in the same ring as a lower energy ion beam. This configuration has been suggested as a method of obtaining charge distributions of nuclei near the drip line (46). The ring in this case is also an accumulator of a radioactive beam formed of spallation products. Adequate luminosity is a potential problem, especially for short-lived ion species.

EXTERNAL VS INTERNAL TARGETS   When a beam is passed once through a target and then dumped, the probability per beam particle of a nuclear interaction in the target is the interesting quantity. If the probability is small, as is normally the case at low energies, a thicker target will give a higher event rate. Interaction of a charged particle beam with electrons in the target atoms causes energy loss and a broadening of the beam energy distribution. These effects set an upper bound to the useful target thickness, except at very high energies, where the thickness may be increased until the probability of a nuclear reaction approaches unity. The lore of single-pass targetry is deeply ingrained in every experimentalist: thin targets give lower event rates and better energy resolution; rates increase with target thickness until either the deterioration of beam quality becomes unacceptable or all the beam particles interact. Loss of beam flux through interactions such as large angle scattering is seldom more than a minor side effect.

The situation is quite different when a stored beam interacts with an internal target. If the target is thin, a beam particle may well survive a single encounter with the target material, and the event rate per beam particle will scale with the mean number of revolutions before loss occurs. With a suitable storage ring design, a luminosity plateau can occur, the same event rate being obtained over a range of target thicknesses provided the loss probability is linear in target thickness. The thickness plateau has a lower edge because a finite beam lifetime is observed in the limit of zero target thickness, as residual gas in the ring acts as a distributed target. The upper edge of the plateau is reached when loss probabilities increase more rapidly than linearly, for example because of an emittance increase when beam heating by the target overwhelms the cooling process. This plateau has been observed in proton+proton scattering at the IUCF Cooler, where at energies near 200 MeV it covers a target thickness range from 10^{14} to 5×10^{15} atoms cm^{-2}.

STORED BEAM LIFETIME   The event rate for a given cross section is the luminosity L, given by the product of beam flux $N \cdot f$, where N particles orbit at frequency f, in particles per second, and target thickness x_t (nuclei cm^{-2}). For a stored beam, a class of events of particular importance is one that removes a particle from the beam. The rate of such events, with loss cross section σ_{loss}, determines the beam lifetime τ:

$$1/\tau = -1/N \cdot \langle dN/dt \rangle_{loss} = L \cdot \sigma_{loss}/N = f \cdot x_t \cdot \sigma_{loss}.$$

If the ring is continually refilled at a fixed average rate $\langle dN/dt \rangle_{fill}$, an equilibrium stored current will be reached when

$$0 = \langle dN/dt \rangle_{fill} - \langle dN/dt \rangle_{loss}.$$

This gives the simple expression $L = \langle dN/dt \rangle_{fill}/\sigma_{loss}$, in which neither the stored current $I/q = N \cdot f$ nor the target thickness x_t appears explicitly. Note the central role played by the quantity $1/\sigma_{loss} = (f \cdot x_t \cdot \tau)$ in determining the luminosity. The expression is only valid on the thickness plateau mentioned above, and only if the increasing stored current obtained by reducing the target thickness remains below stability limits.

The physics determining the dominant loss process is specific to a given operating regime. For electron beams, hard collisions with target electrons may dominate. Small angle scattering may dominate for lighter ions, as discussed below. For low energy, heavy ion beams, charge exchange may be dominant. For very high energy ions, the nuclear reaction cross section σ_r may itself set the loss rate, which implies very efficient use of the stored beam. One may in fact recognize that the efficiency of beam usage is simply σ_r/σ_{loss}.

For a stored ion beam and internal target, beam loss can occur through a single nuclear scattering event if the resultant transverse momentum exceeds the containment limit of the storage ring. The target would normally be placed at or near a beam waist to maximize tolerance to scattering. The parameters giving the largest angle θ for which a beam particle is still contained following a single scattering are given by the equation $A = \pi \cdot \beta \cdot \theta^2$ where A, the ring acceptance, and the envelope function β, defined previously, have dimensions of length. Typical values of θ are only a few milliradians, so Coulomb scattering dominates, and the integrated loss cross section may then be shown to be proportional to $\beta \cdot \pi/A$. A ring designed to minimize this loss has large acceptance and a small β at the target waist.

For a light ion, the maximum momentum transfer in a collision with an electron in a target atom is bounded. Direct loss arising from such a δ-ray electron knock-on event may be avoided by a large enough longitudinal acceptance in the ring. However, a loss can still result from the dispersion-coupling mechanism. Momentum lost to an electron in a region of high dispersion shifts the equilibrium orbit away from the ion position. The particle then executes betatron motion with respect to the new equilibrium orbit. If the amplitude of this oscillation exceeds the transverse acceptance, the particle will be lost. Unfortunately, the condition (large dispersion and small beam size) that would give high resolution through dispersion matching (47) with an internal target is the same condition that leads to enhanced dispersion-coupling loss. Thus one of the common techniques for improving resolution in external beam experiments is less useful with internal targets.

As ion energy increases, transverse scattering decreases but the fractional momentum transferred to an electron increases, so the luminosity limit shifts to longitudinal loss processes.

STORED BEAM EQUILIBRIUM   The phase space distribution of the beam under the combined action of target heating and beam cooling tends toward a time-independent form although not a true thermodynamic equilibrium. Projections of this distribution give the energy resolution, time spread, beam size, and divergence at the target. The heating of the beam caused by scattering from the nuclei and electrons of the target atoms tends to produce a tail on the equilibrium distribution in addition to contributing to the width of the central core. The distribution only approximates the multiple scattering form that would be obtained in a single pass through a target of thickness given by the internal target thickness multiplied by the number of passes in a cooling time. The nonlinearity of the cooling force, the tail truncation by finite acceptance, and the betatron precession all generate new features in the equilibrium distribution.

A number of methods have been used to model the equilibrium distribution. Meyer (48) reported a Monte Carlo simulation predicting that the momentum distribution with electron cooling will consist of two components: a narrow peak of width given by the electron temperature and a broad tail with an area proportional to the target thickness. Hinterberger & Prazuhn (49) reported an approximate analytic method for predicting beam distributions. For simulations to predict correctly the beam lifetime, considerable care must be taken to reproduce the shape of the transverse tail down to densities of order 10^{-6} of the central density. These tails may also be expected to play a crucial role in background generation.

Luminosity Domains

THE CHARGE EXCHANGE BOUNDARY   A normal storage ring accepts a limited range of magnetic rigidities (momentum/charge). If a beam ion gains or loses an electron in passing through a target, the change in magnetic rigidity leads to a particle loss. For a fully stripped ion beam, only electron capture need be considered. At high enough energies, Schlacter et al (50) derived scaling laws showing that the capture cross section σ_c depends strongly on atomic numbers Z_b and Z_t of beam and target, and on beam energy E in MeV/amu. With the approximation of integer exponents, the value of σ_c is of order $\sigma_c = 10^{-22}(Z_b Z_t)^4/(E/A)^5$ cm^2. For a reasonable beam usage efficiency in a nuclear physics experiment, the loss cross section should not exceed perhaps 100 barns. For Ne + Ne, for example, a kinetic energy above 40 MeV/amu is needed to reduce the loss by electron capture to this level.

Rings can be designed with a wide enough magnetic rigidity range to accept an adjacent charge state for a heavy ion beam, as has been demonstrated recently at TSR (44). For such a ring, the threshold energy for carrying out nuclear scattering experiments with internal targets is lowered, because a beam particle can survive a single electron pickup and is likely to be stripped on succeeding passes through the target before another pickup occurs. The ring must have zero dispersion at the target to avoid a large increase in transverse emittance accompanying each change of magnetic rigidity. A more extreme lattice configuration was proposed by Cramer (51) as a method of constructing an extremely efficient stripper. Such a microtron-like ring with several paths for different rigidities could extend nuclear studies with heavy ions to much lower energies. The flexibility to accommodate all possible rigidity ratios would be a challenging design task. Efficient beam utilization in a storage environment is particularly relevant for rare ions such as radioactive species (^{6}He, ^{11}Li, ^{13}N, ...). It may be worthwhile to go to considerable lengths to achieve the potential benefit. The pickup loss could be avoided with a target free of electrons, but neither beams in rings nor ions in traps yet have the target density needed for useful luminosity.

COOLING BOUNDARIES   Synchrotron radiation damping is effective only at highly relativistic energies, restricting its use to electron rings. Electron cooling has a technical upper limit of order 1 GeV/amu set by voltage holding in the electrostatic electron accelerator, although pressurized systems may in future go to higher energies. At low energies, electron beam currents are reduced by space charge effects, so cooling is reduced. Stochastic cooling works for any beam and energy. Laser cooling is beam-species specific.

FACILITY EXAMPLES

Electron Facilities

Coincidence experiments limited by accidental rates take longer to perform in inverse relation to the "duty factor" (which is roughly the fraction of the time in which beam is present). Stretcher rings are motivated by the desire to increase the duty factor of a pulsed accelerator. Pulsed accelerators use less power. The disadvantage of a low duty factor to experiments may be overcome by injecting the accelerator output into a storage ring to remove the time macrostructure. The internal beam retains time microstructure, imposed by the rf cavity used to compensate energy lost to synchrotron radiation, and it may in addition have time structure at the orbit frequency from gaps or intensity nonuniformities introduced by the injection process. Slow extraction makes much of the improvement in duty factor available at an external target. Current examples of external beam rings are located at Tohoku (52), Saskatoon (53), and Bonn (54).

Rings under construction at MIT Bates (55) and NIKHEF-K (56) in Amsterdam are configured for both internal and external target operation. The South Hall Ring (SHR) at Bates economically employs recycled dipoles from the Princeton-Penn accelerator and a layout that passes through the existing south target hall to accommodate internal target experiments. Interest in polarized targets and longitudinally polarized electron beams is part of the motivation for the design. Simulations for the Amsterdam AmPS ring indicate that the dominant beam loss mechanism is bremsstrahlung in collisions with an internal hydrogen target above 700 MeV, while at lower energies, Møller and single scattering are dominant. A beam lifetime of 1–5 minutes with luminosity above 10^{34} cm^{-2} s^{-1} at 900 MeV is predicted.

Internal targets can be added to electron rings that have been developed for other purposes, although the lattice configuration may not be optimal for access to reaction products. Popov (57) reported electronuclear experiments with luminosity of 10^{32} cm^{-2} s^{-1} with an internal beam in VEPP-2 at Novosibirsk and a plan for a new ring (NEP) that would permit an increase to 10^{34} cm^{-2} s^{-1}. Holt and coworkers (58) are developing a polarized internal target for use in the VEPP-3 ring in Novosibirsk. The storage ring ADONE at Frascati uses an internal jet target (59) to produce tagged photons. An example at higher energy is the positron-electron-proton (PEP) storage ring at SLAC, where nuclear interactions have been observed (60) by introducing gas target material in an interaction region where an existing detector could be exploited. The PEGASYS facility (61) was a proposed upgrade.

Light Ion Facilities

There are a number of examples in thin internal targets being used in proton synchrotrons, for example in the nuclear studies at the Fermi National Accelerator Laboratory (FNAL) (62) and in a recent fiber target test (63) at Saturne in Saclay. Saturne now employs the storage ring

MIMAS (64) as an accumulator to increase injected current. At anti-proton facilities, both the past decade of external beams (65) at the low energy antiproton ring (LEAR) and the current experiments with internal targets at LEAR (66) and at the FNAL accumulator (67) are noted.

A recent trend has been the construction of rings designed explicitly to exploit cooled beams. These include the IUCF Cooler (68) in Indiana, the CELSIUS ring in Uppsala (69), the TARN II ring (70) at Institute for Nuclear Studies (INS) in Tokyo, and the CoSy project (71) under con-struction at KFA in Jülich. These facilities share the common features of injection from an existing cyclotron and operation both as synchrotrons and as electron-cooled storage rings. Protons are injected by stripping of a H_2^+ beam. The total running experience with cooled beams at the three operating facilities is only about four ring-years at the time of writing, so there is much yet to be learned about the ultimate performance of these facilities. Before cooling techniques became established, there was a sep-aration between the good beam quality available only at low energy and the beams of higher energy with poorer beam characteristics. Cooling, combined with synchrotron acceleration, allows excellent beam quality to be obtained over an extended energy range, with useful luminosity for light ions on internal targets. This has motivated the recent trend.

Users of the IUCF Cooler have reported first experiments with polarized and unpolarized protons employing a hydrogen jet target of thickness 0.5×10^{16} atoms/cm^2. A study near threshold of the pp $\rightarrow$ ppπ^0 reaction (72) reports time-average luminosities near 300 MeV of about 0.5×10^{29} cm^{-2} s^{-1}. In more recent runs, the luminosity has increased by a further factor of 30 (H.-O. Meyer, private communication). The luminosity is expected to exceed 10^{31} cm^{-2} s^{-1} before reaching a stability limit of the cooled beam intensity determined by the H_2^+ injection energy of 45 MeV.

A study at 185 MeV at IUCF of asymmetry in elastic p-p scattering in the Coulomb-nuclear interference region has been reported (73). The luminosity achieved with a stored, cooled, and polarized proton beam was the same as for the first unpolarized pion experiment. Polarized beam luminosity was increased by a factor of 100 during 1990 by improvements made in the cooled stacking injection method (9). Measurements (74) with the zero gradient synchrotron (ZGS) established a lower bound of several minutes for the lifetime of polarization of a stored beam. The limit has been extended to several hours for an electron-cooled beam (H.-O. Meyer, private communication).

The IUCF Cooler has accelerated protons to 476 MeV (about 98% of design maximum rigidity), and it has reached stored intensities of cooled coasting beam of 4 mA (2.5×10^{10} protons) at 45 MeV, 0.6 mA of cooled bunched beam after acceleration to 292 MeV, and 0.3 mA of polarized protons at 185 MeV. In addition, deuterons have been accelerated to 200 MeV and cooled, $^3\text{He}^{2+}$ ions have been cooled, and H_2^+ ions have been stored. Cooling reduces energy spread and emittance by about one order of magnitude relative to the cyclotron beam.

The CELSIUS ring (75) has accelerated proton beams to 1.13 GeV and $^4\text{He}^{2+}$ to 0.6 GeV, has stored 4×10^{10} protons and 10^{10} α particles uncooled at 50 MeV, and has reached an uncooled luminosity of 2×10^{30} cm^{-2} s^{-1} with a hydrogen cluster jet target of 3×10^{14} atoms/cm^2. Protons of 48 MeV have been electron cooled. Preparations for π^0 and η^0 rare decay experiments are well advanced (76), with some detector tests having been carried out in the accelerated beam of CELSIUS, and the first pellets have been obtained from a prototype of the frozen H_2 pellet target.

The TARN II ring at INS Tokyo (77) began electron cooling of stored beams in September 1989. It has been used to study cooling of 20-MeV protons and of 10-MeV/amu d and α particles. Provision has been made for fast and slow extraction and for operation with internal targets. Acceleration has been demonstrated to about 30% of the design rigidity of 6.1 T·m, limited temporarily by a power supply. Observation of electron pickup by stored ions interacting with the electron beam has been reported (78) for $^1\text{H}^+$, H_2^+, and $^3\text{He}^+$ ions.

The CoSy ring (79) under construction at KFA Jülich has a design rigidity of 11 T·m, provision for ramping at 7 T·m/s, and an emphasis on extracted beam operation. There is provision for electron cooling at the injection energy and for stochastic cooling after acceleration. Testing is scheduled to begin in 1992 and first experiments in 1993. In many respects the research program goals are similar to those of Saturne.

In addition to these rings now in operation and under construction, there are proposals for future additions to the family. A new separated-sector cyclotron is being constructed at the Research Center for Nuclear Physics (RCNP) in Osaka and the building contains a large experimental hall for a recirculator ring (80) associated with a magnetic spectrometer. This ring may eventually evolve into a cooled storage ring. The INS group at Kiev has proposed (81) a ring extension to their cyclotron facility. A design has been prepared and funds are in hand for beginning the laboratory building expansion. At IUCF there has been exploratory planning for a future cooled "spin synchrotron," injected from the present Cooler, and to be used for internal target polarized beam experiments (J. Cameron, private communication).

Heavy Ion Facilities

The distinction between light ion and heavy ion facilities is primarily one of emphasis, since all of the rings mentioned in the previous section have some plans for operating with heavier ions, and the rings described in this section are capable of light ion operation. Nuclear studies with heavier ions in cooling rings are possible at intermediate energies where charge-changing beam losses may become small enough for useful beam lifetime in the presence of target material. However, the motivation for a family of small rings with cooled beams of low energy heavy ions is the opening up of new areas of atomic physics research by this technology. Stored beams may interact with photons, electrons, atoms, or other ions. Cross sections are large enough that useful luminosity is easier to obtain than for nuclear physics applications. An adequate beam lifetime for heavy ions of low energy demands extremely good vacuum (in the nanoPascal range) in these rings.

The TSR ring at the Max-Planck-Institut (MPI) Heidelberg (82) began cooled beam operation in November of 1988. Beam is injected from an MP tandem and linac postaccelerator, making a very wide range of ions and injection energies available. Cooled 21-MeV proton beams of 1.1 mA and a mean lifetime of 36 hours have been reported. The highest stored current to date is 15 electrical milliAmperes (emA) of $^{12}C^{6+}$ at 73 MeV (2.5×10^{10} particles). The TSR ring has demonstrated (83) simultaneous storage of three charge states of the same ion species ($Cu^{24+,25+,26+}$), possible because of the unusually broad rigidity acceptance of the lattice. The TSR ring has also demonstrated simultaneous storage of two different ion species of similar rigidity ($^2H^+$ and $^{16}O^{8+}$), accomplished by retuning the injector ion source after storing the longer-lived species.

The TSR research program in its first two years has emphasized machine studies (equilibrium momentum spreads, lifetimes, collective phenomena) and atomic physics, with the first demonstration (25) of laser cooling in July 1989, and measurements of radiative, dielectronic (41), and laser-assisted recombination. In 1991, a polarized target (84) will be installed to demonstrate the gradual polarization of an initially unpolarized proton beam through the spin dependence of beam-target scattering processes. This technique has potential future application for polarization of stored antiprotons.

The ASTRID ring in Århus University (85) began operation in the Spring of 1990. This ring will operate about half the time as a synchrotron radiation source, filled with electrons from a microtron injector. There is provision for electron cooling of stored ions and for future injection from an existing tandem electrostatic accelerator. The electron system is now being used for dielectronic recombination studies in the tandem beam (86).

Poulsen (87) reported a number of intriguing results from the first months of ASTRID operation. The low energy injected beam from a mass separator is already very cold because of the longitudinal compression associated with electrostatic acceleration, and it is observed to heat itself through intrabeam scattering. Gaps left in the coasting beam by injection kickers appear to propagate like solitons rather than closing up as expected. Laser cooling at Århus of the metastable fraction of a $^7Li^+$ beam to a longitudinal temperature of 0.5 mK is observed. It is not yet clear why the cooled metastable fraction does not exchange heat more strongly with the uncooled ground-state fraction of the stored beam.

The ESR at GSI Darmstadt (88) came into operation in 1990 and has demonstrated electron cooling for 92- and 164-MeV/amu Ar ions injected from the new SIS 18-T·m synchrotron. The ESR is designed to accept all ions from the synchrotron, including 0.5-GeV/amu fully stripped uranium ions and radioactive ions made by spallation reactions, the latter selected by a fragment separator. The ESR has provision also to serve as a general purpose beam processor, accumulating beams that after cooling can be extracted and sent back to the SIS for further acceleration, or on to the external target halls. A wide-ranging research program (89) is planned for the ESR. Atomic physics with few-electron heavy ion species, inverse kinematics nuclear reactions with heavy stored beams impinging on light internal targets, mass measurements near the drip line, and charge transfer in crossing beam collisions are examples of experiments under consideration.

The CRYRING at MSI Stockholm (90) achieved a first turn of beam in the last days of 1990 and began operation as a storage ring during 1991. The existing CRYSIS source followed by a radiofrequency quadrupole (RFQ) serves as injector. Provision is made for rapid acceleration in CRYRING after injection at low energy, which limits the time for losses by charge changing. Vacuum requirements are nevertheless severe. The electron cooling system is designed to ramp its fields during synchrotron acceleration. The research plans for this ring include a number of topics in atomic physics, including ion-atom, ion-ion, ion-electron, and ion-photon interactions. The accelerated beam may also be used for nuclear physics research.

There have been a number of additional proposals for heavy ion rings. Among these are the HISTRAP (91) at Oak Ridge National Laboratory (ORNL), a ring for atomic and nuclear physics, and the K-4/K-10 (T·m) ring pair proposed (92) at the Joint Institute for Nuclear Research (JINR) at Dubna. The possibility of a 22.5-T·m ring at Legnaro has been introduced (93). For completeness, the relativistic heavy ion collider (RHIC) (94) to be constructed at BNL should be mentioned. Cooling is less critical at such a high energy, but is still useful to counteract intrabeam scattering, which is important in limiting the luminosity. A collider has been proposed (95) for Au + Au at 7 GeV/amu as part of the Japanese Hadron Facility. The geometries of nine of the cooling ring for light and heavy ions described above are sketched in Figure 2. Some of their properties are outlined in Tables 1 and 2.

Cooling Ring Evaluation

Are the new cooling rings going to deliver a significant improvement in the conduct of nuclear research? How is the design of these machines limiting their performance, and what changes are needed for the next generation?

A real improvement of about one order of magnitude in energy resolution for an electron-cooled beam in the presence of an internal target has been demonstrated (eg. 42 keV full width half maximum at 306 MeV at the IUCF Cooler, averaged over a one-hour run duration at an average luminosity of 10^{30} cm^{-2} s^{-1}). However, the record kinetic energy of 400 MeV to date for electron-cooled proton beams, while well above the pion production thresholds, is still far below the thresholds for strange particle production. This is one example of research in which the good resolution would be important. The performance of the higher rigidity rings with beams cooled and then accelerated to higher energy for an experiment without further cooling has yet to be established.

Emittance reduction is another demonstrated property of a cooled beam that is retained in the presence of a target. Values below 0.1π μm have been observed. The smaller and more parallel beam may be used to define better the incident four-momentum in a reaction, although a conclusive demonstration of the benefit of this property is not yet at hand. The small emittance does not mean that the ring acceptance may be reduced proportionally, as a reasonable beam lifetime requires space for the distribution tail generated by multiple scattering. Tests at the IUCF Cooler have shown that injection, acceleration, and internal target operation may be carried out with the acceptance restricted to about 10π μm by fixed slits at a nondispersed location. This acceptance will permit operation of a small diameter windowless polarized target cell with good luminosity.

The internal target environment creates difficulties. Rings with a single experimental area for internal targets face sheduling bottlenecks. Even an open space of several meters is not long enough to accommodate all experiments. The optimum β function at the target depends on energy, target, and ion species, favoring a lattice design with some flexibility to tailor the beam envelope to specific experiments. Dispersion at the target is usually not helpful. There is a conflict between the vacuum requirements

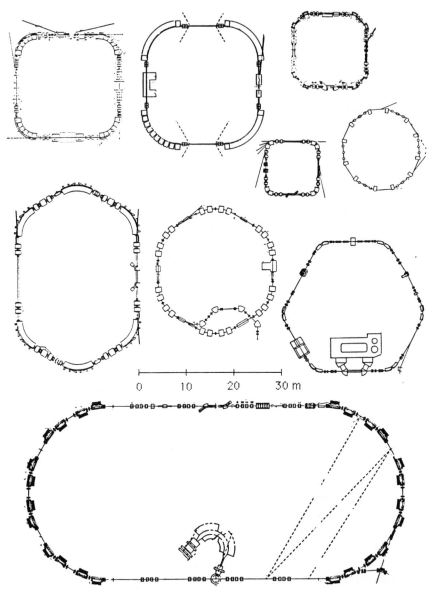

Figure 2   Lattices of the nine cooling rings of Tables 1 and 2, reduced to a common scale for comparison. From left to right, top to bottom are LEAR, CELSIUS, TSR; ASTRID, CRYRING; ESR, TARN-II, IUCF Cooler; and CoSy. On this scale, the diverse lattice details are barely discernible, but the relatively small fraction of the ring circumference available for experimental use is quite evident. Note the common characteristic of either polygonal symmetry or arc segments in a racetrack geometry. Further details are given in the text.

Table 1 Some properties of light ion cooling rings, listed in the order in which electron cooling began

Ring name	LEAR	Cooler	TARN-II	CELSIUS	CoSy
Laboratory	CERN	IUCF	INS	TSL	KFA(FZ)
Location	Geneva	Indiana	Tokyo	Uppsala	Jülich
Country	Switzerland	USA	Japan	Sweden	Germany
Length (m)	78.0	86.8	77.8	82.0	184
Experiment length (m)	3.7	6.1, 6.1 5.2, 2.9	4.2	5.2	6.5, 4.3
Magnetic rigidity $(T \cdot m)^a$	6.7	3.6	5.8	7.0	12
Date started: e-cooling	Nov. 1987	Apr. 1988	Sept. 1989	May 1990	(1992–1993)
Electron kinetic energy (keV) achieved (design)[b]	27 (40)	250 (270)	70 (110)	26 (300)	(100)
Particle species: Principal Other[c] (Future)	antiprotons p, ^{16}O (S, Pb)	polarized p p, d, ^{3}He (pol'd d, Li)	p, d, ^{4}He H_2^+ ^{3}He$^+$ (Li, C, Ne)	p, ^{4}He (pol'd p) (C, Ne)	(p, d) (pol'd p, d) (Ne, Ar, Kr)
Injector type	synchrotron linac	cyclotron	cyclotron	cyclotron	cyclotron
Targets[c]	cluster jet	jet dust fiber cell (pol'd cell)	electrons (jet)	jet fiber (pellet)	
References	66	68, 72, 73	70, 77, 78	69, 75, 76	71, 79

[a] Magnetic rigidity (in Tesla-meters) determines the maximum stored beam momentum per unit charge (1.00 T · m = 299.79 MeV/c).
[b] The maximum electron kinetic energy determines the maximum beam kinetic energy per nucleon that may be electron-cooled. Entries in parentheses are design values or goals.
[c] Polarized = pol'd.

for long beam lifetime and the real needs of the experimenter to install gassy objects such as a proportional chamber with thin windows within the ring vacuum. Access for detectors to reaction products at zero degrees requires forethought or later lattice modifications. All ring designs make compromises that benefit particular research interests. A few more years of operating experience will be needed before one can critically evaluate the diverse design choices made for the current generation of rings.

Table 2 Some properties of heavy ion cooling rings, listed in the order in which electron or laser cooling began[a]

Ring name	TSR	ASTRID	ESR	CRYRING
Laboratory	MPI	Univ.	GSI	MSI
Location	Heidelberg	Århus	Darmstadt	Stockholm
Country	Germany	Denmark	Germany	Sweden
Length (m)	55.4	40.0	108.4	51.6
Experiment length (m)	3.8	3.0	9	2.7, 2.7
Magnetic rigidity (T·m)	1.5	2.0	10.0	1.4
Electron kinetic energy (keV)[b]	12 (20)	(2.5)	95 (330)	
Date started:				
e-cooling	Nov. 1988	—	May 1990	—
laser-cooling	1989	May 1990	—	—
Particle species:				
Principal	C, H, Li$^+$, Be$^+$	Ar, Li$^+$	Ar	p
Other	O, Si, S, Cu	electrons		
(Future)	(I)		(Ne-U)	(Ne, Kr)
Injectors	tandem linac	separator microtron (tandem)	synchrotron fragment-separator	RFQ CRYSIS
References	25, 82, 83 41, 84	85–87	88, 89	90

[a] The layouts of the nine rings described in Tables 1 and 2 and in the text are shown in Figure 2.
[b] Entries in parentheses are design values or goals.

CONCLUSION

A storage ring provides a mode for studying the interaction of beams with target material that is complementary (96) to the more traditional methods. Whether the stored beam mode is advantageous is a quantitative question whose answer depends on the field of investigation. The smaller rings for atomic physics are tapping a rich vein of new research possibilities and will greatly enrich this field over the next few years.

The study of phenomena with smaller nuclear cross sections requires a large increase in luminosity over that required for atomic physics research. Useful performance in the internal target mode is now being obtained for light ions at intermediate energies. A goal of 10^{32} cm^{-2} s^{-1} seems feasible for electron-cooled beams, although not yet demonstrated. Whether even higher luminosity is needed is not clear, for the question of the useful luminosity range is tied to the design of detectors. A large solid-angle array can be saturated with luminosities already achieved.

Slower and heavier ions have severe problems with the charge-changing beam loss mechanism. The TSR has shown that a few charge states can be stored simultaneously, which is an important step in reducing this loss. New kinds of rings (51) that can store a set of different magnetic rigidities may be a logical next step.

External use of extracted cooled beams has the intrinsic difficulty that cooling lowers the storage capacity and takes time. The average output current in particles per second (I/Qe) cannot exceed the number of particles stored divided by the processing (accumulation/acceleration/cooling) time. This ratio may correspond to a few particle nanoAmperes (I/Q).

Internal target use requires unconventional lattice designs with careful provision for adequate space for detectors, access to forward reaction products, multiple target stations to circumvent scheduling problems, etc. Fortunately it would appear that there are no fundamental restrictions to such experiment-friendly designs. The IUCF Cooler breaks most of the traditional design rules and functions nevertheless.

The internal target mode has real advantages that make it particularly attractive in some experiments. The luminosity is not linearly dependent on target thickness, and the optimum thickness overlaps the realm of technical feasibility for windowless jet targets. Such targets may be pure, insensitive to radiation damage, and transparent to reaction products. Recirculation greatly improves the efficiency with which beam is used, so a larger fraction of the particles interact usefully with the target; this in turn makes possible better utilization of rare ion species and a reduction of background by eliminating the beam dump.

One of the attractive features of a cooled beam is its insensitivity to properties of the injector. In the approach to thermal equilibrium, all memory of initial conditions is lost. The beam quality is determined by the equilibrium in the ring and is independent of the character of the injector. A wider range of ion source and accelerator technology becomes accessible to serve as ring injector. A low duty factor is even an advantage because the peak current over one orbit period in the ring determines the stored current. It appears contradictory that most of the new cooling rings have been built in laboratories with injectors of much better beam quality than ring injection requires. But these are just the sites where the better quality of the cooled beam is most appreciated and likely to be exploited. Future rings at other locations can be filled from less elaborate injectors.

BIBLIOGRAPHIC NOTE

The subject spanned by this overview is so broad as to make impractical a full survey of the literature. I have chosen to make a somewhat subjective sampling, including some of the earlier papers that I found helpful, and more recent work containing reference to the broader literature. For absent favorites, the fault is mine alone.

ACKNOWLEDGMENTS

This work was supported in part by the US National Science Foundation under grant NSF PHY 87-14406 and by Indiana University. The patience of my family and my colleagues as I struggled to meet a rigid deadline is much appreciated.

Literature Cited

1. Courant, E. D., Snyder, J. S., *Ann. Phys.* 3: 1–48 (1958)
2. Ohnuma, S., *AIP Conf. Proc.* 123: 415–23 (1984)
3. Douglas, D. R., *AIP Conf. Proc.* 153: 390–473 (1987)
4. Steffan, K., *Part. Accel.* 24: 45–52 (1988)
5. Goodwin, J. E., Meyer, H.-O., Minty, M. G., et al., *Phys. Rev. Lett.* 64: 2779–82 (1990)
6. Wei, J., Lee, S. Y., Ruggiero, A. G., *Part. Accel.* 24: 211–22 (1989)
7. Hofmann, I., *Part. Accel.* 34: 211–20 (1990)
8. Colton, E., Neuffer, D., Thiessen, H. A., et al., *AIP Conf. Proc.* 176: 302–7 (1988)
9. Pei, X., In *Proc. 1989 IEEE Part. Accel. Conf.*, IEEE Conf. Rep. 89CH2669-0 (1989), pp. 666–68
10. Anderson, O. A., Soroka, L., In *Proc. 1987 IEEE Part. Accel Conf.*, IEEE Conf. Rep. 87CH2387-9 (1987), pp. 1043–45
11. Robinson, K. W., *Phys. Rev.* 111: 373–80 (1958)
12. Skrinsky, A. N., Parkhomchuk, V. V., *Sov. J. Part. Nucl.* (AIP Transl. of *Fiz. Elementarnykh Chastits i Atomnogo Yadra*) 12: 223–47 (1981)
13. Cole, F. T., Mills, F. E., *Annu. Rev. Nucl. Part. Sci.* 31: 295–335 (1981)
14. Cooper, R. K., Lawrence, G. P., *IEEE Trans. Nucl. Sci.* 22: 1916–19 (1975)
15. Frieze, W. E., Gidley, D. W., Lynn, K. G., *Phys. Rev.* B31: 5628–33 (1985)
16. Renieri, A., *Theoretical Aspects of the Behavior of Beams in Accelerators and Storage Rings*, CERN Rep. 77-13, pp. 82–110 (1977)
17. Budker, G. I., Dikansky, N. S., Kudelainen, V. I., et al., *Part. Accel.* 7: 197–211 (1976)
18. Bell, M., Chaney, J., Herr, H., et al., *Nucl. Instrum. Methods* 190: 237–55 (1981)
19. Forster, R., Hardek, T., Johnson, D. E., et al., *IEEE Trans. Nucl. Sci.* NS-28: 2386–8 (1981)
20. Derbenev, Y. S., Skrinsky, A. N., *Part. Accel.* 8: 235–43 (1978)
21. Ogino, T., Ruggiero, A. G., *Part. Accel.* 10: 197–205 (1980)
22. Sørensen, A. H., Bonderup, E., *Nucl. Instrum. Methods* 215: 27–54 (1983)
23. Poth, H., *Phys. Rep.* 196: 135–297 (1990)
24. Steck, M., Bisoffi, G., Blum, M., et al., *Nucl. Instrum. Methods* 287: 324–27 (1990)
25. Möhl, D., Petrucci, G., Thorndahl, L., van der Meer, S., *Phys. Rep.* 58: 73–119 (1980)
26. Schröder, S., Klein, R., Boos, S., et al., *Phys. Rev. Lett.* 64: 2901-4 (1990)
27. Bjorken, J. D., Mtingwa, S. K., *Part. Accel.* 13: 115–43 (1983)
28. Pollock, R. E., In *Proc. Workshop on Electron Cooling and Related Applications*, ed. H. Poth, KfK Rep. 3846 (1985), pp. 109–21
29. Kilian, K., See Ref. 28, p. 376; CERN Intern. Rep. EP 84-05
30. Weng, W. T., *AIP Conf. Proc.* 153: 348–89 (1987)
31. Wangler, T. P., Crandall, K. R., Mills,

R. S., Rieser, M., *IEEE Trans. Nucl. Sci.* NS-32: 2196–2200 (1985)
32. Lawson, J. D., *The Physics of Charged Particle-Beams*, Fig. 4-11. Oxford: Clarendon (1977), 462 pp.
33. Schiffer, J., In *Proc. Workshop on Crystalline Beams*, ed. R. W. Hasse, I. Hofmann, D. Liesen, GSI Rep. 89-10 (1989), pp. 2–32
34. Bollinger, J. J., Gilbert, S. L., Wineland, D. L., See Ref. 33, pp. 321–40
35. Hofmann, A., Risselada, T., *IEEE Trans. Nucl. Sci.* NS-30: 2400-2 (1983)
36. Neil, V. K., Judd, D. L., Laslett, L. J., *Rev. Sci. Instrum.* 32: 267–76 (1961)
37. Laslett, L. J., Neil, V. K., Sessler, A. M., *Rev. Sci. Instrum.* 36: 436–48 (1965)
38. Neil, V. K., *IEEE Trans. Nucl. Sci.* NS-14: 522–28 (1967)
39. Cocher, S., Hofmann, I., *Part. Accel.* 34: 189–210 (1990)
40. Tennyson, J. L., *AIP Conf. Proc.* 87: 345–94 (1982)
41. Kilgus, G., Berger, J., Blatt, P., et al., *Phys. Rev. Lett.* 64: 737–40 (1990)
42. Sauer, J. R., Milburn, R. H., Sinclair, C. K., *IEEE Trans. Nucl. Sci.* NS-16: 1069–72 (1969)
43. Jones, K. W., In *Proc. Workshop on Atomic Physics with Stored Cooled Heavy Ion Beams*, ORNL Conf.-860144 (1986), pp. D1–40
44. Jaeschke, E., In *Proc. 19th INS Symp. on Cooler Rings and Their Applications*, Tokyo, Nov. 1990. Singapore: World Scientific, in press (1991)
45. Franzke, B., Schmelzer, Ch., *IEEE Trans. Nucl. Sci.* NS-32: 2678–80 (1985)
46. Ando, A., Katayama, T., See Ref. 44
47. Martin, S. A., Hardt, A., Meissburger, J., et al., *Nucl. Instrum. Methods* 214: 281–303 (1983)
48. Meyer, H.-O., *Nucl. Instrum. Methods* B10/11: 342–6 (1985)
49. Hinterberger, F., Prazuhn, D., *Nucl. Instrum. Methods* A279: 413–22 (1989)
50. Schlacter, A. S., Stearns, J. W., Graham, W. G., et al., *Phys. Rev.* A27: 3372–4 (1983)
51. Cramer, J. G., *Nucl. Instrum. Methods* 130: 121–23 (1975)
52. Tamae, T., Sugawara, M., Yoshida, K., et al., *Nucl. Instrum. Methods* A264: 173–85 (1988)
53. Dallin, L. O., See Ref. 9, pp. 22–26
54. Althoff, K. H., von Drachenfels, W., Dreist, A., et al., *Part. Accel.* 27: 101–6 (1990)
55. Flanz, J. B., *AIP Conf. Proc.* 176: 290–95 (1988)
56. Luijckx, G., Boer Rookhuisen, H., Bruinsma, P. J. T., et al., In *Proc. 2nd Eur. Part. Accel. Conf.*, ed. P. Marin, P.
Mandrillon. Gif-sur-Yvette: Ed. Frontières (1990), pp. 589–91
57. Popov, S. G., In *Proc. Conf. Electro-Nucl. Phys. with Internal Targets*, ed. R. G. Arnold, Singapore: World Scientific (1989), pp. 37–47
58. Young, L., Coulter, K., Gilman, R. A., et al., See Ref. 57, pp. 125–31
59. Taiuti, M., Anghinolfi, M., Bianchi, N., et al., See Ref. 57, pp. 140–45
60. Dietrich, F. S., Melnikoff, S. O., van Bibber, K. A., *AIP Conf. Proc.* 150: 378–81 (1986)
61. van Bibber, K., See Ref. 57, pp. 189–94
62. Hirsch, A. S., Bujak, A., Finn, J. E., et al., *Phys. Rev.* C29: 508–25 (1984)
63. Koch, H. R., Riepe, G., Hamacher, A., et al., *Nucl. Instrum. Methods* A271: 375–82 (1988)
64. Olivier, M., Chamouard, Pa., Rommel, G., Tkatchenko, A., *Part Accel.* 32: 161–66 (1990)
65. Allen, D., Asseo, E., Baird, S., et al., See Ref. 10, pp. 814–18
66. Baird, S., Chanel, M., Möhl, D., Tranquille, D., *Part. Accel.* 26: 223–28 (1990)
67. Seth, K. K., See Ref. 57, pp. 211–26
68. Pollock, R. E., See Ref. 9, pp. 17–21; See Ref. 44.
69. Ekström, C., Fransén, E., Gajewski, K., et al., *Phys. Scripta* T22: 256–68 (1988)
70. Katayama, T., Andou, A., Chida, K., et al., See Ref. 56, pp. 577–79
71. Maier, R., Pfister, U., Theenhaus, R., See Ref. 56, pp. 131–33
72. Meyer, H.-O., Ross, M. A., Pollock, R. E., et al., *Phys. Rev. Lett.* 65: 2846–49 (1990)
73. Pitts, W. K., Haerberli, W., Knutson, L. D., et al., In *Proc. 9th Int. Symp. on High Energy Spin Physics*, Bonn (1990), in press
74. Cho, Y., Martin, R. L., Parker, E. F., et al., *IEEE Trans. Nucl. Sci.* NS-24: 1509–11 (1977)
75. Reistad, D., See Ref. 56, pp. 128–30
76. Kullander, S., See Ref. 44.
77. Katayama, T., Chida, K., Hattori, T., et al., *Part. Accel.* 32: 105–12 (1990)
78. Tanabe, T., See Ref. 44
79. Maier, R., Martin, S., Pfister, U., *Part. Accel.* 32: 91–96 (1990)
80. Ando, A., Hosono, K., Katayama, I., Ikegami, H., In *Sixth Symp. on Accel. Sci. and Technology.* Tokyo: Ionics (1987), pp. 310–12
81. Rudchik, A. T., See Ref. 44
82. Grieser, M., Blum, M., Habs, D., et al., See Ref. 56, pp. 620–22
83. Habs, D., See Ref. 44
84. Graw, G., In *Physics with Polarized Beams on Polarized Targets*, ed. J. So-

winski, S. E. Vigdor, Singapore: World Scientific (1990), pp. 328–48

85. Möller, S., In *Proc. First Eur. Part. Accel. Cont.*, ed. S. Tazzari, Singapore: World Scientific (1989), pp. 112–14
86. Andersen, L. H., See Ref. 44
87. Poulsen, O., See Ref. 44
88. Franzke, B., Beckert, K., Eickhoff, H., et al., See Ref. 56, pp. 46–48
89. *Proc. Workshop on Experiments and Experimental Facilities at SIS/ESR*, GSI Rep. 87-7 (1987)
90. Rensfelt, K.-G., See Ref. 56, pp. 623–25
91. Olsen, D. K., Alton, G. D., Datz, S., et al., *Nucl. Instrum. Methods* B24/25: 26–37 (1987)
92. Oganessian, Y. T., Ter-Akopian, G. M., *Heavy Ion Storage Rings with Electron Cooling*, Intern. Rep. I. V. Kurchatov Inst. of Atomic Energy, Moscow, pp. 65–77 (1990)
93. Ruggiero, A. G., In *Proc. Workshop on Electron Cooling*, Legnaro, May 15–17, ed. R. Calabrese. Singapore: World Scientific (1991), pp. 100–14
94. Hahn, H., *Part. Accel.* 32: 75–82 (1990)
95. Yoshii, M., See Ref. 44
96. Pollock, R. E., *Comments Nucl. Part. Phys.* 12: 73–84 (1983)

Annu. Rev. Nucl. Part. Sci. 1991. 41: 389–428

THE STANFORD LINEAR COLLIDER[1]

John T. Seeman

Stanford Linear Accelerator Center, Stanford, California 94309

KEY WORDS: accelerator, particle, emittance, luminosity, polarization

CONTENTS

1. INTRODUCTION

In the early 1980s the world community of high energy physics faced the challenge of building an electron accelerator to provide direct observations and detailed studies of the leptonic and hadronic decays of the then hypothesized Z^0 boson. If the Z^0 boson existed, annihilations of electrons with their antiparticles (positrons) at center-of-mass energies of over 90 GeV/c^2

---

[1] The US Government has the right to retain a nonexclusive royalty-free license in and to any copyright covering this paper.

should produce the highly sought after Z^0 decays in large quantities and with very low detector backgrounds. From this challenge emerged two electron-positron colliding-beam accelerators.

The first accelerator, called the Stanford Linear Collider (SLC), was started at the Stanford Linear Accelerator Center (SLAC) in California (1–7). This linear collider could be built relatively inexpensively by using as a basis the existing two-mile accelerating structure at SLAC, albeit with many modifications and additions from beginning to end. Because of the existing facilities at SLAC, this first-of-its-kind collider had the potential of producing data at the Z^0 mass before any other collider and, more importantly, of becoming a prototype of larger and more energetic linear colliders of the future (8–12). Furthermore, it was expected that during the design and commissioning of the SLC numerous advances in accelerator physics and beam diagnostics of high energy, high brightness beams would be made (13–15). These advances would have applications not only for the next generation of linear colliders but also for synchrotron radiation sources, free electron lasers, high brightness beam sources, and ultra-low-emittance storage rings. As with any new frontier accelerator, the risks were high but so too were the rewards.

The second accelerator, the large electron-positron project (LEP) (16, 17), was started in Geneva, Switzerland, at the CERN laboratory. This large circular collider was a new facility. CERN, however, had the advantage of a twenty-year history of storage ring technology and thus was expected to commission LEP rapidly. It is widely accepted that the cost of this style of collider increases approximately quadratically with beam energy (18), while the cost of a linear collider is expected to grow linearly with beam energy. Thus, the fiscal possibility of building a future electron-positron circular collider larger than LEP is remote at best.

A decade later, in the summer of 1989, the SLC was completed; it produced refined measurements of the mass (91.14 ± 0.12 GeV/c^2) and width ($2.42^{+0.45}_{-0.35}$ GeV/c^2) of the Z^0 boson and determined the number of neutrino families (three) (19, 20). Later that year LEP completed an initial data run and provided greatly improved statistical errors on the above results. Today LEP continues to produce numerous Z^0 decays each day, checking the foundations of the Standard Model. The SLC has turned its efforts to studying Z^0 decays with polarized electrons and studying the requirements for the next linear collider. There are few, if any, studies at the moment anywhere in the world studying electron-positron circular colliders more energetic than LEP, whereas, many institutions around the globe are designing, constructing, and testing new devices and techniques for future high energy, electron-positron linear colliders (21).

This report discusses the new accelerator physics concepts developed and tested while making the SLC a working research tool. At present, there are more interesting accelerator physics tests being proposed each day than there is accelerator time to perform them. This pace is expected to continue into the mid 1990s. Some of the issues being explored include very small emittance bunches, beams of extremely high power densities, multiple bunch beams, beam-beam interaction, positron target design, electron gun design, low emittance damping rings, wakefield compensation, chromatic correction, halo collimation, reliability, instrumentation, and controls. In this report implications for future accelerators are indicated throughout the discussions.

This summary of SLC accelerator advances cannot possibly cover in detail the thousands of design documents, SLC Collider Notes, logbooks of observations, publications of accelerator physics and technology, and individual reports of investigations generated throughout the history of the SLC. The references contained in the bibliography are a good source for further study.

2. DESIGN CRITERIA FOR THE SLC

The SLC was designed to collide single bunches of electrons (e^-) and positrons (e^+) head-on at a single interaction point (IP) with single-beam energies, E, up to 70 GeV. The design concept for the SLC was prefaced by exploratory scenarios for lepton colliders using beams from two linear accelerators (22–24). Rather than construct a second linear accelerator (linac) at SLAC to aim at the existing linac, a decision was made to accelerate both bunches in the same linac and collide them after they pass through oppositely curving arcs. The SLC collider is the highest energy accelerator that can use the arc system because synchrotron radiation losses for each particle, which grow rapidly with energy (E^4/ρ), soon become unmanageable.

The usefulness of a collider is gauged by the time-integrated luminosity with conditions acceptable for the physics detector at the IP. The luminosity L of the SLC determines the event rate R, where R is the product of L and the physics cross section. The luminosity can be calculated from the parameters of the beams:

$$L = N^+ N^- f/(4\pi\sigma_x\sigma_y), \qquad\qquad 1.$$

where N^+ is the average number of positrons per bunch, N^- the corresponding number of electrons, f the collision rate, and σ_x and σ_y are horizontal and vertical bunch sizes at the IP. The bunch size is determined by the emittance ε of the beam and the betatron function β, $\sigma = (\varepsilon\beta)^{1/2}$ (25).

The emittances throughout the SLC are usually stated in energy-invariant units given by $\gamma\varepsilon$, where γ is the relativistic energy factor E/mc^2.

The highest luminosity is obtained when the bunch charges are the highest, the repetition rate is as large as possible, and the transverse bunch sizes are as small as possible. Furthermore, the physics detector at the IP desires a minimum amount of debris or backgrounds from the beam so as not to cloud or prevent data collection. However, all of these goals push the technological limits and are often strongly interrelated.

The normal collision cycle for the SLC can be followed with Figure 1. A positron bunch and two electron bunches are extracted from their respective damping rings. The positron bunch and first electron bunch are accelerated to 47 GeV in the SLAC linac, with a gradient of about 18 MeV/m. After they pass through the two arcs and are reduced to a small size by the final focus system, they collide at the IP. The spent beams are discarded after the collision. On the same acceleration cycle, the second electron bunch after being accelerated to about 30 GeV in the linac (2 km) is extracted and made to strike a water-cooled tungsten target. Positrons emerge from the target with energies near 1 MeV. The positrons are carefully collected while being accelerated to 200 MeV. They are then transported to the beginning of the linac in a quadrupole lattice. Upon the arrival of the positron bunch, the first section of the linac (100 m) is pulsed with radiofrequency (rf) power and used to accelerate the positron and two new electron bunches, which are made by either a thermionic or a polarized gun. These bunches are injected into the two 1.15-GeV damping rings, where radiation damping reduces the emittances to values required for small beam sizes at the final focus. Throughout this cycle a second positron bunch has remained in the positron damping ring, where two damping cycles are required to reduce its naturally large emittance. This complex cycle is repeated at 120 Hz. The inherent instabilities of linacs in general have been compensated in the SLC by the use of slow (one minute) and fast (every pulse) feedback systems, all of which are computer driven. Nearly 100 measured beam parameters are actively controlled. In the interaction region the MKII detector has been in place from the fall of 1987 through November 1990 (26). The SLD detector has been installed and will be ready for beams by the summer of 1991 (27).

Selected design parameters of the SLC are listed in Table 1 (28). Also included in Table 1 are the best individually achieved parameter values (while not maintaining the others), the simultaneously achieved values during the best MKII data collection period, and the projected parameters for collisions with the SLD. In part, the technological challenges of the SLC come from the large range of beam dimensions at various locations in the accelerator. A schematic view of various beam dimensions in differ-

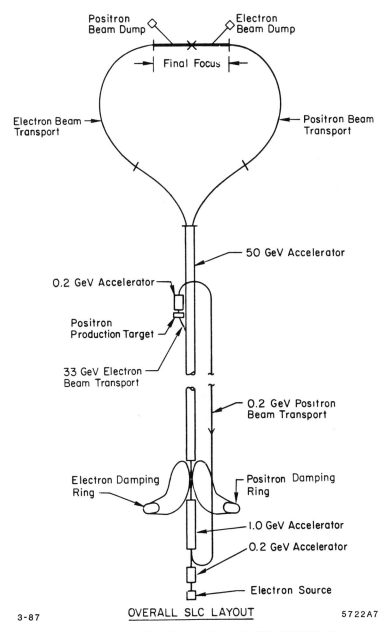

Figure 1 A schematic overview of the Stanford Linear Collider (SLC) (3). The accelerator length, including the arcs and final focus, is about 4 km.

Table 1 SLC accelerator parameters

Accelerator parameter	Units	1984 Design	Best independently achieved	Simultaneous during MKII collisions	Goal for SLD collisions
Beam energy	GeV	50	53	46.6	46.6
Repetition rate	Hz	180	120	120	120
Energy spectrum	%	0.25	0.2	0.3	0.3
N^- at IP	10^{10}	7.2	3.3	2.7	5.0
N^+ at IP	10^{10}	7.2	1.9	1.6	4.0
N^- in linac	10^{10}	7.2	5.3	3.8	6.0
N^+ in linac	10^{10}	7.2	3.5	2.1	5.0
N^- (210 MeV)	10^{10}	8.0	7.0	4.5	8.0
N^+ (210 MeV)	10^{10}	14	10.0	9.0	10.0
Ring damping τ_x	ms	3.5	4.0	4.0	3.5
Target e^+ yield	e^+/e^-	4.0	4.5	4.0	4.0
$\gamma\varepsilon^+_{x(y)}$ in linac	10^{-5} r-m	3 (3)	2.0 (3.0)	2.1 (3.7)	3.5 (3.5)
$\gamma\varepsilon^-_{x(y)}$ in linac	10^{-5} r-m	3 (3)	4.0 (3.0)	7.5 (3.2)	3.5 (3.5)
$\gamma\varepsilon^{+/-}_x$ at IP	10^{-5} r-m	4.0	6.0	7.4	5.4
$\gamma\varepsilon^{+/-}_y$ at IP	10^{-5} r-m	3.2	4.0	4.4	3.6
IP divergence	μrad	300	275	220	300
IP β^*	mm	5	10	15	5
σ^*_x	μm	2.07	3.0	3.5	2.6
σ^*_y	μm	1.65	2.6	2.7	1.6
Bunch length (σ_z)	mm	0.5–1.5	0.5–12.0	1.0	1.2
Pinch enhancement		2.2	1.0	1.0	1.1
Luminosity	10^{29}/cm^2 s	60.0	0.45	0.45	5.0
Luminosity	Z^0/hr	650.0	4.9	4.9	54.0
Efficiency	%	100	90	25	50
Polarization	%	39	0	0	34

ent parts of the SLC is shown in Figure 2. For example, the size of the positron beam changes by four orders of magnitude going from just after the production target, through the various systems, and on to the IP in a few milliseconds. Furthermore, the horizontal to vertical size aspect ratio changes from unity to 250 and back, concurrently. Orchestrating this orderly decrease in beam sizes through the various systems and transitions while minimizing undesired emittance growth requires constant vigilance. The strong focusing quadrupole lattices throughout the collider make this possible to a large degree. A particle that has a transverse displacement of one beam radius, say 300 μm, just after the damping ring will be displaced 4 km downstream at the IP by at most one radius (3 μm) (if we ignore transverse wakefields for the moment) and will remain in collision.

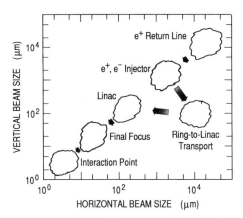

Figure 2 The transverse dimensions of an SLC beam are reduced up to four orders of magnitude from generation through collision. The horizontal-to-vertical aspect ratio also changes by two orders of magnitude. These variations are routine during operation.

3. ELECTRON AND POSITRON SOURCES

3.1 *Electron Source*

The electron source (29–31) consists of a 40-MeV injector and a 1.1-GeV, 2856-MHz accelerator. A schematic of the injector is shown in Figure 3. There are two electron guns: one is a thermionic gridded cathode driven by two high voltage pulsers, and the other is a polarized GaAs photoemitter excited by a laser. The polarized source is discussed further in Section 10. The gun must produce two bunches separated by about 60 ns. After leaving one of these guns, the bunches go through a "Y" bend and into two rf prebunchers operating at 178 MHz. The prebunchers, combined with the final s-band buncher at 2856 MHz, reduce the 60 cm long 150-keV beam pulse into a 3 mm long 2-MeV bunch through velocity-position correlations. The particle density in these bunches is large, requiring special computer simulation codes (32, 33) be used to calculate the expected emittance enlargement from space charge, errors, and wakefields. The enlargement from the bunching process dominates the emittance generated from the cathode (34). The design of an ideal high current gun remains an active subject. The focus of this research is for future low emittance, high current sources for the next collider and for gun structures of klystrons. Bunch charges over 1.3×10^{11} e$^-$ are routinely made with invariant emittances below 5×10^{-4} radian-meters (r-m).

After length compression, the two e$^-$ bunches are accelerated in the injector linac to 210 MeV, where they are joined by the positron bunch coming from the target. These three bunches are then accelerated to 1.15

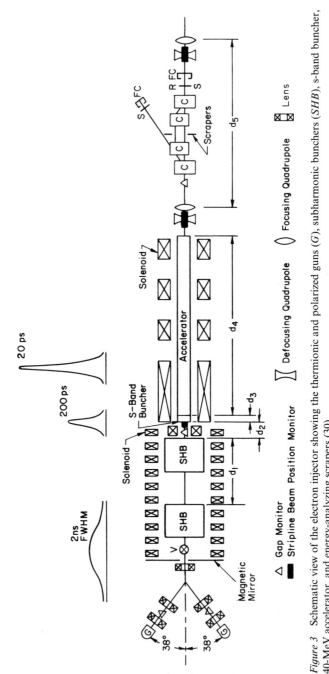

Figure 3 Schematic view of the electron injector showing the thermionic and polarized guns (*G*), subharmonic bunchers (*SHB*), s-band buncher, 40-MeV accelerator, and energy-analyzing scrapers (30).

GeV and injected into their respective damping rings. The intense charge of each bunch requires the compensation of longitudinal beam loading (energy loss) produced by the irises in the rf accelerating structure (35). The phase of each bunch on the rf wave ($\lambda = 10.4$ cm) is adjusted to remove the linear part of the head-to-tail internal energy loading. The long-range longitudinal loading between bunches is compensated by timing changes of the rf so that sufficient refilling of the structure occurs within the bunch separation (60 ns).

The intense charge in each bunch also requires the control of transverse wakefields, which cause off-axis bunches to generate internal forces (36). These forces are such that the longitudinal head of the bunch drives trailing (core and tail) particles to ever increasing amplitudes. There are several methods to control transverse wakefields (see Sections 5 and 9). Here, the lattice has short quadrupole spacings, which make a very short focal length to keep the beams near the axis and thus make the response to wakefields smaller. The alignment of the accelerator, quadrupoles, and the beam position monitors (BPM) needs to be at the level of 100 μm (37).

3.2   Positron Source

The positron source (38–40) is designed to produce as many positrons on every pulse as possible so that ultimately the e^+ bunch in its damping ring has as much charge as its electron partner does in the other ring. The process is as follows. The third accelerated bunch (30 GeV e^-) in the main linac is extracted, passed through an extraction transport line, and made to strike a target. Emerging from the target (which is made of tungsten-rhenium and is six radiation lengths long) is a vast spectrum of low energy electrons, positrons, and photons. A portion of the positron spectrum is carefully focused by a pulsed solenoid (flux concentrator) (41) just downstream of the target, accelerated, and bunched by a 1.4-m "capture" accelerator with a gradient of 40 MeV/m (42). The energy of the positron bunch is then raised to 210 MeV in the "booster region," whereupon it enters the return line to be transported back to the injector linac 2 km upstream. Internal to the booster region, the particle charge is collimated to remove unwanted captured electrons. After the positrons reach 210 MeV, transverse and energy collimators remove particles that are far off axis and far off energy. At this point the positron yield (number of 210 MeV e^+ per number of incident 30 GeV e^-) has been measured to be 2.5, as expected.

The most difficult task in building the positron source has been to design a target that does not crack under intense high power incident bunches. Each 30-GeV electron bunch at 6×10^{10}, with dimensions 0.3 mm diameter and 1 mm bunch length (gaussian sigma), contains 290 joules. Much of this energy is absorbed in the target in a few picoseconds and the rest is absorbed in components downstream. The incoming beam diameter has been enlarged by a factor of two by using a scattering foil a meter in front of the target. Without the foil the electromagnetic shower from the beam would have a small volume and heat the target to above the temperature where the pressure shock cracks the material in a single pulse. The remaining problem is the average power deposition of the beam arriving at 120 Hz. Many solutions to the heating problem have been studied. The best solution is to make a transversely rotating target that distributes the load evenly over a 2-cm circle. The engineering required to make this device reliable in the heavy radiation environment was difficult (43, 44). For example, a nearly automatic removal system was built with quick cable and vacuum disconnects, a remote crane, and special elevator. This remote system can remove a highly radioactive target safely in about one hour with little exposure to personnel (45). Targets for a future collider will use this design as a basis, but an order of magnitude higher beam power is expected.

The length of the positron bunch produced is important because long bunch lengths yield large energy spreads in the injector accelerator that are not accepted easily into the damping ring. A system has been built to monitor bunch length by examining the high frequency voltages induced in a cavity through a ceramic window (A. Kulikov, unpublished). This diagnostic device is ideal for a noninvasive length monitor in a potential future collider.

Finally, the positron bunch that is transported to the injector has a large transverse size. The small dodecapole fields in the quadrupole magnets in the return transport line have been shown to enlarge the transverse emittance if the betatron phase advance in the lattice is near ninety degrees per cell (46). Lowering the phase advance per cell improved the emittance, an indication that small errors over many lattice cells can accumulate significantly.

4. DAMPING RINGS

The damping rings of the SLC have been designed to reduce rapidly the transverse emittance of two bunches spaced nearly equally around the 35-m circumference (47–49). In order to make the damping time short compared to the beam cycle time, the horizontal betatron tune ($v = 8.3$) was made very high compared with a normal storage ring of that size. As a consequence, the vacuum chamber apertures were made quite restricted to allow for the maximum possible dipole fields (20 kG). Furthermore, the ring is operated on the coupling resonance to make the horizontal and vertical emittances equal. The measured emittance reduction with damping time is as expected. Even though new lower emittance damping rings have been designed for a next collider and for synchrotron radiation sources, the SLC damping rings have retained the low emittance record for the past decade.

The bunches are injected on axis into the ring using a kicker with a fast fall time (60 ns) and extracted from the ring a few milliseconds later with a fast rise time kicker (50). The positron ring kickers must have fast rise and fall times so as to not disturb the already stored bunch. The extraction kickers are required to have amplitude stability at the 0.05% level to make the beam trajectory stable in the main linac and thereby avoid transverse wakefields. Furthermore, the timing stability of the kicker pulse must be well below a nanosecond as the kicker pulses do not have perfectly flat tops. The design of these kickers has evolved with time: the latest involves fine control of high power thyratrons, efficient ferrite magnets, radiation hard magnet insulation, subnanosecond timing feedback systems, and perturbative prepulsers (51). This experience will be applied to the next generation collider, where tighter tolerances are expected.

The bunch length σ_z in the damping ring is 6 mm at low currents and lengthens toward 10 mm at high currents (52). This length must be reduced to the 0.5 to 1.5 mm needed in the linac. A radiofrequency accelerator (32 MeV) operating at the zero phase crossing is located in the ring-to-linac transport line and generates a head-tail energy difference in the bunch. When this difference is combined with the chromatic path length difference designed into the transport line, the bunch length is shortened (53). Because of the large energy spectrum ($\sigma_E/E = 1\%$) produced and the resulting beam aspect ratio ($\sigma_x/\sigma_y = 250$), this transport line requires strong corrections to the second-order optics, corrections that are made with sextupoles. The transport line was designed using second-order achromates in which all the second-order chromatic and geometric transport terms were made small (54). This design works well. The concept of second-order achromats is now used worldwide (55). Nevertheless, subtle field errors in the extraction magnets and anomalous magnetic errors in the transport line require first- and second-order betatron, coupling, and dispersion adjustments of the beam at the entrance to the linac.

Transverse instabilities have not been observed in the damping rings, but, three longitudinal effects have: longitudinal bunch lengthening (mentioned above) due to vacuum chamber impedances, longitudinal dipole oscillations (fixed by rf feedback), and coupled bunch longitudinal (π mode) instabilities (56). The π mode instability is temporarily controlled with rf cavity temperature while a feedback cavity is being designed and built. Studies here apply directly to a future collider in which smaller emittances and larger number of bunches in the damping ring are foreseen.

A jitter of the injection position (x,y) or angle (x',y') into the linac causes wakefield growth and filamentation of the beam during acceleration. The launch jitter comes primarily from amplitude instabilities of sensitive dipole magnets in the damping ring system, including the extraction kicker and septum. Ring trajectory control at the 30 μm level is also needed. Constant improvement of the power supplies has reduced the oscillation amplitudes to their present values: 100 μm horizontally $(\sigma_x/3)$ and 30 μm vertically $(\sigma_y/10)$ as observed in the linac. The present jitter reduction activities are concentrated on searching for the 0.25 to 0.5 gauss-meters of unstable field with frequencies below 10 Hz and on applying pulse-by-pulse feedback to the launch variables. Field stabilization and rapid feedback are necessities of all future colliders and storage rings.

5.   LINEAR ACCELERATOR

The main linear accelerator at Stanford has a length of 2946 m and is powered by 230 klystrons (57). The accelerating gradient is about 20 MeV/m. A strong focusing lattice consists of 282 quadrupoles used to maintain the transverse beam size. A pair of x-y correction dipoles and a stripline beam position monitor are associated with each quadrupole for trajectory correction. High resolution profile monitors—both screen (video) and wire scanner types—are located along the linac approximately every factor of three in energy. Monitors for the energy, energy spectrum, and emittance enlargement are placed near the end of the linac to allow either automatic or operator correction during SLC operations.

The primary goal of the linac is to transform the six-dimensional phase-space volume of a low emittance, low energy bunch to high energy without significant phase-space enlargement. Acceleration by itself reduces the absolute emittance of the bunches by increasing the longitudinal velocities of the particles while leaving the transverse velocities constant. However, deleterious effects such as transverse wakefields, rf deflections, chromatic filamentation, and injection errors can increase the emittance, if left unchecked. A technique called BNS damping (58) is used to reduce the effects of transverse wakefields: one adds a head-tail energy spread along a bunch by back-phasing early klystrons and forward-phasing later klystrons. BNS damping works very well. More details of these techniques are discussed in Section 9.

The klystrons are the key to acceleration (59). Each produces 67 MW at 2856 MHz for 3.5 μs. The power is compressed using a SLED system (60). The power is then evenly divided among four 3-m constant gradient accelerating sections on a support girder. At 20 MeV/m, each klystron is thus capable of providing 250 MeV to each particle. The phase and amplitude of each klystron are monitored and adjusted using a new control system that maintains the phase and power tolerances at 0.2° and 0.2% respectively over the 3-km linac length (61). The phase of each klystron can be determined to about 4° absolute and remains close to the desired value for months. The rate at which each klystron produces an errant pulse is well below one for every 10,000 pulses on the average. Thus at a 120-Hz rate with 230 klystrons, errant acceleration cycles occur well below one every forty pulses. The mean klystron lifetime is well over 20,000 hours, which means that one klystron per week needs replacing. Finally, a power limitation in the modulators restricts the repetition rate to 120 Hz.

The energy spectrum of a bunch is determined by its current, its bunch length, and the rf parameters. Intrabunch longitudinal wakefields cause a longitudinal position-dependent deceleration, which is mostly cancelled by moving the average phase of the bunch ahead of the voltage crest of the rf wave. For example, at 5×10^{10} particles and a bunch length of 1 mm, the tail is decelerated by 2 GeV. A compensated energy spectrum has a complex shape, as shown in Figure 4. Both spectra have widths less than 0.2% root mean square (rms) when 1% energy collimation is made. The "double-horned" spectrum at high bunch intensities results from the non-linear longitudinal wakefields and the curvature of the rf waveform (62). The energy spectra are measured in an energy-dispersive region at the beginning of the arcs in a location where the betatron function is low. A

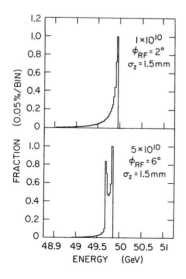

Figure 4 Simulated energy spectrum of a bunch at the end of the linac (47 GeV) at low and high intensities (57). At high intensity the nonlinearity in the rf field and the longitudinal wakefields create the complex shape.

noninteracting spectrum monitor (63) has been built using a vertical wig-gler magnet to generate a vertical synchrotron radiation stripe (3 MeV critical energy). The x-ray stripe strikes an off-axis fluorescent screen viewed by a video camera. The width of the stripe is a measure of the spectrum. A resolution of 0.07% is routine.

With the spectra of both positron and electron bunches adjusted using their respective overall linac phases, the energies of the two bunches are set to the desired values. First, the SLED timing is used to fix the proper energy difference between bunches, and then the appropriate number of klystrons are applied to make the energy sum correct (64). The absolute energy of the two bunches changes rapidly over time up to a percent or so. A pulse-by-pulse feedback system keeps the electron (and positron) energy fixed using opposite-phase adjustments of two eight-klystron "sectors" in the linac (65). This system works well, keeping the rms energy variations at about 0.1%. Energy differences between the bunches grow very slowly and an operator or a slow feedback (minutes) can track them.

Bunch current variations from the injector will cause beam loading changes in the linac, which in turn result in energy errors at the arc entrance and at the target extraction line. This problem is severe enough that at high currents one missing positron bunch can place the third bunch (e⁻) out of the extraction line acceptance and thus stop all future production of positrons. This event is called the "bootstrap" problem, referring to the method of recovery. A fast energy feed-forward system (66) has just been implemented; it measures the current changes in both damping rings and adjusts the klystron phases in an energy feedback system to compensate for the anticipated energy offset in advance of bunch extraction. Lessons learned here are crucial for any future collider, which likely will need many feed-foward systems.

The trajectories of positrons and electrons can be corrected using several steering algorithms (67). Oppositely charged beams can both be steered to nearly the same trajectory even though they are traveling in the same direction because the response of each beam to a given dipole field depends on the betatron function at the dipole. In the lattice, quadrupoles alternate in sign. A given quadrupole focuses horizontally and defocuses vertically, or vice versa. The betatron function alternates high to low from focus to defocus quadrupoles and also for oppositely charged beams. The basic unit of this type of correction is called the "magic beam bump" (68). The steering algorithm that works best in the linac is called "one-to-one" in which the electrons are steered to the centers of the position monitors in their focusing quadrupoles and the positrons are steered to their focusing quadrupoles thus using alternate magnets and position monitors. This algorithm has been shown to be more robust than others against an occasional broken corrector or position monitor. Finally, using a computer program similar to that for trajectory correction, a measured beam oscillation can be used to check and diagnose errors in the lattice phase advance or the local beam energy.

The alignment of the accelerator is important for maintaining the beam emittances. The accelerator structure is supported on girders 12 m long and aligned with a laser system (69). Internal to the girder, the components are aligned using a local optical telescope. In addition, trajectory measurements for both beams using several quadrupole lattice settings are used to make a beam-based determination of the offsets (70). These determinations result in a more accurate check on the alignment than is now possible by mechanical means. Using these methods in combination, the alignment errors of the quadrupoles have been reduced to about 100 microns rms, the position monitors to about 75 microns, and the accelerating structure about 250 microns. The California earthquake of October 1989 produced a 1-cm transverse offset in the linac tunnel housing, shearing the east half of SLAC from the west (71). A gentle s-bend over 200 metres was placed in the linac to correct for this shear. The effect on the emittances from this gentle bend are negligible.

6. ARC TRANSPORT

The arcs must transport two high energy bunches from the linac to the final focus without significant emittance dilution (72, 73). The arc system consists of a short matching region at the end of the linac leading into a very strong focusing array of combined function magnets including dipole, quadrupole, and sextupole fields. Each lattice cell has a betatron phase advance of 108° and a length of 5 m. Given the length constraints of the site, the arc bending radius had to be 279 m. As a result, each 47-GeV particle loses about 1 GeV to synchrotron radiation in a single pass through the system. Ten pairs of magnets form a second-order achromat that can safely transport beams with an energy spectrum of 0.5%. There are 23 [22] achromats in the north [south] arc. At the end of each F [D] magnet within an achromat a beam position monitor is attached to sense the x [y] beam position with an accuracy of about 25 μm. To maximize the bending radius, the magnets and the position monitors occupy all of the longitudinal distance except for a few matching sections.

The existing surface elevation of the SLAC site is irregular, so the arcs were designed to follow the terrain surface. Achromat units (20 magnets) as groups were rotated around the beam axis up to 10° to provide the needed vertical bending. The vertical elevation spans a range of about 75 feet. The beams easily passed through these twisted transport lines to the final focus. However, there were initial matching problems. Over the length of an achromat, slight differences of the actual betatron phase advance from the design caused unwanted coupling of vertical and horizontal beam motion at the roll boundaries. Also, slight magnet misalignments contributed. Unwanted emittance enlargements resulted. A combination of several solutions has now produced acceptable results: tapered rolls at the achromat boundaries, betatron phase corrections using backleg coils on the magnets, trajectory steering by magnet movers, harmonic corrections at certain spatial frequencies, and 3π trajectory bumps for skew correction are all distributed along the arcs (74–76). These corrections are made after exhaustive oscillation data are taken throughout the arcs starting in the linac. After constructing the actual first-order and (soon) second-order transport matrices R_{ij} and T_{ijk}, one can implement the appropriate corrections. Overall, we learn that, with such adjustments and the current knowledge of achromatic systems, terrain-following accelerators can be made to work, should they be necessary because of limited funds or difficult terrain.

The sextupole fields in the magnets introduce strong nonlinearities to beam transport if misalignments of the arc magnets are too large. Complex surveying procedures have been devised to align these three-dimensionally undulating arcs to a precision of about 150 μm (77). Here also, beam-based alignment studies can resolve offsets; for example, those made by the motion of the tunnel floor.

The emittance excitation from synchrotron radiation is kept to a minimum by reducing the horizontal dispersion function to about 35 mm using the compactness of combined function magnets. The expected emittance growths are $\Delta\gamma\varepsilon = 1.3 \times 10^{-5}$ r-m horizontally and 0.5×10^{-5} r-m vertically at 50 GeV. The vertical enlargement comes from the twisted achromats in the arcs. Several other effects introduce small emittance enlargements. Resistive wall wakefields increase the emittance by a few percent given the bunch charge, shape, and the small vacuum chamber diameter (12 mm) (78). Centrifugal forces from the bunches bending away from their electromagnetic fields also add a few percent (79). Radiation anti-damping adds a few percent, as well. All these emittance studies of the SLC arcs are relevant to a future collider, in which two bunch-length-compression sections (80) are likely to be needed. These sections must bend high energy, very-low-emittance bunches through large angles.

In the arcs, misdirected beams can damage components quickly, as they have small dimensions ($\sigma_x < 100$ μm). These sizes, when combined with the high energy, can place a power of 70 kW in a very small volume of nearby material. Long-term heat dissipation is a major concern for vacuum chambers, beam collimators, and profile monitoring instruments. Furthermore, the energy deposition in thick material in a single pulse can be large enough to cause shock waves and instantaneous cracking (81). Several solutions for these heating problems have been successful. All sensitive equipment should be shielded by upstream collimation or by an active beam shut off that measures beam loss. Thin spoilers are used to scatter the high power beam so that the energy density downstream is much lower in a high power absorbing device such as a collimator. Finally, average heating losses can be removed by appropriately designed cooling circuits.

7. FINAL FOCUS

The final focus system of the SLC must perform four functions: it must focus the two opposing beams to small sizes at the collision point; it must steer the two beams into head-on collision; it must collimate and mask errant particles and synchrotron radiation to make acceptable detector backgrounds; and it must safely transport the disrupted 70-kW beams to respective dumps (82–85).

A schematic layout of the component of one arm of the final focus is shown in Figure 5. The beam first enters a correction region to remove dispersion entering from upstream. Then the beam passes through the first demagnifying telescope where the x and y planes have demagnifications of 8.5 and 3.1, respectively. Betatron mismatches and x-y coupling arriving from upstream are also corrected in this region. Next, the chromatic correction section, which contains gentle bends and sextupoles, is used to correct the trajectories of different-energy particles so that they focus at the same longitudinal position at the interaction point. The final telescope provides the last demagnification to make the smallest spots possible and to make the vertical and horizontal spots of equal size. After passing through the IP, each beam traverses through the opposing beam's transport and is deflected into an extraction line to a high power dump. The final focus as built for the MKII detector uses conventional iron quadrupoles as the final triplet near the IP. These magnets raised the minimum possible β^* that could be produced because the distance from the IP to the first quadrupole was increased. New superconducting quadrupoles are being installed with the SLD detector (86). These quadrupoles are closer to the IP, which allows a reduced β^* and a doubling of the achievable luminosity.

The minimum spot sizes at the IP depend on the incoming beam emittances, the maximum allowed divergence angles at the IP, and chromatic corrections. The design values for the incoming beam emittances ($\gamma\varepsilon$) are 4.5×10^{-5} r-m horizontally and 3.5×10^{-5} r-m vertically. However, during current operation the actual emittances are about 50% higher. The angular

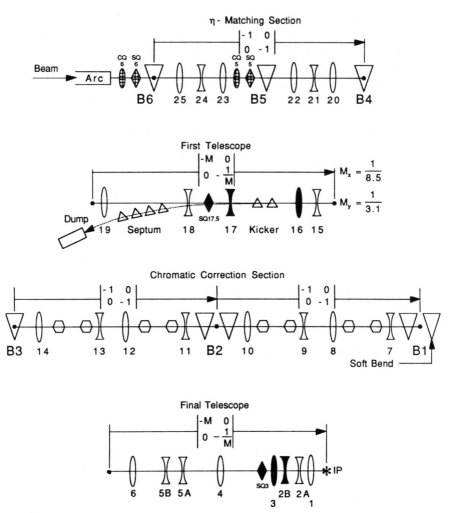

Figure 5 A schematic view of one arm of the final focus system (83). The various sections from top to bottom match the beam phase space from the arcs to the interaction point (IP), demagnify the beam dimensions, and chromatically correct the focusing of off-energy particles.

divergence is limited by synchrotron radiation coming from the strong focusing quadrupoles near the IP. The masking near the detector limits the angular divergence to 250 μrad for the MKII and 300 μrad for the SLD. Finally, the chromatic corrections (87, 88) are made by eliminating the unwanted first-, second-, and third-order matrix elements: R_{ij}, T_{ijk}, U_{ijkl}, respectively, where the subscripts $(1, 2, \ldots 6)$ refer respectively to $(x, x', y, y', \Delta l, \Delta E/E)$.

Chromatic corrections are accomplished by carefully designed symmetries in the final focus. Here are several examples. The first-order dispersions $(R_{16}$ and $R_{36})$ are corrected by placing the final focus dipole magnets in equal-strength pairs 180° apart in betatron space. The second-order dispersions $(T_{166}$ and $T_{266})$ are corrected by providing two identical bending modules that have identical dispersion functions but out-of-phase betatron trajectories. The elements $(T_{126}$ and $T_{346})$ are corrected by using four sextupole pairs in the energy dispersive regions. After the first- and second-order terms are corrected, the third-order terms dominate. By adjusting the ratio of the dipole to sextupole strengths in inverse proportion, one can minimize the third-order terms without changing the second-order correction. The best expected spot sizes for the SLC after these corrections are made are shown in Figure 6 as a function of the first-order betatron function β^* at the IP. In the actual accelerator these matrix terms must be minimized by real-time adjustments. Effective, though elaborate, tuning procedures have been developed for this minimization (88).

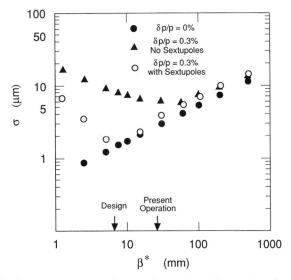

Figure 6   The beam spot size σ (μm) at the interaction point as a function of the linear optics β^* at the collision point (87). The solid circles represent the sizes produced if only first-order optics are calculated. If second- and third-order optics are included, the triangles are obtained. With the use of sextupoles in the final focus to correct the higher order terms, the open circles represent the improvement in the minimum size.

The art of chromatic correction presents a challenge for the SLC and especially for a next collider. The new effects of asymmetric emittances, large energy spreads (σ_E/E), large angular divergences, energy loss from synchrotron radiation in strong quadrupoles (89), multiple bunches, wakefield steering, and high power density beams have all led to innovative designs, one of which will be tested in the Final Focus Test Beam at SLAC (D. Burke, unpublished, 1990).

The beams are very dense at the collision point and can exert large transverse forces on each other, referred to as beam-beam deflections (90). As the two beams are steered through each other, the beam-beam deflection first adds and then subtracts from the bending angle. An example of a measured beam-beam deflection is shown in Figure 7. Beam-beam deflections can be measured in the horizontal, vertical, and skew planes. From the observed deflections, many beam properties can be derived. The beam centroid offsets are determined from the place where the deflection crosses zero. This offset is removed by using nearby dipole magnets to bring the beams into head-on collision. The shape of the deflection curve indicates the size of the combined two-beam system and is a good indicator when upstream components have changed the beam parameters. Jitter in the deflection measurements often indicates pulse-by-pulse position changes.

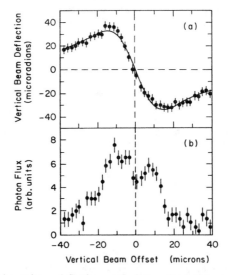

Figure 7   Observed beam-beam deflections at the interaction point are shown in the upper plot (90), and a beamstrahlung radiation signal (radiated photons from the beam-beam interaction) is shown in the lower plot (91).

Present jitter is about one third of the beam size, and results in only a small loss in average luminosity.

The particles bent in the beam-beam interaction radiate a form of synchrotron radiation called "beamstrahlung" (91). This radiation travels forward and is detected using a gas Čerenkov detector about 40 m downstream. The radiation from one beam is most intense where the particle density of the other beam is changing most rapidly, at about one transverse beam sigma. The integrated signal over the bunch can be measured as the beams are steered through each other. An SLC beamstrahlung measurement is shown in Figure 7. The signal shape can have a single peak or two peaks, depending on the initial sizes of the two beams. An on-line computer determination of the inferred spot shapes from the beamstrahlung measurements will soon be tested (92).

An exciting feature of the beam-beam interaction is that the beams will focus each other. This mutual focusing is characterized by a parameter called disruption D (93):

$$D = Nr_e\sigma_z/\gamma\sigma_x^2 \qquad\qquad 2.$$

for round beams, where r_e is the classical electron radius. With strong disruption the beams can be markedly pinched over the length of the bunch and result in an increased luminosity. There are many additional effects of disruption: mutual alignment of the beams, kink instabilities, pair production backgrounds from photon interaction with the opposing bunch, and necessarily large outgoing chamber apertures. Disruption plays a major role in the design of the next linear collider. The SLC will not see significant disruption effects until the bunch densities at the IP increase by about a factor of two over the present values.

Collimation of errant particles generated upstream and synchrotron radiation from the final focus quadrupoles has been difficult for the SLC. The aim was to provide backgrounds similar to circular storage rings. However, equilibrium conditions in a ring are much less noisy than transient conditions in a linear collider. Significant progress has been made (94) through many separate contributions. Collimation, both primary and secondary, of particles with transverse offsets are made at 47 GeV at the end of the linac. Both beams pass through eight 1-mm^2 holes spaced over 150 m with only 10 to 20% losses. Just downstream, the off-energy particles are collimated in a dispersive region early in the arcs. Internal to the arcs and in the early final focus there are tertiary collimators for transverse particle offsets coming primarily from scattering from the edges of the upstream collimators. In the final focus there is also a secondary energy cut. Several masks near the detector shield the IP from locally generated synchrotron radiation emanating from the final focus quadrupoles and dipoles. Finally, two magnetized steel toroids on each side of the final focus deflect muons produced in upstream collimators away from the detector. Experience from the SLC indicates that collimation considerations must be dealt with early in the design of a linear collider and that background-resistant particle detectors are important.

The energies of the SLC beams are measured very precisely with noninterfering spectrometers in the beam dump lines of the final focus (95). These devices are modeled after the energy spectrum monitor at the end of the linac. Two horizontal dipoles are separated by a precisely known vertical bend. The synchrotron radiation stripes downstream are thus separated in angle by the bend of the vertical dipole. Video cameras view the interaction of the radiated x rays with a phosphor-coated screen. The image is then digitized and analyzed. These noninterfering x-ray monitors have been very useful, and a future machine will use them extensively.

Since the two colliding bunches take different paths from the linac to the final focus, it is not obvious that the magnetic path lengths can be adjusted accurately enough to center the longitudinal collision point in the final focus. The bunches can be moved relative to each other in the linac but only in steps of one or more complete rf wavelengths (10.4 cm). There is some adjustment in the focal length of the final quadrupoles. However, the larger the adjustment the more difficult the overall operation of the final focus will be. A test of the arrival times of the two bunches in the final focus was made using a Čerenkov radiator and a streak camera (R. E. Erickson, et al., unpublished, 1987). The resulting path difference between the two arcs and final foci (1400 m each) was about 1 mm: a triumph for the survey and alignment group and a significant contribution to the operation of the SLC final focus.

8. BEAM OBSERVATION AND CONTROL

Control and monitoring of the SLC hardware and beams have significant consequences for stable operations and advancement of accelerator physics. Therefore, a large degree of flexibility has been built into the control system and considerable redundancy into the beam diagnostics.

A VAX mainframe computer interacts through about 75 microcomputers located throughout the accelerator complex to control subunits of the collider (96, 97). The microcomputers interact with the hardware through CAMAC hardware and have sufficient memory and computational power to handle reasonably complicated tasks such as magnet trimming, timing changes to pulsed devices, and the reading and averaging of beam position data. Operators interact with the VAX through control consoles. Each console can operate the entire SLC, and dozens of consoles can be operational at any one time. Reasonable care eliminates most interferences between consoles.

The SLAC linac can in principle produce 360 beam pulses every second (98). The control system allows 256 different beam code definitions, which can be used in any order and at any rate (99). The SLC uses several dozen beam codes to produce colliding beams at 120 Hz. More codes are used for SPEAR, PEP, and nuclear physics beams. With a few new pulsed magnets, all beams could run simultaneously in a time sharing mode. Pulsed devices can be controlled with a timing resolution of about 0.5 ns. At 120 Hz the accelerator acts differently for the two kinds of beam pulses placed oppositely on the 60-Hz alternating current line, which is somewhat asymmetric. Pulsed devices and these flexible beam codes allow correction of most of these "time-slot separation" effects (100).

There are approximately 3000 magnets in the SLC. They are calibrated, standardized, and then set to the desired field strengths using field-current polynomials. Several magnets in the damping ring, arcs, and final focus require regulation at the level of one part in ten thousand. When the beam energy changes—for example, when the linac klystron population changes or the experimenters request a different IP energy—the magnets are scaled appropriately. Because the klystron population changes often, the linac magnets are scaled by a special application code that allows completion in well under a minute, including operator interaction.

The VAX computer has several "watch dog" functions in operation constantly to monitor magnets, klystron phases and amplitudes, temperatures, beam trajectories, and total beam pulses. It alerts the operations staff immediately if a parameter exceeds its tolerances.

Beam position monitors (BPM) are stripline devices with four electrodes (101). The signal from each electrode is sent through a cable to an electronic digitizer, where the position is determined by subtracting the voltage of opposite electrodes and dividing by the sum for normalization. The subtraction is sometimes done before and sometimes after digitization. There are eight different designs for the 2000 BPMs located throughout the SLC, tailored to the local beam properties: for example, high order mode heating must be reduced in the damping ring, centering ability is important in the linac, good accuracy is desired in the arcs, and large diameter and two-beam capability are needed in the final focus. The range of resolutions of the BPMs is 10 to 30 microns, limited mostly by digitization. Averaging of multiple beam pulses improves the resolution by a factor of two to three.

The two types of beam profile monitors used most often in the SLC are phosphor-coated screens viewed by video cameras (102) and wire scanners (103). Screen profile monitors place relatively thick plates (100 to 500 μm) into the beam. The screens interfere with downstream operation, but they are very useful in observing real-time shape changes and providing detailed information of transverse tail formation. The emittance of a beam can be measured using a screen profile monitor and an adjustable quadrupole upstream. Emittance resolutions of about $\Delta\varepsilon = 1 \times 10^{-10}$ r-m (absolute) are routine. Screen profile monitors were irreplaceable during the early commissioning phase of the SLC when beam jitter and asymmetrical beam shapes occurred frequently. However, with the need for a finer resolution and observed damage to the screen material with higher density beams, wire scanning monitors were developed. Wire scanning profile monitors are now used throughout the SLC.

In the final focus, thin carbon fibers 4, 7, or 30 μm in diameter can be inserted into the vacuum chamber near the IP (104). The beam is moved over the wire using a dipole. Both a bremsstrahlung signal downstream and a secondary emission signal from the wire are measured. This device works very well, achieving a resolution of about one third the wire diameter. However, two effects limit operation at the IP. Above 1×10^{10} particle per pulse and with beam sizes below 4 μm, the beam density is high enough to break the wire from thermal shock on a single pulse. Calculations (D. Walz, unpublished) gave predictions of that threshold. Beam observations on the breakage of several wires confirm the calculations to within a factor of two. Secondly, at somewhat lower intensities the fields generated by the beam itself reach the level of ten volts per angstrom and the secondary emission signal changes drastically. The fields from the positron beam pull electrons out of the wire, even though the beam passes by outside. Fields from the electron bunch also affect the emission. Such carbon filament monitors can be used only for low current studies.

Moving an intense beam also leads to particle loss. Thus all the wire scanning monitors away from the IP are moving devices that scan a stationary beam. Measurements taken upstream of the collimators at the end of the linac can also be obtained during normal colliding operations. Upstream of the IP, wire diameters of 40 to 100 microns are used. As a wire is moved through the beam, scattered radiation is measured by either a photomultiplier tube or a fast ion chamber on every pulse. Secondary emission signals are too difficult to use because the intense beam introduces a "position monitor" signal as well as an emission signal on the wire that must be carefully filtered. In a typical arrangement, four scanners are located near each other, separated by about 45° in betatron phase. The readings from the four scanners made in sequence can be used to determine the emittance and phase-space dimensions of the beam (105). Resolutions of $\Delta\varepsilon = 0.3 \times 10^{-10}$ r-m (absolute) have been achieved with bunches of

3.5×10^{10} e$^-$ at 120 Hz. A computer-automated program scans the wires at the beginning and end of the linac every half hour and records a history of the measured emittances and phase-space dimensions for later diagnosis.

Feedback systems are needed throughout the SLC to keep the naturally unstable beams within acceptable tolerances (106). Several pulse-to-pulse feedback systems (mostly independent of the SLC control system) for energy and position at the end of the linac and for position at the IP were commissioned several years ago and have been used with excellent results. Recently, a new fast feedback system, which allows the use of hardware of the SLC control system and takes advantage of modern matrix control theory, has been completed (107). It is now being implemented throughout the project and shows great promise for providing a new level of stable beams.

9. EMITTANCE CONTROL

The absolute transverse emittances ε_x and ε_y of the electron [positron] beam just prior to injection into its damping ring are approximately 1.3×10^{-7} r-m [4×10^{-7} r-m]. This emittance must be reduced by a factor of 265 [800] along its path to the IP. As accounted for in the design, the damping rings store the electron [positron] beam for one [two] acceleration cycles making an emittance reduction of a factor of 10 [30]. Acceleration in the linac reduces the emittance a factor of 41 [41]. The arcs and final focus increase the emittance through unavoidable radiation effects by a factor of about 1.5 [1.5]. The goal for the individual SLC systems is to prevent any additional effects from increasing the emittance. There are many possible enlargement effects (108), and the four different SLC systems are affected by them differently.

The equilibrium emittance of the as-built damping rings is smaller than designed. A change in the method of chromaticity compensation using permanent magnet sextupoles allowed a modification to the dipoles that gave a reduced equilibrium emittance. Operating with a coupled beam, the output emittance is 0.6 of the design at 120 Hz. The observed bunch lengthening in the ring does not affect the transverse emittance except for a small lengthening of the linac bunch. It also probably moves the threshold for transverse mode coupling to beyond our operating intensity. The longitudinal π mode instability mentioned in the section on damping rings causes unwanted random trajectory changes downstream, where there is finite dispersion. Feedback on the extracted bunch reduces this unwanted effect. Finally, the extraction septum has very nonlinear fields near the conductor blades, as all septa do. If an extracted beam trajectory is moved toward the septum, the vertical betatron function first becomes mismatched from the extra quadrupole field and later the emittance increases directly from higher order fields (109). Care in the placement of the extracted trajectory is therefore important.

The bunch in the transport line from the ring to the linac is shortened in length by adding a large energy spread, which introduces chromatic focusing effects. First- and second-order matrix elements involving the energy must be properly adjusted or the enlarged beam size entering the linac will undergo filamentation downstream. Quadrupole and sextupole adjustments are required to remove these errors using measured trajectories, emittances, and betatron functions in the early part of the linac as observables. Since the beam aspect ratio in this transport line is as large as 250, small skew quadrupoles are needed to compensate for rotation errors.

In the linear accelerator many emittance enlargement effects may occur: some result from injection errors, some arise from accumulation of errors along the linac, and some occur at the end. The injection errors include betatron mismatches, dispersion mismatches, x-y coupling, static injection offsets, and launch jitter. Accumulating errors include transverse wakefields, misaligned quadrupoles and position monitors, rf deflections, and component vibration. Effects at the end of the linac include collimator wakefields and x-y coupling.

Betatron mismatches occur when the injection bunch has a phase-space orientation (β, α) that does not match the linac lattice (110). The linac lattice cannot be chromatically corrected because it is straight. Therefore, particles with different energies have different oscillation frequencies. Since the injected beam has an internal energy spread that changes during acceleration, the trajectories of the different energy portions of the beam rotate in phase space at different speeds and soon undergo filamentation. Given a beam β ($\alpha_0 = \alpha = 0$) that is mismatched from the lattice design β_0, the emittance enlargement after filamentation is $\varepsilon/\varepsilon_0 = (\beta/\beta_0 + \beta_0/\beta)/2$. Standard emittance measurements can give the measured betatron functions and indicate the corrections needed.

Dispersion mismatches are similar (111, 112). A dispersion η error means that there is a transverse position-energy correlation, $x = \eta \Delta E/E$. This correlation adds to the apparent size and emittance of the beam: $\sigma^2 = \varepsilon\beta + \eta^2(\Delta E/E)^2$. Given the chromatic lattice, the beam particles displaced by dispersion undergo filamentation if allowed and the real emittance grows. Minimization of the measured emittance early in the linac by using dispersion adjustments correctly removes the dispersion (113).

Emittance growth from x-y coupling occurs when particle trajectories in one plane, say x, are rotated into the other plane (y) by skew transport elements such as rotated quadrupoles or off-axis sextupoles (114). This problem is of more concern for a future collider, where flat beams are needed. Coupling can increase both effective emittances (x and y) or move emittance from one plane to the other. Filamentation of the mixed beam can cause further growth.

A static launch error of the beam injected into the linac generates a betatron oscillation. Standard trajectory correction restores the proper launch. However, if the launch of the beam jitters more rapidly than feedback can correct it, other methods of control are needed. The transport equation of motion for various particles in the bunch is used to study oscillations along the linac.

$$\frac{d^2}{ds^2}x(z,s)+k^2(z,s)x(z,s) = \frac{r_e}{\gamma(z,s)}\int_z^\infty dz'\rho(z')W_\perp(z'-z)x(z',s), \qquad 3.$$

where s is the distance along the accelerator; z is the distance internally along the bunch; k is the quadrupole focusing term (which varies along the linac and along the bunch because of energy changes); ρ is the line density; W is the transverse wakefield due to the accelerating structure; and r_e is the classical electron radius. The left-hand side of Equation 3 represents the form of a betatron oscillation including (slow) acceleration. The right-hand side indicates the forces from transverse wakefields on a particle generated by position errors of all the preceding particles in the accelerating structure. The transverse wakefields for the SLC structure increase in strength with the distance between the leading and the trailing particles (115). Thus, the particles at the end of the bunch, in general, see the largest forces. The displacements grow exponentially with distance along the linac (116). An experiment from the SLC showing this effect is shown in Figure 8. Photographs of beam enlargement from oscillations are shown in Figure 9.

The strongly forced oscillation of the tail by the head in Equation 3 can be ameliorated by changing the energy spectrum along the bunch so that the head is higher in energy than the tail; this can be accomplished with rf phasing adjustments. The wakefield forces, which act like a defocusing force on the tail of the bunch, can be mostly cancelled by the increased focusing of the tail by the quadrupole lattice. This effect is called BNS damping. BNS damping has been studied at the SLC (117) and has been shown to be so effective that all linac operations now use it.

This wakefield cancellation of forces may be exploited further. By careful arrangement of the bunch charge density based on knowledge of the local beam energy, lattice, and rf structure, all particles in the bunch can be made to follow exactly the same trajectory. The conditions for this behavior can be derived by substituting an identical oscillation into Equation 3 for all

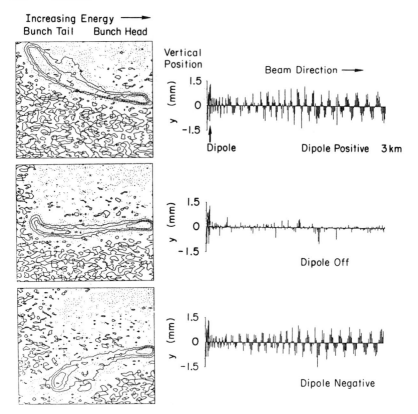

Figure 8 Experimental results demonstrating that the bunch head drives the tail to larger amplitudes during a betatron oscillation due to transverse wakefields (116). (*Right side*) Three trajectories (transverse positions along the linac) are shown. The upper plot has a "positive" oscillation produced by a small change in a dipole magnet early in the linac, the center shows "no" oscillation, and the bottom has a "negative" oscillation. (*Left side*) The three resulting beam shapes are shown measured on a profile monitor. The accelerator conditions are such that the longitudinal head of the bunch is located on the right side of the measured profile and the longitudinal tail is on the left side. Note that the position of the bunch head is independent of the oscillation but that the tail of the bunch moves vertically with the sign of the oscillation. This example shows that leading particles in a bunch drive trailing particles to large amplitudes via transverse wakefields.

particles, and cancelling position terms on both sides (118). This condition is called autophasing (119). However, simulations and experimental attempts to match this condition in the actual operation of the SLC have not been fruitful to date, though studies continue. If this condition can be

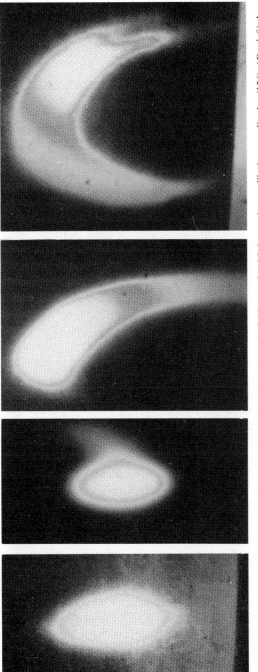

Figure 9 Images of an electron bunch on a profile monitor showing wakefield growth with increasing oscillation amplitudes (136). (*Far left*) An image for a well-steered beam. (*Middle left*) An image produced by an oscillation amplitude of about 0.2 mm. (*Middle right*) The image resulting from a 0.5 mm oscillation. (*Far right*) An image showing a 1-mm oscillation. The beam intensity is 2×10^{10} electrons. The core sizes σ_x and σ_y are about 120 μm.

satisfied in a future collider, the beam intensity threshold for wakefield emittance growth will increase significantly.

Misalignments of quadrupoles, position monitors, and accelerating structures in the linac cause each beam (after correction) to have a trajectory that is neither straight nor centered in the accelerating structure. These offsets generate dispersion and wakefield emittance growth as described above. There are several methods to deal with these errors. (a) They can be found mechanically and fixed, although the required accuracy is well below 100 μm (120). (b) Calculations using knowledge of the beam trajectory as a function of the quadrupole lattice strength can determine the relative quadrupole and position monitor errors to about 75 to 100 μm (121). The misalignments can then be mechanically corrected. (c) A dispersion-reducing trajectory correction may be tried (122). (d) Mechanical movers of the rf structure in the tunnel can be used in a betatron harmonic correction scheme to reduce the final emittance (123). (e) Betatron oscillations can be forced onto the beam at various locations along the linac to cancel effects of the existing absolute trajectory and thus minimize the final emittance (124). (f) Finally, harmonic changes can be added to the quadrupole lattice to cancel random and systematic errors in the quadrupole field strengths (125). The best combination of the available solutions depends on the particular errors involved. All have been tried on the SLC linac with various degrees of success (126).

The rf structures in the linac have several small asymmetries that generate transverse fields (127). These transverse fields can be in or out of phase with the accelerating fields and deflect (slightly) all or portions of each bunch. The accumulation of these kicks along the linac can enlarge the emittance. There are three effects: (a) The average deflection component can be removed by trajectory correction. However, changes in the rf amplitude of a klystron cause changes in the deflecting fields and thus position jitter downstream (128). (b) The steady-state portion of the transverse rf kick causes local trajectory changes, even with steering, and produces off-center beams leading to dispersion and wakefield growth (129). (c) Finally, the magnitude of an rf deflection can vary over the length of the bunch, which causes direct emittance growth and further filamentation downstream (130). Care must be taken during construction of the rf structure to minimize these asymmetries. They cannot be fixed after construction except possibly by cancellation using appropriate pairing.

Mechanical vibration of quadrupoles in the SLC causes trajectory jitter in the beams, resulting in dispersion and wakefield emittance growth. Studies in the SLAC tunnels of quadrupoles with solid supports show vibrations at the 0.05 μm level, mostly with frequencies below 10 Hz (131). Quadrupoles centered on long girders (12 m) show 0.5 μm vibration levels at the resonant frequency of the girder (about 8 Hz) (132). These levels are adequate for the SLC but study is needed for the next collider. Low solid supports are most likely adequate for most of the next accelerator but active vibration control is needed in the final focus.

The intense small beams at the end of the linac are passed through small collimators to remove unwanted halos. If the core of the bunch is slightly off axis in these collimators, the bunch experiences deflecting forces that vary over the bunch (133). The SLC conditions are such that this effect is reasonably small. However, for the next collider special collimation sections must be designed (134).

The emittance growth due to radiation from the dipole magnets in the arcs and final focus was minimized in the original design. However, anomalous dispersion and x-y coupling are major effects that are measured and minimized by using the techniques described in their respective sections (135).

In summary, the many studies of emittance in the SLC have made significant progress toward our desired goal, even with increased beam currents. A time evolution of the gains in emittance and beam currents is shown in Figure 10 (136).

10.   POLARIZATION

A polarized electron source with a three-electrode photocathode is being prepared for use in late 1991 to provide longitudinally polarized electrons at the IP for a precision measurement of the Weinberg angle ($\sin^2 \theta_w$) (137). The emitted polarized electron bunches are handled exactly the same as unpolarized bunches downstream of the gun, with the exception of injection and extraction from the damping ring. The transport line from the linac to the damping ring is designed to have the proper energy and spin precession angle so that a longitudinal superconducting solenoid (6.4 tesla-meters) in that line rotates the spin into the vertical direction for injection into the ring (138). The vertical spin remains polarized during the damping cycle if care is taken to avoid depolarizing resonances. On extraction, the bunch passes through two similar solenoids, one in the ring-to-linac transport line and the other in the early linac. These solenoids are used to align the spin in the precise orientation to make longitudinally polarized electrons at the IP after many precession cycles, both horizontally and vertically in the SLC arcs. The polarization direction of the electrons from the gun will be changed by 180° pulse by pulse randomly in order to remove systematic errors in the data.

The circularly polarized light that irradiates the gun cathode comes from a pulsed dye laser. Pumped by a flash lamp, the dye laser produces about

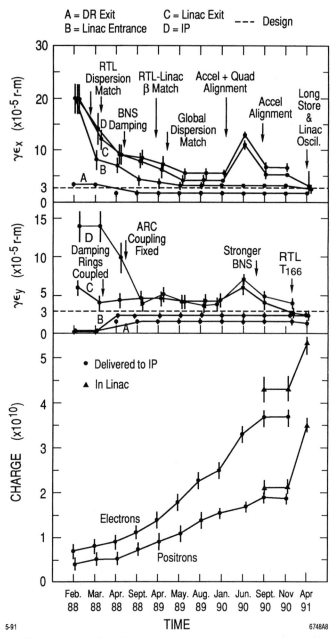

Figure 10   Time evolution of the SLC linac beam emittances and intensities (136) showing when significant corrections occurred. Note the large increase in beam brightness with time.

70 kW of peak power in a 600-ns pulse at up to 120 Hz. Additional opto-electronic devices will be used to divide this pulse into two 3-ns micropulses separated by the 60-ns spacing required by the SLC electron bunches. The stability of the timing and intensity of the light source are important for efficient SLC operation. Tunable solid-state lasers are interesting possi-bilities for a future upgrade.

A new gun with an improved design is under construction (139). The details are shown in Figure 11. A beam of polarized electrons is emitted from a specially prepared GaAs cathode after a pulsed optical beam illuminates the surface. The GaAs cathode is biased negatively and emits electrons with a quantum efficiency of about 1% and a polarization of 40 to 50%. Gun gap voltages of well over 100 kV are needed to reach SLC single-bunch intensities.

There are three polarimeters to allow measurement of the spin direction. One Moller polarimeter is installed at the end of the linac and a second one in the transport line to the electron dump in the final focus. A Compton polarimeter is also installed in the final focus. Even with the often noisy

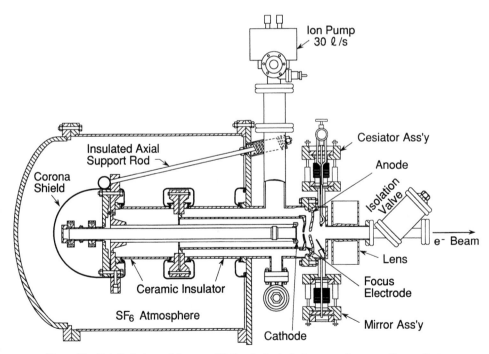

Figure 11   Detailed view of the new SLC polarized electron gun incorporating a GaAs cathode processed with cesium (139).

backgrounds of the transient beams in the SLC, these polarimeters have seen clean signal-to-noise ratios and stand ready for the coming of polarized electrons.

The depolarizing effects of the accelerator rf fields in a linac were shown long ago to be small (140). However, with higher gradients and the bunches being off the rf crest because of BNS damping, a new calculation was done (141). Again, the effects were small but nonzero. Since future linear colliders will also use polarization, experimental studies on the SLC are crucial for planning their spin systems.

Studies of new materials for photocathodes that will produce nearly 100% polarization are under active investigation (142). The motivation for a higher polarization is that the effective luminosity for the physics process being studied at the IP is directly proportional to the square of the polarization.

11. HISTORY, COMMISSIONING, AND RESULTS

The luminosity of the SLC has grown steadily during commissioning and the first collision experiments. The peak instantaneous luminosity recorded over the life of the SLC is shown in Figure 12. The luminosity is determined

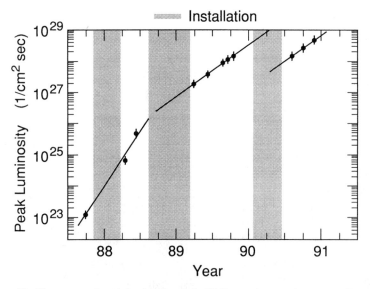

Figure 12   The measured peak luminosity of the SLC over the past three years. Recently, the luminosity has increased at the rate of an order of magnitude for every six months of time devoted to accelerator physics studies (143).

by (a) measuring separately all the parameters entering Equation 1 and (b) guaranteeing that the beam-beam defections at the interaction point are of the proper size, which indicates centered collisions. The rise in luminosity follows an exponential curve with an e-folding time of about one month for the first period and two months in the past two years. The growth is highly correlated with accelerator physics studies and new hardware (143). Hardware improvements made during the installation period yielded immediate gains early in the project, but later gains required more commissioning time as the accelerator operations became more involved. In addition, the peak luminosity growth rate decreased with time because the accelerator time was being used extensively for collisions, as it should be.

The determination of the expected integrated luminosity from the peak luminosity is complicated by many factors. There are several inefficiencies: (a) The accelerator has downtimes, mostly random, due to broken hardware (10%) and the necessary tuning recovery time afterwards (10%). (b) The luminosity does not remain at its peak because of drifting accelerator parameters requiring active operator intervention (5%). (c) Backgrounds often become too severe from drifting accelerator parameters and require frequent retuning (20%). (It is natural to push the performance of the accelerator until the tuning time is about 25%.) (d) The beams do not remain colliding head-on without active control (5%). (e) The particle physics detector has intrinsic "dead" time from overlapping events and electronic event readout dead time (20%) and from system downtime as well (10%). Thus, the overall efficiency compared to the peak luminosity is the product of the efficiencies, or about 40%. This efficiency is comparable with that of other (circular) colliders. (f) Whether to spend time on accelerator physics or on collisions is a difficult decision; collisions add to the total Z^0 count but accelerator physics studies increase the luminosity. (g) Finally, the number of observed events depends on whether the experimenter has the beam energy set to the peak cross section of the Z^0 resonance or not. In total the SLC has integrated about a thousand Z boson equivalents.

The number of active components in the SLC, including power supplies, klystrons, kickers, vacuum pumps, computers, controls, and instrumentation, is nearly an order of magnitude larger than that of recently built circular electron colliders. Since a very large fraction of these components must be operational in order for the accelerator to function, the reliability of each component must be greater. Much effort is spent at SLAC to maintain reliable active components. A future linear collider, as well as any new large circular collider, will have a still larger number of components. Reliability engineering must be an ongoing study.

12. SUMMARY AND CONCLUSIONS

The Stanford Linear Collider has gone from a conceptual idea to an active research tool in just over a decade. Many new accelerator techniques have been proven on the SLC and some hard realities discovered. The accelerator physics of linear colliders has advanced tremendously. In addition, a high energy physics program explored the properties of the Z^0 boson and continues today with the near-term implementation of polarized electrons to the interaction region.

There have been many accelerator physics successes (too many to discuss them all here). The controlled collision of opposing $3\mu m$ diameter beams, the transport of very low emittance beams, the alignment of the accelerator components at the 100 μm level, and acceleration and energy control of multiple bunches with 0.1% stability are crucial steps toward a next collider. The proven ideas of second-order achromats for emittance preservation, the control of 10,000 active devices, and the operation of a 70-kW high energy-density positron target are also necessary milestones. Advances have been made in the production and transport of polarized electrons in intense bunches. The use of active pulse-by-pulse feedback throughout the accelerator to control over one hundred beam variables is unprecedented. These key steps along with many others achieved at the SLC confidently point toward the design of a future collider.

Several operational issues of the SLC illustrate the unavoidable fact that all large accelerators will have inefficiencies. The SLC is an interrelated complex of smaller accelerators: injector, damping ring, linac, positron source, arcs, final focus, and physics detector. Each of these subparts has its own difficulties but, nevertheless, must function with a high degree of reliability in order for the whole to operate efficiently. A fault or tuning problem in one system impedes progress in all other systems. The SLC has had its share of difficulties: detector backgrounds from errant particles, kicker reliability, long-term maintenance of 25-year-old equipment, constant trajectory control, recovery time after a downtime, lack of redundancy in several diagnostic systems, pulse-by-pulse jitter, time pressure, and aggressive schedules for technological advances. Many of these issues of the SLC have been either solved or reduced significantly over the three years of operation. Improvements in the handling of high power, high repetition rate beams remain to be done. Progress has been very steady. Overall, no accelerator physics issue has been discovered that prevents our reaching the goals of Table 1, but it will take time.

Finally, here are several general comments on accelerator design and commissioning learned through observations. (*a*) All potential problems with accelerator operation surmised during the design stage should be thought through carefully, if not acted on, so that a solution can be easily incorporated in the as-built machine. As much flexibility as possible should be incorporated. (b) Commissioning of the subparts of the accelerator should be done in parallel so that all groups can be fully utilized and solutions can be developed in parallel. (c) Tuning improvements must be attempted even with less than ideal beam conditions. General progress is then obtained more rapidly. (d) Most importantly, if a beam problem cannot be measured quantitatively, it cannot be fixed. Beam parameters entering and exiting an accelerator subpart must be measured carefully. The transition regions between subparts are as important as the repetitive regions within a subpart.

Looking toward the future, work is under way at SLAC and around the world to make a detailed design of the next linear collider. Much exciting work remains for this design. The accelerator results and lessons learned from the SLC are playing a key role in shaping realistic plans.

ACKNOWLEDGMENTS

The design, construction, and commissioning of the Stanford Linear Collider required the dedication, insight, hard work, and perseverance of many people at the Stanford Linear Accelerator Center and from associated universities and laboratories. Many technical groups have worked closely together for years to make the SLC a viable accelerator; groups focusing on accelerator physics, operations, controls, instrumentation, vacuum, detector, polarization, mechanical engineering, power conversion, rf engineering, software, and facilities. Also, many visitors from around the world have contributed to the SLC. It is to the credit of the over 1000 people involved that tremendous progress has been made in understanding the theoretical and experimental accelerator physics needed for linear colliders and that the first observations of the hadronic decays of the Z^0 boson were completed. This work was supported by the US Department of Energy, contract DE-AC03-76SF00515.

Literature Cited

1. Stanford Linear Accelerator Center, *SLAC Linear Collider Conceptual Design Report*, SLAC-Rep. 229. Stanford: SLAC (1980)
2. Richter, B., In *Proc. 11th Int. Conf. on High Energy Accel.*, ed. W. S. Newman. Geneva: Birkhauser Verlag (1980), pp. 168–88
3. Erickson, R., ed., *SLC Design Handbook*. Stanford: SLAC (1984)
4. Ecklund, S., *IEEE Trans. Nucl. Sci.* NS-32: 1592–95 (1985)
5. Stiening, R., *IEEE Trans. Nucl. Sci.* 87CH2387-9: 1–7 (1987)
6. Sheppard, J., see Ref. 21, pp. 1–34
7. Seeman, J. T, In *Proc. 1990 Linear Accel. Conf.*, Los Alamos Natl. Lab. Rep. LA-12004-C, ed. C. Beckmann (1990), pp. 3–7
8. Wilson, P., In *Proc. 1988 Linear Accel.*

Conf., CEBAF-Rep. 89-001, ed. C. Leemann (1988), pp. 700–5

9. Guignard, G., See Ref. 7, pp. 8–12
10. Takata, K., See Ref. 7, pp. 13–17
11. Skrinsky, A., In *Proc. 12th Int. Conf. on High Energy Accel.*, ed. F. T. Cole, R. Donaldson. Batavia, Ill: Fermi Natl. Accel. Lab (1983), p. 104
12. Siemann, R. H., *Annu. Rev. Nucl. Part. Sci.* 37: 243–66 (1987)
13. Seeman, J. T., *IEEE Trans. Nucl. Sci.* 89CH2669-0: 1736–40 (1989)
14. Stiening, R., *AIP Conf. Proc.* 105: 281 (1982)
15. Phinney, N., *IEEE Trans. Nucl. Sci.* 32: 2117 (1985)
16. Picasso, E., In *Proc. EPAC 88 Conf.*, ed. S. Tazzari. Singapore: World Scientific (1988), p. 3
17. Bachy, G., et al., *Part. Accel.* 26: 19–31 (1990)
18. Richter, B., *Nucl. Instrum. Methods* 47: 136 (1976)
19. Abrams, G. S., et al., *Phys. Rev. Lett.* 63: 724–27 (1989)
20. Abrams, G. S., et al., *Phys. Rev. Lett.* 63: 2173–76 (1989)
21. Kurokawa, S., Nakayama, H., Yoshioka, M., eds., *Proc. 2nd Int. Workshop on Next-Generation Linear Collider*, KEK Rep. 90-22. Tsukuba, Japan: KEK (1990)
22. Tigner, M., *Nuovo Cimento* 37: 1228–31 (1965)
23. Amaldi, U:, *Phys. Lett.* B61: 313–15 (1976)
24. Skrinsky, A., In *Proc. 6th Natl. Accel. Conf.* Dubna (1978). 1: 19
25. Sands, M., SLAC-Rep. 121. Stanford: SLAC (1970)
26. Abrams, G. S., et al., *Nucl. Instrum. Methods* A281: 55 (1989)
27. Breidenbach, M., et al., *SLAC-PUB-3798.* Stanford: SLAC (1985)
28. Breidenbach, M., et al., *SLAC-SLC-1991.* Stanford: SLAC (1990)
29. James, M. B., Miller, R. H., *IEEE Trans. Nucl. Sci.* 28: 3461 (1981)
30. Clendenin, J. E., et al., *IEEE Trans. Nucl. Sci.* 28: 2452 (1981)
31. Sheppard, J. C., et al., *IEEE Trans. Nucl. Sci.* 87CH2387-9: 43–46 (1987)
32. Herrmannsfeldt, W. B., *SLAC-TN-68-36.* Stanford: SLAC (1968)
33. Herrmannsfeldt, W. B., *SLAC-PUB-4623.* Stanford: SLAC (1988)
34. Ross, M. C., et al., *IEEE Trans. Nucl. Sci.* 32: 3160 (1985)
35. Bane, K. L. F., *SLAC-CN-16.* Stanford: SLAC (1980)
36. Bane, K. L. F., *AIP Conf. Proc.* 153: 971 (1985)
37. Stiening, R., *SLAC-CN-40.* Stanford:

38. Clendenin, J. E., et al., See Ref. 8, p. 568
39. Bulos, F., et al., *IEEE Trans. Nucl. Sci.* 32: 1832–34 (1985)
40. Clendenin, J. E., et al., *Part Accel.* 30: 85–90 (1990)
41. Kulikov, A., et al., *SLAC-PUB-5473.* Stanford: SLAC (1991)
42. Ecklund, S. D., *AIP Conf. Proc.* 184: 1592 (1988)
43. Ecklund, S. D., *SLAC-CN-128.* Stanford: SLAC (1981)
44. Reuter, E., et al., *SLAC-PUB-5370.* Stanford: SLAC (1991)
45. Reuter, E., et al., *SLAC-PUB-5369.* Stanford: SLAC (1991)
46. Kulikov, A., et al., *SLAC-PUB-5474.* Stanford: SLAC (1991)
47. Hutton, A., et al., *IEEE Trans. Nucl. Sci.* 32: 1659 (1985)
48. Rivkin, L. Z., et al., *IEEE Trans. Nucl. Sci.* 32: 1659 (1985)
49. Fischer, G. E., et al., See Ref. 11, p. 37
50. Mattison, T. S., et al., *Part. Accel.* 30: 115–20 (1990)
51. Mattison, T. S., et al., *SLAC-PUB-5462.* Stanford: SLAC (1991)
52. Bane, K. L. F., *SLAC-PUB-5177.* Stanford: SLAC (1990)
53. Wiedemann, H., *SLAC-CN-57.* Stanford: SLAC (1981)
54. Fieguth, T. H., Murray, J. J., See Ref. 11, p. 401
55. Brown, K., *SLAC-PUB-2257.* Stanford: SLAC (1979)
56. Corredoura, P. L., et al., *IEEE Trans. Nucl. Sci.* 89CH2669-0: 1879 (1989)
57. Seeman, J. T., Sheppard, J. C., In *Proc. 1986 Linear Accel. Conf.*, SLAC-Rep. 303, ed. G. Loew. Stanford: SLAC (1986), p. 214
58. Balakin, V., et al., See Ref. 11, p. 119
59. Allen, M. A., et al., *IEEE Trans. Nucl. Sci.* 87-CH2387-9: 1713–15 (1987)
60. Farkas, Z. D., et al., *SLAC-PUB-1561.* Stanford: SLAC (1975)
61. Jobe, R. K., et al., *IEEE Trans. Nucl. Sci.* 87CH2387-9: 735–37 (1987)
62. Bane, K. L. F., *SLAC-AP-76.* Stanford: SLAC (1989)
63. Seeman, J. T., et al., See Ref. 57, p. 441.
64. Seeman, J. T., et al., *SLAC-PUB-5438.* Stanford: SLAC (1991)
65. Abrams, G. S., et al., *Part. Accel.* 30: 91–96 (1990)
66. Hsu, I., et al., See Ref. 7, p. 662.
67. Sheppard, J. C., *IEEE Trans. Nucl. Sci.* 32: 2180 (1985)
68. Seeman, J. T. *SLAC CN-251.* Stanford: SLAC (1983)
69. Herrmannsfeldt, W. B., et al., *Appl. Optics* 7: 996 (1968)

70. Levine, T. L., et al., *SLAC-PUB-4720*. Stanford: SLAC (1988)
71. Fischer, G. E., *SLAC-Rep. 358*. Stanford: SLAC (1990)
72. Fischer, G. E., et al., *IEEE Trans. Nucl. Sci.* NS-32-5: 3657–59 (1985)
73. Weng, W. T., et al., *IEEE Trans. Nucl. Sci.* NS-32-5: 3660–62 (1985)
74. Weng, W. T., Sands, M., *SLAC-CN-339*. Stanford: SLAC (1986)
75. Barklow, T. L., et al., *Part. Accel.* 30: 121–26 (1990)
76. Bambade, P., Hutton, A., *SLAC-CN-370*. Stanford: SLAC (1989)
77. Pitthan, R., et al., *IEEE Trans. Nucl. Sci.* 87-CH2387-9: 1402 (1987)
78. Stiening, R., *SLAC-CN-20*. Stanford: SLAC (1980)
79. Sand, M., *SLAC-CN-326*. Stanford: SLAC (1986)
80. Kheifets, S. A., et al., *Part. Accel.* 30: 79–84 (1990)
81. Walz, D., In *Proc. 1st Int. Workshop on Next Generation Linear Collider*. SLAC-Rep. 335, ed. M. Riordan. Stanford: SLAC (1988), p. 262
82. Brown, K. L., *SLAC-PUB-4159*. Stanford: SLAC (1987)
83. Erickson, R. A., *AIP Conf. Proc.* 184: 1553 (1988)
84. Murray, J. J., et al., *IEEE Trans. Nucl. Sci.* 87-CH2387-9: 1331 (1987)
85. Hawkes, C. M., Bambade, P., *SLAC-PUB-4621*. Stanford: SLAC (1988)
86. Erickson, R. E., et al., *IEEE Trans. Nucl. Sci.* 87-CH2387-9: 142 (1987)
87. Breidenbach, M., et al., *SLC Performance in 1991*, SLAC Rep. Stanford: SLAC (1990)
88. Bambade, P., et al., *SLAC-PUB-4776*. Stanford: SLAC (1989)
89. Hirata, K., et al., *Phys. Lett.* B224: 437 (1989)
90. Bambade, P., et al., *Phys. Rev. Lett.* 62: 2949 (1989)
91. Bonvicini, G., et al., *Phys. Rev. Lett.* 62: 2381 (1989)
92. Ziemann, V., *SLAC-PUB-5479*. Stanford: SLAC (1991)
93. Hollebeek, R., *AIP Conf. Proc.* 184: 680 (1988)
94. Jacobsen, R., et al., *SLAC-PUB-5205*. Stanford: SLAC (1990)
95. Levi, M., *SLAC-PUB-4922*. Stanford: SLAC (1989)
96. Ross, M. C., *IEEE Trans. Nucl. Sci.* 87-CH2387-9: 508 (1987)
97. Phinney, N., Shoaee, H., *IEEE Trans. Nucl. Sci.* 87-CH2387-9: 789 (1987)
98. Neal, R. B., *The Stanford Two-Mile Accelerator*. New York: Benjamin (1968)
99. Thompson, K., Phinney, N., *IEEE Trans. Nucl. Sci.* 32: 2123–25 (1985)
100. Ross, M. C., *SLC Experiment Note 202*. Stanford: SLAC (1990)
101. Pellegrin, J. L., et al., *IEEE Trans. Nucl. Sci.* 87-CH2387-9: 673 (1987)
102. Ross, M. C., et al., *IEEE Trans. Nucl. Sci.* 32: 2003 (1985)
103. Ross, M. C., *SLC Experiment Note 179*. Stanford: SLAC (1990)
104. Field, R. C., et al., *Nucl. Instrum. Methods* A274: 37 (1989)
105. Ross, M. C., et al., *IEEE Trans. Nucl. Sci.* 87CH2387-9: 725–28 (1987)
106. Jobe, R. K., et al., *IEEE Trans. Nucl. Sci.* 87CH2387-9: 713–15 (1987)
107. Himel, T. H., *SLC Experiment Rep. 210*. Stanford: SLAC (1991)
108. Seeman, J. T., *IEEE Trans. Nucl. Sci.* 89-CH2669-0: 1736 (1989)
109. Emma, P., et al., *SLAC-CN-381*. Stanford: SLAC (1991)
110. Spence, W. L., et al., *SLAC-PUB-5276*. Stanford: SLAC (1990)
111. Sheppard, J. C., *SLAC-CN-298*. Stanford: SLAC (1985)
112. Seeman, J. T., *SLAC-CN-330*. Stanford: SLAC (1986)
113. Merminga, N., et al., *SLC Experiment Rep. 172*. Stanford: SLAC (1991)
114. Servranckx, R., Brown, K., *SLAC-CN-350*. Stanford: SLAC (1986)
115. Bane, K. L. F., Wilson, P., See Ref. 2, p. 592
116. Seeman, J. T., Sheppard, J. C., In *Proc. of Workshop on New Developments on Part. Accel. Techniques*, Orsay, CERN-87-11, ed. S. Turner. Geneva: CERN (1987), p. 122.
117. Seeman, J. T., *SLAC-PUB-4968*. Stanford: SLAC (1991)
118. Seeman, J. T., Merminga, N., See Ref. 7, p. 387
119. Balakin, V., See Ref. 81, p. 56
120. Seeman, J. T., *SLC Experiment Rep. 221*. Stanford: SLAC (1991)
121. Adolphsen, C. E., et al., *SLAC-PUB-4902*. Stanford: SLAC (1989)
122. Raubenheimer, T., Ruth, R. D., *SLAC-PUB-5355*. Stanford: SLAC (1991)
123. Seeman, J. T., See Ref. 7, p. 390
124. Chao, A., et al., *Nucl. Instrum. Methods* 178: 1 (1980)
125. Stiening, R., *SLAC-CN-161*. Stanford: SLAC (1982)
126. Seeman, J. T., et al., *SLAC-PUB-5437*. Stanford: SLAC (1991)
127. Seeman, J. T., et al., *IEEE Trans. Nucl. Sci.* 32: 2629 (1985)
128. Stiening, R., *SLAC-CN-181*. Stanford: SLAC (1982)
129. Seeman, J. T., *IEEE Trans. Nucl. Sci.* 87-CH2387-9: 1267 (1987)

130. Seeman, J. T., *Part. Accel.* 30: 73–78 (1990)
131. Stiening, R., *SLAC-CN-42.* Stanford: SLAC (1981)
132. Fischer, G. E., *AIP Conf. Proc.* 153: 1047 (1984)
133. Bane, K. L. F., Morton, P., See Ref. 57, p. 490
134. Merminga, N., et al., *SLAC-PUB-5436.* Stanford: SLAC (1991)
135. Toge, N., et al., *IEEE Trans. Nucl. Sci.* 89-CH2669-0: 1844 (1989)
136. Seeman, J. T., et al., *Part. Accel.* 30: 91–96 (1990)
137. Clendenin, J. E., et al., *SLAC-PUB-5368.* Stanford: SLAC (1991)
138. Moffeit, K. C., In *Proc. Minneapolis Spin Conf.*, ed. S. Smith. Minneapolis: Univ. Minn. (1988), p. 901
139. Clendenin, J. E., *SLAC-PUB-5368.* Stanford: SLAC (1990)
140. Chao, A., *SLAC-CN-29.* Stanford: SLAC (1980)
141. Panofsky, W. K. H., *SLAC-CN-383.* Stanford: SLAC (1991)
142. Maruyama, T., et al., *SLAC-PUB-5420.* Stanford: SLAC (1990)
143. Seeman, J. T., *SLAC-CN-377.* Stanford: SLAC (1991)

Annu. Rev. Nucl. Part. Sci. 1991. 41: 429–68

QUASI-ELASTIC HEAVY-ION COLLISIONS[1]

K. E. Rehm

Physics Division, Argonne National Laboratory, Argonne, Illinois 60439

KEY WORDS: heavy-ion reactions

CONTENTS

1.   INTRODUCTION

Collisions between two heavy nuclei at energies above the Coulomb barrier produce a diverse spectrum of different reaction modes, a spectrum much wider than that observed for light-ion-induced reactions. For the latter case, two processes are observed: direct reactions and compound nucleus formation. Studies of heavy-ion reactions have identified additional pro-

---

cesses such as deep-inelastic reactions, incomplete fusion, and quasi-fission reactions. Although the boundaries between the various proesses are not well defined, it is generally accepted that with increasing impact parameter the interaction evolves gradually from fusion- and fission-like reactions, which require a substantial overlap of the two nuclei through deep-inelastic processes, to quasi-elastic reactions, which are associated with the most grazing-type collisions.

The term quasi-elastic reaction, which entails inelastic scattering and few-nucleon transfer reactions populating low-lying states, first emerged in the early 1970s in order to distinguish peripheral collisions associated with small energy losses from the more violent deep-inelastic reactions, which showed substantial amounts of energy and mass transfer. A clear distinction between quasi-elastic and deep-inelastic reactions with regard to the amount of energy or mass transfer, however, does not exist. This is illustrated in Figure 1. It is evident that only reaction products with masses in the immediate vicinity of the projectile ^{58}Ni have considerable yields around $Q = 0$ and therefore qualify as candidates for quasi-elastic reactions. With increasing mass and charge transfer, the centroids of the energy distributions move to more negative Q values, i.e. toward deep-inelastic reactions. Figure 1 also demonstrates that no clear separation exists

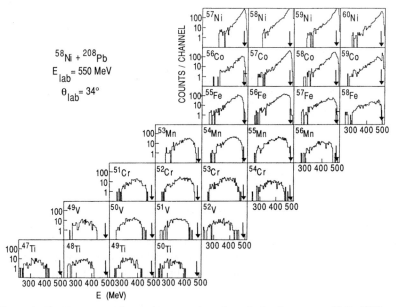

Figure 1 Energy spectra for various reaction channels in the system ^{58}Ni + ^{208}Pb at $E_{lab} = 550$ MeV and $\theta_{lab} = 34°$. The arrows correspond to $Q = 0$.

between the quasi-elastic and the deep-inelastic reaction modes. All experiments with good mass, charge, and energy resolution performed so far indicate that deep-inelastic reactions evolve gradually from quasi-elastic processes. Different prescriptions have been given to separate quasi-elastic reactions from deep-inelastic processes. One involves an arbitrary energy cut and another separates the two processes on the basis of their different angular distributions (see Section 4.2.6). In the following, "quasi-elastic" reactions are only loosely defined as processes in which the projectile emerges after the interaction more or less intact, only slightly modified in mass, charge, and energy.

The more central collisions leading to fusion and deep-inelastic scattering have been studied extensively in the past, both experimentally and theoretically. For these processes the particles in the outgoing channel are sufficiently different in mass, charge, or energy from the particles in the entrance channel, and thus can more easily be distinguished from the elastic channel. Since many nucleons are involved in these reactions, various macroscopic and statistical assumptions can be made in a theoretical treatment. Several review articles have already been published covering deep-inelastic reactions (1–4) or heavy-ion-induced fusion reactions (5–7).

Quasi-elastic reactions induced by heavy ions with masses $A < 20$ have been reviewed previously (8–11). The discussion of quasi-elastic reactions in this article is restricted to reactions induced by heavier projectiles with masses $A > 20$. High-resolution studies with heavier projectiles started about ten years ago when detection techniques with sufficient mass, charge, and energy resolution had been developed. Various techniques are now available for these investigations, and experiments up to the heaviest systems $(U + U)$ have been performed.

A comprehensive theoretical treatment of quasi-elastic reactions is complicated since usually only a few degrees of freedom are involved and thus statistical assumptions may not be justified. On the other hand, one-step calculations are too simplistic because in many cases a particular state can be populated via several routes requiring a more complex coupled-channels treatment. Quasi-elastic reactions therefore cover the full range from simple (ordered) one-step transitions to complicated (chaotic) multistep reactions, which dominate the more central collisions.

Although considerable progress in coupled-channels calculations has been made during the last few years, we are still far from a complete theoretical understanding of these reactions. New techniques have been developed to handle complicated multichannel calculations, but only very simple systems have been investigated so far.

New insight has been gained into the influence of quasi-elastic channels on various other reaction modes. Quasi-elastic heavy-ion reactions have been the topic of several recent conferences or workshops, and additional information can be found in the conference proceedings (12–16).

The interactions between two heavy nuclei ($A > 20$) exhibit distinct differences if compared to reactions induced by "lighter" heavy ions. Figure 2 shows angular distributions for elastic and inelastic scattering populating the lowest-lying excited state measured with ^{16}O, ^{28}Si, ^{58}Ni, and ^{86}Kr projectiles on ^{208}Pb (17–20). The solid lines are the results of distorted wave Born approximations or coupled-channels calculations (discussed in Section 4.1). The cross-hatched bars represent the angle-integrated inelastic yields, which increase from 23 mb for ^{16}O to 1770 mb for ^{86}Kr. Most of this huge increase is caused by the long-range Coulomb potential. Nevertheless, the large yields for peripheral inelastic scattering can strongly influence fusion reactions, which require a strong nuclear overlap. Studies of quasi-elastic reactions have for this reason attracted increased interest during recent years (see Section 5.2).

While the strong rise in the inelastic scattering yields is expected because of the increased Coulomb excitation probability, quasi-elastic transfer reactions (which require a larger overlap of the two colliding nuclei) also exhibit larger cross sections in heavier systems. Figure 3 shows the contributions of quasi-elastic transfer reactions to the total reaction yield for ^{16}O-induced (21, 22) and ^{58}Ni-induced (23, 24) reactions on ^{120,122}Sn and ^{208}Pb, respectively. These reactions were studied at bombarding energies that are about 25% above the respective Coulomb barriers. It is obvious that the contributions of the quasi-elastic transfer reactions to the total reaction cross sections change considerably with increasing mass of the system. While the systems ^{16}O+^{122}Sn and ^{16}O+^{208}Pb are still

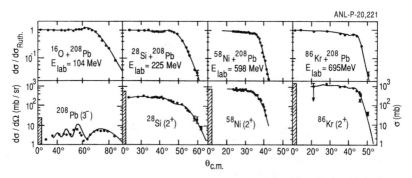

Figure 2 Angular distributions for elastic scattering and inelastic excitation of the lowest state measured with ^{16}O, ^{28}Si, ^{58}Ni, and ^{86}Kr projectiles on ^{208}Pb. The bombarding energies are indicated for each reaction. The solid lines are the results of DWBA or coupled-channel calculations. The cross-hatched bars represent the angle-integrated cross sections.

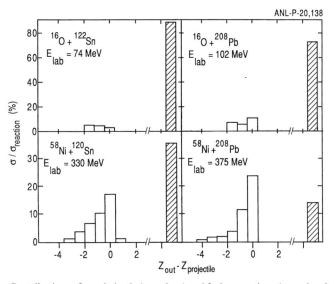

Figure 3 Contributions of quasi-elastic (*open bars*) and fusion reactions (*cross-hatched bars*) to the total reaction cross sections for ^{16}O- and ^{58}Ni-induced reactions on $^{120,122}Sn$ and ^{208}Pb.

dominated by fusion reactions (shown as cross-hatched bars in Figure 3), fusion is less important for the heavier systems $^{58}Ni + {}^{120}Sn$, ^{208}Pb. (In the following, fusion is defined as the sum of compound nucleus formation and fission.) In these heavier systems, transfer reactions are the dominant reaction modes.

The dominance of the quasi-elastic reaction channels is especially large at energies in the vicinity of the Coulomb barrier. Figure 4 shows, as an example, the energy dependence of the various reaction modes in the system $^{58}Ni + {}^{124}Sn$, where a complete set of cross sections has been measured (23, 25–27). The dotted line for inelastic scattering in Figure 4 includes Coulomb and nuclear excitation, the dot-dashed line includes nuclear excitation only. Coulomb excitation takes place mainly at large internuclear distances and with partial waves that are not involved in nuclear reactions (9). Quasi-elastic transfer reactions (shown as solid dots in Figure 4) are the most important nuclear processes at energies close to the barrier and, together with inelastic scattering, are the dominant reaction modes at low energies.

This article discusses the experimental results of quasi-elastic reaction studies performed with heavy projectiles and their interpretations within various (mainly macroscopic) models. The experimental techniques avail-

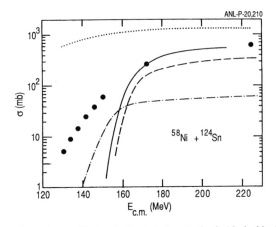

ANL-P-20,210

Figure 4   Energy dependence of fusion (*solid line*), deep-inelastic (*dashed line*), quasi-elastic transfer (*circles*), and inelastic scattering (*dotted* and *dot-dashed lines*) of the lowest 2^+ and 3^- states measured for the system ^{58}Ni + ^{124}Sn (23).

able in studies of quasi-elastic reactions are described in Section 2. Section 3 presents a short summary of the theoretical methods available for the analysis of heavy-ion-induced inelastic scattering and transfer reactions. Section 4 summarizes the experimental results obtained so far. Section 5 discusses the influence of quasi-elastic reactions on other reaction modes. Open questions to be investigated in future experiments are discussed in Section 6.

2.   EXPERIMENTAL TECHNIQUES

The large number of reaction products produced in collisions between two heavy nuclei (see Figure 1) requires a detection system with good energy, mass, and charge resolution as well as a large dynamic range. Various experimental techniques have been developed to meet these requirements, all of which have advantages in certain areas and deficiencies in others.

Some of the early high-resolution experiments utilized Si surface-barrier detector telescopes for particle identification (28). Although Si detectors are capable of achieving excellent time resolutions (29), the flight paths for time-of-flight measurements become prohibitively long for experiments with heavier particles, thus resulting in very small solid angles (28). For this reason Si detector telescopes have been used less frequently in recent studies and are not discussed here in more detail. Kinematic coincidence techniques, i.e. detecting both particles in the outgoing channel with large-area detectors, have been used successfully in some experiments (30, 31).

These techniques are best suited for systems in which the projectile and target have about equal mass. A disadvantage of this method is, however, that complete mass and Z determination has so far been achieved only for lighter systems.

The majority of the more recent studies of quasi-elastic reactions use one of the following three detection techniques:

1. particle identification with magnetic spectrographs or recoil mass separators;
2. measurements of the characteristic γ rays of the reaction products; or
3. chemical separation techniques, followed by a measurement of the characteristic γ rays.

2.1 *Magnetic Spectrographs*

In the past, magnetic spectrographs were used mainly as high-resolution instruments for light-ion-induced reactions. Together with modern focal plane detectors (32–36) they can also serve as effective heavy-ion detector systems with excellent mass, charge, and energy resolution. Spectrographs have reasonably large solid angles, especially if magnetic quadrupole elements are included in their ion-optical design. They separate spatially the elastic peak from events associated with inelastic scattering. A disadvantage is the sometimes large separation of individual charge states in the focal plane, which requires either large focal plane detectors or measurements with different magnetic field settings in order to determine the complete charge-state distribution. The strong kinematic energy shift observed in reactions between two heavy nuclei can be compensated. New spectrographs have been designed to incorporate variable dispersion and perpendicular angle of incidence in the focal plane (37) as well as good kinematic compensation (37, 38). However, some of the older spectrographs [for example, the Enge split-pole (39) and Q3D (40) spectrographs] have also been used successfully for many studies of quasi-elastic reactions.

The heavy-ion focal plane detectors used with magnetic spectrographs typically have active lengths of 50–100 cm, and they measure position and angle-of-incidence in the focal plane along with time-of-flight, energy, and the energy loss or the height of the Bragg maximum for the incident particles (32–36). Complete mass and charge separation for nuclei with masses up to about $A = 150$ has been achieved. Figure 5 shows mass and charge spectra measured with a 1200-MeV ^{208}Pb beam in the system ^{144}Sm + ^{208}Pb at GSI in Darmstadt (41). The mass and charge resolutions obtained in this experiment are about $m/\delta m = 200$ and $Z/\delta Z = 78$. Magnetic spectrographs have been used in studies of even heavier systems up to ^{208}Pb + ^{208}Pb (42).

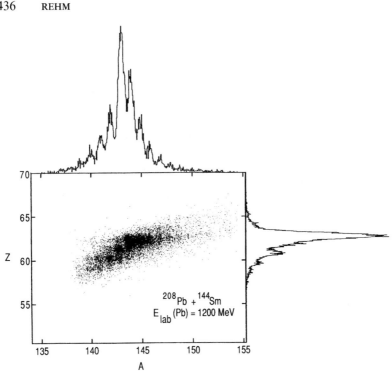

Figure 5 Mass and nuclear charge spectra measured with a position-sensitive ionization chamber at GSI for the system ^{208}Pb + ^{144}Sm (41).

One complication in all experiments using particle identification techniques originates from the possibility of particle decay of the reaction products. Since the time-of-flight through the spectrometers is typically longer than 100 ns, only secondary particles are detected in the focal plane. This difficulty increases at higher bombarding energies. Evaporation calculations have been performed, which in an iterative process generate the primary mass distribution (43).

In addition to the good particle separation, magnetic spectrographs can also provide compensation for the angle dependence of the energy for the scattered particles, which can be a severe problem for high-resolution experiments with very heavy projectiles. The limits for the kinematic constant $k = 1/p \, \delta p/\delta \theta$ with older spectrographs are typically around $k = 0.3$. New spectrographs can compensate up to $k = 1.15$, which makes studies of inverse reaction kinematics (heavy projectile on a light target) feasible (38). Figure 6 shows an energy spectrum for the reaction ^{58}Ni(^{90}Zr, ^{90}Zr)^{58}Ni measured at the Japanese Atomic Energy Research Institute

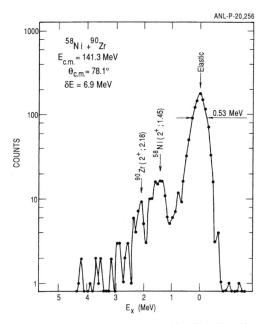

Figure 6   Energy spectrum measured for the reaction ^{58}Ni(^{90}Zr, ^{90}Zr)^{58}Ni with the ENMA spectrograph at JAERI (44).

(JAERI), where for a kinematic constant $k = 0.89$, an energy resolution $\Delta E = 530$ keV has been obtained (44). Without kinematic compensation the energy resolution at an angular acceptance of $\Delta\theta = 0.77°$ would have been 6900 keV.

2.2   *Recoil Mass Separators*

Studies of quasi-elastic reactions at energies in the vicinity of or below the Coulomb barrier are of particular interest since at these energies quasi-elastic reactions are the only reaction modes (see Figure 4) and therefore only a few degrees of freedom are involved. There are, however, experimental difficulties arising from the relatively small cross sections for these processes. In addition, the low energies of the projectile-like particles emerging at backward angles impose limits on the mass and charge resolution that can be achieved under these conditions. These difficulties can be overcome if an inverse reaction kinematics is used, i.e. bombarding a lighter target with a heavier projectile. The target-like reaction products from close collisions are then emitted at forward angles with high energies, which simplifies the particle identification procedure. The results in Figure 5 were obtained using the inverse kinematics reaction ^{144}Sm(^{208}Pb, ^{144}Sm)^{208}Pb.

Another approach is the use of recoil mass separators to detect target-like residual nuclei at forward angles in a reaction with normal kinematics (13, 45). Recoil mass separators combine electric and magnetic elements to separate the beam from the reaction products emitted at small scattering angles (27, 46–48).

The target-like reaction products from a subbarrier transfer reaction are emitted at forward angles and can be identified using time-of-flight and ΔE-E techniques. An example of a mass spectrum measured for the system ^{58}Ni + ^{161}Dy at 0° with the Daresbury recoil spectrometer is shown in Figure 7 (48). It demonstrates the excellent mass resolution that can be achieved with these instruments. Because of the low energy of the recoil products, no Z information was obtained. Q-value and binding-energy arguments as well as the results of particle-γ experiments (27) indicate that transfer reactions at subbarrier energies are dominated by neutron transfer reactions. A disadvantage of recoil mass separators is their relatively small dynamic range in velocity and charge acceptance; one can usually measure no more than four charge states simultaneously. The detection of target-like recoil particles furthermore requires highly enriched targets in order to avoid having elastically scattered target impurities contaminate the yields for the particle of interest (48). Despite these difficulties, recoil mass spectrometers have proven to be important tools for the study of quasi-elastic transfer reactions, especially at subbarrier energies.

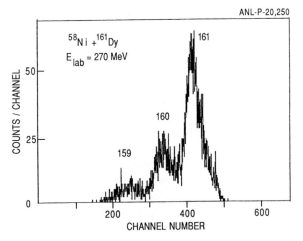

Figure 7 Mass spectrum obtained with the Daresbury recoil separator for the system ^{58}Ni + ^{161}Dy (48).

2.3   Detection of Characteristic γ Rays

The detection of characteristic γ rays from the reaction products has been used since the early days of heavy-ion physics as a high-resolution means of studying quasi-elastic reactions in the vicinity of the Coulomb barrier (49). The use of large detector arrays such as the Spin Spectrometer at Oak Ridge National Laboratory (ORNL) (50) or the Darmstadt-Heidelberg Crystal Ball (51) has considerably increased the efficiency of such measurements. The main advantage of this technique lies in the excellent energy resolution of the Ge detectors. The good energy resolution in turn results in mass and charge resolutions superior to those obtained in experiments using particle detection techniques. The detection of γ rays in coincidence with scattered particles allows one to select grazing trajectories, and it also provides the angle information necessary to correct the Doppler shift of the γ rays emitted from the moving nuclei. Figure 8 shows a γ spectrum measured for the system ^{206}Pb + ^{232}Th (52). Transitions corresponding to one- and two-neutron transfer reactions populating states in ^{231}Th and ^{230}Th are clearly visible in the spectrum.

While the channel resolution achieved with this technique is unsurpassed, a cross-section measurement requires a detailed knowledge of the decay scheme of the final nucleus of interest. Another deficiency of the

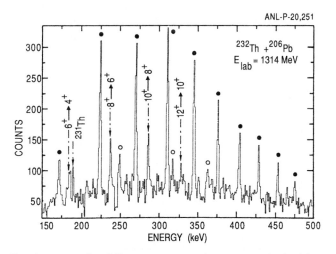

Figure 8   Doppler-corrected and Compton-suppressed γ spectrum obtained from particle-γ coincidence measurements for the system ^{206}Pb + ^{232}Th (52). The Pb projectiles are scattered into center-of-mass angles between 114° and 146°. Transitions in ^{230}Th resulting from the two-neutron transfer reaction are indicated by the appropriate spins. The full and open dots correspond to transitions in the ground and octupole band of ^{232}Th, respectively.

method is that transitions to the ground state and to low-lying states, which decay mainly via internal conversion, cannot be detected. Coincidence measurements between scattered particles and conversion electrons have also been performed (53). More details about the γ-ray technique can be found in the literature (54, 55).

2.4 *Radiochemical Methods*

In radiochemical techniques, the residual nuclei are stopped in a catcher foil, which after irradiation, undergoes a chemical separation process that is sensitive to specific elements or elemental groups. The α and γ activity of the various fractionations is then analyzed and provides very good mass and charge determination (56, 57). Since the catcher foils cover a large area, high sensitivities can be achieved. In addition, the low background in off-line counting allows measurement of very small cross sections. Contrary to particle detection techniques, this method cannot determine the production cross sections for all nuclides, because stable isotopes and isotopes with very short half-lives or unfavorable decay properties cannot be detected. Furthermore, only a very crude measurement of the energy spectra using the stacked-foil technique can be performed.

Despite these deficiencies radiochemical methods are very useful for reactions with very heavy nuclei as well as for studies of reactions with very small cross sections. Figure 9 shows the nuclei identified with radiochemical techniques in ^{238}U + ^{238}U (58). The figure not only illustrates the selectivity of the radiochemical method, but also the high sensitivity, which is emphasized by the fact that very exotic transfer reactions (2p9n transfer reactions with a cross section of only 29μb) have been detected.

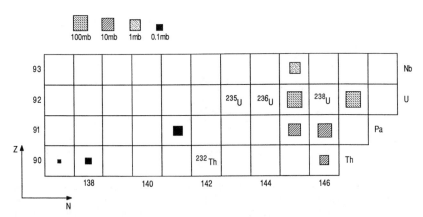

Figure 9   Cross sections measured with radiochemical techniques for the system ^{238}U + ^{238}U at E_{lab} = 6.2 MeV/u (58). The size of the squares is proportional to the cross section.

3. THEORETICAL ANALYSIS OF QUASI-ELASTIC REACTIONS

Quasi-elastic reactions induced by light ions are usually analyzed within the framework of distorted wave theories (59). In these theories it is assumed that elastic scattering is the most important channel. The removal of flux into other channels is treated by the introduction of an imaginary potential, with no individual channel contributing significantly to the absorption. The cross section is then given by

$$\frac{d\sigma}{d\Omega}(\theta) = \frac{k_f}{k_i} \sum_{lm} |T_{lm}(\theta)|^2. \qquad 1.$$

The T-matrix element is calculated from the integral

$$T_{lm} = \int \chi_f^*(r_f k_f) \langle \Psi_B \Psi_b | V_{eff} | \Psi_A \Psi_a \rangle \chi_i(r_i k_i) \, dr_i \, dr_f, \qquad 2.$$

where $\chi_{i,f}$ are the distorted wave functions describing the elastic scattering in the entrance and exit channel, respectively. They are calculated from an optical potential whose parameters have been adjusted to fit data for elastic scattering at the same energy. Distorted wave theories have been very successful in light-ion-induced reactions (59) and have also been applied to reactions induced by heavier projectiles (8–11, 60).

Not all of the assumptions given above are valid for reactions induced by medium-mass projectiles. It was shown in Figure 3 that the influence of fusion reactions (which for light ions are the main contributors to the imaginary part of the optical potential) is reduced in reactions induced by heavier projectiles, with a corresponding increase in the strength of quasi-elastic processes. Elastic scattering is not necessarily the strongest channel; inelastic scattering can be so strong that perturbation theory is no longer adequate. In these cases, inelastic excitation of low-lying states must be included explicitly in a coupled-channels treatment. The effects of channel coupling are clearly observable in inelastic scattering, as discussed in Section 4.1. A treatment of transfer reactions in a similar coupled reaction channels (CRC) formulation has only been attempted in a few cases (61).

Strong inelastic channels affect not only the elastic channel; they may also influence (via two-step processes) various transfer reactions, for example if they precede a particle transfer reaction. Two-step processes are also important for two-particle transfer reactions, where the transfer of a pair of nucleons can be treated either as a cluster transfer or sequentially in two successive one-particle transfer reactions. The complete T-matrix element for this process consists of three terms (62):

$$T = T_{sim}^{(1)} + T_{seq}^{(2)} + T_{NO}^{(2)}, \qquad 3.$$

where $T_{sim}^{(1)}$ is the first-order simultaneous contribution, $T_{seq}^{(2)}$ is the sequential part, and $T_{NO}^{(2)}$ is a nonorthogonal term that for small residual interactions V_{12} cancels the first-order simultaneous term (63). Most two-step calculations have been limited to reactions induced by lighter heavy ions, and clear evidence for the influence of two-step processes has been observed (10).

All calculations for quasi-elastic reactions induced by heavy ions include many partial waves in the sum of Equation 1. Various computer codes have been developed using elaborate interpolation routines for the matrix elements T_{lm}, which change smoothly with angular momentum (64–67).

While the large number of partial waves can complicate numerical calculations, it also allows one to simplify the description of heavy-ion reactions. Replacing the sum in Equation 1 by an integral and using l-dependent T-matrix elements are two of the basic ingredients in many so-called semiclassical or semiquantal treatments. They are especially useful for the analysis of global quantities such as total reaction cross sections and average Q values. These models are described extensively in the literature (68–70).

4.   EXPERIMENTAL RESULTS

High-resolution studies with heavy projectiles populating individual states have been performed only in a few cases. Elastic scattering, a necessary ingredient for obtaining optical potential parameters, has been measured mainly for systems involving nuclei in the vicinity of closed neutron or proton shells (18–20, 23, 30, 31, 44, 71–76). Even for these systems, the energy resolution is in most cases barely sufficient to separate the elastic channel from inelastic excitations. For the system S + Ni, conflicting results have been reported (71, 72).

4.1 *Inelastic Scattering*

Inelastic scattering of medium-weight nuclei provides an ideal testing ground for the study of multistep processes such as multiple Coulomb excitation. Severe experimental limitations restrict, however, the choice of projectile and targets to nuclei in the vicinity of closed shells. The heaviest system investigated so far with particle detection techniques is $^{86}Kr + {}^{208}Pb$ (20, 43, 74) and cross sections for the excitation of the 2^+ level in ^{86}Kr of about 1400 mb/sr have been observed (see Figure 2).

The large excitation probabilities observed in many inelastic scattering reactions require a coupled-channels treatment in the theoretical analysis. Figure 10 shows a comparison between experiment and coupled-channels calculations for elastic and inelastic scattering in the system $^{28}Si + {}^{208}Pb$

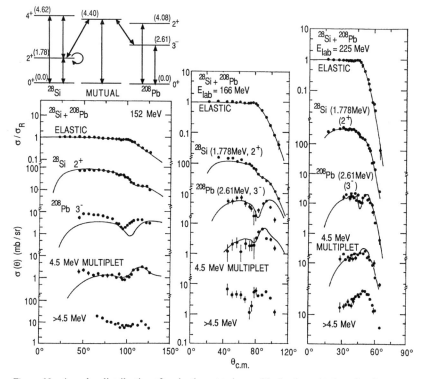

Figure 10   Angular distributions for elastic scattering and inelastic excitations for the system ^{28}Si + ^{208}Pb at three bombarding energies (18, 75). The solid lines are the result of coupled-channel calculations, as discussed in the text. The coupling scheme is indicated.

measured at three energies ranging from 152 to 225 MeV (18, 75). The coupling scheme included a total of six states in the projectile and target and is also shown in Figure 10. Later measurements for the same system extended the energy range to 420 MeV (76). The agreement between experiment and theory is quite remarkable (considering that there is no free parameter in the calculation) once the optical potential has been chosen to describe the angular distribution for elastic scattering. [The $B(E2)$ values were all taken from the literature (see 18, 75 for details).]

^{28}Si + ^{208}Pb is the only heavy system studied with good energy and mass resolution over a large energy range (see Figure 11). While the cross section for populating the 2^+ state in ^{28}Si reaches a maximum at about $E_{lab} = 225$ MeV, a strong increase in the yields to higher-lying (unresolved) states ($E_x > 4.5$ MeV) is observed at higher bombarding energies.

High-resolution inelastic scattering experiments using particle detection

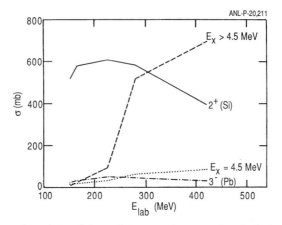

Figure 11   Energy dependence of the angle-integrated cross sections for inelastic excitation of various levels in the system ^{28}Si + ^{208}Pb.

techniques face major difficulties for reactions involving medium-weight projectiles or deformed target nuclei. In these cases, the high resolution of Ge detectors can be used to identify the reaction products via their characteristic γ rays. These methods are especially useful at bombarding energies in the vicinity of or below the Coulomb barrier (49). Excitation functions for inelastic scattering have been measured for several systems (77–79). For Ar- and Kr-induced reactions (77, 78) it was shown that both quantum mechanical (80) and semiclassical (77) calculations can describe the experimental results satisfactorily.

All measurements of heavy-ion-induced inelastic scattering performed at energies above and below the barrier can be reproduced within coupled-channels or semiclassical models, if the optical model parameters and the coupling matrix elements are known. This has important consequences since inelastic scattering is an important doorway to fusion reactions (5–7) (see Section 5.2) and must be included in any theoretical analysis of subbarrier fusion reactions. Since $B(EL)$ values for the most important transitions are known in most cases, coupled-channels calculations can be used for a reliable prediction of the coupling strengths in these reactions.

4.2   Transfer Reactions

Next to inelastic scattering, one-nucleon transfer reactions—especially one-neutron transfer reactions—have the largest cross sections among the quasi-elastic processes. Several experiments have shown that transfer reactions, like inelastic scattering, can also influence other reaction modes such as elastic scattering (81–83) and fusion reactions (5–7, 84, 85). For this reason, a knowledge of the transfer yields for these processes is required. Experimental data for heavy-ion-induced transfer reactions to individual states, however, are even more sparse than data for inelastic scattering, and some examples are discussed in Section 4.2.1. The majority of heavy-ion-induced transfer reactions deal with energy-integrated cross sections (Section 4.2.2).

4.2.1 TRANSITIONS TO INDIVIDUAL LEVELS

As for inelastic scattering, the energy resolution achievable in transfer reactions with heavy ions restricts the choice for projectile and target to lighter nuclei around closed shells. Systems in which transfer reactions to individual states have been investigated include ^{28}Si + ^{40}Ca (86), ^{28}Si + ^{48}Ca (87), ^{32}S + ^{40}Ca (30), ^{37}Cl + ^{40,44,48}Ca (73), and ^{36}S + ^{92}Mo (88). Except for the ^{32}S + ^{40}Ca system, magnetic spectrographs were used in all measurements. Figure 12 shows angular distributions for the one-neutron and one-proton transfer reactions ^{40}Ca(^{28}Si, ^{27}Si)^{41}Ca and ^{40}Ca(^{28}Si, ^{27}Al)^{41}Sc populating several low-lying states in the mass 27 and 41 nuclei (81).

With the exception of a slight shift of the theoretical angular distributions toward larger angles, which is well known from studies with

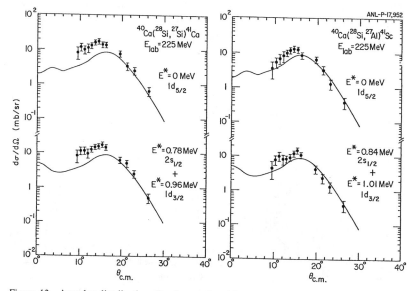

Figure 12 Angular distributions for the one-particle transfer reactions ^{40}Ca(^{28}Si, ^{27}Si)^{41}Ca and ^{40}Ca(^{28}Si, ^{27}Al)^{41}Sc populating various levels in the mass 27 and 41 nuclei (86). The solid lines are the result of distorted wave Born approximation (DWBA) calculations with spectroscopic factors taken from light-ion studies and optical model parameters fitted to the elastic scattering data (86).

lighter heavy ions (60, 89, 90), one-step distorted wave Born approximation (DWBA) calculations predict the shape of the cross sections and also the absolute magnitude for both reactions within a factor of ~2. Considering that the parameters of the binding potentials are subject to some uncertainties (60), the agreement between experiment and theory is quite remarkable. It should be mentioned that the 2^+ state in ^{28}Si is excited with cross sections of 500 mb/sr in the angular region where the transfer reactions show their maximum yields and could therefore influence the transfer via a two-step reaction (transfer following inelastic scattering) (10). Furthermore, polarization effects of the orbitals involved in the transfer, which have been discussed in the literature for reactions with lighter ions (10), should be even more severe in reactions involving heavier nuclei. The good agreement between the experimental data and simple one-step DWBA calculations is thus even more remarkable. Similar good agreement between theory and experiment has been observed in other systems as well (30, 73, 86–89).

4.2.2 ENERGY-INTEGRATED ONE-NEUTRON TRANSFER REACTIONS Transitions to individual levels in heavier systems can no longer be resolved. Figure 13 shows energy spectra for the systems ^{28}Si + ^{208}Pb (18) and ^{58}Ni + ^{208}Pb (91). They are compared with a theoretical spectrum obtained by folding the experimental energy resolution to the results of DWBA calculations for various states. The spectroscopic factors were taken

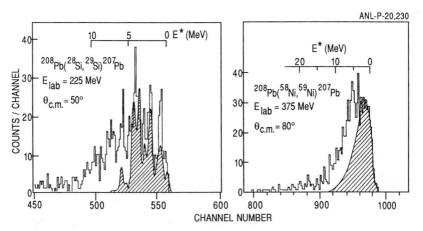

Figure 13 Energy spectra measured for the one-neutron transfer reactions ^{208}Pb(^{28}Si, ^{29}Si)^{207}Pb and ^{208}Pb(^{58}Ni, ^{59}Ni)^{207}Pb at E_{lab} = 225 and 375 MeV, respectively. The cross-hatched area is a spectrum obtained by folding the experimental energy resolution to the results of DWBA calculations.

from light-ion data. Typically 20–30 combinations of states in projectile and target have been taken into account in these calculations. With increasing mass of the nuclei, these calculations become more complex (92). At low excitation energies, where sufficient spectroscopic information is available, the agreement between experiment and theory is generally quite satisfactory. The energy spectra, however, extend to excitation energies of about 10–15 MeV. In this energy region, problems with mutual excitation of ejectile and residual nucleus, as well as the lack of spectroscopic information, make a detailed comparison impossible. For this reason only energy-integrated cross sections are discussed here.

The angular distributions for energy-integrated cross sections are generally bell-shaped, similar to the results obtained for transitions to individual states. Figure 14 illustrates this feature for various neutron transfer reactions (24, 93). The widths of the angular distributions decrease in going from ^{48}Ti to ^{80}Se. However, if the cross section $d\sigma/d\Omega$ is transformed into $d\sigma/dR$, where R is the distance of closest approach for a given θ, and plotted as function of the reduced distance of closest approach $d_0 = R/(A_1^{1/3} + A_2^{1/3})$ (see Figure 14), it is clear that all neutron transfer reactions

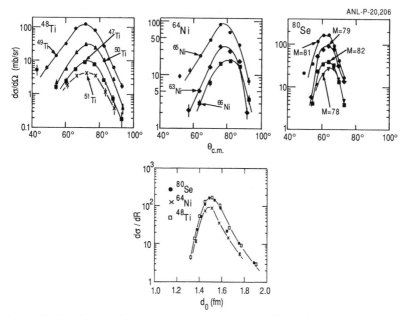

Figure 14 Angular distributions for the energy-integrated neutron transfer reactions induced by ^{48}Ti, ^{64}Ni, and ^{80}Se projectiles on ^{208}Pb at 300, 380, and 525 MeV, respectively. The solid lines serve to guide the eye. See text for a description of the lower plot.

(only one-neutron pickup is shown for simplicity) occur with maximum probability at an internuclear distance corresponding to a radius parameter $d_0 = 1.5$ fm.

The angle-integrated cross sections for neutron transfer reactions are generally quite large and contribute considerably to the total quasi-elastic as well as to the total reaction cross sections (91). The contribution of the neutron transfer to the quasi-elastic yields (excluding inelastic scattering) exhibits an interesting dependence on the projectile mass. Figure 15 shows the relative contributions of the neutron transfer reactions (sum of one- and two-neutron transfer yields) to the total quasi-elastic cross sections (excluding inelastic scattering) measured with various heavy projectiles on ^{208}Pb targets (24, 42). The bombarding energy in these experiments was typically 10–25% above the Coulomb barrier. As can be seen, the contributions from neutron transfers to the total quasi-elastic yield increase from 40% for ^{16}O to 100% for the heaviest projectile (^{208}Pb). This behavior is caused by the increasing influence of the Coulomb potential, which prevents close collisions for the high-Z systems and thereby increases peripheral reactions favoring the exchange of weakly bound particles (see 42 for details).

The energy- and angle-integrated yields for individual one-neutron transfer reactions are strongly system dependent. For the pickup reaction ^{40}Ca(^{28}Si, ^{29}Si)^{41}Ca, an integrated yield of 20 mb was obtained (86). This value increases to 265 mb for the pickup reaction ^{208}Pb(^{58}Ni, ^{59}Ni)^{207}Pb (24). In contrast, the corresponding yield for the stripping process

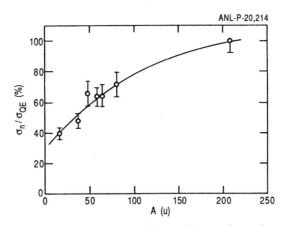

Figure 15 Contributions from neutron transfer reactions to the total quasi-elastic yield (excluding inelastic scattering) for reactions induced by various projectiles with mass A on ^{208}Pb.

^{208}Pb(^{58}Ni, ^{57}Ni)^{209}Pb is only 11 mb (24). The strong variations of these individual cross sections are, to a large extent, caused by changes in the available phase space that is smaller for reactions with negative ground-state Q values. The Q-value dependence of the one-neutron transfer cross sections is illustrated in Figure 16 (24, 94). The one-neutron transfer yield increases by almost two orders of magnitude within a Q-value range of 10 MeV. DWBA calculations can reproduce this Q-value dependence, but the calculations for energy-integrated spectra can become quite complex. A simpler method based on semiclassical matching conditions is described below.

Reactions induced by heavy projectiles can be described in semiclassical models by treating the two interacting nuclei as particles moving on classical trajectories (95). In these models, the observed Q-value preference of heavy-ion transfer reactions is the result of a "matching condition" requiring that the entrance and the exit trajectory evolve smoothly into one another (95, 96). Similar results have been obtained from a quantum-mechanical analysis (97). All matching conditions predict (95–99) that heavy-ion transfer reactions populate states in the final nuclei that are located in a Gaussian-shaped Q window $\exp[-(Q-Q_{opt})^2/\Gamma^2]$ with a centroid given by

$$Q_{opt} = E_{cm}\left(\frac{Z'z'}{Zz} - 1\right) \qquad\qquad 4.$$

and a variance Γ^2

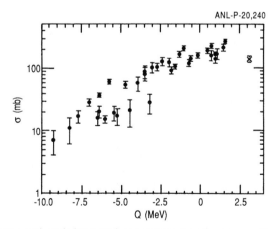

ANL-P-20,240

Figure 16   Energy- and angle-integrated one-neutron transfer cross sections induced by projectiles from ^{28}Si to ^{86}Kr on targets between ^{58}Ni and ^{232}Th, plotted as function of the ground-state Q value.

$$\Gamma^2 = \frac{4\alpha(2E_{cm} - V)\hbar^2}{r_B m_{aA}}$$ 5.

(recoil effects are neglected for simplicity). Here Zz and $Z'z'$ are the nuclear charges in the entrance and exit channels, respectively, E_{cm} is the center-of-mass energy, α is the slope of the transfer form factor, r_B is the radius of the residual nucleus, m_{aA} is the reduced mass in the entrance channel, and V is the Coulomb barrier.

For neutron transfer reactions ($Z = Z'$, $z = z'$), the centroid of the Q window is at $Q = 0$, independent of the bombarding energy, and the width Γ decreases with decreasing bombarding energy. DWBA calculations confirm this simple semiclassical picture (24). The strong ground-state Q-value dependence exhibited in Figure 16 can now be understood from Equations 4 and 5. Figure 17 shows the Q window (obtained from a DWBA calculation) and some of the states that can be populated in the neutron transfer reaction $^{208}Pb(^{58}Ni, ^{59}Ni)^{207}Pb$ with a ground-state Q value of $Q_{gg} = 1.63$ MeV. It is clear from Figure 17 that neutron transfer reactions with negative Q_{gg} will have lower energy-integrated cross sections than reactions with positive Q_{gg}.

The energy-integrated cross section can be obtained by integrating the Gaussian-shaped Q distribution from Equations 4 and 5:

$$\sigma \propto \int_{-\infty}^{Q_{gg}} \exp\left[-(Q - Q_{opt})^2/\Gamma^2\right] dQ$$ 6.

$$\sigma \propto \Gamma\left[1 + \mathrm{erf}\left(\frac{Q_{gg} - Q_{opt}}{\Gamma}\right)\right],$$ 7.

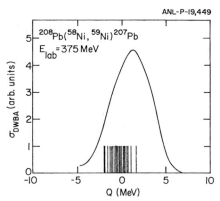

Figure 17 Q window obtained from DWBA calculations for the reaction $^{208}Pb(^{58}Ni, ^{59}Ni)^{207}Pb$. The vertical lines correspond to various known single-particle states, which can be populated in the outgoing channel.

where erf(x) is the error function (24, 94). Improvements to this simple picture incorporate binding-energy effects for the transferred neutron (reactions involving weakly bound neutrons have larger cross sections than those involving tightly bound particles) and effects associated with the increasing number of single-particle states accessible for heavier nuclei. These modifications are discussed in the literature (24).

This model describes the general trend of the one-neutron transfer yields quite well. Figure 18 compares experimental cross sections (corrected for the different binding energies) for a variety of transfer reactions induced by projectiles ranging in mass from ^{28}Si to ^{86}Kr on target nuclei from ^{58}Ni to ^{208}Pb with the semiempirical formula of Equation 7. Only even-even projectile and target nuclei are included in this comparison. For nuclei with odd neutron numbers, the pairing energy has to be considered in the calculation of the optimum Q value (see 24, 94 for details). Good agreement between theoretical and experimental cross sections is observed.

The simple semiclassical formula for the energy-integrated transfer cross section also describes remarkably well the variations in the transfer yields within an isotopic chain of nuclei without any free parameter. Figure 19 compares the experimental and theoretical yields for the one-neutron transfer reactions (^{58}Ni, ^{57,59}Ni) and (^{64}Ni, ^{63,65}Ni) measured on various ^{A}Sn isotopes ($A = 112, 116, 120, 124$) (23). The deviations are typically less than 40%.

Deviations from the general systematics have been observed for some nuclei in the mass 100 region (100, 101). They are probably caused by the small number of low-spin single-particle neutron states available in these

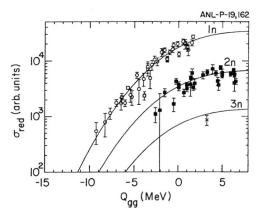

Figure 18 Binding-energy-corrected one-, two-, and three-neutron transfer reactions induced by various projectiles plotted as function of Q_{gg}. The solid lines are obtained by integration over a Gaussian-shaped Q window (see Equation 7).

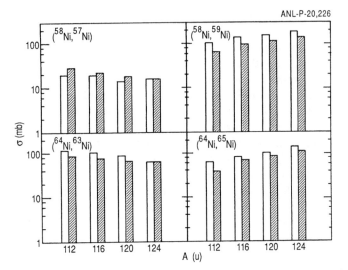

Figure 19   Comparison of the experimental integrated transfer cross sections (*open bars*) in the systems ^{58,64}Ni $+$ ^{112,116,120,124}Sn with the results obtained from Equation 7 (*cross-hatched bars*).

nuclei. In general, however, the energy- and angle-integrated one-neutron transfer yields can be predicted with an accuracy of about 50% using the simple model given above. Nuclear structure effects, on the other hand, are still observable but require a more detailed treatment.

4.2.3   TWO-NEUTRON TRANSFER REACTIONS   Two-neutron transfer reactions have been studied extensively during the past few years. It has been argued that two-neutron transfer reactions, because of their mostly positive ground-state Q values, represent important doorway channels toward fusion, which may explain the large subbarrier fusion enhancement observed in several systems (102–104). Two-neutron transfer reactions have also been investigated in order to obtain information about correlated pair transfer, which was predicted to show large enhancement effects in reactions involving "superfluid" Sn isotopes (the nuclear Josephson effect) (99, 105–112). The analogy between two-neutron transfer reactions and the Josephson effect, however, is not obvious because the pairing energy is usually smaller than the one-neutron binding energy. In addition, the transfer of two particles at low bombarding energies proceeds mainly via two-step processes. These aspects of two-nucleon transfer reactions were discussed in a recent review in this series (54). In the following, only macroscopic properties of the two-neutron transfer reactions are discussed.

Figure 20 shows energy spectra for two-neutron transfer reactions induced by ^{58}Ni on ^{64}Ni and ^{208}Pb targets (91, 113). It is evident that for these systems the ground states are only weakly populated, while the main yield goes to states at high excitation energies (E_x = 5–10 MeV). This is to be expected from the matching conditions discussed in Section 4.2.2, which predicts the maximum of the Q window at an excitation energy $E_x = Q_{gg}$.

The angular distributions for two-neutron transfer reactions are generally bell-shaped, with maxima at the same angles as those observed for the corresponding one-neutron transfer (Figure 14). The total angle- and energy-integrated transfer yields are smaller by about a factor of 4–5 than the yields for a one-neutron transfer cross section with the same ground-state Q value (see Figure 18).

4.2.4   ENERGY DEPENDENCE OF THE ONE-NEUTRON TRANSFER CROSS SECTIONS
At subbarrier energies, neutron transfer occurs via a tunneling process (95) that exponentially increases the neutron transfer cross sections as function of the bombarding energy:

$$\sigma \propto \frac{1}{E_{cm}} \exp\left[-(2\alpha Zze^2/E_{cm})\right]. \qquad 8.$$

Here Z,z are the nuclear charges in the entrance channel and α describes the falloff of the wave function of the transferred neutron. It can be calculated from the binding energy B:

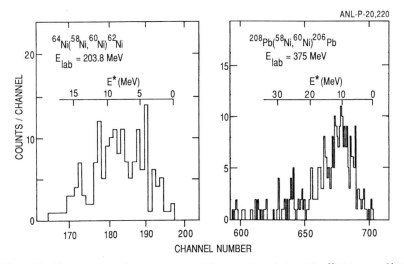

Figure 20 Energy spectra for two-neutron pickup reactions induced by ^{58}Ni ions on ^{64}Ni and ^{208}Pb targets.

$$\alpha = \sqrt{\frac{2\mu B}{\hbar^2}}, \qquad\qquad 9.$$

with μ the reduced mass of the transferred nucleon or cluster. At energies above the barrier, the overlap of the two nuclei is sufficiently large to allow neutrons to be easily exchanged between the two nuclei. The energy dependence in this region is given by the single-particle level densities and by the width of the Q window, as described by Equation 5. Since Γ varies only slightly with energy, a weaker energy dependence is expected from Equation 7 at energies above the barrier.

Figure 21 exhibits the energy dependence of the one-neutron transfer reaction (^{58}Ni, ^{59}Ni) measured for ^{124}Sn and ^{208}Pb targets (23, 25, 27, 91, 92, 114, 115). A good agreement between experiment and theory is observed. Similar results have been obtained for the system ^{28}Si + ^{208}Pb, where one-neutron transfer yields were measured in the energy range 152–420 MeV (18, 75, 76).

4.2.5   PROTON TRANSFER REACTIONS   Because of the higher binding energy for protons and the associated steeper falloff of the wave functions at large distances, a larger overlap of the two interacting nuclei is required to exchange charged particles. Q-matching conditions and mostly negative ground-state Q values favor proton stripping reactions, (^{58}Ni, ^{57}Co), (^{48}Ti, ^{47}Sc).... Proton pickup reactions occur (with smaller yields) in

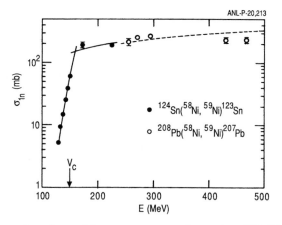

Figure 21   Energy dependence of the one-neutron transfer reactions ^{124}Sn(^{58}Ni, ^{59}Ni)^{123}Sn and ^{208}Pb(^{58}Ni, ^{59}Ni)^{207}Pb. The lines were calculated using Equations 7 and 8, and were normalized to the data.

systems involving neutron-rich projectiles. One-proton transfer reactions to individual states were discussed above in Section 4.2.1.

Energy-integrated proton transfer reactions display a behavior similar to that of neutron transfer reactions. The energy spectra for charged-particle transfer reactions are centered at more negative Q values than the ones observed for neutron transfer processes (see Figure 1), and it is sometimes not clear which reactions should be associated with quasi-elastic processes. The angular distributions are bell-shaped, with maxima located at slightly larger angles than for the corresponding one-neutron transfer case (93). The energy- and angle-integrated one- and two-proton transfer cross sections may be treated within the semiclassical framework discussed above (see 24 for details).

It is interesting to note that the ratio between the energy- and angle-integrated yields for one- and two-proton transfer reactions for a given ground-state Q value is only about 2, while it was 4–5 for the neutron transfer reactions. Heavy-ion-induced two-proton transfer reactions have recently received renewed interest. It has been argued that they represent the best analogue of the Josephson effect in nuclear physics since, at large distances, the transferred particles still feel the Coulomb barrier between the two interacting nuclei (116, 117). The pairing energy, however, is again smaller than the one-proton separation energy, which makes the analogy somewhat doubtful.

The energy dependence of proton transfer reactions has not been studied extensively. Figure 22 shows the energy dependence of the cross sections for the proton transfer reaction ^{208}Pb(^{58}Ni, Co) (integrated over the Co

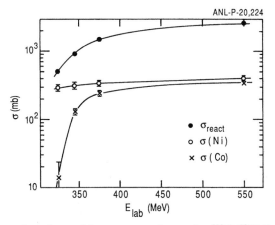

Figure 22   Energy dependence of the proton transfer reaction ^{208}Pb(^{58}Ni, Co) compared to the neutron transfer reaction ^{208}Pb(^{58}Ni, Ni).

isotopes $^{55-59}$Co) in comparison with the results from the neutron transfer reactions ^{208}Pb(^{58}Ni, Ni) (integrated over the nickel isotopes ^{57,59,60,61}Ni) and the total reaction cross sections (excluding inelastic excitations) taken from coupled-channels calculations. While the neutron transfer cross sections remain fairly constant between 325 and 550 MeV, the yield for the proton transfer reaction increases to about $E_{lab} = 380$ MeV and saturates for higher energies. Because of the smaller cross sections, only a few proton transfer reactions have been studied in heavy systems. If future experiments confirm the analogy between proton pair transfer reactions and the Josephson effect, heavy-ion-induced charged-particle transfer reactions will be an important tool for the study of particle correlations in finite systems.

4.2.6 MULTIPARTICLE TRANSFER REACTIONS   Transfer reactions induced by heavy projectiles produce a large spectrum of multiparticle transfer processes (Figure 1). As discussed above for the proton transfer reactions, the extent to which these complex multiparticle transfer reactions can be labeled "quasi-elastic" has not yet been resolved. For the system ^{58}Ni + ^{208}Pb, only the one- and two-particle transfer reactions show appreciable yields at $Q = 0$. The angular distributions for multiparticle transfer reactions exhibit a change from a bell-shaped structure associated with quasi-elastic scattering to a forward peaking, characteristic of deep-inelastic reactions. This is demonstrated in Figure 23 for the system ^{58}Ni + ^{208}Pb for a variety of transfer reactions.

The evolution of multiparticle transfer processes can be studied through their energy dependences. Figure 24 shows the relative contributions of the different quasi-elastic transfer reactions to the total reaction cross section in the system ^{58}Ni + ^{208}Pb (115). In order to define quasi-elastic reactions independent of the bombarding energy, the yields of the Gaussian part of the angular distributions (see Figure 23) have been associated with quasi-elastic processes (93). The yields for each element Z are integrated over all isotopes populated in the reaction. The data are typical for heavy-ion-induced transfer reactions. At energies close to the Coulomb barrier, neutron transfer ($Z = 28$) is the dominant reaction process, contributing almost 60% to the total reaction cross section at $E_{lab} = 325$ MeV. Quasi-elastic charge transfer reactions contribute less than 5%, and fusion and deep-inelastic processes make up the rest of the total reaction cross section at the lowest energy. The exchange of charged particles requires a larger nuclear overlap and sets in at somewhat higher energies. Multiparticle transfer reactions exhibit a strong energy dependence at low energies, similar for the $Z = 23–27$ channels. After sufficient overlap has been established, the relative contributions of the multiparticle transfer reactions remain practically unchanged. This is in contrast to deep-inelastic reactions, which increase strongly with bombarding energy.

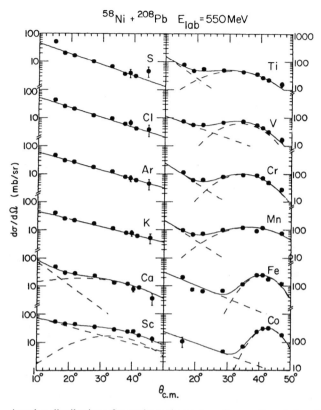

Figure 23 Angular distributions for various charge transfer reactions induced by ^{58}Ni projectiles on ^{208}Pb at $E_{lab} = 550$ MeV.

For these complex multiparticle transfer processes, no microscopic calculations have been performed yet. Multiparticle transfer reactions, however, are an interesting field for the investigation of the transition from deterministic one-step reactions to complex statistical processes in nuclear systems.

5. QUASI-ELASTIC REACTIONS AND THEIR INFLUENCE ON OTHER REACTION MODES

Contrary to light-ion-induced reactions, elastic scattering and compound nucleus formation are not the dominant processes in heavy-ion reactions near the Coulomb barrier. In heavy-ion-induced collisions, quasi-elastic

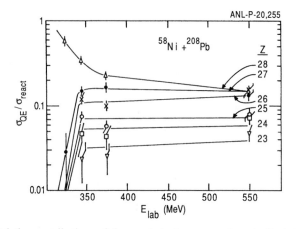

Figure 24 Relative contributions of the quasi-elastic cross sections to the total reaction cross section measured at various bombarding energies for the system ^{58}Ni + ^{208}Pb. The yields have been integrated over the various isotopes populated for a given element. The solid lines serve to guide the eye.

reactions successfully compete in reaction strength and can influence other reaction modes in a variety of ways that are discussed briefly in this section.

5.1   *Quasi-elastic Reactions and Elastic Scattering*

The effect of quasi-elastic channels on elastic scattering can take various forms. As illustrated in Figure 2, strong inelastic channels can lead to deviations of the angular distribution for elastic scattering from the Rutherford prediction, even at angles forward of the grazing angle. This effect is well known from many studies of heavy-ion-induced inelastic scattering (9, 118). Transfer reactions influence elastic scattering mainly at larger angles; this effect has been observed only recently in the systems ^{16}O + ^{208}Pb (81, 82) and ^{28}Si + ^{64}Ni (83).

Another, more subtle, effect of quasi-elastic reactions arises from the energy dependence of quasi-elastic processes (discussed in Section 4.2.4). The cross sections for quasi-elastic reactions, which dominate at low energies, increase strongly in the vicinity of the barrier (Figure 4). Since these channels are the main contributors to the imaginary part of the optical potential W, $W(E)$ experiences a sharp increase at energies near the barrier, as shown schematically in the top part of Figure 25. Based on causality arguments, it has been shown (119, 120) that the real part V and the imaginary part W of the optical potential are related by a dispersion relation

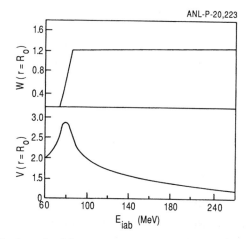

Figure 25 (*Top*) A schematic of the energy dependence of the imaginary part of the optical potential. (*Bottom*) Resulting energy dependence of the real part of the potential obtained from Equation 10.

$$V(E) = V_0 + \frac{P}{\pi} \int \frac{W(E')}{E' - E} \, dE', \qquad\qquad 10.$$

where V_0 is an energy-independent real potential. With an energy-dependent imaginary potential, a peak in the real potential $V(E)$ is obtained, as shown in the bottom part of Figure 25. Experimental data for elastic and inelastic scattering in various systems have confirmed this relationship between the real and imaginary part of the optical potential (121–124).

5.2 Quasi-elastic Reactions and Subbarrier Fusion Enhancement

The large deviations of heavy-ion-induced fusion cross sections from the predictions of simple barrier-penetration calculations at energies at and below the Coulomb barrier have been studied extensively during the last ten years, and several review articles can be found in the literature (5–7). Various theoretical explanations including macroscopic (125, 126) (neck formation) or microscopic treatments (127–131) (coupled-channels effects) have been developed. The main channels responsible for the enhancement of subbarrier fusion cross sections that must be included in a coupled-channels description are the inelastic and transfer channels discussed in this article.

Inelastic scattering gives the dominant contribution in most cases (5–7, 84), but does not account for all features observed in subbarrier fusion reactions. Figure 26 shows fusion yields and quasi-elastic transfer cross sections for the systems ^{32}S, ^{58}Ni + ^{58,64}Ni (5, 132, 133). There are large differences in the fusion cross sections at low energies, with the more neutron-rich target nuclei exhibiting larger yields. Since the $B(E2)$ values do not change appreciably among the various Ni isotopes, inelastic scattering alone cannot explain the neutron-number dependence of the fusion yields. It has been suggested (101–103) that two-neutron transfer processes are responsible for these differences, since the ground-state Q value for the ^{64}Ni(^{58}Ni, ^{60}Ni)^{62}Ni reaction is positive ($Q = 3.887$ MeV), while it is -2.083 MeV for ^{58}Ni(^{58}Ni, ^{60}Ni)^{56}Ni. Similar observations hold for the S + Ni system. The energy gain for a channel with a positive ground-state Q value can increase the fusion probability considerably, especially at energies below the barrier. Q-matching conditions, however, are not favorable for neutron transfer reactions with positive ground-state Q values. Two-neutron transfer reactions favor an optimum Q value of $Q_{opt} = 0$, which means that the states with the most positive Q value are only weakly populated, while the main transfer strength goes to states at $E_x = Q_{gg}$ (see Figure 20). It has been shown, however, that states with negative Q values can also contribute to the fusion enhancement, if the coupling to the ground state is sufficiently strong (114). This means that other quasi-elastic transfer reactions must also be included in a coupled-channels treatment.

For the Ni + Ni and S + Ni systems, the differences in the subbarrier fusion enhancement are correlated with a stronger quasi-elastic yield for the neutron-rich target ^{64}Ni (113, 134). These processes were included in a coupled-channel treatment for a few cases (84, 85, 135), and the agreement between experiment and theory has improved considerably. For the Ni + Ni system, however, the fusion cross sections at low energies are still underpredicted by a factor of ~ 5 and even larger discrepancies are

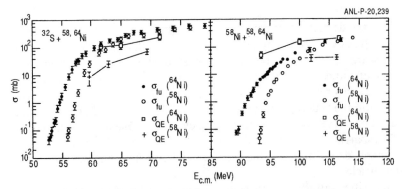

Figure 26 Fusion yields and quasi-elastic cross sections measured for the systems ^{32}Si + ^{58,64}Ni and ^{58}Ni + ^{58,64}Ni.

observed in heavier systems (84). This clearly reveals that subbarrier fusion reactions with heavy ions are not yet fully understood.

5.3   Quasi-elastic Reactions and the Limit of Fusion for Heavy Nuclei

While the nuclear structure effects in heavy-ion-induced fusion reactions discussed in Section 5.2 seem to be limited to nuclei with $A < 100$, another correlation between quasi-elastic reactions and fusion occurs in heavier systems, where a dynamical hindrance for the fusion channel has been observed. Experiments in very heavy systems have revealed that fusion cross sections for nuclei with a combined nuclear charge larger than $Zz = 1600$ are smaller than expected on the basis of systematics (136, 137). This hindrance of the fusion process is caused by dissipation in the system before fusion can occur and has led to the introduction of the so-called extra push energy, which is needed in order to achieve sufficient overlap between the interacting nuclei (138).

The reduced fusion yield at low energies has to reappear in other reaction channels. At energies at or below the Coulomb barrier, where the contributions from deep-inelastic collisions are small, one may therefore expect an increase in the quasi-elastic reaction strength for very heavy systems ($Zz > 1600$).

Because of the strong Q value and energy dependence discussed in Section 4.2.2, the data must either be selected to have the same ground-state Q values or bombarding energy and Q-value corrections must be made.

In several experiments using radiochemical methods as well as measurements with a magnetic spectrograph, the quasi-elastic cross section for a series of nuclei with a Zz product between 1152 and 2000 have been studied (139, 140). Figure 27 presents the quasi-elastic cross sections for these systems, together with the results from other studies (mainly at low energies) plotted as function of Zz (140). The Q values for the neutron transfer reactions for these systems are typically -3.5 MeV. A jump in the cross sections is observed around $Zz = 1600$, with large yields observed for systems above 1600 where there is considerable dynamical hindrance for fusion. For lighter systems, the low-energy quasi-elastic yield is smaller since no hindrance of the fusion channel exists for these systems.

5.4   Transfer Probabilities for Quasi-elastic Reactions

In many analyses of quasi-elastic reactions, particularly at low bombarding energies, the data have been transformed into so-called transfer probabilities. The probability for a transfer reaction with a cross section $d\sigma_{tr}$ originating from a certain impact parameter b can be written as

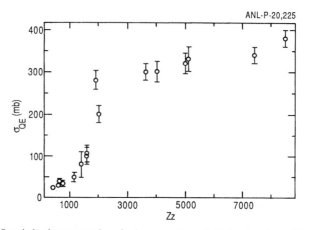

ANL-P-20,225

Figure 27 Quasi-elastic cross sections for heavy systems plotted as function of Zz. A strong increase is observed at $Zz = 1600$, where the effects of dynamical fusion hindrance set in.

$$P_t = d\sigma_{tr}/2\pi b \, db. \qquad\qquad 11.$$

Since the impact parameter b depends on the scattering angle θ via the relation

$$b = \hbar\eta/(mv)\cot(\theta/2), \qquad\qquad 12.$$

Equation 11 can easily be transformed into

$$P_t = d\sigma_{tr}/d\Omega/d\sigma_{Rutherford}/d\Omega. \qquad\qquad 13.$$

When the overlap of the two nuclei is small, the wave function of the transferred particles can be approximated by a Hankel function (for neutron transfer reactions) to obtain a semiclassical expression for P_t (95):

$$P_t \approx \sin(\theta/2)\exp(-\alpha D). \qquad\qquad 14.$$

In Equation 14, D is the distance of closest approach calculated by assuming pure Rutherford trajectories, and α is calculated from the binding energy of the transferred particle (see Equation 9).

Transfer probabilities are particularly useful, since they require no knowledge of the absolute cross section, but only the relative yields between transfer process and elastic scattering. There are two experimental methods to vary the internuclear distance D in Equation 14. One approach measures excitation functions at subbarrier energies at $\theta = 180°$, to change D according to the formula $D = Zze^2/E_{cm}$. The other makes use of the angular dependence of D:

$$D = \frac{Zze^2}{2E_{cm}} \left(1 + \frac{1}{\sin \dfrac{\theta}{2}} \right)$$ 15.

in a measurement of $d\sigma/d\Omega$ at a fixed energy above the barrier. For measurements at energies above the barrier, care must be taken that the assumption of pure Rutherford trajectories is justified at the measured scattering angles (141).

Figure 28 shows the transfer probability $P_t/\sin(\theta/2)$ as function of the reduced distance of closest approach measured (58) for the one-neutron transfer reaction $^{238}\text{U}(^{238}\text{U}, {}^{237}\text{U})^{239}\text{U}$ using radiochemical techniques. A least-squares fit to the data gives a slope of $\alpha = 1.07$ fm^{-1}, which is very close to the value $\alpha = 1.03$ fm^{-1} calculated from Equation 9. Similar good agreement between the theoretical and experimental slopes has been obtained for a large number of one-neutron transfer reactions in different systems (142).

Deviations of the slopes from their theoretical values indicate that some of the assumptions made in the semiclassical analysis might not be valid. Several recent experiments have examined possible deviations from the tunneling picture.

In the system $^{48}\text{Ti} + {}^{104}\text{Ru}$, which leads to the superdeformed nucleus ^{152}Dy, possible correlations between quasi-elastic scattering and the super-

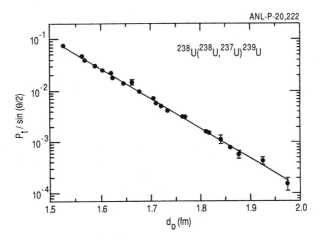

Figure 28   Transfer probability $P_t/\sin(\theta/2)$ as function of the reduced distance of closest approach d_0 measured for the reaction $^{238}\text{U}(^{238}\text{U}, {}^{237}\text{U})^{239}\text{U}$. The solid line is a least-squares fit to the data with a slope of $\alpha = 1.07$ fm^{-1}.

deformed minimum in ^{152}Dy have been investigated (100). The falloff of the angular distribution at forward angles could be followed over three orders of magnitude (limited by charge-exchange collisions with the residual gas atoms) and no deviations from an exponential falloff were observed. In a more recent experiment using radiochemical techniques, possible time delays in collisions between very heavy ions have been studied (143). Time delays can occur when the two interacting nuclei stick together for a finite amount of time. These delays have been discussed in connection with the sharp positron-electron coincidence peaks observed in collisions between high-Z ions (144, 145). Time delays in heavy-ion collisions lead to deviations from pure Rutherford trajectories. In the system ^{238}U + ^{197}Au, the transfer probability was measured over eight orders of magnitude and no deviation from the exponential falloff with the correct slope was observed.

Anomalous slopes have been seen in several cases involving mainly two-neutron transfer reactions (146, 147). In some cases, oscillations were observed in the transfer probability at large internuclear distances (148). This was linked to interference effects in the transfer process (149). As shown by Wuosmaa et al, however, two-neutron transfer reactions are usually associated with very localized form factors and the assumption of semiclassical trajectories may not be valid in all cases, especially at energies above the Coulomb barrier (88).

6. SUMMARY

The last ten years have witnessed a significant increase in the number of experimental and theoretical studies of quasi-elastic reactions. A variety of questions have been answered but many problems still remain.

The main difference between heavy-ion-induced reactions and light-ion collisions is the increased importance of the coupling between the various reaction modes. As a result of this coupling, reaction modes associated with central collisions, such as fusion, can be influenced by peripheral quasi-elastic reactions. These reactions (inelastic scattering and few-nucleon transfer reactions) depend through their $B(E\lambda)$ values and spectroscopic factors strongly on the structure of the interacting nuclei. Fusion reactions were once thought to be mainly a macroscopic process, but channel coupling makes them strongly dependent on nuclear structure. A complete theoretical understanding of these microscopic nuclear structure effects has not yet been achieved.

A complete set of measurements (i.e. experiments measuring all reaction modes at various energies) exists for only a few systems. For inelastic heavy-ion scattering, experiment and theory are usually in good agreement.

One-neutron transfer reactions are also well described by one-step DWBA calculations, although the quality of the data is in most cases not good enough to rule out completely the influence of the multistep processes that have been found to occur in reactions with lighter projectiles. Experiments with improved resolution are urgently needed in this particular field.

Many areas have barely been touched in the quasi-elastic studies performed because in most cases only the strongest channels have been measured. With improved detection techniques, more exotic transfer reactions leading to new nuclei or states that cannot be reached with light-ion reactions will be investigated. The new recoil mass separators, with their large efficiencies and excellent mass separation, will be essential for these experiments.

The transition from simple one-step to complex multiparticle transfer reactions has not been investigated in detail, neither experimentally nor theoretically. Multistep transfer reactions populate complex multiparticle-multihole states, which can be identified with the new Compton-suppressed Ge detector arrays. First experiments have already been performed with the existing detector systems and encouraging results have been obtained. The next generation of devices such as GAMMASPHERE and EUROGAM, with their higher efficiencies, will significantly improve these experiments.

Quasi-elastic reactions are important doorway states to fusion, and the correlation between both reaction modes is well established, even though a full quantitative understanding has not yet been achieved. Similar correlations between quasi-elastic reactions and deep-inelastic scattering have not yet been investigated, mainly because of difficulties with the detection techniques for heavy systems where deep-inelastic processes become more important.

The question of transfer from excited states has been studied in a few cases. Future experiments will provide information on the interplay between inelastic excitation and transfer from high-spin states in deformed nuclei. Interesting interference effects are expected to occur, but the experiments performed so far, do not have sufficient statistics to show these effects clearly. The next generation of Compton-suppressed arrays will change this situation.

Transfer at large internuclear distances is an active area of research. Radiochemical methods have been used to measure transfer probabilities down to 10^{-9}, and neutron transfer has been observed at distances 10 fm larger than the strong interaction radius. If the analogy between a superconducting solid-state junction and two nuclei separated at large distances is valid, these reactions can give important information about particle correlations in finite systems.

The next generation of quasi-elastic reaction studies is thus expected to give valuable new information, not only in the fields of nuclear structure and nuclear dynamics, but also more generally in the area of finite manybody quantal systems.

ACKNOWLEDGMENT

This work was supported by the US Department of Energy, Nuclear Physics Division, under contract W-31-109-ENG-38.

Literature Cited

1. Volkov, V. V., *Sov. J. Part. Nucl.* 6: 420–45 (1976)
2. Schröder, W. U., Huizenga, J. R., *Annu. Rev. Nucl. Part. Sci.* 27: 465–547 (1977)
3. Lefort, M., Ngô, Ch., *Ann. Phys.* 3: 5–114 (1978)
4. Volkov, V. V., *Phys. Lett.* C44: 93–157 (1978)
5. Beckermann, M., *Phys. Lett.* C129: 145–223 (1985)
6. Steadman, S. G., Rhoades-Brown, M. J., *Annu. Rev. Nucl. Part. Sci.* 36: 649–81 (1986)
7. Beckermann, M., *Rep. Prog. Phys.* 51: 1047–1103 (1988)
8. von Oertzen, W., In *Nuclear Spectroscopy and Reactions*, Part B, ed. J. Cerny. New York, London: Academic (1974), pp. 279–342
9. Landowne, S., Vitturi, A., In *Treatise on Heavy Ion Science*, Vol. I, ed. D. A. Bromley. New York: Plenum (1985), pp. 355–460
10. Ascuitto, R. J., Seglie, E. A., See Ref. 9, pp. 463–614
11. Hodgson, P. E., *Nuclear Heavy-Ion Reactions.* Oxford: Clarendon (1978)
12. *Proc. Beijing Int. Symp. on Physics at Tandem*, ed. J. Chenglie, Singapore: World Scientific (1986)
13. *Proc. Symp. Heavy Ion Interactions around the Coulomb Barrier*, Legnaro, Italy, ed. C. Signorini et al. (1988); *Lecture Notes in Physics*, Vol. 317. Berlin, Heidelberg: Springer-Verlag (1988)
14. *Proc. Int. Symp. Heavy Ion Reaction Dynamics in Tandem Energy Region*, Hitachi, Japan, ed. Y. Sugiyama et al. (1988); Universal Academy Press (1989)
15. *Proc. Workshop Heavy Ion Collisions at Energies Near the Coulomb Barrier*, Daresbury, UK, ed. M. A. Nagarajan. Bristol, Philadelphia, New York: Inst Phys. Press (1991), Vol. 110
16. *Proc. Workshop Interface between Nuclear Structure and Heavy Ion Reaction Dynamics*, Univ. Notre Dame, USA, ed. R. R. Betts, J. J. Kolata. Bristol, Philadelphia, New York: Inst. Phys. Press (1991), Vol. 109
17. Becchetti, F. D., et al., *Phys. Rev.* C6: 2215–18 (1972)
18. Kolata, J. J., et al., *Phys. Rev.* C30: 125–33 (1984)
19. Beckermann, M., et al., *Phys. Rev.* C36: 657–66 (1987)
20. Cheng-Lie, J., et al., *Phys. Rev. Lett.* 47: 1039–41 (1981)
21. Videbaek, F., et al., *Phys. Rev.* C15: 954–71 (1977)
22. Spieler, H., et al., *Z. Phys.* A278: 241–50 (1976)
23. van den Berg, A. M., et al., *Phys. Rev.* C37: 178–86 (1988)
24. Rehm, K. E., et al., *Phys. Rev.* 42: 2497–2507 (1990)
25. Wolfs, F. L. H., et al., *Phys. Lett.* 196: 113–16 (1987)
26. Lesko, K. T., et al., *Phys. Rev. Lett.* 55: 803–6 (1985); *Phys. Rev.* C34: 2155–64 (1986)
27. Pass, C. N., et al., *Nucl. Phys.* A499: 173–99 (1989)
28. Pontoppidan, S., et al., *Phys. Rev.* C28: 2299–2311 (1983)
29. Spieler, H., *IEEE Trans. Nucl. Sci.* 29: 1142–58 (1982)
30. Bilwes, B., See Ref. 15, pp. 51–61
31. Stefanini, A. M., et al., *Phys. Lett.* B240: 306–10 (1990)
32. Schüll, D., In *Proc. Hans Geiger Symp. Detectors in Heavy Ion Reactions*, Berlin, 1982, pp. 80–91. *Lecture Notes in Physics*. Berlin, Heidelberg: Springer-Verlag (1983), Vol. 178

33. Hynes, M. V., et al., *Nucl. Instrum. Methods* 224: 89–96 (1984)
34. Cunningham, R. A., et al., *Nucl. Instrum. Methods* A234: 67–80 (1985)
35. Takekoshi, E., et al., *Nucl. Instrum. Methods* A237: 512–22 (1985)
36. Rehm, K. E., Wolfs, F. L. H., *Nucl. Instrum. Methods* A273: 262–72 (1988)
37. Walcher, Th., *Experimental Methods in Heavy Ion Physics.* Berlin, Heidelberg: Springer-Verlag (1978)
38. Sugiyama, Y., et al., *Nucl. Instrum. Methods* 187: 25–35 (1981)
39. Spencer, J. E., Enge, H. A., *Nucl. Instrum. Methods* 49: 181–93 (1967)
40. Drentje, A. G., Enge, H. A., Kowalski, S. B., *Nucl. Instrum. Methods* 122: 485–90 (1974)
41. Speer, J., et al., *Phys. Lett.* B259: 422 (1991)
42. Holzmann, R., et al., *GSI Ann. Rep.* (1988), GSI-89-1, and to be published
43. Sohlbach, H., et al., *Z. Phys.* A328: 205–17 (1987)
44. Sugiyama, Y., et al., To be published (1992)
45. Cormier, T., *Annu. Rev. Nucl. Part. Sci.* 37: 537–65 (1987)
46. Betts, R. R., et al., *Phys. Rev. Lett.* 59: 978–81 (1987)
47. Herman, M. G., et al., See Ref. 13, pp. 137–42
48. Lilley, J. S., See Ref. 15, pp. 63–76
49. Bohn, H., et al., *Phys. Rev. Lett.* 29: 1337–40 (1972)
50. Jääskeläinen, M., et al., *Nucl. Instrum. Methods* 204: 385–405 (1983)
51. Metag, V., Habs, D., Schwalm, D., *Comments Nucl. Part. Phys.* 16: 213–29 (1986) and references cited therein
52. de Boer, F. W. N., et al., *Z. Phys.* A325: 457–66 (1986)
53. Himmele, G., et al., *Nucl. Phys.* A404: 401–12 (1983)
54. Wu, C. Y., et al., *Annu. Rev. Nucl. Part. Sci.* 40: 285–326 (1990)
55. Gerl, J., et al., *Z. Phys.* A334: 195–206 (1989)
56. Kratz, J. V., et al., *Nucl. Phys.* A357: 437–70 (1981)
57. Kratz, J. V., et al., *Phys. Rev.* C13: 2347–65 (1976)
58. Wirth, G., et al., *Phys. Lett.* B177: 282–86 (1986)
59. Satchler, G. R., *Direct Nuclear Reactions.* New York: Oxford (1983)
60. Pieper, S. C., et al., *Phys. Rev.* C18: 180–204 (1978)
61. Thompson, I. J., See Ref. 15, pp. 165–76 and references therein
62. Broglia, R. A., et al., *Phys. Lett.* C29: 291–362 (1977); Broglia, R. A., et al., *Phys. Lett.* B73: 401–4 (1978)
63. Broglia, R. A., et al., *Phys. Lett.* B45: 23–26 (1973)
64. Tamura, T., Low, K. S., *Comp. Phys. Commun.* 8: 349–76 (1974)
65. Macfarlane, M. H., Pieper, S. C., *PTOLEMY: A Program for Heavy-Ion Direct-Reaction Calculations,* Rep. ANL-76-11, Rev. 1. Argonne National Lab. (1976)
66. Raynal, J., *Phys. Rev.* C23: 2571–85 (1981)
67. Thompson, I. J., *Comp. Phys. Rep.* 7: 167–212 (1988)
68. Strutinski, V. M., *Sov. Phys. JETP* 19: 1401–5 (1964)
69. Frahn, W., See Ref. 9, pp. 135–290 and references therein
70. Broglia, R. A., Winther, A., *Heavy Ion Reactions.* Redwood City, Calif: Addison-Wesley (1991)
71. Stefanini, A. M., et al., *Phys. Rev. Lett.* 59: 2852–55 (1987)
72. Tighe, R. J., et al., *Phys. Rev.* C42: 1530–39 (1990)
73. Wilpert, Th., PhD Thesis, Hahn-Meitner-Inst., Berlin (1989) (unpublished)
74. Roussel-Chomaz, P., et al., *Phys. Lett.* B209: 187–91 (1988)
75. Vojtech, R. J., et al., *Phys. Rev.* C35: 2139–45 (1987)
76. Dixit, S., et al., *Phys. Rev. C.* To be published (1992)
77. Guidry, M. W., et al., *Phys. Rev. Lett.* 40: 1016 (1978)
78. Neese, R. E., et al., *Phys. Lett.* B85: 201–5 (1979)
79. Macchiavelli, A. O., et al., *Nucl. Phys.* A432: 436–50 (1985)
80. Rhoades-Brown, M. J., et al., *Phys. Rev.* C24: 2747–50 (1984)
81. Thompson, I. J., et al., *Phys. Lett.* B157: 250–54 (1985)
82. Pieper, S. C., Rhoades-Brown, M. J., Landowne, S., *Phys. Lett.* B162: 43–46 (1985)
83. Sugiyama, Y., et al., *Phys. Rev. Lett.* 62: 1727–30 (1989)
84. Landowne, S., See Ref. 15, pp. 121–32
85. Ebensen, H., Landowne, S., *Nucl. Phys.* A492: 473–92 (1989)
86. Vineyard, M. F., et al., *Phys. Rev.* 33: 1325–32 (1986)
87. Hoath, S. D., et al., *Phys. Lett.* B154: 33–36 (1985)
88. Wuosmaa, A. H., et al., *Phys. Lett.* B255: 316–20 (1991)
89. Brendel, C., et al., *Nucl. Phys.* A477: 162–88 (1988)
90. Tamura, T., Low, K. S., *Phys. Rev. Lett.* 31: 1356–58 (1973)
91. Rehm, K. E., et al., *Phys. Rev. Lett.* 51: 1426–29 (1983)

92. Beckermann, M., et al., *Phys. Rev. Lett.* 58: 455–58 (1987)
93. Rehm, K. E., et al., *Phys. Rev.* C37: 2629–46 (1988)
94. van den Berg, A. M., et al., *Phys. Lett.* B194: 334–37 (1987)
95. Bass, R., *Nuclear Reactions with Heavy Ions.* Berlin, Heidelberg: Springer-Verlag (1980)
96. Brink, D. M., *Phys. Lett.* B40: 37–40 (1972)
97. Buttle, P. J. A., Goldfarb, L. J. B., *Nucl. Phys.* A176: 299–320 (1971)
98. Broglia, R. A., Pollarolo, G., Winther, A., *Nucl. Phys.* A361: 307–25 (1981)
99. von Oertzen, W., et al., *Z. Phys.* A326: 463–81 (1987)
100. Sanders, S. J., et al., *Phys. Rev.* C37: 1318–21 (1988)
101. Rehm, K. E., See Ref. 14, pp. 95–104
102. Broglia, R. A., et al., *Phys. Rev.* C27: 2433–75 (1983)
103. Pengo, R., et al., *Nucl. Phys.* A411: 255–74 (1983)
104. Broglia, R. A., et al., *Phys. Lett.* B133: 34–38 (1983)
105. Goldanski, V. I., Larkin, A. I., *Sov. Phys. JETP* 26: 617–19 (1968)
106. Dietrich, K., *Phys. Lett.* B32: 428–30 (1970); *Ann. Phys.* 66: 480–508 (1971)
107. Hara, K., *Phys. Lett.* B35: 198–200 (1971)
108. Kleber, M., Schmidt, H., *Z. Phys.* A245: 68–80 (1971)
109. Broglia, R. A., et al., *Phys. Lett.* B73: 401–4 (1978)
110. Weiss, H., *Phys. Rev.* C19: 834–47 (1979)
111. von Oertzen, W., et al., *Z. Phys.* A313: 189–95 (1983)
112. Landowne, S., Pollarolo, G., Dasso, C. H., *Nucl. Phys.* A486: 325–34 (1988)
113. Rehm, K. E., et al., *Phys. Rev. Lett.* 55: 280–83 (1985)
114. Henning, W., et al., *Phys. Rev. Lett.* 58: 318–21 (1987)
115. Wolfs, F. L. H., Rehm, K. E., et al., *Phys. Rev. C.* To be published (1992)
116. Künkel, R., et al., *Phys. Lett.* B208: 355–60 (1988)
117. Künkel, R., et al., *Z. Phys.* A336: 71–89 (1990)
118. Thorn, C. E., et al., *Phys. Rev. Lett.* 38: 384–86 (1977)
119. Nagarajan, M. A., Mahaux, C., Satchler, G. R., *Phys. Rev. Lett.* 54: 1136–38 (1985)
120. Nagarajan, M. A., Satchler, G. R., *Phys. Lett.* B173: 29–33 (1986)
121. Baeza, A., et al., *Nucl. Phys.* A419: 412–28 (1984)
122. Lilley, J. S., et al., *Phys. Lett.* B151: 181–84 (1985)
123. Fulton, B. R., et al., *Phys. Lett.* B162: 55–58 (1985)
124. Fulton, B. R., See Ref. 15, pp. 15–28 and references therein
125. Aguiar, C. E., et al., *Phys. Lett.* B201: 22–24 (1988); *Phys. Rev.* C38: 541–42 (1988)
126. Stelson, P. H., *Phys. Lett.* B205: 190–4 (1988); Stelson, P. H., et al., *Phys. Rev.* C41: 1584–99 (1990)
127. Esbensen, H., *Nucl. Phys.* A352: 147–56 (1981)
128. Dasso, C. H., Landowne, S., Winther, A., *Nucl. Phys.* A405: 381–96 (1983)
129. Jacobs, P., Smilansky, U., *Phys. Lett.* B127: 313–16 (1983)
130. Landowne, S., Pieper, S. C., *Phys. Rev.* C29: 1352–57 (1984)
131. Rhoades-Brown, M. J., Braun-Munzinger, D., *Phys. Lett.* B136: 19–23 (1984)
132. Stefanini, A. M., et al., *Nucl. Phys.* A456: 509–34 (1986)
133. Beckermann, M., et al., *Phys. Rev.* C25: 837–49 (1982)
134. Stefanini, A. M., et al., *Phys. Lett.* 185: 15–19 (1987)
135. Scarlassara, F., et al., *Z. Phys.* A338: 171–82 (1991)
136. Toke, J., et al., *Nucl. Phys.* A440: 327–65 (1985)
137. Lützenkirchen, K., et al., *Nucl. Phys.* A452: 351–80 (1986)
138. Swiatecki, W. J., *Phys. Scr.* 24: 113–22 (1981); *Nucl. Phys.* A376: 275–91 (1982)
139. Bellwied, R., et al., See Ref. 13, pp. 125–30
140. Bellwied, R., PhD thesis, Univ. Mainz, Germany (unpublished) (1988)
141. Vigezzi, E., Winther, A., *Ann. Phys.* 192: 431–86 (1989)
142. Rehm, K. E., In *Proc. XII Workshop on Nuclear Physics*, Iguazu Falls, Argentina, 1989. Singapore: World Scientific (1990), pp. 212–31
143. Wirth, G., et al., *Phys. Lett.* B253: 28–32 (1991)
144. Cowan, T., et al., *Phys. Rev. Lett.* 56: 444–47 (1986)
145. Koenig, W., et al., *Z. Phys.* A328: 129–45 (1987)
146. Juutinen, S., et al., *Phys. Lett.* B192: 307–11 (1987)
147. Kim, H. J., et al., *Phys. Rev.* C38: 2081–85 (1988)
148. Wu, C. Y., et al., *Phys. Rev.* C39: 298–301 (1989)
149. Landowne, S., Price, C., Esbensen, H., *Nucl. Phys.* A484: 98–116 (1988)

Annu. Rev. Nucl. Part. Sci. 1991. 41: 469–509

QUANTITATIVE TESTS OF THE STANDARD MODEL OF ELECTROWEAK INTERACTIONS[1]

William J. Marciano

Brookhaven National Laboratory, Upton, New York 11973

KEY WORDS: precision measurements, radiative corrections

CONTENTS

[1] This manuscript has been authored under contract number DE-AC02-76CH0016 with the US Department of Energy. Accordingly, the US Government retains a nonexclusive, royalty-free license to publish or reproduce the published form of this contribution, or allow others to do so, for US Government purposes.

469

1. INTRODUCTION

Over the past several decades, tremendous theoretical and experimental advances have been made in elementary particle physics. A renormalizable quantum field theory of strong and electroweak interactions based on an underlying local $SU(3)_c \times SU(2)_L \times U(1)_Y$ gauge invariance has emerged (1, 2). It beautifully describes all known particles and their observed interaction properties (modulo gravity) while incorporating the proven symmetries and successes of the quark model, quantum electrodynamics, and the old four-Fermi vector–axial vector (V-A) theory. It even correctly predicted weak neutral currents (3), as well as the existence and properties of gluons, $W^{\pm}$ and Z bosons. Those impressive successes have earned for the $SU(3)_c \times SU(2)_L \times U(1)_Y$ theory its title as the "Standard Model," a designation that describes its acceptance as a proven standard against which future experimental results and alternative theories are to be compared.

As an outline of the standard model, I have illustrated in Table 1 its minimal spectrum of particles and some of their basic properties. The fermions are grouped into three generations (or families) of leptons and

Table 1 Elementary particles

Particle	Symbol	Spin	Charge	Color	Mass (GeV)	
Electron neutrino	ν_e	1/2	0	0	$< 0.94 \times 10^{-8}$	
Electron	e	1/2	-1	0	0.51×10^{-3}	First
Up quark	u	1/2	2/3	3	5×10^{-3}	generation
Down quark	d	1/2	$-1/3$	3	9×10^{-3}	
Muon neutrino	ν_μ	1/2	0	0	$< 0.25 \times 10^{-3}$	
Muon	μ	1/2	-1	0	0.106	Second
Charm quark	c	1/2	2/3	3	1.25	generation
Strange quark	s	1/2	$-1/3$	3	0.175	
Tau neutrino	ν_τ	1/2	0	0	< 0.035	
Tau	τ	1/2	-1	0	1.78	Third
Top quark	t	1/2	2/3	3	> 89	generation
Bottom quark	b	1/2	$-1/3$	3	4.3	
Photon	γ	1	0	0	0	
W boson	$W^{\pm}$	1	± 1	0	80.14 ± 0.31	Gauge
Z boson	Z	1	0	0	91.17 ± 0.02	bosons
Gluon	g	1	0	8	0	
Higgs scalar	H	0	0	0	$48 \lesssim m_H \lesssim 1000$	

quarks spanning an enormous mass range. Experiments are consistent with massless neutrinos as required by the minimal standard model (i.e. with no right-handed neutrino fields and only a Higgs scalar doublet). However, observation of a nonzero neutrino mass could easily be accommodated. There are currently some hints of very small neutrino masses and mixings from solar neutrino experiments and dark matter speculations. The bounds I quote come from direct kinematical measurements (4).

The top quark remains undiscovered and the CDF bound (5) on m_t in Table 1 has become quite high. We are confident that the top quark exists, since experimental studies indicate that the bottom quark must have an $SU(2)_L$ isodoublet partner. Renormalizability also requires the top quark (or some complicated alternative). Indeed, global studies of electroweak radiative corrections and precision measurements suggest $m_t \simeq 130 \pm 40$ GeV, which is in good accord with the CDF bound. Why is the top quark so heavy? That question serves to highlight the more general problem of why Nature chose to repeat the generation structure three times and give fermions their observed pattern of masses and mixings. It may be easier to understand the large top quark mass than the relatively small masses of other fermions.

Quarks and leptons interact by exchanging gauge bosons. Eight massless gluons couple to the color $SU(3)_c$ charge of quarks and mediate strong interactions, while the $W^\pm$, Z, and γ of the $SU(2)_L \times U(1)_Y$ sector are responsible for weak and electromagnetic interactions. If $SU(2)_L \times U(1)_Y$ were an exact symmetry, all fermion and gauge boson masses would be zero. However, a Higgs complex isodoublet breaks the symmetry down to $U(1)_{em}$ and provides particles with their masses. Fermion masses are essentially free parameters determined by experiments. They are a measure of the particles' couplings to the Higgs doublet. The $W^\pm$ and Z masses are related to the gauge couplings and thereby interconnected. The precise measurements of m_Z and m_W at CERN and Fermilab constitute one of the most important confirmations of the standard model, even at the quantum loop level.

Three of the components of the Higgs isodoublet field have been discovered as longitudinal components of $W^\pm$ and Z bosons. (They enter as massless Goldstone bosons.) The fourth component, called the Higgs scalar, remains elusive. Its mass is unspecified, except for the recent LEP bounds $m_H \gtrsim 48$ GeV (6) and the theoretical consistency arguments that imply $m_H \lesssim 1$ TeV (7, 8). Much of the motivation for building the high-energy Superconducting Super Collider (SSC) pp collider with $\sqrt{s} = 40$ TeV comes from the Higgs scalar. Either the Higgs scalar will be discovered as a fundamental remnant of the symmetry breakdown or "new physics" will take its place.

Given our lack of understanding of masses and mixings, it seems quite possible that some richer underlying new dynamics is actually waiting in place of the Higgs scalar; but so far no completely compelling alternative exists. One potentially attractive scenario that replaces the fundamental Higgs doublet by $t\bar{t}$ condensation is described below (9).

At the present time, there are no solid experimental results that cannot be accounted for by the standard model. That accomplishment is very impressive when one considers the wealth of precise experimental data the model must confront. Indeed, as discussed below, new weak interaction experimental tests have reached the percent or better mark and now probe the standard model at the level of its quantum loop corrections. Nevertheless, there is a rather general expectation that new physics must emerge as higher energies are probed. That conviction stems from the intellectual drive to further unify the fundamental forces and from a dissatisfaction with the $SU(2)_L \times U(1)_Y$ electroweak sector. Unlike the $SU(3)_c$ quantum chromodynamics (QCD) part of the model, which has but one parameter (the mass scale $\Lambda_{\overline{MS}}$), the electroweak sector has a minimum of 17 independent masses and couplings. One would hope that the truly fundamental theory of everything will have no arbitrary free parameters. That shortcoming points to currently missing physics—perhaps in the form of heavy fermions, additional gauge bosons, supersymmetry, new dynamics, etc—that will complement the standard model and guide us to the ultimate theory of everything.

My aim in this review is to describe some of the precision tests of the standard model's electroweak sector. In that regard, it is an update of an earlier review in this series (10). Those precision tests serve not only to confirm the standard model but also to constrain allowable new physics that may be appended to it.

The review is organized as follows. In Section 2, I survey precise charged current tests of unitarity and describe how they relate to new physics. A description is given of some important parameters such as the Fermi constant, quark mixing angles, and pseudoscalar decay constants, including effects of radiative corrections. Predictions for tau decay rates are also presented. Section 3 provides a discussion of some weak neutral current phenomenology. A renormalized weak mixing angle $\sin^2 \theta_W(m_Z)_{\overline{MS}}$ is defined and used to parametrize weak amplitudes, including standard model loop corrections. A few specific processes in which precision measurements have been made, such as neutrino scattering and atomic parity violation, are discussed. Brief remarks about the potential of polarized electron scattering asymmetries are also given. The spectacular advances in measurements of $W^{\pm}$ and Z gauge boson properties are described in Section 4. Mass and decay measurements are shown to provide important tests of the standard model at the quantum loop level. They also constrain the possibilities of new physics, such as a fourth generation or heavy Z' boson mixing. Constraints on the top quark mass from direct searches and a favored value from electroweak radiative corrections are reviewed in Section 5. There I also describe the relatively new $t\bar{t}$ condensation scenario as an alternative to the fundamental Higgs doublet. Section 6 discusses constraints on the Higgs scalar mass as well as expected signatures of its presence at future high-energy colliders. In Section 7, I outline the S and T parametrization of gauge boson radiative corrections popularized by Peskin & Takeuchi (11). Constraints on those harbingers of new physics are reviewed and future sensitivities discussed. That formalism is particularly powerful for high mass scale loop effects such as technicolor models. In fact, it is shown that experimental data already provides a severe constraint for technicolor scenarios. A brief discussion of the state of grand unified theories is also given. Finally, I conclude in Section 8 with some comments on how far we have come in testing the standard model and with an outlook on the future.

2. WEAK CHARGED CURRENTS

Weak charged current interactions are described by the interaction Lagrangian

$$\mathscr{L}_{\text{int}}^{\text{CC}} = -\frac{g_{2_0}}{\sqrt{2}} W^{\mu}(x)J_{\mu}^{\text{CC}}(x) + \text{hermitian conjugate}, \qquad 1.$$

where g_{2_0} is the bare $SU(2)_L$ gauge coupling, W^{μ} is the W boson field, and

$$J_{\mu}^{\text{CC}} = \sum_{\ell=e,\mu,\tau} \bar{v}_{\ell_L}\gamma_{\mu}\ell_L + \sum_{\substack{q=u,c,t \\ q'=d,s,b}} \bar{q}_L V_{qq'}\gamma_{\mu}q'_L. \qquad 2.$$

Only the left-handed components $\psi_L = \frac{1}{2}(1-\gamma_5)\psi$ participate and V is a 3×3 unitary matrix, the Cabibbo-Kobayashi-Maskawa (CKM) matrix (12):

$$V = \begin{pmatrix} V_{ud} & V_{us} & V_{ub} \\ V_{cd} & V_{cs} & V_{cb} \\ V_{td} & V_{ts} & V_{tb} \end{pmatrix}. \qquad 3.$$

The $V_{qq'}$ are finite measurable quantities (after renormalization) that result from diagonalization of quark mass matrices. As such, they must contain valuable information about the underlying mass-generating mechanism. In addition, they provide the mixing and phase necessary to explain CP violation in the kaon system and predict anticipated CP violation in B decays. Therefore, we would like to learn as much about those important parameters as possible. To that end, I present current values of the $V_{qq'}$, focusing on the cornerstone, V_{ud}, and the role of electroweak radiative corrections. Using measured values in conjunction with unitarity tests, one can bound or search for hints of new physics beyond the standard model. Some examples are subsequently discussed. The CKM elements can also be used, together with meson decay rates, to obtain pseudoscalar decay constants f_π, f_K, etc. A prescription for extracting those constants is given and the resulting f_π checked for consistency in the Goldberger-Treiman relation and $\Gamma(\pi^0 \to \gamma\gamma)$ rate predicted by PCAC. Finally, the status of tau decay theory and experiment are discussed.

2.1 *Muon Decay—G_μ Definition*

The Fermi constant, G_μ, obtained from muon decay is one of the most precisely determined weak interaction parameters. As such, it is used to normalize low-energy, weak charged and neutral current amplitudes. Also, taken together with the W boson mass, m_W, it provides a physical definition of the renormalized SU(2)$_L$ coupling g_{2_R}. Given its important role, let me describe how G_μ is extracted from muon decay.

The very precisely measured muon lifetime (4)

$$\tau_\mu = 2.197035 \pm 0.000040 \times 10^{-6}\,\text{s} \qquad\qquad 4.$$

is related to G_μ via the defining equation (13)

$$\tau_\mu^{-1} = \Gamma(\mu \to \text{all}) = \frac{G_\mu^2 m_\mu^5}{192\pi^3} f\!\left(\frac{m_e^2}{m_\mu^2}\right)\!\left(1 + \frac{3m_\mu^2}{5m_W^2}\right)\!\left[1 + \frac{\alpha(m_\mu)}{2\pi}\left(\frac{25}{4} - \pi^2\right)\right]$$

$$5a.$$

$$f(x) = 1 - 8x + 8x^3 - x^4 - 12x^2 \ln x \qquad\qquad 5b.$$

$$\alpha^{-1}(m_\mu) = \alpha^{-1} - \frac{2}{3\pi}\ln(m_\mu/m_e) + \frac{1}{6\pi} = 136. \qquad\qquad 5c.$$

In Equation 5a, the traditional radiative corrections in the old V-A theory have been explicitly factored out in defining G_μ. From that expression one finds

$$G_\mu = 1.16637 \pm 0.00002 \times 10^{-5}\,\text{GeV}^{-2}. \qquad\qquad 6.$$

A renormalized SU(2)$_L$ coupling g_{2_R} can also be defined using

$$\frac{G_\mu}{\sqrt{2}} = \frac{g_{2_R}^2}{8m_W^2}. \qquad\qquad 7.$$

The measured W mass, $m_W = 80.14 \pm 0.31$ GeV, then implies

$$g_{2_R} = 0.6510 \pm 0.0025. \qquad\qquad 8.$$

Of course, G_μ and g_{2_R} are merely defined quantities. To utilize their precision, one must know the complete $O(\alpha)$ quantum loop corrections for both muon decay and the process to which they are applied. It is, therefore, sometimes more convenient to work with a nonphysical running coupling such as $g_2(\mu)_{\overline{MS}}$ defined by modified minimal subtraction, $\overline{MS}$, with μ a sliding mass scale (14). That renormalized coupling is defined by a renormalization prescription rather than tied to a specific physical process. As such, weak amplitudes can be renormalized in terms of $g_2(\mu)_{\overline{MS}}$ without direct knowledge of the radiative corrections to muon decay. The prescription is subtraction of $[1/(n-4)+(\gamma/2)-\ln\sqrt{4\pi}]$ terms. Of course, knowing the one-loop radiative corrections to muon decay allows one to relate $g_2(\mu)_{\overline{MS}}$ and g_{2_R}. One finds

$$g_{2_R}^2 = g_2^2(\mu)_{\overline{MS}}\left[1 + \frac{g_2^2(\mu)_{\overline{MS}}}{16\pi^2}\left[-8\ln\left(\frac{m_Z}{\mu}\right)+6+\left(\frac{7}{2}-6s^2\right)\frac{\ln c^2}{s^2}\right]\right.$$

$$\left.+\left.\frac{\mathrm{Re}\,\Pi_{WW}(m_W^2)-\Pi_{WW}(0)}{m_W^2}\right|_{\overline{MS}}\right], \qquad 9.$$

where $s^2 = \sin^2\theta_W$, $c^2 = \cos^2\theta_W$, $\Pi_{WW}(q^2)$ is the W boson self-energy, and $|_{\overline{MS}}$ means all $[1/(n-4)+(\gamma/2)-\ln\sqrt{4\pi}]$ contributions are subtracted (15). There are subtleties in the $\overline{MS}$ subtraction when masses in the W loops are greater than μ. We encounter that problem below when we take $\mu = m_Z$ and allow $m_t > m_Z$. Note that whereas g_{2_R} is renormalization group invariant $\mu(d/d\mu)g_{2_R} = 0$, $g_2(\mu)_{\overline{MS}}$ satisfies $\mu(d/d\mu)g_2(\mu)_{\overline{MS}} = \beta(g_2)$ with β the SU(2)$_L$ beta function.

2.2 CKM Matrix

The CKM quark mixing matrix in Equation 3 is constrained by unitarity to satisfy

$$\sum_i V_{ij}^* V_{ik} = \sum_i V_{ji}^* V_{ki} = \delta_{jk}. \qquad 10.$$

In addition, relative quark phases are unobservable. Therefore, V can be parametrized in terms of four independent quantities. Sometimes three angles and one phase are chosen. Alternatively, four moduli can be employed to parametrize the others. By measuring many different charged current transition rates, one thus tests the unitarity of V and searches for deviations because of new physics. With that goal in mind, I update the values of the CKM matrix elements and discuss some inferences.

$|V_{ud}| = 0.9750 \pm 0.0007$ The most precise determination of $|V_{ud}|$ comes from comparing superallowed nuclear beta decays ($0^+ \rightarrow 0^+$ Fermi transitions) with muon decay. Because those beta decays proceed through a conserved vector current (in the limit $m_d - m_u = 0$), strong interactions do not renormalize the hadronic amplitude except through small loop-induced $O(\alpha)$ corrections, which can be estimated.

After absorbing all divergences into the renormalization of $g_{2_0}^2/m_W^{0^2}$ via G_μ, one finds that the ft values of superallowed decays provide a determination of $|V_{ud}|$ through

$$|V_{ud}|^2 = \frac{\pi^3 \ln 2}{ft G_\mu^2 m_e^5 (1+\delta_{RC})} = \frac{2984.4 \pm 0.1 \text{ s}}{ft(1+\delta_{RC})} \qquad 11.$$

where δ_{RC} represents finite $O(\alpha)$ radiative corrections, which are quite large (~ 3–4%) and extremely important for unitarity checks (10, 16). There are also corrections due to final-state Coulomb interactions, nuclear isospin mixing, atomic screening and mismatch, nuclear size effects, etc, which must be properly accounted for. I do not discuss those complications but instead start with ft values that have already been corrected and reflect their associated uncertainties (16).

The radiative corrections δ_{RC} divide into two pieces traditionally called outer and inner corrections, $\delta_{RC} = \Delta_{outer} + \Delta_{inner}$ (17). The outer corrections depend on the nucleus considered, while the inner corrections are universal. That separation gives rise to a useful nucleus-independent quantity

$$Ft = ft(1+\Delta_{outer}) \qquad 12.$$

such that

$$|V_{ud}|^2 = \frac{2984.4 \pm 0.1 \text{ s}}{Ft(1+\Delta_{inner})}. \qquad 13.$$

The universal Δ_{inner} has been carefully scrutinized, including a summation of leading short-distance logarithms. Updating earlier work (18) for $m_t > 89$ GeV, one obtains

$$\Delta_{inner} = 0.0234 \pm 0.0012, \qquad 14.$$

where the error comes primarily from the uncertainty in matching radiative corrections in the effective nucleon theory with short-distance corrections.

The estimated error was reduced by recent calculations of some nuclear structure loop effects that I have included in ft (19).

In Table 2, an up-to-date survey of Ft values is given for various superallowed transitions. If one accepts the small Ft errors, the agreement is not very good. On that basis, it has been argued that nuclear corrections may have introduced a spurious Z dependence in the Ft values (20). Empirical studies hint at an additional correction that reduces those values by $2 \times 10^{-4}\ Z$. To illustrate that point, I give in Table 2 values of $|V_{ud}|$ obtained from individual decays along with the effect of a Z-dependent correction. It is clear that the Z-dependent correction provides better consistency.

Given the present quandary about Z-dependent effects, I have chosen to use the lowest Z decay (without any extra Z-dependent correction)

$$|V_{ud}| = 0.9750 \pm 0.0007 \qquad (^{14}\text{O beta decay}) \qquad 15.$$

as the current "best" value. One may, however, wish to be conservative and include an additional error of between ± 0.0005 and ± 0.0010 to reflect the present nuclear confusion.

An alternative way to determine $|V_{ud}|$ is by measuring both the neutron lifetime, τ_n, and g_A (from neutron decay correlations). Together, those measurements give $|V_{ud}|$ via

$$|V_{ud}|^2 = \frac{4904.0 \pm 5.0\ \text{s}}{\tau_n (1 + 3g_A^2)}, \qquad 16.$$

where the error corresponds to a rough (conservative) estimate of the

Table 2 Ft and $|V_{ud}|$ for various superallowed beta decays after correcting for nuclear effects, Coulomb, and outer radiative corrections (16)[a]

Nucleus	Z	Ft (s)	$|V_{ud}|$	$|V_{ud}| \times (1 + 10^{-4}Z)$
^{14}O	7	3067.9 ± 2.4	0.9750 ± 0.0007	0.9756 ± 0.0007
^{26m}AL	12	3071.1 ± 2.6	0.9744 ± 0.0007	0.9756 ± 0.0007
^{34}CL	16	3074.2 ± 3.1	0.9740 ± 0.0008	0.9755 ± 0.0008
^{38m}K	18	3071.5 ± 3.2	0.9744 ± 0.0008	0.9762 ± 0.0008
^{42}Sc	20	3077.1 ± 2.9	0.9735 ± 0.0008	0.9754 ± 0.0008
^{46}V	22	3078.7 ± 3.2	0.9732 ± 0.0008	0.9754 ± 0.0008
^{50}Mn	24	3073.2 ± 5.2	0.9741 ± 0.0010	0.9765 ± 0.0010
^{54}Co	26	3075.1 ± 3.7	0.9738 ± 0.0008	0.9764 ± 0.0008
Average		3073.0 ± 1.1	0.9741 ± 0.0006	0.9757 ± 0.0006

[a] Z represents the charge of the daughter nucleus. The last column includes an empirical correction suggested by fits to the data (20).

uncertainty in the radiative corrections (21). Using the average of the two most recent (and most precise) measurements of τ_n (22)

$$\tau_n = 889.1 \pm 2.6 \text{ s} \qquad\qquad 17.$$

along with the single best g_A determination (23)

$$g_A = 1.262 \pm 0.005 \qquad\qquad 18.$$

leads to

$$|V_{ud}| = 0.9770 \pm 0.0005 \pm 0.0014 \pm 0.0032 \qquad \text{(neutron decay)}, \qquad 19.$$

where the errors come from radiative corrections, τ_n, and g_A respectively.

The value of $|V_{ud}|$ in Equation 19 is quite consistent with the nuclear beta decay results, but the error is still relatively large. It is, however, anticipated that τ_n and g_A determinations should be significantly improved with the advent of new dedicated cold neutron experiments.

A final way to determine $|V_{ud}|$ involves the pion decay rate $\Gamma(\pi^+ \to \pi^0 e^+ \nu_e)$, which has a relatively small uncertainty in its radiative corrections. At present

$$|V_{ud}| = 0.968 \pm 0.018 \qquad \text{(pion beta decay)}, \qquad 20.$$

which is consistent with other measurements (10, 24). Unfortunately, the small branching ratio $BR(\pi^+ \to \pi^0 e^+ \nu_e) \simeq 10^{-8}$ makes a precision experiment very difficult. Nevertheless, a new experiment at the Paul Scherrer Institute (PSI) aims for a $\pm 0.25\%$ measurement of $|V_{ud}|$ and there has been some discussion of a $\pm 0.05\%$ measurement at the proposed LAMPF PILAC (pion accelerator) (W. K. McFarlane, private communication). Such measurements would be theoretically very clean and could thus be decisive in pinpointing $|V_{ud}|$.

Values of the other CKM elements are also obtained by normalizing relevant amplitudes in terms of G_μ and separating out residual corrections. Unfortunately, those determinations are often plagued by uncertainties in hadronic matrix elements. Below I summarize current values of the CKM elements and briefly comment on their determinations.

$|V_{us}| = \mathbf{0.220 \pm 0.002}$ This value comes from averaging the K_{e3} determination (25) 0.2196 ± 0.0023 with the hyperon decay value (26) $0.220 \pm 0.001 \pm 0.003$. K_{e3} decays are cleaner theoretically (the theory error in $|V_{us}|$ is ± 0.0018). There is still some room for improvement in K_{e3} experiments. Given the importance of $|V_{us}|$ and the existence of many dedicated kaon decay detectors, a new round of K_{e3} experiments seems warranted. Rendering the experimental uncertainty in $|V_{us}|$ negligible might spur new theoretical studies and a reduction in the overall error.

$|V_{cd}| = \mathbf{0.215 \pm 0.016}$ That value is obtained by averaging CDHS and

Fermilab dimuon production rates in deep inelastic ν_μ and $\bar{\nu}_\mu$ scattering (27). A proposed Fermilab emulsion experiment (28) would be capable of measuring $|V_{cd}|$ to about $\pm 3\%$.

$|V_{cs}| = 0.98 \pm 0.12$ That value follows from averaging a $|V_{cs}|$ obtained from dimuon production (27) in $\overset{(-)}{\nu}_\mu$ scattering with one obtained from D_{e3} decays.

$|V_{cb}| = 0.046 \pm 0.005$ That value is obtained by averaging $|V_{cb}|$ determinations from $\tau_b = 1.18 \times 10^{-12}$ s and $\Gamma(b \to c\ell\nu)$ (29). There is still considerable controversy regarding the extraction of $|V_{cb}|$ from those measurements. In particular, the semileptonic branching ratios of B mesons are smaller than expected. That may be caused by an overestimate of the $B\bar{B}$ production rate (assumed to be 100%) in $\Upsilon(4S)$ decays or enhancements in the nonleptonic B^0 decays.

$|V_{ub}| = 0.005 \pm 0.002$ That value comes from the CLEO and ARGUS measurements (30) of $\Gamma(b \to u\ell\nu)$, which imply $|V_{ub}|/|V_{cb}| = 0.10 \pm 0.03$, where the error reflects theoretical uncertainties in the decay spectrum shape.

Since the top quark has not been discovered, no direct measurements of $|V_{ij}|$ exist. However, using three-generation unitarity, one infers from the above values

$$|V_{tb}| = 0.9989 \pm 0.0005$$

$$|V_{ts}| < 0.052$$

$$|V_{td}| < 0.021. \qquad\qquad 21.$$

It is ironic that $|V_{tb}|$ is the best-determined matrix element even though it has never been directly measured.

From the first row of the CKM matrix one finds (experimentally)

$$|V_{ud}|^2 + |V_{us}|^2 + |V_{ub}|^2 = 0.9991 \pm 0.0016. \qquad\qquad 22.$$

That result is a beautiful confirmation of three-generation unitarity and the standard model at the level of its radiative corrections. [Without proper accounting of radiative corrections (31), one would have obtained ~ 1.04, an apparent violation of unitarity.]

The result in Equation 22 can be used to limit heavy neutrino mixing (16), constrain fourth-generation mixing (18), and set bounds on extra gauge boson masses (32), supersymmetry (33), etc. A few of those constraints are described below.

Fourth Generation LEP measurements of the Z boson width are consistent with three light neutrinos. Therefore, if a fourth generation (ν_L, L, t', b') exists, its members must all be quite massive. Correspondingly, fourth-generation mixing with other quarks, except perhaps the top quark, is expected to be small. The current CKM values combined with unitarity require

$$|V_{ub'}| \leq 0.054$$

$$|V_{cb'}| < 0.54$$

$$|V_{t'd}| \leq 0.11$$

$$|V_{t's}| \leq 0.54. \qquad\qquad 23.$$

These constraints are useful but not very stringent. If a heavy fourth generation exists, it would better manifest itself in loop effects such as CP violation or rare flavor-changing decays. The B meson system would seem to be a good probe of such effects, particularly $\bar{B}_s^0$-B_s^0 oscillations and $B_s^0 \to \mu^+\mu^-$.

Z' Bosons   The existence of an extra Z' gauge boson would modify the radiative corrections to muon and beta decays. That modification would shift each of the $|V_{uj}|^2$ in Equation 22 by $1+\Delta$, where Δ is a calculable $O(\alpha)$ correction. For example, in the SO(10) model there is one extra neutral gauge boson Z_χ that gives (32)

$$\Delta(Z_\chi) = -0.0048\,\frac{\ln X}{X-1}, \qquad X = m_{Z_\chi}^2/m_W^2. \qquad 24.$$

Comparing that effect with Equation 22 gives

$$m_{Z_\chi} > 265\ \text{GeV} \qquad (90\%\ \text{CL}). \qquad\qquad 25.$$

That bound is not quite as good as direct bounds on m_{Z_χ} from p$\bar{\text{p}}$ collider experiments or neutral current experiments such as atomic parity violation, but it does illustrate the utility of charged current unitarity tests.

Other unitarity constraints on new physics can be found in the literature (10, 34).

2.3   Pseudoscalar Decay Constants

The pseudoscalar meson decay constants f_π and f_K are important scale-setting parameters in chiral symmetry expansions. In addition, f_π enters the famous Goldberger-Treiman relation (35) $f_\pi g_{\pi pn} \simeq (m_n + m_p) g_A/\sqrt{2}$, which is exact in the chiral limit, $m_\pi = 0$. Deviations provide a measure of chiral symmetry-breaking effects (36), which in the standard model are due to the current quark masses m_u and m_d and thus are expected to be small, $O(1-2\%)$. Heavier meson decay constants f_D and f_B have not been directly measured, but are important parameters for weak decay and oscillation studies.

To extract f_π and f_K from $\pi_{\mu 2}$ and $K_{\mu 2}$ decays, one must take into account electroweak radiative corrections (37). Those corrections are, unfortunately, dependent on hadronic structure uncertainties and therefore are somewhat ambiguous. That arbitrariness can be absorbed into the definitions of f_π and f_K and be viewed as an inherent uncertainty in some applications. Other structure-independent (process-dependent) corrections should, however, be factored out in the extraction of f_π and f_K. Those include the short-distance loop effects described in Section 2.2 as well as QED corrections specific to the particular decay modes used to obtain those parameters.

With the above goals in mind, f_π is defined by

$$\Gamma[\pi \to \mu\nu(\gamma)] = \frac{G_\mu^2 f_\pi^2}{8\pi} |V_{ud}|^2 m_\pi m_\mu^2 \left(1 - \frac{m_\mu^2}{m_\pi^2}\right)^2 \left[1 + \frac{\alpha}{\pi} R\left(\frac{m_\mu}{m_\pi}\right)\right],$$ 26a.

where

$$R(\mu) = 2\ln\frac{m_Z}{m_\rho} - \frac{3}{2}\ln\frac{m_\rho}{m_\pi} + \frac{13}{8} - \frac{3\mu^2}{4(1-\mu^2)} + \frac{6 - 20\mu^2 + 11\mu^4}{2(1-\mu^2)^2}\ln\mu$$

$$- 2\left(\frac{1+\mu^2}{1-\mu^2}\ln\mu + 1\right)\ln(1-\mu^2) + 2\frac{1+\mu^2}{1-\mu^2} L(1-\mu^2)$$ 26b.

$$L(x) = \int_0^x \frac{dt}{t}\ln(1-t).$$ 26c.

The first term in Equation 26b contains the short-distance correction induced by the use of G_μ to normalize the decay amplitude. The other terms are taken from Kinoshita's effective field theory calculation (38) of the low-energy QED corrections, employing m_ρ in place of his cutoff scale. The hadronic structure uncertainty can be viewed as residing in the $-3/2 \ln m_\rho/m_\pi + 13/8$ term, which depends on how short- and long-distance calculations are matched. Since there is an ambiguity of $O(\alpha)$, one should view applications of f_π as having an inherent uncertainty of about $\frac{1}{4}\%$.

Using $\Gamma[\pi \to \mu\nu(\gamma)] = 2.5284 \pm 0.0023 \times 10^{-14}$ MeV, $G_\mu = 1.16637 \times 10^{-5}$ GeV^{-2}, $|V_{ud}| = 0.9750 \pm 0.0007$, and appropriate particle masses, one finds from Equation 26

$$f_\pi = 130.7 \pm 0.1 \text{ MeV.}$$ 27.

This value is numerically close to the result found by Holstein (39), even though the one-loop radiative corrections given there do not contain quite the same μ dependence as Equation 26. (They have, for example, a different μ dependence in the limit $\mu \to 0$.)

The formula in Equation 26 can be applied to $K_{\mu 2}$ decays (40) with the replacement $f_\pi \to f_K$, $V_{ud} \to V_{us}$, $m_\pi \to m_K$, and $m_\rho \to m_{K^*}$. Using $\Gamma[K \to \mu\nu(\gamma)] = 3.38 \pm 0.01 \times 10^{-14}$ MeV and $|V_{us}| = 0.220 \pm 0.002$ gives

$$f_K = 159.9 \pm 1.5, \qquad\qquad 28.$$

where the main uncertainty comes from $|V_{us}|$.

The value of f_π can be tested in the Goldberger-Treiman relation. Using $g_A = 1.2646 \pm 0.0018$ (from averaging τ_n and direct g_A measurements) and the most recent (41) (still controversial) $g_{\pi pn} = 13.04 \pm 0.06$, one finds

$$\Delta_\pi = 1 - \frac{(m_n + m_p) g_A}{\sqrt{2} f_\pi g_{\pi pn}} = 0.015 \pm 0.005, \qquad\qquad 29.$$

which is in excellent agreement with expectations (36). I note that the earlier standard (42) $g_{\pi pn} = 13.4 \pm 0.1$ gives a much less acceptable deviation (4.3%) from the chiral prediction (36). One might contemplate using the Goldberger-Treiman relation as a definition of $g_{\pi pn}$.

Another test of f_π is provided by the PCAC prediction $\Gamma(\pi^0 \to \gamma\gamma) = \alpha^2 m_{\pi^0}^3 / (32\pi^3 f_\pi^2)$ (43). Using the f_π in Equation 27 gives $\Gamma(\pi^0 \to \gamma\gamma) = 7.73$ eV, whereas the world average, $\Gamma(\pi^0 \to \gamma\gamma)_{exp} = 7.50 \pm 0.17$ eV (4), deviates from that prediction by only 3%. That deviation is acceptable given the anticipated effects of chiral symmetry breaking (43). The best single π^0 decay experiment gives $\Gamma(\pi^0 \to \gamma\gamma) = 7.25 \pm 0.23$, which would imply a harder to explain 6% deviation from the PCAC prediction (44).

The above consistency checks further validate $f_\pi = 130.7$ MeV as a good value in chiral expansions. (Usually $f_{\pi^0} = f_\pi / \sqrt{2} = 92.4$ MeV is employed.) Even though there is some ambiguity in the hadronic contributions to radiative corrections, the definition of f_π in Equation 26 seems to work rather well in both the Goldberger-Treiman relation and the PCAC prediction for $\Gamma(\pi^0 \to \gamma\gamma)$. However, given the lingering controversy regarding $g_{\pi pn}$ and $\Gamma(\pi^0 \to \gamma\gamma)$ experiments, both tests should continue to be scrutinized.

2.4   Tau Decays

Tau decay measurements are approaching an interesting level of precision. Future high-luminosity e^+e^- facilities (tau factories) should make further significant advances. Already we are confronted by a puzzle, the "missing decays problem" (45), that needs experimental resolution. (See subsequent discussion.) In addition, there is hope that accurate measurements of the τ lifetime and leptonic branching ratios can be compared with perturbative QCD predictions to obtain a very precise determination of $\Lambda_{\overline{MS}}$ (46–48). With those issues and prospects in mind, I give a comparison of tau decay predictions with experiment.

Including electroweak radiative corrections, the leptonic decay rates are predicted to be (48)

$$\Gamma(\tau \to \ell \bar{v}_\ell v_\tau) = \frac{G_\mu^2 m_\tau^5}{192\pi^3} f\left(\frac{m_\ell^2}{m_\tau^2}\right)\left(1 + \frac{3}{5}\frac{m_\tau^2}{m_W^2}\right)\left[1 + \frac{\alpha(m_\tau)}{2\pi}\left(\frac{25}{4} - \pi^2\right)\right], \qquad 30.$$

where ℓ = e or μ, $f(x)$ is given in Equation 5b and $\alpha^{-1}(m_\tau) \simeq 133.3$. Employing the tau mass, $m_\tau = 1784.2 \pm 3.2$ MeV, then gives

$$\Gamma(\tau \to ev\bar{v}) = 4.115 \pm 0.037 \times 10^{-13} \text{ GeV}$$

$$\Gamma(\tau \to \mu v \bar{v}) = 4.003 \pm 0.036 \times 10^{-13} \text{ GeV}, \qquad 31.$$

with the $\pm 0.9\%$ uncertainty due to the error in m_τ.

Electroweak radiative corrections to semihadronic τ decays are larger. Analogous to the discussion in Section 2.2, there is a short-distance enhancement factor (48)

$$1 + \frac{2\alpha}{\pi}\ln\frac{m_Z}{m_\tau} \simeq 1.02 \qquad 32.$$

for all hadronic rates normalized with G_μ^2. Including that leading correction, one finds the branching ratio predictions in Table 3.

Comparing the theoretical predictions for single-charge-prong tau decays with the experimental averages in Table 3, one finds theory to be systematically higher. That difference may arise from my use of too long a τ lifetime; however, the particle data tables quote an even longer lifetime, $\tau_{\text{tau}}^{\text{ave}} = 3.03 \pm 0.08 \times 10^{-13}$ s (4). Alternatively, it may indicate problems with the experimental normalizations. There are indications that the latter is more likely. Indeed, adding the experimental single-charged-prong branching ratios in Table 3 gives $\simeq 80\%$ compared with the theory pre-

Table 3 Experimental measurements of branching ratios for one charged particle (4) compared with theoretical predictions (34) normalized using $\tau_{\text{tau}} = 2.98 \times 10^{-13}$ s

Decay mode	Experiment (%)	Theory (%)
$\tau \to ev\bar{v}$	17.7 ± 0.4	18.65
$\tau \to \mu v \bar{v}$	17.8 ± 0.4	18.14
$\tau \to \pi^- v$	11.0 ± 0.5	11.2
$\tau \to \pi^- \pi^0 v$	22.7 ± 0.8	23.5
$\tau \to K^- X$	1.7 ± 0.2	2.5
$\tau \to \pi^- \pi^0 \pi^0 v$	7.5 ± 0.9	7.5
Other	< 2.7	~ 4

diction of 85.5%. The theoretical prediction agrees with direct measurement of tau decays into one charged prong $\approx 86.1 \pm 0.3\%$ obtained from topological analyses (4, 49), while the sum of experimental branching ratios falls short. That shortcoming is sometimes called the "missing decays problem" (45). A further reason for believing the theory predictions comes from QCD perturbation theory, as we discuss next.

An important ratio R_H

$$R_H \equiv \frac{\Gamma(\tau \to \nu_\tau + \text{hadrons})}{\Gamma(\tau \to e\nu\bar{\nu})} \qquad\qquad 33.$$

was shown by Braaten to be calculable in QCD perturbation theory (46). One finds (for four effective quark loop flavors)

$$R_H \simeq 3(1.02)(|V_{ud}|^2 + |V_{us}|^2)\left[1 + \frac{\alpha_s}{\pi} + 4.8\left(\frac{\alpha_s}{\pi}\right)^2 + \cdots\right], \qquad 34.$$

where the 2% short-distance electroweak correction from Equation 32 has been included and $\alpha_s = \alpha_3(m_\tau)_{\overline{\text{MS}}}$ is the QCD $\overline{\text{MS}}$ coupling. A recent calculation of the $O(\alpha_s^3)$ correction (50, 51) leads to $+19.6(\alpha_s/\pi)^3$, which is neglected here. The important thing is that the coefficient be small enough to justify the truncation in Equation 34. A subtlety involves my use of four flavors in the $(\alpha_s/\pi)^2$ coefficient. Fortunately, for three flavors $4.8 \to 5.2$, not much of a change.

R_H can be obtained from the tau lifetime,

$$\tau_{\text{tau}}^{-1} = \Gamma(\tau \to \text{all}) = (1.9728 + R_H)\Gamma(\tau \to e\nu\bar{\nu}) \qquad 35.$$

or from either leptonic branching ratio

$$BR(\tau \to e\nu\bar{\nu}) = 1.028\,BR(\tau \to \mu\nu\bar{\nu}) = (1.9728 + R_H)^{-1}. \qquad 36.$$

Using averages for those quantities, one finds

$$\tau_{\text{tau}} = 2.98 \pm 0.08 \times 10^{-13}\,\text{s} \to R_H = 3.39 \pm 0.15 \qquad 37a.$$

$$BR(\tau \to e\nu\bar{\nu}) = 0.177 \pm 0.004 \to R_H = 3.68 \pm 0.13 \qquad 37b.$$

$$BR(\tau \to \mu\nu\bar{\nu}) = 0.178 \pm 0.004 \to R_H = 3.49 \pm 0.12. \qquad 37c.$$

Comparing the R_H values with Equation 34 gives (52)

$$R_H = 3.39 \pm 0.15 \to \alpha_s = 0.25 \pm 0.09 \to \Lambda_{\overline{\text{MS}}}^{(4)} = 163^{+152}_{-135} \qquad 38a.$$

$$R_H = 3.68 \pm 0.13 \to \alpha_s = 0.40 \pm 0.06 \to \Lambda_{\overline{\text{MS}}}^{(4)} = 436^{+93}_{-111} \qquad 38b.$$

$$R_H = 3.49 \pm 0.12 \to \alpha_s = 0.30 \pm 0.06 \to \Lambda_{\overline{\text{MS}}}^{(4)} = 266^{+112}_{-134}. \qquad 38c.$$

The errors on $\Lambda_{\overline{MS}}^{(4)}$ are still quite large; however, the central value in Equation 38a is in the best accord with other measurements of that QCD parameter. Turning the argument around, for $\Lambda_{\overline{MS}}^{(4)} \simeq 150_{-50}^{+100}$ MeV, one expects $\tau_{tau} = 2.99_{-0.05}^{+0.03} \times 10^{-13}$ s. It will be interesting to compare these values with future experimental results. If those efforts provide a consistent R_H, then a 1% measurement (at that point, theoretical uncertainties start to become an issue) will determine $\Lambda_{\overline{MS}}^{(4)}$ to about ± 30 MeV. That precision is worth the experimental effort.

3. WEAK NEUTRAL CURRENTS

The standard model's neutral current interaction Lagrangian is given by

$$\mathscr{L}_{int}^{NC} = -e_0 A^\mu(x) J_\mu^{em}(x) - \frac{g_{2_0}}{\cos \theta_W^0} Z^\mu(x) J_\mu^{NC}(x), \qquad 39.$$

where A^μ and Z^μ are photon and Z boson gauge fields and

$$e_0 = g_{2_0} \sin \theta_W^0 \qquad 40a.$$

$$J_\mu^{em} = \sum_f Q_f \bar{f} \gamma_\mu f \qquad 40b.$$

$$J_\mu^{NC} = \sum_f (T_{3f} \bar{f}_L \gamma_\mu f_L - \sin^2 \theta_W^0 Q_f \bar{f} \gamma_\mu f). \qquad 40c.$$

The summations in Equation 40 are over all fermion fields with $Q_f = (0, -1, 2/3, -1/3)$ and $T_{3f} = (\frac{1}{2}, -\frac{1}{2}, \frac{1}{2}, -\frac{1}{2})$ for $f = (\nu_e, e, u, d)$ along with identical assignments for higher generations.

The bare weak mixing angle θ_W^0 is of central importance. It is first introduced via diagonalization of the W_μ^3-B_μ mass matrix such that

$$A_\mu = B_\mu \cos \theta_W^0 + W_\mu^3 \sin \theta_W^0$$

$$Z_\mu = W_\mu^3 \cos \theta_W^0 - B_\mu \sin \theta_W^0. \qquad 41.$$

As such, it describes the mixing between $SU(2)_L$ and $U(1)_Y$ sectors and the relative strengths of their gauge couplings. Besides occurring in the fermion weak neutral current (see Equation 40c) and the normalization of e_0 relative to g_{2_0}, the weak mixing angle also relates $W^\pm$ and Z masses

$$m_W^0 = m_Z^0 \cos \theta_W^0 \qquad 42.$$

for Higgs doublet symmetry breaking. Thus one way to test the standard model is to measure $\sin^2 \theta_W$ in as many different ways as possible. A deviation in the value obtained from one experiment as compared with others would signal new physics. Of course, in any high-precision comparison, standard model radiative corrections must be properly accounted for and not confused with new physics.

In comparing different experiments, the natural relation

$$\sin^2 \theta_W^0 = e_0^2/g_{2_0}^2 = 1 - (m_W^0/m_Z^0)^2 \qquad 43.$$

is extremely useful. Reexpressed in terms of measurable renormalized quantities, that relation is modified by finite calculable loop effects (53). So tests of that relation probe the standard model at the level of its electroweak radiative corrections and give a constraint on m_t. In addition, they are sensitive to "new physics" (see Section 7). Of course, any precise extraction of $\sin^2 \theta_W$ from experiment requires a renormalization prescription and complete $O(\alpha)$ calculation of radiative corrections. Below, a specific definition of the renormalized mixing angle is advocated and its utility illustrated for a few experimental processes.

3.1 $sin^2 \theta_W(m_Z)_{\overline{MS}}$

A very convenient renormalized weak mixing angle can be defined by modified minimal subtraction (54, 55). It is given by

$$\sin^2 \theta_W(\mu)_{\overline{MS}} \equiv e^2(\mu)_{\overline{MS}}/g_2^2(\mu)_{\overline{MS}}, \qquad 44.$$

where the couplings $e^2(\mu)_{\overline{MS}}$ and $g_2^2(\mu)_{\overline{MS}}$ are individually defined by $\overline{MS}$. When μ, the running mass unit in dimensional regularization, is greater than m_t, the $\overline{MS}$ prescription is very simple. It corresponds merely to subtracting all $[1/(n-4)+\gamma/2-\ln\sqrt{4\pi}]$ loop corrections to those couplings. For the standard model with three generations and one Higgs doublet, that translates to

$$\sin^2 \theta_W(\mu)_{\overline{MS}} = \sin^2 \theta_W^0$$
$$\times \left[1 + \frac{e_0^2 \mu^{n-4}}{8\pi^2} \left(\frac{11}{3} + \frac{19}{6\sin^2\theta_W^0} \right) \left(\frac{1}{n-4} + \frac{\gamma}{2} - \ln\sqrt{4\pi} \right) + \cdots \right], \quad 45.$$

where . . . indicates higher orders (54). The running $\sin^2 \theta_W(\mu)_{\overline{MS}}$ satisfies the renormalization group equation (for $\mu > m_t$)

$$\mu \frac{\partial}{\partial \mu} \sin^2 \theta_W(\mu)_{\overline{MS}} = \frac{\alpha}{2\pi} \left(\frac{11}{3} \sin^2 \theta_W + \frac{19}{6} \right) + \cdots \qquad 46.$$

For most low-energy processes as well as measurements at the Z pole, it is convenient to choose $\mu = m_Z$ as a fixed scale. That causes a slight complication since it appears very likely that $m_t > m_Z$. To deal with that problem within the $\overline{MS}$ rules, one must subtract an extra

$$\frac{\alpha}{\pi}\left(\frac{1}{3} - \frac{8}{9}\sin^2\theta_W\right)\ln\frac{m_t}{\mu} \qquad\qquad 47.$$

from Equation 45 when $\mu < m_t$ (56). (The two definitions match at $\mu = m_t$.) That additional subtraction keeps the γ-Z mixing independent of $\ln m_t$; but $\alpha \ln m_t$ and even αm_t^2 effects can arise in other calculations. Numerically, the effect of Equation 47 is quite insignificant. For example, if $m_t = 140$ GeV and $\mu = m_Z$, it reduces $\sin^2\theta_W(m_Z)_{\overline{MS}}$ by 0.0001. Nevertheless, it is important to adhere to a strict $\overline{MS}$ formalism so new physics with mass scales exceeding m_Z can be similarly dealt with.

Choosing $\sin^2\theta_W(m_Z)_{\overline{MS}}$ and G_μ as renormalized expansion parameters allows one to deal with electroweak radiative corrections to weak neutral current processes in a straightforward manner. The weak amplitude is modified from its generic lowest-order form by the replacements $g_{2_0}^2/8m_Z^{0^2}\cos^2\theta_W^0 \to \rho G_\mu/\sqrt{2}$ and $\sin^2\theta_W^0 \to \kappa\sin^2\theta_W(m_Z)_{\overline{MS}}$ where ρ and κ are $1 + O(\alpha)$ process-dependent quantities (57). Complete $O(\alpha)$ calculations of ρ and κ have been carried out for most processes of interest and thus allow $\sin^2\theta_W(m_Z)_{\overline{MS}}$ to be extracted with precision. Some examples are described below.

3.2 Neutrino Scattering

Precise neutrino scattering experiments have been performed for (a) deep inelastic ${}^{(-)}_{\nu_\mu}$N scattering, (b) elastic ${}^{(-)}_{\nu_\mu}$p scattering, and (c) ${}^{(-)}_{\nu_\mu}$e scattering. The importance and current status of those studies are briefly described below, and the radiative corrections to νe scattering are also discussed.

After including electroweak radiative corrections and charm quark threshold effects, one finds from measurements of $R_\nu \equiv \sigma_{NC}(\nu_\mu N \to \nu_\mu X)/\sigma_{CC}(\nu_\mu N \to \mu X)$ (using $m_H \simeq 100$ GeV)

$$\sin^2\theta_W(m_Z)_{\overline{MS}} \simeq 0.230 \pm 0.003 \pm 0.005 + 0.0018\left(\frac{m_t}{m_W}\right)^2, \qquad 48.$$

where the first error is experimental and the second comes mainly from charm production uncertainties (58, 59). The sensitivity to m_t comes about because neutral and charged current cross sections are directly compared and $\sin^2\theta_W(m_Z)_{\overline{MS}}$ is employed. The result in Equation 48 currently provides an important constraint on m_t (see Section 5).

Elastic ${}^{(-)}_{\nu_\mu}$p scattering does not provide a competitive measure of $\sin^2\theta_W(m_Z)_{\overline{MS}}$. (It gives roughly 0.210 ± 0.033.) Instead, the utility of those experiments is their sensitivity to axial-isoscalar neutral current effects. In particular, they probe the strange quark content of the proton through $\langle p|\bar{s}\gamma_\mu\gamma_5 s|p\rangle$ matrix elements (60). An experiment at Brookhaven National

Laboratory shows evidence for a substantial strange quark contribution (61). To confirm and quantify that finding in a meaningful way would require new high statistics measurements of $d\sigma[\overset{(-)}{\nu}_\mu p \to \overset{(-)}{\nu}_\mu p]/dQ^2$ over a range of Q^2.

The purely leptonic $\overset{(-)}{\nu}_\mu e$ measurements are free of hadronic uncertainties (at the tree level). Unfortunately, the cross sections are very small and therefore hard to measure. One can easily incorporate electroweak radiative corrections. They modify the cross sections such that (57, 62)

$$\sigma(\nu_\mu e \to \nu_\mu e) = \frac{\rho_{\nu_\mu e}^2 G_\mu^2 m_e E_\nu}{2\pi}$$

$$\times \left[1 - 4\kappa_{\nu_\mu e}(q^2) \sin^2 \theta_W(m_Z)_{\overline{MS}} + \frac{16}{3} \kappa_{\nu_\mu e}^2(q^2) \sin^4 \theta_W(m_Z)_{\overline{MS}} \right]$$

$$\sigma(\bar{\nu}_\mu e \to \bar{\nu}_\mu e) = \frac{\rho_{\nu_\mu e}^2 G_\mu^2 m_e E_\nu}{2\pi}$$

$$\times \left[\frac{1 - 4\kappa_{\nu_\mu e}(q^2) \sin^2 \theta_W(m_Z)_{\overline{MS}} + 16\kappa_{\nu_\mu e}^2(q^2) \sin^4 \theta_W(m_Z)_{\overline{MS}}}{3} \right]. \quad 49.$$

In addition, there are QED corrections that depend on the specific experimental setup (62). The radiative corrections in Equation 49 are given by

$$\rho_{\nu_\mu e} = 1.0056 + 0.0006 \left[\frac{3}{4\sin^2 \theta_W} \frac{m_t^2}{m_W^2} + G(\xi) \right]$$

$$G(\xi) = \frac{3}{4s^2} \xi \left[\frac{\ln(c^2/\xi)}{c^2 - \xi} + \frac{1}{c^2} \frac{\ln \xi}{1 - \xi} \right] \quad 50.$$

$s^2 = \sin^2 \theta_W$, $c^2 = \cos^2 \theta_W$, $\xi = m_H^2/m_Z^2$, and

$$\kappa_{\nu_\mu e}(q^2 = 0) \simeq 0.9977 \pm 0.0025,$$

where the uncertainty in $\kappa_{\nu_\mu e}$ comes from hadronic loop effects (63). For $q^2 \neq 0$, see the paper by Sarantakos et al (62).

Note that $\kappa_{\nu_\mu e}$ is independent of m_t and m_H because we employed $\sin^2 \theta_W(m_Z)_{\overline{MS}}$. That means that measurements of the ratio $R \equiv \sigma(\nu_\mu e \to \nu_\mu e)/\sigma(\bar{\nu}_\mu e \to \bar{\nu}_\mu e)$ are insensitive to those still unknown parameters and provide a direct determination of $\sin^2 \theta_W(m_Z)_{\overline{MS}}$. From the average of R measurements, one finds

$$\sin^2 \theta_W(m_Z)_{\overline{MS}} = 0.2325 \pm 0.010 \pm 0.0006 \quad \overset{(-)}{\nu}_\mu e \text{ average}, \quad 51.$$

where the second error comes from hadronic loop uncertainties. The good agreement between Equation 51 and other measurements of $\sin^2 \theta_W$ provides interesting constraints on new physics. For example, comparing Equations 48 and 51 gives $m_t \lesssim 250$ GeV (90% CL). They can also be used to bound Z' masses and heavy fermion loop effects (see Section 7).

A possibility exists to measure

$$R' \equiv \frac{\sigma(v_\mu e \rightarrow v_\mu e)}{\sigma(v_e e \rightarrow v_e e) + \sigma(\bar{v}_\mu e \rightarrow \bar{v}_\mu e)} \qquad 52.$$

to about $\pm 2\%$ at the LAMPF pion beam stop (64). Such a measurement would determine $\sin^2 \theta_W(m_Z)_{\overline{MS}}$ to about 1% and provide useful constraints on Z' bosons (65), neutrino oscillations (66), etc.

3.3 Atomic Parity Violation

Atomic parity violation experiments with heavy atoms are becoming very precise (67, 68). Those experiments generally measure the so-called weak charge $Q_W(Z, A)$ where Z is the atomic number and $A - Z$ is the number of neutrons in the nucleus (68, 69). Normalizing the neutral current amplitude in terms of G_μ leads to a weak charge given by (70)

$$Q_W(Z, A) = \rho'_{PV}[2Z - A - 4Z\kappa'_{PV}(0) \sin^2 \theta_W(m_Z)_{\overline{MS}}], \qquad 53.$$

where

$$\rho'_{PV} = 0.9799 + \frac{\alpha(m_Z)}{2\pi} \left[\frac{3}{8s^2} \frac{m_t^2}{m_W^2} + \frac{3\xi}{8s^2} \left(\frac{\ln c^2/\xi}{c^2 - \xi} + \frac{1}{c^2} \frac{\ln \xi}{1 - \xi} \right) \right]$$

$$\kappa'_{PV}(0) = 1.003 \pm 0.0025 \qquad 54.$$

with $\xi = m_H^2/m_Z^2$. The uncertainty in Equation 54 comes from hadronic corrections to γ-Z mixing, which is approximated by a dispersion relation and $e^+e^- \rightarrow$ hadrons data (63).

For $m_t = 130$ GeV and $m_H = 100$ GeV, one finds $\rho'_{PV} = 0.9849$. Measurements of Q_W can then be used to extract $\sin^2 \theta_W(m_Z)_{\overline{MS}}$. Alternatively, a measurement of $\sin^2 \theta_W(m_Z)_{\overline{MS}}$ from another process such as Z decay asymmetries can be used in conjunction with Q_W to constrain m_t. Finally, if m_Z is used to determine $\sin^2 \theta_W(m_Z)_{\overline{MS}}$ (see Section 4), the m_t dependence induced by that procedure tends to cancel the m_t dependence from Equation 54. In that case Q_W predictions are insensitive to present m_t uncertainties (56, 71).

At present, the most precise atomic parity violation experiment uses $^{133}_{55}$Cs. One then predicts (56) for $m_t = 130$ GeV, $m_H = 100$ GeV, and $m_Z = 91.17$ GeV [these values correspond to $\sin^2 \theta_W(m_Z)_{\overline{MS}} = 0.2326$]

$$Q_W(^{133}_{55}\text{Cs}) = -73.20 \pm 0.13. \qquad\qquad 55.$$

That prediction is to be compared with

$$Q_W(^{133}_{55}\text{Cs})^{\text{exp}} = -71.04 \pm 1.58 \pm 0.88, \qquad\qquad 56.$$

where the first error is experimental (67) (mainly statistical) and the second comes from atomic theory uncertainties (72). An ongoing experiment plans to reduce the first error by a factor of five (C. Wieman, private communication). As discussed in Section 7, such a precise measurement has important implications for new physics beyond the standard model. (It is important to stress that the prediction in Equation 55 is insensitive to m_t and therefore is a direct probe of new physics.)

3.4 Polarized Electron Scattering

An interesting test of the standard model involves elastic scattering of polarized electrons on a spinless isoscalar target such as ^{12}C or ^4He (73). Experiments measure the parity-violating asymmetry

$$A \equiv \frac{d\sigma_R - d\sigma_L}{d\sigma_R + d\sigma_L}, \qquad\qquad 57.$$

which is predicted to be

$$A = \frac{-G_\mu q^2}{\sqrt{2\pi\alpha}} \rho'_{\text{PV}} \kappa'_{\text{PV}}(q^2) \sin^2 \theta_W(m_Z)_{\overline{\text{MS}}}, \qquad\qquad 58.$$

where ρ'_{PV} and $\kappa'_{\text{PV}}(0)$ are given in Equation 54. For $q^2 \neq 0$ there are modifications of κ'_{PV} and QED effects must be included. A recently completed BATES experiment on carbon gave (74)

$$\sin^2 \theta_W(m_Z)_{\overline{\text{MS}}} = 0.200 \pm 0.051. \qquad\qquad 59.$$

The error is relatively large; however it is mainly statistical. New experiments with carbon or helium targets could in principle reduce the error on $\sin^2 \theta_W(m_Z)_{\overline{\text{MS}}}$ to about $\pm 1\%$ (P. A. Souder, private communication). (At that level, nuclear theory uncertainties enter.) It has been argued that such experiments also provide a probe of $\langle N|\bar{s}\gamma_\mu s|N\rangle$ matrix elements at $Q^2 \neq 0$ and thus measure the strange content of the nucleus (75).

4. INTERMEDIATE VECTOR BOSONS

The discovery of the $W^\pm$ and Z bosons has been followed by explorations of their properties at $p\bar{p}$ and e^+e^- colliders. Measurements of their masses and decay properties now constitute some of the best tests of the standard model. In fact, m_Z determinations at LEP have become so precise that m_Z

is now a standard for comparison with other neutral current measurements. In this section a status report on $W^\pm$ and Z properties and their utility is presented. More detail on the LEP results can be found in the accompanying review by Burkhardt & Steinberger (76).

4.1 $W^\pm$ and Z Masses

Any discussion of gauge boson masses starts with the natural lowest-order relation in Equation 43. Reexpressed in terms of finite renormalized parameters, that relation becomes (15, 57, 77, 78)

$$\frac{\pi\alpha}{\sqrt{2}G_\mu m_W^2} = \left(1 - \frac{m_W^2}{m_Z^2}\right)(1 - \Delta r) = \sin^2\theta_W(m_Z)_{\overline{\text{MS}}}[1 - \Delta r(m_Z)_{\overline{\text{MS}}}], \qquad 60.$$

where $\alpha = 1/137.036$, $G_\mu = 1.16637 \times 10^{-5}\,\text{GeV}^{-2}$, m_W and m_Z are physical masses (real part of the propagator pole), and Δr and $\Delta r(m_Z)_{\overline{\text{MS}}}$ represent finite electroweak radiative corrections. The combination $1 - m_W^2/m_Z^2$ is often called $\sin^2\theta_W$ and used in physical renormalization prescriptions. The large value of m_t now makes such a prescription awkward, so I have not followed that approach.

The one-loop radiative corrections in Δr and $\Delta r(m_Z)_{\overline{\text{MS}}}$ have the following features. Each contains about a $+7\%$ vacuum polarization correction (54). That effect is large because there are many fermions in the standard model and each contributes $O[(2\alpha/3\pi)Q_f^2\ln(m_Z/m_f)]$. One can view that 7% as representing the running of α from $1/137.036$ at zero momentum to $\alpha(m_Z)_{\overline{\text{MS}}} \simeq 1/127.8$ at $\mu = m_Z$. The top quark also makes an important contribution to Δr (for $m_t^2 \gg m_W^2$) (57, 79):

$$\Delta r(m_t) \simeq -\frac{3\alpha(m_Z)_{\overline{\text{MS}}}}{16\pi}\frac{c^2}{s^4}\frac{m_t^2}{m_W^2} \simeq -0.007\frac{m_t^2}{m_W^2}. \qquad 61.$$

That negative correction becomes very substantial for $m_t > 200$ GeV. By measuring Δr via Equation 60, one can constrain m_t. The $\overline{\text{MS}}$ correction $\Delta r(m_Z)_{\overline{\text{MS}}}$ has a very small dependence on m_t. A large Higgs mass can also shift Δr by

$$\Delta r(m_H) \simeq \frac{11\alpha(m_Z)_{\overline{\text{MS}}}}{24\pi s^2}\ln\frac{m_H}{m_Z}, \qquad 62.$$

which is too small to have a significant effect on present experiments (54, 57). It does, however, lead to an uncertainty in the extraction of m_t from precision measurements.

Using $m_Z = 91.17$ GeV and assuming $m_H \simeq 100$ GeV allows one to predict m_W, Δr, $\Delta r(m_Z)_{\overline{\text{MS}}}$, and $\sin^2\theta_W(m_Z)_{\overline{\text{MS}}}$ as a function of m_t via Equation 60. The results of such an analysis by Degrassi, Fanchiotti &

Sirlin (80) are given in Table 4. [I have modified $\Delta r(m_Z)_{\overline{MS}}$ and $\sin^2 \theta_W(m_Z)_{\overline{MS}}$ because of a slight difference in definition; see Equation 47.] Note the sensitivity of Δr to m_t changes. In contrast, $\Delta r(m_Z)_{\overline{MS}}$ exhibits little dependence on m_t.

From Table 4, one sees the importance of m_W measurements for pin-pointing m_t. At present, $m_W = 80.14 \pm 0.31$ GeV (4) corresponds to $m_t \simeq 130 \pm 50$ GeV with an additional uncertainty of about $^{+25}_{-10}$ GeV from Higgs mass uncertainties. Future measurements of m_W to ± 50 MeV will determine m_t to ± 10 GeV modulo Higgs effects. Of course, we expect m_t to be directly measured in the future. Comparing its measured mass with the loop-inferred value will test the standard model and perhaps even provide an indication of what m_H range is preferred. It will also constrain or provide a hint of new physics. Pushing measurements of m_W as far as possible should clearly be a high priority.

4.2 Z Decay Rates and Asymmetries

The four LEP experiments have accumulated more than 700,000 Z decays. With those statistics, interesting comparisons between theory (81) and experiment (76, 82) become possible. For example, in Table 5, the total and leptonic Z widths are compared with standard model expectations.

Table 4 Standard model predictions (80) as a function of m_t, using $m_Z = 91.17$ GeV, $\alpha = 1/137.036$, and $G_\mu = 1.16637 \times 10^{-5}$ GeV^{-2} as input and assuming $m_H \simeq 100$ GeV

m_t (GeV)	$10^2 \Delta r$	$10^2 \Delta r(m_Z)_{\overline{MS}}$	m_W (GeV)	$\sin^2 \theta_W(m_Z)_{\overline{MS}}$
90	6.08	6.84	79.91	0.2336
100	5.76	6.88	79.97	0.2334
110	5.45	6.91	80.03	0.2331
120	5.13	6.94	80.08	0.2329
130	4.79	6.96	80.14	0.2326
140	4.44	6.98	80.20	0.2323
150	4.07	6.99	80.27	0.2319
160	3.68	7.01	80.33	0.2316
170	3.27	7.02	80.40	0.2312
180	2.83	7.03	80.48	0.2309
190	2.37	7.04	80.55	0.2304
200	1.87	7.05	80.63	0.2300
210	1.35	7.06	80.71	0.2296
220	0.80	7.07	80.80	0.2291
230	0.22	7.08	80.88	0.2286
240	−0.40	7.09	80.98	0.2282
250	−1.05	7.09	81.07	0.2276

Table 5 Comparison of experimental decay rates found from averaging the four LEP experiments (82) with standard model predictions (76)

Experiment (MeV)	Theory[a] (MeV)
$\Gamma(Z \to \text{all}) = 2487 \pm 9$	2490
$\Gamma(Z \to \ell^+\ell^-) = 83.3 \pm 0.4$	83.6
$\Gamma(Z \to \text{invisible}) = 493 \pm 10$	500

[a] The theoretical rates employ $m_Z = 91.17\,\text{GeV}$ and $\sin^2 \theta_W(m_Z)_{\overline{\text{MS}}} = 0.2326$, which corresponds to $m_t \simeq 130\,\text{GeV}$ and $m_H \simeq 100\,\text{GeV}$.

The agreement is impressive. Indeed, there is very little room left for new physics effects such as Z-Z' mixing or even additional loop corrections (see Section 7). Unfortunately, we have reached a point where the theoretical uncertainties in extracting Γ_Z are becoming comparable to the errors in Table 5; anticipated integrated luminosity increases will probably not significantly improve the situation.

Decay asymmetry measurements of the Z boson have also been carried out at LEP. The forward-backward (FB) decay asymmetries of various fermion pairs ($\mu^+\mu^-$, $\tau^+\tau^-$, $b\bar{b}$, etc) have been observed and collectively found to yield (76, 82)

$$\sin^2 \theta_W(m_Z)_{\overline{\text{MS}}} = 0.230 \pm 0.003 \qquad (\text{from } A_{\text{FB}}). \qquad 63.$$

Note that LEP results are often quoted using $\sin^2 \bar{\theta}_W \simeq \sin^2 \theta_W (m_Z)_{\overline{\text{MS}}} + 0.0001$ (83). The error in Equation 63 is expected to decrease to about ± 0.001 by the end of 1993. That will be extremely interesting, since the value of $\sin^2 \theta_W(m_Z)_{\overline{\text{MS}}}$ given above is slightly below the world average $\sin^2 \theta_W(m_Z)_{\overline{\text{MS}}} = 0.2326$. (I note, however, that it is not clear whether all radiative corrections have been properly accounted for in the asymmetry studies.)

Left-right (LR) polarization asymmetries are also being studied. The ALEPH collaboration used their measurements of the τ polarization in $Z \to \tau^+\tau^-$ to obtain (82)

$$\sin^2 \theta_W(m_Z)_{\overline{\text{MS}}} = 0.231 \pm 0.007 \qquad (\text{from } A_{\text{LR}}). \qquad 64.$$

The error in Equation 64 may ultimately decrease by about a factor of five at LEP. To do better seems to require polarized e^- beams. An effort to construct such a facility is under way at the Stanford Linear Collider (SLC). In principle, the uncertainty in $\sin^2 \theta_W(m_Z)_{\overline{\text{MS}}}$ could be reduced to about ± 0.0004. Unfortunately, about one million Z decays are needed.

(Alternatively, very high polarization could decrease the required statistics.) Reaching such high statistics at SLC will be extremely difficult, but worth the effort.

From global fits to all LEP Z decay data, one finds (76, 82)

$$\sin^2\theta_W(m_Z)_{\overline{MS}} = 0.2325 \pm 0.0014 \qquad \text{(All LEP data).} \qquad 65.$$

That value is in excellent accord with neutrino scattering, W boson mass measurements, etc. Thus it appears that there are no great surprises so far at the Z pole, only spectacular confirmation of the standard model.

4.3   Neutrino Counting

The invisible width of the Z in Table 5 is obtained by subtracting measured hadronic and charged lepton decay rates from the total Z width. Dividing the invisible width by $\Gamma(Z \rightarrow \nu\bar{\nu}) \simeq 1.99\Gamma(Z \rightarrow e^+e^-) = 165.8 \pm 0.8$ MeV then gives the number of light neutrino species, N_ν

$$N_\nu = 2.97 \pm 0.06. \qquad 66.$$

That result is consistent with the three known species ν_e, ν_μ, and ν_τ. It rules out the possibility of a fourth generation with the usual couplings and $m_\nu \lesssim 40$ GeV and it severely constrains models with other light weakly interacting particles that can be emitted in Z decays. So, it seems that any new spectroscopy is likely to reside only at high energies.

5.   TOP QUARK PHYSICS

Direct experimental searches have so far failed to discover the top quark. The CDF collaboration at Fermilab has published the most prohibitive bound, $m_t > 89$ GeV (5). That finding raises the question: Why is the top quark so heavy? Indeed, in comparison all other known fermions are almost massless. Is there something special about the top quark?

There were a number of early indications that the top quark might be very heavy. The larger than expected B_d^0-$\bar{B}_d^0$ oscillations and smaller than expected CKM elements $|V_{bc}|$ and $|V_{bu}|$ pointed to a need for a large value of m_t in flavor-changing loops. Those constraints are unfortunately clouded by hadronic uncertainties, so it is difficult to obtain a real determination of m_t. On the other hand, neutrino scattering, m_W, Z widths, etc allow a rather precise determination of m_t from quantum loop corrections to W and Z boson propagators. I review those bounds below.

Given the reality of a large m_t, a number of authors have speculated that it may be a signal of dynamical symmetry breaking via $t\bar{t}$ condensation (9). Such a scenario is the minimal alternative to the standard Higgs mechanism. Although incomplete, that model has some testable consequences. For example, a high value of $m_t \gtrsim 200$ GeV appears necessary in the minimal $t\bar{t}$ condensation model. Aspects of that requirement are subsequently discussed and contrasted with experimental constraints.

5.1 Mass Constraints

At present, the best loop constraints on m_t come from comparing α, G_μ, and m_Z with (a) deep inelastic $\nu_\mu N$ scattering, (b) m_W, and (c) Z decay widths and asymmetries. Each of those measurements has been discussed in this review. Here we put those constraints together and check for consistency. We assume throughout that the only missing pieces of the standard model are the top quark and the Higgs scalar. To begin we assume m_H is 100 GeV and then allow its mass to vary from 48 GeV to 1 TeV. To facilitate comparison with standard model predictions, we use Table 4 as our starting point. Individual constraints on m_t are given in Table 6. Those determinations of m_t are quite consistent. Taken together with all other data and with the m_H uncertainty, we obtain

$$m_t \simeq 130 \pm 40 \text{ GeV} \qquad\qquad 67.$$

as the current "best bet" range. Since the standard deviation in Equation 67 is not linear, it is perhaps better to quote a bound of $m_t \lesssim 180$ GeV.

During the next Fermilab $p\bar{p}$ collider run, the potential for discovering the top quark is expected to extend up to about 140–150 GeV. Equation 67 implies better that a 50-50 chance for its discovery during that run. Ultimately, one should be able to find the top quark up to about 200–250 GeV at Fermilab with anticipated upgrades. Beyond that will require the higher energy of the SSC or LHC. In the meantime, the error on m_W is expected to be lowered to about ± 100 MeV after the next Fermilab run. That alone should reduce the error in Equation 67 to about $\pm 15^{+25}_{-10}$ GeV. Combined with LEP data and $\nu_\mu N$ scattering one can anticipate $\pm 5^{+25}_{-10}$.

Table 6 Constraints on m_t from various measurements[a]

Measurement	m_t Constraint
$R_\nu \equiv \sigma_{NC}(\nu_\mu N)/\sigma_{CC}(\nu_\mu N)$	90^{+50}_{-90} GeV
$m_W = 80.14 \pm 0.31$ GeV	130^{+45}_{-50} GeV
All LEP data	135 ± 50 GeV
All the above	120^{+25}_{-30} GeV

[a] A central value of $m_H \simeq 100$ GeV has been employed. For the range $m_H \simeq 48$–1000 GeV, an additional uncertainty of about $^{+25}_{-10}$ GeV should be appended, with a larger m_H corresponding to a heavier m_t.

5.2 $t\bar{t}$ Condensation

The possibility of replacing the explicit Higgs mechanism with a dynamical $t\bar{t}$ condensation has gained popularity with increasing m_t. The basic idea is that dynamical generation of a top quark mass via condensation, $\langle 0|t\bar{t}|0 \rangle \neq 0$, will spontaneously break the $SU(2)_L \times U(1)_Y$ gauge symmetry and give the gauge bosons mass. There will be three pseudoscalar massless Goldstone bosons with the quantum numbers of $t\bar{t}$, $t\bar{b}$, and $\bar{t}b$ that become the longitudinal components of Z and $W^\pm$ bosons. The natural relation $m_W^{0^2} = m_Z^{0^2} \cos^2 \theta_W^0$ is preserved (84). There is also a tightly bound $t\bar{t}$ state with 0^{++} quantum numbers that replaces the Higgs scalar (9). In all respects, this scenario looks just like the minimal standard model. Its one advantage is that m_t (and to some extent m_H) are predicted. A drawback is that there is no "new physics" all the way up to the condensation scale.

Solving the renormalization group equations, one finds (roughly)

$$m_t^2 \simeq \frac{4}{9} \frac{\alpha_3(m_t)}{\alpha_2(m_t)} \frac{\alpha_3^{1/7}(m_t)}{\alpha_3^{1/7}(m_t) - \alpha_3^{1/7}(\Lambda)} m_W^2, \qquad\qquad 68.$$

where α_3 and α_2 are running $SU(3)_c$ and $SU(2)_L$ gauge couplings, $\alpha_i = g_i^2/4\pi$, and Λ is the scale of $t\bar{t}$ condensation (9, 85). Using $\alpha_3(m_t) \simeq 0.1$, $\alpha_2(m_t) \simeq 0.033$ and allowing Λ to vary gives the predictions in Table 7. There are, of course, higher-order effects that could shift those predictions by $O(10\%)$, but they are nevertheless rather stable. Since one expects $\Lambda \lesssim 10^{19}$ GeV (quantum gravity becomes important at shorter distances), this scenario seems to suggest $m_t \gtrsim 200$ GeV. From our previous discussion in Section 5.1, we know that such large top quark masses are disfavored by electroweak loop phenomenology. One can reduce m_t by adding more fermions, such as a fourth generation (85), with large condensates (including a heavy neutrino) or supersymmetry (86). That relaxes the predictability somewhat but suggests a richer spectroscopy waiting to be discovered. Further advances in this type of model require top discovery,

Table 7   Predictions for m_t as a function of Λ, the scale of dynamical symmetry breaking, in the minimal $t\bar{t}$ condensation model

Λ (GeV)	m_t (GeV)	Λ (GeV)	m_t (GeV)
∞	98	10^{12}	250
10^{19}	200	10^3	500
10^{16}	210		

measurement of its mass and decay properties, and scrutiny of any new phenomenology that accompanies the top quark.

6. HIGGS SCALAR

The standard $SU(2)_L \times U(1)$ electroweak model requires the existence of a spin-zero Higgs scalar, a remnant of spontaneous symmetry breaking. Much of the motivation for building the SSC lies in its potential for finding the Higgs scalar or whatever new physics takes its place. How the Higgs scalar will be discovered depends crucially on its mass, since both Higgs production and decay rates are highly mass dependent. Below, a discussion of various experimental and theoretical constraints on m_H is given and current discovery strategies are outlined.

6.1 Mass Constraints

The LEP experiments have searched for $Z \to H\ell^+\ell^-$ and $Z \to H\nu\bar{\nu}$ with negative results. The ALEPH collaboration has given the best bound (87)

$$m_H \gtrsim 48 \text{ GeV} \qquad\qquad 69.$$

but the other experiments have similar sensitivity. As higher statistics are accumulated, those bounds are expected to reach 55–60 GeV (or a discovery might be made). LEP II can extend the search up to about $m_H \simeq 80$ GeV via $e^+e^- \to HZ$. Masses beyond 80 GeV are reserved for the next generation of pp colliders (see subsequent discussion).

What does theory tell us about m_H? Unfortunately, very little. In the minimal standard model, perturbative unitarity and triviality arguments suggest $m_H \lesssim 1$ TeV (7, 8). For larger masses, it is quite likely that additional structure will appear at high energies, and the enormous Higgs width implies that it ceases to be a well-defined resonance. Requiring the Higgs (running) self-coupling to remain finite up to scales $O(10^{19} \text{ GeV})$ implies $m_H \lesssim 150$ GeV (88), so there is some prejudice that m_H should not be too heavy if there is a "great desert" up to 10^{19} GeV. The $t\bar{t}$ condensation scenario also suggests $m_h < 2m_t$ (perhaps much less) (9).

6.2 Higgs Discovery Signatures

If the mass of the Higgs scalar exceeds about 80 GeV, its discovery is reserved for the next generation of pp colliders, the SSC with $\sqrt{s} = 40$ TeV and an initial design luminosity $\mathscr{L} = 10^{33}$ cm^{-2} s^{-1} or the Large Hadron Collider (LHC) at CERN with $\sqrt{s} = 16$ TeV and a possible design luminosity $\mathscr{L} = 10^{34}$ cm^{-2} s^{-1}.

It is generally agreed that the region $m_H \simeq 140$–800 GeV can be covered at the SSC by detecting two e^+e^- or $\mu^+\mu^-$ pairs either from $H \to ZZ$ at higher masses or from H → ZZ* (Z* = virtual Z) below the ZZ threshold. At higher and lower masses, the best Higgs discovery strategies have been less clear. For very large m_H, one may have to resort to a careful partial wave analysis of W_L-W_L scattering. At the other end of the scale, $m_H \simeq 80$–150 GeV, a well-defined discovery mode H → $\gamma\gamma$ now seems feasible (particularly with associated $t\bar{t}$ production); so I give a brief discussion.

The production mechanism gluon+gluon → H has quite a large cross section of about 100 pb for $m_H \approx 100$ GeV. The rare but distinctive decay H → $\gamma\gamma$ has a branching ratio of $1 \sim 2 \times 10^{-3}$ (89). Therefore, one expects at the SSC (for $m_H \simeq 100$ GeV)

$$\sigma(pp \to H \to \gamma\gamma) \simeq 1.5 \times 10^{-1} \text{ pb.} \qquad 70.$$

An SSC year, 10^7 seconds, corresponds to an integrated luminosity of 10^4 pb^{-1}, which implies 1500 raw H → $\gamma\gamma$ decays per SSC year. Cuts and detector efficiencies are expected to reduce the H → $\gamma\gamma$ signal to about 500 events per year. Unfortunately, the background from $q\bar{q} \to \gamma\gamma$ leads to $10^7\gamma\gamma$ pairs per year. To reduce that background requires extraordinary $\gamma\gamma$ mass resolution ($\Delta m_{\gamma\gamma}/m_{\gamma\gamma} \simeq \frac{1}{2}\%$) (90). Even then, the signal-to-background ratio is not overwhelming, particularly at the lower masses $m_H \simeq 80$–100 GeV. To explore that region thoroughly seems to require higher luminosity or longer runs.

An alternative idea is to search for the associated production of a Higgs scalar and decay H → $\gamma\gamma$ with $W^\pm$ (91, 92) or $t\bar{t}$ pairs (93). The decay W → ℓv_ℓ (ℓ = e or μ) or t → Wb, W → ℓv_ℓ with the requirement that the charged lepton be isolated provides a significant background-reducing tag. Hence the decay H → $\gamma\gamma$ becomes observable. Of the two, the associated production with $t\bar{t}$ appears more feasible since the cross section $\sigma(pp \to t\bar{t}H)$ is roughly two to three times larger at the SSC than $\sigma(pp \to W^\pm H)$. In addition, since either t or $\bar{t}$ can produce an isolated lepton, one gains another factor of two. The extra overall factor of five or six makes the $t\bar{t}H$ signal rather robust whereas $W^\pm H$ turns out to be marginal.

Before cuts and efficiencies, one expects

$$\sigma(pp \to Ht\bar{t}X \to \gamma\gamma\ell v_\ell X)_{SSC} \simeq 5.2 \text{ fb}$$

$$\sigma(pp \to Ht\bar{t}X \to \gamma\gamma\ell v_\ell X)_{LHC} \simeq 0.67 \text{ fb} \qquad 71.$$

for $m_H \simeq 80$–130 GeV and somewhat smaller cross sections for 130–150 GeV (93). Including isolation cuts and typical detector efficiencies leads to 21 events per year at the SSC and 3 events per year at the LHC for 10^{33} cm^{-2} s^{-1} luminosity. Backgrounds to this process have not been thoroughly scrutinized, but good $\gamma\gamma$ mass resolution is expected to eliminate the most likely background from pp → $t\bar{t}\gamma\gamma$. A signal of 21 events would allow a measurement of the H $\rightarrow \gamma\gamma$ branching ratio and provide a nice consistency check on the standard model. Associated production with $t\bar{t}$ and the resulting isolated lepton tag may also provide a window to new physics at the SSC.

7. NEW PHYSICS

Given the high precision already attained by electroweak experiments, one can search for hints of new physics in tree level amplitudes or even in loop effects. A nice formalism for studying heavy physics effects on loop corrections has been introduced by Peskin & Takeuchi (11), the S and T parametrization. In the next sections, I describe the implementation of that formalism and current experimental constraints that it implies.

7.1 S and T Radiative Corrections

Heavy new fermions with generic mass m_F enter low-energy phenomenology through gauge boson self-energies. In the electroweak sector those include the vacuum polarization functions $\Pi_{\gamma\gamma}(q^2)$, $\Pi_{\gamma Z}(q^2)$, $\Pi_{ZZ}(q^2)$, and $\Pi_{WW}(q^2)$. Using $\overline{MS}$ subtraction removes heavy particle contributions to the first two, $\Pi_{\gamma\gamma}$ and $\Pi_{\gamma Z}$, up to terms of the form $\alpha q^2/m_F^2$, which are neglected (56). That prescription absorbs the heavy particle effects into the definitions of $\alpha(\mu)_{\overline{MS}}$ and $\sin^2\theta_W(\mu)_{\overline{MS}}$ for $\mu < m_F$. There are, however, residual effects in Π_{ZZ} and Π_{WW} that remain after $\overline{MS}$ subtraction and are potentially observable as corrections to the natural relation in Equation 43.

In general, the observable heavy fermion loop effects can be parametrized in terms of three observables S_W, S_Z, and T defined by (56)

$$\left.\frac{\Pi_{WW}^{new}(m_W^2) - \Pi_{WW}^{new}(0)}{m_W^2}\right|_{\overline{MS}} = \frac{\alpha(m_Z)_{\overline{MS}}}{4\sin^2\theta_W(m_Z)_{\overline{MS}}} S_W \qquad 72a.$$

$$\left.\frac{\Pi_{ZZ}^{new}(m_Z^2) - \Pi_{ZZ}^{new}(0)}{m_Z^2}\right|_{\overline{MS}} = \frac{\alpha(m_Z)_{\overline{MS}}}{\sin^2 2\theta_W(m_Z)_{\overline{MS}}} S_Z \qquad 72b.$$

$$\frac{\Pi_{WW}^{new}(0)}{m_W^2} - \frac{\Pi_{ZZ}^{new}(0)}{m_Z^2} = \alpha(m_Z)_{\overline{MS}} T, \qquad 72c.$$

where the superscript "new" means only heavy new particle loops are included, $|_{\overline{MS}}$ means a modified minimal subtraction is made (including ln m_F/μ terms), and $\alpha(m_Z)$ has been factored out. Holdom (94) introduced the notation $U = S_W - S_Z$ such that $S = S_Z$, T, and U can be redefined as the independent parameters. U and T primarily contain isospin-violating corrections. In general $U \ll T$; so to a good approximation one can assume $U = 0$ and take $S_W \simeq S_Z \simeq S$ to be the isospin-conserving parameter originally introduced by Peskin & Takeuchi (11). Here I will use S_W, S_Z, and T.

Even without new physics, S_W, S_Z, and T provide a convenient way to approximate large m_t and m_H loop effects. Normalizing the standard model radiative corrections at $m_t = 130$ GeV and $m_H = 100$ GeV, one finds that deviations from those values are given by (56, 95)

$$S_W \simeq \frac{1}{6\pi} \ln\left(\frac{m_H}{100 \text{ GeV}}\right) + \frac{2}{3\pi} \ln\left(\frac{m_t}{130 \text{ GeV}}\right) \qquad \text{73a.}$$

$$S_Z \simeq \frac{1}{6\pi} \ln\left(\frac{m_H}{100 \text{ GeV}}\right) - \frac{1}{3\pi} \ln\left(\frac{m_t}{130 \text{ GeV}}\right) \qquad \text{73b.}$$

$$T \simeq \frac{3}{16\pi s^2}\left(\frac{m_t^2 - (130 \text{ GeV})^2}{m_W^2}\right) - \frac{3}{8\pi c^2} \ln\left(\frac{m_H}{100 \text{ GeV}}\right). \qquad \text{73c.}$$

Note that $U = 1/\pi \ln(m_t/130 \text{ GeV}) \ll T$ for large m_t.

S_W, S_Z, and T are measured as follows. Employing α, G_μ, and m_Z as input and assuming $m_t = 130$ GeV, $m_H = 100$ GeV completely specifies the standard model's predictions modulo new physics or deviations in m_t or m_H. For example, $\sin^2\theta_W(m_Z)_{\overline{\text{MS}}}$ is obtained from (56)

$$\sin^2 2\theta_W(m_Z)_{\overline{\text{MS}}} = \frac{4\pi\alpha}{\sqrt{2G_\mu m_Z^2(1 - \Delta\hat{r})}} \qquad \text{74.}$$

where

$$\Delta\hat{r} = 0.0632 + \frac{\alpha(m_Z)_{\overline{\text{MS}}}}{4s^2c^2} S_Z - \alpha(m_Z)_{\overline{\text{MS}}} T \qquad \text{75.}$$

or

$$\sin^2\theta_W(m_Z)_{\overline{\text{MS}}} = 0.2326 + 0.00365 S_Z - 0.00261 T. \qquad \text{76.}$$

A deviation in $\sin^2\theta_W(m_Z)_{\overline{\text{MS}}}$ from 0.2326 as measured, say, in Z decay asymmetries could signal a nonvanishing S_Z or T. Similarly, the W mass is predicted to be

$$m_W = 80.14 + 0.45T - 0.63 S_Z + 0.34 S_W \text{ GeV.} \qquad \text{77.}$$

It is the only observable that depends on $U = S_W - S_Z$ in this approach. The dependence of all other observables on S_Z and T is found by replacing

$\sin^2\theta_{\rm W}(m_Z)_{\overline{\rm MS}}$ by Equation 76 and multiplying all neutral current G_μ factors by $\rho(0)^{\rm new}$, where

$$\rho(0)^{\rm new} = 1 + \alpha(m_Z)_{\overline{\rm MS}}T = 1 + 0.00782T. \qquad 78.$$

That procedure leads to the $S_{\rm W}$, S_Z, and T dependences given by Marciano & Rosner (56) and illustrated in Table 8. The results exhibit several interesting features. Existing measurements are rather consistent with $S \simeq T \simeq 0$. Several experiments are already sensitive to $O(1)$ S and T effects and the projected future sensitivity may go to ± 0.1.

Some measurements are particularly sensitive to T. Those provide good constraints on $m_{\rm t}$ (see Equation 73). Taking $S = 0$ in Table 8, one finds $T \simeq -0.09 \pm 0.23$, the value that is roughly expected since $T = 0$ corresponds to $m_{\rm t} = 130$ GeV (the global average). Taking $T = 0$ and setting $S_{\rm W} = S_Z = S$ leads to

$$S \simeq -0.10 \pm 0.47, \qquad 79.$$

which is also consistent with no new physics. (For arbitrary T, $S \simeq -0.10 + 1.64T \pm 0.47$.) Note that atomic parity violation experiments with heavy atoms such as Cs have almost no T dependence (56); so, they give a direct measurement of S_Z. There is at present a small deviation in S_Z from zero coming from $Q_{\rm W}({\rm Cs})^{\rm exp}$. It will be interesting to follow an ongoing measurement of $Q_{\rm W}({\rm Cs})$ as it lowers the error on S_Z to $\pm 0.4 \pm$ atomic theory.

One can measure $S_{\rm W}$ directly by comparing $m_{\rm W}$ with $\sin^2\theta_{\rm W}(m_Z)_{\overline{\rm MS}}$ obtained from Z decay asymmetries or $\sigma(\nu_\mu e)/\sigma(\bar\nu_\mu e)$:

Table 8 Current constraints on S and T from various experiments and projected future sensitivities[a]

Experiment	Current constraint	Future sensitivity
$m_{\rm W} = 80.14 \pm 0.31$ GeV	$T - 1.4S_Z + 0.76S_{\rm W} = 0 \pm 0.75$	± 0.13
$Q_{\rm W}({\rm Cs}) = -71.04 \pm 1.58 \pm 0.88$	$S_Z + 0.006T = -2.7 \pm 2.0 \pm 1.1$	± 0.5
$\Gamma(Z \to {\rm all}) = 2487 \pm 9$ MeV	$T - 0.36S_Z = -0.11 \pm 0.34$	± 0.3
$\Gamma(Z \to \ell^+\ell^-) = 83.3 \pm 0.4$ MeV	$T - 0.23S_Z = -0.39 \pm 0.51$	± 0.45
$A(Z)_{\rm FB}$ (LEP)	$S_Z - 0.69T = -0.71 \pm 0.81$	± 0.3
$A(Z)_{\rm LR}$ (ALEPH)	$S_Z - 0.69T = -0.43 \pm 1.88$	± 0.1
$R_\nu \equiv \sigma(\nu_\mu{\rm N})_{\rm NC}/\sigma(\nu_\mu{\rm N})_{\rm CC}$	$T - 0.37S_Z = -0.37 \pm 0.62$	± 0.24
$R_{\bar\nu}$	$T - 0.02S_Z = 1.4 \pm 1.3$	± 0.65
$\sigma(\nu_\mu e)/\sigma(\bar\nu_\mu e)$	$S_Z - 0.69T = 0.01 \pm 2.7$	± 1.4
$\sigma(\nu_\mu e)/\sigma(\nu_e e) + \sigma(\bar\nu_\mu e)$	$T - 0.8S_Z$	± 0.3
Polarized e on carbon	$S_Z - 0.19T = -8.76 \pm 13.75$	± 0.63

[a] This analysis follows that of Marciano & Rosner (56), but uses $m_{\rm t} = 130$ GeV and $m_{\rm H} = 100$ GeV.

$$S_W \simeq 118 \left(2 \frac{m_W - 80.14\,\text{GeV}}{80.14\,\text{GeV}} + \frac{\sin^2 \theta_W (m_Z)_{\overline{\text{MS}}} - 0.2326}{0.2326} \right).$$ 80.

Using $m_W = 80.14 \pm 0.31$ GeV and $\sin^2 \theta_W (m_Z)_{\overline{\text{MS}}} = 0.2305 \pm 0.0027$ obtained from averaging Z decay asymmetries and $\overset{(-)}{\nu}_\mu e$ scattering gives

$$S_W = -1.08 \pm 0.91 \pm 1.37.$$ 81.

(Again, I caution the reader that the forward-backward asymmetry measurements at LEP may not have been completely corrected for QED effects. Since Equation 81 is dominated by those measurements, it is still very tentative.) Averaging that result with the atomic parity violation constraint in Table 8 gives

$$S = -1.6 \pm 1.3 \quad \text{(independent of } T\text{).}$$ 82.

Since that constraint is independent of assumptions about T, it represents a better characterization of the state of S. It is interesting that S appears to be negative and somewhat removed from 0. Of course, the error is too large to get excited about a possible new physics signal in S. If S stays negative, it could be an indication of heavy Z' bosons (56).

T parametrizes isospin-breaking loop effects. The data indicate nothing present in T other than a heavy top quark. Why is S, the isospin-conserving correction, interesting? It has been shown to provide a measure of new high-energy spectroscopy (11). For example, each heavy fermion degenerate doublet contributes $1/6\pi$ to S (79, 96). So, a heavy degenerate fourth generation is expected to give

$$S \simeq \frac{4}{6\pi} \simeq 0.2 \quad \text{(heavy fourth generation).}$$ 83.

More interesting are technicolor models, which can have many new doublets. Indeed, those models give (11, 97)

$$S \simeq (0.05\text{--}0.10) N_T N_D + 0.12 \quad \text{(technicolor),}$$ 84.

where the range 0.05–0.10 represents a rough estimate of the uncertainty (97), N_T is the number of technicolors, and N_D is the number of technidoublets. For example a generic one-generation model with $N_D = 4$ and $N_T = 4$ is expected to give

$$S \simeq 0.9\text{--}1.7 \quad \text{(one generation, } N_T = 4 \text{ technicolor).}$$ 85.

That prediction is already contradicted by the constraints in either Equation 79 or 82. Such a contradiction poses a serious problem for technicolor advocates. Even models such as the color sextet symmetry-breaking scenario (98) give $S \simeq 0.5 \sim 0.8$, which is contradicted by the above constraints on S. Note that most heavy fermion models seem to suggest a positive S whereas experiment favors a negative value.

In the future, individual S measurements should reach a sensitivity of about $\pm 0.2 \sim \pm 0.3$ (independent of T). If a positive deviation is not found at that level, it will not bode well for technicolor models. Confirmation of a negative value will likely stimulate new ideas. It should be noted that supersymmetry models and minimal $t\bar{t}$ condensation scenarios tend to give $S \simeq 0$, but a thorough study has not been carried out.

7.2 Grand Unified Theories

The $SU(3)_c \times SU(2)_L \times U(1)_Y$ gauge couplings g_i, $i = 3, 2, 1$ are now known with high precision. Employing the $\overline{MS}$ definition $\alpha_i(\mu)_{\overline{MS}} = g_i^2(\mu)_{\overline{MS}}/4\pi$, one finds

$$\alpha_3(m_Z)_{\overline{MS}} = 0.106 \pm 0.006$$

$$\alpha_2(m_Z)_{\overline{MS}} = \alpha(m_Z)_{\overline{MS}}/\sin^2\theta_W(m_Z)_{\overline{MS}} = 0.03364 \pm 0.00009$$

$$\alpha_1(m_Z)_{\overline{MS}} = 5\alpha(m_Z)_{\overline{MS}}/3\cos^2\theta_W(m_Z)_{\overline{MS}} = 0.01699 \pm 0.00003$$

$$\alpha(m_Z)_{\overline{MS}} = 1/127.8 \pm 0.2$$

$$\sin^2\theta_W(m_Z)_{\overline{MS}} = 0.2326 \pm 0.0005. \qquad 86.$$

Using those initial values and evolving the $\alpha_i(\mu)$ to higher energies, one finds that they fail to meet at a point (see Figure 1). That implies either that new physics at some threshold modifies the evolutions so they do eventually unify, or that we must give up the beautiful idea of grand unification of the standard model in some compact simple group such as $SU(5)$, $SO(10)$, E_6, etc. Another way of illustrating the problem is to assume unification at scale m_X with no new physics between m_Z (or m_t) and m_X. Then using $\alpha_3(m_Z)_{\overline{MS}}$ and $\alpha(m_Z)_{\overline{MS}}$ as input, one can predict m_X and $\sin^2\theta_W(m_Z)_{\overline{MS}}$. One finds (99, 100)

$$m_X \simeq 2^{+2}_{-1} \times 10^{14} \text{ GeV}$$

$$\sin^2\theta_W(m_Z)_{\overline{MS}} \simeq 0.215 \pm 0.003. \qquad \text{(great desert GUTS).} \qquad 87.$$

Those predictions hold for many grand unified theories (GUTS) with a "great desert" between the standard model's spectrum and the unification mass m_X. An example of such a model is the minimal $SU(5)$ theory of Georgi & Glashow (101). The prediction for $\sin^2\theta_W(m_Z)_{\overline{MS}}$ disagrees with experiment by almost six standard deviations. Even worse, in such theories the proton can decay. The partial lifetime for $p \rightarrow e^+\pi^0$ is predicted (100) to be about

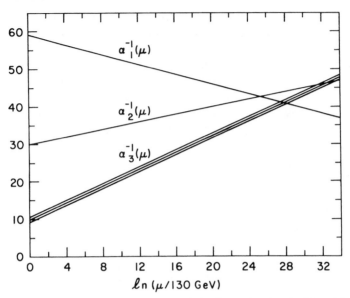

Figure 1 Evolution of the standard model gauge couplings using the $\alpha_i(m_Z)$ from Equation 86 as input and assuming a "great desert" for $\mu > m_Z$.

$$\tau(p \to e^+\pi^0) \simeq 4 \times 10^{29 \pm 0.7} (m_X/2 \times 10^{14} \text{ GeV})^4 \text{ yr}. \qquad 88.$$

That lifetime is well below the IMB bound $\tau(p \to e^+\pi^0) \gtrsim 5.5 \times 10^{32}$ yr (M. Goldhaber, private communication).

Given that minimal SU(5) and most other GUTS with "great desert" scenarios are ruled out by proton decay experiments and $\sin^2\theta_W(m_Z)_{\overline{\text{MS}}}$ measurements, should they be abandoned? Probably not, since new physics thresholds can easily rectify both problems. For example, if supersymmetry at mass scale m_{SUSY} is appended to the SU(5) model, it changes the evolution of couplings between m_{SUSY} and m_X. Assuming two light Higgs doublets (required in SUSY models), one finds

$$m_X \simeq m_Z \exp\left\{\frac{\pi}{2}\left[\frac{1}{\alpha_2(m_Z)_{\overline{\text{MS}}}} - \frac{1}{\alpha_3(m_Z)_{\overline{\text{MS}}}}\right]\right\} \qquad 89.$$

independent of m_{SUSY} (102–104). Using the α_3 and α_2 couplings in Equation 86 then gives

$$m_X \simeq 6.4^{+8}_{-4} \times 10^{15} \text{ GeV}$$
$$\tau(p \to e^+\pi^0) \simeq 1.4 \times 10^{35 \pm 0.7 \pm 1.5} \text{ yr} \qquad \text{[SUSY SU(5)]}. \qquad 90.$$

The predicted proton lifetime is safely beyond the experimental bound. In the case of the weak mixing angle, the prediction depends on m_{SUSY}:

$$\sin^2 \theta_{\text{W}}(m_Z)_{\overline{\text{MS}}} = 0.237 \pm 0.004 - 1.25 \times 10^{-3} \ln (m_{\text{SUSY}}/m_Z). \qquad 91.$$

For $m_{\text{SUSY}} \simeq 1$ TeV, that prediction is in beautiful agreement with experiment. In fact, the finding in Equation 91 is the only present evidence in support of supersymmetry (see Figure 2). Unfortunately, the underlying uncertainty in the spectrum of SUSY masses that I have parametrized by one mass scale, m_{SUSY}, makes a precise determination of the required supersymmetry masses impossible. It will be interesting to see if the agreement between Equation 91 and experiment is an accident or a harbinger of supersymmetry at SSC energies.

8. CONCLUSION AND OUTLOOK

The standard model has withstood intense experimental scrutiny. It has been tested at the level of its electroweak radiative corrections both in charged and neutral current processes. To date, there are no significant deviations that can be interpreted as a strong hint of new physics. Those measurements have, however, proved to be a powerful probe of the top

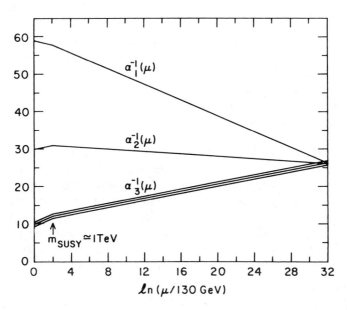

Figure 2 Evolution of the gauge couplings assuming a supersymmetry threshold at $m_{\text{SUSY}} \simeq 1$ TeV and two Higgs doublets.

quark mass through its contributions to quantum loops, which exhibit an $\alpha m_t^2/m_W^2$ behavior. Current constraints suggest $m_t \simeq 130 \pm 40$ GeV, a value in excellent accord with the direct CDF bound $m_t > 89$ GeV and with the requirements from CP violation and B_d^0-$\bar{B}_d^0$ oscillations, which also point to a large m_t. The advent of a larger than originally expected m_t has motivated speculations that $t\bar{t}$ condensation may provide a dynamical alternative to the usual Higgs mechanism. Such developments suggest that the top quark is distinct from other quarks. It may be the key to understanding electroweak symmetry breaking and mass generation. Precision measurements will continue to constrain m_t, but what is really needed is the discovery of the top quark and a direct measurement of its mass. Perhaps the experimenters at Fermilab can accommodate that request during the next few years.

Higgs scalar effects are too small to be seen in quantum loops. Finding that particle or its dynamical alternative is likely to require new high-energy e^+e^- and pp colliders. Fortunately, it now appears that clear discovery signatures are observable at future facilities for m_H values all the way up to ~ 800 GeV.

In pursuing the elusive Higgs particle, it is hoped that new physics will be found. There are many speculations as to what the new physics might be, but there is no strong positive evidence for any of the conjectures. Generic technicolor models are well motivated, but it is difficult to construct a complete theory. In addition, recent constraints on the isospin-conserving loop correction, S (see Equation 82), further diminish their viability. Similarly, there is no direct evidence for extra gauge bosons, a fourth generation, supersymmetry, etc. However, supersymmetric grand unified theories do predict $\sin^2 \theta_W (m_Z)_{\overline{MS}}$ near its experimental value and do not violate proton lifetime bounds. That could be our first indirect hint that a wealth of supersymmetry spectroscopy is waiting at SSC energies.

To make further headway seems to require new high-energy facilities such as the SSC and some lucky breaks in well-motivated "long shot" experiments studying, for example, neutrino oscillations, proton decay, electric dipole moments, etc. The SSC offers the opportunity of thoroughly exploring Nature's short-distance domain all the way down to about 3×10^{-18} cm. It should provide the answer or at least important clues regarding the source of mass generation and electroweak symmetry breaking. "Long shot" experiments are not as clearly motivated, but they can have huge payoffs if a positive effect is uncovered.

Great progress has been made in deciphering Nature's laws. The success of the standard model is a testament to that fact. However, many puzzles still remain. Why is the $SU(2)_L$ symmetry left-handed? What is the fundamental underlying cause of symmetry breaking and *CP* violation? Why are there three generations? Those deep questions serve to whet our intellectual appetites and stimulate our scientific imaginations. They make our long inner space odyssey a fantastic adventure.

Literature Cited

1. Weinberg, S., *Phys. Rev. Lett.* 19: 1264 (1967); Salam, A., In *Elementary Particle Theory*, ed. N. Svartholm. Stockholm: Almquist & Wiksells (1968), p. 367
2. Fritzsch, H., Gell-Mann, M., Leutwyler, H., *Phys. Lett.* B47: 365 (1973); Weinberg, S., *Phys. Rev. Lett.* 31: 494 (1973); Gross, D., Wilczek, F., *Phys. Rev.* D8: 3633 (1973)
3. Hasert, F. J., *Phys. Lett.* B46: 121 (1973)
4. Review of Particle Properties, *Phys. Lett.* B239: 1 (1990)
5. Sliwa, K. (CDF Collaboration), In *Proc. XXV Recontre de Moriond*, ed. J. Tran Thanh Van. Singapore: World Scientific (1990)
6. Decamp, D., et al. (ALEPH collaboration), *Phys. Lett.* B246: 306 (1990)
7. Dicus, D., Mathur, V., *Phys. Rev.* D7: 3111 (1973); Lee, B. W., Quigg, C., Thacker, H., *Phys. Rev.* D16: 1519 (1977)
8. Callaway, D. J. E., *Phys. Rep.* 167: 241 (1988); Dashen, R., Neuberger, H., *Phys. Rev. Lett.* 50: 1897 (1983)
9. Nambu, Y., Enrico, Fermi Inst. Rep. (1989), unpublished; Miransky, V., Tanabashi, M., Yamawaki, K., *Mod. Phys. Lett.* A4: 1043 (1989); Marciano, W., *Phys. Rev. Lett.* 62: 2793 (1989); Bardeen, W., Hill, C., Lindner, M., *Phys. Rev.* D41: 1647 (1990)
10. Marciano, W. J., Parsa, Z., *Annu. Rev. Nucl. Part. Sci.* 36: 171 (1986)
11. Peskin, M. E., Takeuchi, T., *Phys. Rev. Lett.* 65: 964 (1990)
12. Kobayashi, M., Maskawa, K., *Prog. Theor. Phys.* 49: 652 (1973)
13. Roos, M., Sirlin, A., *Nucl. Phys.* B29: 296 (1971)
14. Bardeen, W. A., et al., *Phys. Rev.* D18: 3998 (1978)
15. Fanchiotti, S., Sirlin, A., *Phys. Rev.* D41: 319 (1990)
16. Hardy, J. C., et al., *Nucl. Phys.* A509: 429 (1990)
17. Sirlin, A., *Phys. Rev.* 164: 1767 (1967)
18. Marciano, W. J., Sirlin, A., *Phys. Rev. Lett.* 56: 22 (1986)
19. Jaus, W., Rasche, G., *Phys. Rev.* D41: 166 (1990)
20. Rasche, G., Woolcock, W. S., *Mod. Phys. Lett.* A5: 1273 (1990); Wilkinson, D. H., TRIUMF preprint TRI-PP-90-44 (1990)
21. Wilkinson, D. H., *Nucl. Phys.* A377: 424 (1982)
22. Mampe, W., et al., *Phys. Rev. Lett.* 63: 593 (1989); Byrne, J., et al., *Phys. Rev. Lett.* 65: 289 (1990)
23. Bopp, P., et al., *Phys. Rev. Lett.* 56: 919 (1986)
24. McFarlane, W. K., et al., *Phys. Rev.* D32: 547 (1985)
25. Leutwyler, H., Roos, M., *Z. Phys.* C25: 91 (1984)
26. Donoghue, J., Holstein, B., Klimt, S., *Phys. Rev.* D35: 934 (1987)
27. Abramowicz, H., et al., *Z. Phys.* C15: 19 (1982); Schaevitz, M., Nevis Lab. Rep. 1415 (1989), unpublished
28. Reay, N. W., et al., Fermilab Proposal 803. Batavia, Ill: Fermilab
29. Anjos, J. C., et al., *Phys. Rev. Lett.* 62: 1587 (1989)
30. Fulton, R., et al., *Phys. Rev. Lett.* 64: 16 (1990); Albrecht, H., et al., DESY preprint 89/152 (1989), unpublished
31. Sirlin, A., *Nucl. Phys.* B71: 29 (1974); *Rev. Mod. Phys.* 50: 573 (1978)
32. Marciano, W. J., Sirlin, A., *Phys. Rev.* D35: 1672 (1987)
33. Barbieri, R., et al., *Phys. Lett.* B156: 348 (1985)
34. Marciano, W. J., *Nucl. Phys.* B(Proc. Suppl.)11: 5 (1989)
35. Goldberger, M. L., Treiman, S. B., *Phys. Rev.* 111: 354 (1958)
36. Dominguez, C. A., *Riv. Nuovo Cimento* 8: 1 (1985)
37. Sirlin, A., *Phys. Rev.* D5: 436 (1972)
38. Kinoshita, T., *Phys. Rev. Lett.* 2: 477 (1959); Smorodinskii, Ya. A., Shih-K'e, Hu, *Sov. Phys. JETP* 14: 438 (1962)
39. Holstein, B. R., *Phys. Lett.* B244: 83 (1990)
40. Marciano, W. J., In *Rare Decay Symposium*, ed. D. Bryman, et al. Singapore: World Scientific (1989)
41. Arndt, R. A., et al., *Phys. Rev. Lett.*

65: 157 (1990); Bergervoet, J. R., et al. *Phys. Rev.* C41: 1435 (1990)

42. Dumbrajs, O., et al., *Nucl. Phys.* B216: 277 (1983)
43. Kitazawa, Y., *Phys. Lett.* B151: 165 (1985)
44. Atherton, H. W., *Phys. Lett.* B158: 81 (1985)
45. Truong, T. N., *Phys. Rev.* D30: 1509 (1984); Gilman, F. J., *Phys. Rev.* D35: 3541 (1987); Gilman, F. J., Rhie, S.-H., *Phys. Rev.* D31: 1066 (1985)
46. Braaten, E., *Phys. Rev. Lett.* 60: 1606 (1988); *Phys. Rev.* D39: 1458 (1989)
47. Narison, S., Pich, A., *Phys. Lett.* B211: 183 (1988)
48. Marciano, W. J., Sirlin, A., *Phys. Rev. Lett.* 61: 1815 (1988)
49. Perl, M. L., SLAC-PUB-4632 (1988), unpublished
50. Gorishny, S. G., Kataev, A. L., Larin, S. A., *Phys. Lett.* B212: 238 (1988); B259: 144 (1991)
51. Surguladze, L. R., Samuel, M. A., *Phys. Rev. Lett.* 66: 560 (1991)
52. Marciano, W. J., *Phys. Rev.* D29: 580 (1984)
53. Bollini, G. B., Giambiagi, J. J., Sirlin, A., *Nuovo Cimento* A16: 423 (1973); Marciano, W. J., *Nucl. Phys.* B84: 132 (1975)
54. Marciano, W. J., *Phys. Rev.* D20: 274 (1979)
55. Marciano, W. J., Sirlin, A., *Phys. Rev. Lett.* 46: 163 (1981); See Ref. 100
56. Marciano, W. J., Rosner, J. L., *Phys. Rev. Lett.* 65: 2963 (1990)
57. Marciano, W. J., Sirlin, A., *Phys. Rev.* D22: 2695 (1980); *Nucl. Phys.* B189: 441 (1981)
58. Amaldi, U., et al., *Phys. Rev.* D36: 1385 (1987)
59. Langacker, P., In *Proc. 1990 PASCOS Conf.*
60. Kaplan, D., Manohar, A., *Nucl. Phys.* B310: 527 (1988)
61. Ahrens, L. A., et al., *Phys. Rev.* D35: 785 (1987)
62. Sarantakos, S., Sirlin, A., Marciano, W. J., *Nucl. Phys.* B217: 84 (1983)
63. Degrassi, G., Sirlin, A., Marciano, W. J., *Phys. Rev.* D39: 287 (1989)
64. White, D. H. (spokesman), LAMPF Proposal No. LA-11300-P
65. Godfrey, S., Marciano, W. J., In *Proc. 1987 BNL Neutrino Workshop*, ed. M. Murtagh. Brookhaven Natl. Lab., NY (1987)
66. Marciano, W. J., *Phys. Rev.* D36: 2859 (1987); Rosen, S. P., Kayser, B., *Phys. Rev.* D23: 669 (1981)
67. Noecker, M. C., Masterson, B. P., Wieman, C. E., *Phys. Rev. Lett.* 61: 310

(1988)
68. Bouchiat, M. A., In *Proc. 12th Int. Atomic Physics Conf.* (1990)
69. Commins, E. D., Bucksbaum, P. H., *Annu. Rev. Nucl. Part. Sci.* 30: 1 (1980)
70. Marciano, W. J., Sirlin, A., *Phys. Rev.* D29: 75 (1984); D27: 552 (1983)
71. Sandars, P. G. H., *J. Phys.* B23: L655 (1990)
72. Blundell, S. A., Johnson, W. R., Sapirstein, J., *Phys. Rev. Lett.* 65: 1411 (1990); Dzuba, V., Flambaum, V., Sushkov, O., *Phys. Lett.* A141: 147 (1989)
73. Feinberg, G., *Phys. Rev.* D12: 3575 (1975)
74. Souder, P. A., et al., *Phys. Rev. Lett.* 65: 694 (1990)
75. Beck, D. H., *Phys. Rev.* D39: 3248 (1989)
76. Burkhardt, H., Steinberger, J., *Annu. Rev. Nucl. Part. Sci.* 41: 55–96 (1991)
77. Sirlin, A., *Phys. Rev.* D22: 971 (1980)
78. Sirlin, A., *Phys. Lett.* B232: 123 (1989)
79. Veltman, M., *Nucl. Phys.* B123: 89 (1977)
80. Degrassi, G., Fanchiotti, S., Sirlin, A., *Nucl. Phys.* B351: 49 (1991)
81. Borrelli, A., Maiani, L., Sisto, R., *Phys. Lett.* B244: 117 (1990)
82. Augustin, J., Orsay preprint LAL90-87 (1990)
83. Consoli, M., Hollik, W., In *Z-Physics at LEPI*, CERN 89-08. Geneva: CERN (1989)
84. Carter, A., Pagels, H., *Phys. Rev. Lett.* 43: 1845 (1979)
85. Marciano, W. J., *Phys. Rev.* D41: 219 (1990)
86. Bardeen, W., Clark, T., Love, S., *Phys. Lett.* B237: 235 (1990)
87. Decamp, D., et al., *Phys. Lett.* 246B: 306 (1990)
88. Maiani, L., Parisi, G., Petronzio, R., *Nucl. Phys.* B136: 115 (1978); Marciano, W. J., Valencia, G., Willenbrock, S., *Phys. Rev.* D40: 1725 (1989)
89. Vainshtein, A., et al., *Sov. J. Nucl. Phys.* 30: 711 (1979)
90. Barter, C., et al., In *Proc. Snowmass 88*, ed. S. Jensen. Singapore: World Scientific (1989), p. 98
91. Kleiss, R., Kunszt, Z., Stirling, W. J., *Phys. Lett.* B253: 269 (1991)
92. Han, T., Willenbrock, S., Fermilab-Pub-91/70-T (1991)
93. Marciano, W. J., Paige, F. E., *Phys. Rev. Lett.* 66: 2433 (1991); Gunion, J., UC Davis preprint (1991)
94. Holdom, B., Fermilab preprint 90/263-T (1990)
95. Kennedy, D. C., Langacker, P., *Phys. Rev. Lett.* 65: 2967 (1990)

96. Bertolini, S., Sirlin, A., *Nucl. Phys.* B248: 589 (1984)
97. Golden, M., Randall, L., Fermilab Rep. No. 90/83-T (1990); Holdom, B., Terning, J., *Phys. Lett.* B247: 88 (1990); Dobado, A., Espriu, D., Herrero, M., CERN Rep. No. CERN-TH.5785/90 (1990); Roiesnel, C., Truong, T. N., *Phys. Lett.* B253: 439 (1991)
98. Marciano, W. J., *Phys. Rev.* D21: 2425 (1980)
99. Georgi, H., Quinn, H., Weinberg, S., *Phys. Rev. Lett.* 33: 451 (1974)
100. Marciano, W. J., Sirlin, A., In *The Second Workshop on Grand Unification*, ed. J. Leveille, L. Sulak, D. Unger. Boston: Birkhauser (1981), p. 151
101. Georgi, H., Glashow, S., *Phys. Rev. Lett.* 32: 438 (1974)
102. Dimopoulos, S., Raby, S., Wilczek, F., *Phys. Rev.* D24: 1681 (1981)
103. Marciano, W. J., In *Snowmass Proc. 1986*, ed. R. Donaldson, J. Marx, p. 726
104. Marciano, W. J., Senjanovic, G., *Phys. Rev.* D25: 3092 (1982)

Annu. Rev. Nucl. Part. Sci. 1991. 41: 511–45

CHARM BARYONS: THEORY AND EXPERIMENT

J. G. Körner

Institut für Physik, Johannes Gutenberg-Universität, Staudinger Weg 7, Postfach 3980, D-6500 Mainz, Germany

H. W. Siebert

Physikalisches Institut, Universität Heidelberg, Philosophenweg 12, D-6900 Heidelberg, Germany

KEY WORDS: charm baryon decays, lifetimes, semileptonic decay, nonleptonic decay, heavy flavor symmetry, angular decay distributions

CONTENTS

1. INTRODUCTION AND OUTLINE

A new era in particle physics began with the discovery in 1974 of the J/ψ, a narrow meson resonance of mass 3.1 GeV (1, 2). In the subsequent years

511

0163–8998/91/1201–0511$02.00

this state was successfully interpreted as a bound state composed of the heavy charm quark, with mass $m_c \approx 1.5$ GeV and charge 2/3, and its antiparticle. Soon after the discovery of the so-called hidden charm state J/ψ, further so-called open charm hadron species composed of a charm quark and light (u:up, d:down, s:strange) quarks and antiquarks were found. The first candidate charm baryon states were detected in 1975 in neutrino interactions (3), soon followed by the identification of charm meson states at the SPEAR e^+e^- ring in 1976 (4, 5). In retrospect, charm hadrons had probably made their appearance several years earlier in cosmic ray interactions (6).

The discovery in 1977 of the Υ family of mesons was the first indication of the existence of a fifth quark, the bottom quark b, with mass $m_b \approx 5$ GeV and a charge $-1/3$ (7). Again, open bottom meson states composed of a heavy bottom quark and a light antiquark were identified somewhat later (8, 9), but bottom baryons have not yet been seen. Finally, a third species of heavy flavor quarks is anticipated but not yet identified in the form of the top quark with a mass $m_t \approx 140$ GeV and charge 2/3.

Recent measurements and theoretical calculations have substantially enhanced our understanding of charm meson states, their spectroscopy and decays. Experimental results on charm baryons and their decays are beginning to be good enough to compare what has been learned in the charm meson sector to the charm baryon sector. In addition, several experiments are being proposed to study charm baryons, their masses, lifetimes, and weak decays. It is therefore timely to review what we know now about these states and what we can expect to learn from future experiments. How can we extrapolate theoretical calculations from mesons to baryons, and how do they translate to heavy flavor baryons with bottom quantum numbers?

The heavy charm and bottom quarks and the heavy hadrons composed of them are quite distinct in their properties from the light flavored hadrons composed of u, d, and s quarks. The large mass of the heavy flavored quarks introduces a mass scale greatly exceeding that of the confinement scale $\Lambda \approx 300$ MeV that governs the physics of the light hadrons.

Although heavy hadrons with different heavy flavors weigh in quite differently, they are in some sense similar to one another once the appropriate mass scale is chosen and possible anomalous dimension factors are taken into account. Recently this notion has been described in greater detail and is called the Heavy Quark Effective Theory (HQET) (10–16), patterned after the Bloch-Nordsieck model of soft photon radiation from electrons (16, 17).

In the heavy quark limit, the interaction of the heavy quark with the light quark system becomes spin and flavor independent (at equal velocity).

This independence generates an additional spin and flavor symmetry in current-induced transitions among heavy hadrons (including a normalization condition) not manifest in the original QCD Lagrangian. With this new theoretical ingredient there is now hope that one can turn from the physics at the quark and gluon level (where theory is formulated) to the physics at the hadron level (where one can compare with experiment) in a reasonably model-independent way. In this way one can hope to extract fundamental Standard Model parameters from transitions among heavy hadrons, such as the Kobayashi-Maskawa (KM) matrix elements, the CP-violating phase angle, and possibly Higgs boson couplings, as well as possible non–Standard Model physics.

Charm hadrons, being the lightest of the heavy hadrons, may not be the best candidates to test and apply the predictions of HQET formulated for infinitely heavy quarks. But charm hadrons and their decays will certainly be the best studied experimentally in the next few years, at least as far as baryons are concerned. They are also an ideal laboratory to study the influence of preasymptotic $1/m_Q$ effects to the heavy quark limit $m_Q \rightarrow \infty$, where m_Q is the mass of the heavy quark. And last, but not least, the quality of the b $\rightarrow$ c physics to be extracted from bottom baryon to charm baryon transitions depends on a detailed knowledge of the decay properties of charm baryons, which are the main subject of this review.

We begin in Section 2 with a discussion of the spectrum of the lowest-lying charm baryons. We review the experimental status of the masses of charm baryons and briefly comment on theoretical attempts to understand their spectroscopy. In Section 3 we discuss lifetime measurements and lifetime hierarchies suggested by the interplay of various theoretical mechanisms contributing to the decays. In Section 4 we discuss semileptonic decays of charm baryons. Experimental information on semileptonic decays is quite scant at present, but a wealth of interesting physics lies ahead of us. Section 5 treats exclusive nonleptonic charm baryon decays, where there are more data to be compared to theoretical modeling. Section 6 contains our summary and an outlook on future charm baryon experiments.

Space limitations preclude an exhaustive treatment of charm baryon physics and the many fascinating aspects of heavy hadron physics. Thus we focus on the properties of charm baryons as revealed in their decays. We refer the reader to earlier reviews on heavy hadron physics in this series (18–20) and elsewhere (21–23).

We conclude the introduction by providing a brief compendium on charm baryon production experiments. Charm baryons have been studied since 1975 in about 25 different experiments performed at e^+e^- colliders (SPEAR, DORIS, CESR), at the CERN-ISR pp-collider, and with fixed targets using neutrino, photon, pion, kaon, proton, neutron, and hyperon beams. The large variety of production processes (and energies) in these experiments yields a wealth of experimental data on the production of charm baryons, but these data are not the topic of this review. Neither do we give a complete account of the fascinating story of charm baryon experiments; instead, we limit ourselves to those experiments whose results are important to our discussion of charm baryons. To facilitate this discussion, in Table 1 we list experiments by their code names and give their principal characteristics. A complete list of all experiments concerned can be found, for instance, in the compilations of the Particle Data Group (24).

The Λ_c was first observed in 1975 in a bubble chamber exposed to a neutrino beam (3) and was seen again in 1976 in a photoproduction

Table 1 Experiments on charmed baryons mentioned in the text, identified by their code-name

Collider experiments:		
R 415	$pp, \sqrt{s} = 62$ GeV	CERN
Mark II	$e^+e^-, \sqrt{s} = 4.5 - 6.8$ GeV	SLAC
ARGUS	$e^+e^-, \sqrt{s} = 10.5$ GeV	DESY
CLEO	$e^+e^-, \sqrt{s} = 10.5 - 10.9$ GeV	Cornell
Fixed-target experiments:		
E 687	tagged γ, 100 - 260 GeV	FNAL
NA 14/2	tagged γ, 50 - 200 GeV	CERN
E 687	γ, 50 - 300 GeV	FNAL
NA 32	π^-, 230 GeV	CERN
E 400	$n, E_{\text{peak}} = 600$ GeV	FNAL
BIS - 2	n, 40 - 70 GeV	Serpukhov
WA 62	Σ^-, 135 GeV	CERN
E 769	π, K, p, 250 GeV	FNAL
E 791	π^-, 500, 800 GeV	FNAL
WA 89	Σ^-, 340 GeV	CERN
E 781 [a]	Σ^-, 600 GeV	FNAL

[a] in preparation

experiment (25). Recent experiments have reached statistics of several hundred events. Quite often, however, the Λ_c appears only as a peak in a mass spectrum with a small signal-to-background ratio. Very clean Λ_c samples have been obtained by applying so-called lifetime cuts, that is, by requiring a minimum distance between the charm production and decay vertices. This method was first used in bubble chambers and emulsions, and is now also widely applied using high resolution counters (microstrip devices and CCDs).

The Σ_c was first observed in 1975 in the above-mentioned bubble chamber experiment (3); the observed event had the decay chain $\Sigma_c^{++} \to \Lambda_c^+ \pi^+$, $\Lambda_c^+ \to \Lambda \pi^+ \pi^+ \pi^-$. The decay $\Sigma_c \to \Lambda_c \pi$ is easily identifiable because of the small mass difference (168 MeV), similar to the decay D* $\to$ Dπ. As in the search for D decay, attempts to further clean up Λ_c signals by using cuts on this mass difference have had little success, because only a small fraction (≈ 10–25%) of all Λ_c's come from the decay of Σ_c's.

The Ξ_c^+ was first observed in 1983 in a hyperon beam experiment (26). Its isospin partner, the Ξ_c^0, has been observed more recently in e^+e^- collision and hadroproduction experiments (27–29). The Ω_c has so far been observed only in the above-mentioned hyperon beam experiment (30).

Before leaving production and turning to the spectrum and decay properties of charm baryons, we mention one remarkable theoretical prediction of the HQET concerning the ratio of charm baryon pair production in the e^+e^- interactions close to threshold. One finds $\sigma_{\Lambda_c \bar{\Lambda}_c} : \sigma_{\Sigma_c \bar{\Sigma}_c} : \sigma_{\Sigma_c \bar{\Sigma}_c^* + \Sigma_c^* \bar{\Sigma}_c} : \sigma_{\Sigma_c^* \bar{\Sigma}_c^*} = 27:1:16:10$, where Λ_c production has been related to the other pair production process using a factorizing spin piece in the charm baryon wave functions (31).

2. CLASSIFICATION OF STATES AND MASS MEASUREMENTS

The ground-state charm baryons are classified as usual as members of the SU(4) multiplets 20′ and 20. The $J^P = 1/2^+$ ground-state baryons (containing the ordinary $C = 0$ octet baryons) comprise the 20′ representation, and the $J^P = 3/2^+$ ground-state baryons (containing the ordinary $C = 0$ decuplet baryons) comprise the 20 representation. In Tables 2 and 3 we list the quantum number content and masses of the charm baryon members of the 20′ and 20 representation, where we use the same notation as the Particle Data Group (24). I and I_3 denote the isospin; S and C refer to the strangeness and charm quantum numbers.

Let us first trace the sources of the experimental mass values for the Λ_c^+, Ξ_c^+, Ξ_c^0, Σ_c^{++}, Σ_c^+, and Σ_c^0 baryons given in Table 2. The mass of the

Table 2 Charmed $1/2^+$ baryon states. $[ab]$ and $\{ab\}$ denote antisymmetric and symmetric flavour index combinations

Notation	Quark content	SU(3)	(I, I_3)	S	C	Mass (MeV)
Λ_c^+	$c[ud]$	3^*	$(0, 0)$	0	1	2285.0 ± 0.6
Ξ_c^+	$c[su]$	3^*	$(1/2, 1/2)$	-1	1	2466.2 ± 2.2
Ξ_c^0	$c[sd]$	3^*	$(1/2, -1/2)$	-1	1	2472.8 ± 1.7
Σ_c^{++}	cuu	6	$(1, 1)$	0	1	2453 ± 0.7
Σ_c^+	$c\{ud\}$	6	$(1, 0)$	0	1	2453 ± 3.0
Σ_c^0	cdd	6	$(1, -1)$	0	1	2452.5 ± 0.9
$\Xi_c^{+'}$	$c\{su\}$	6	$(1/2, 1/2)$	-1	1	2561
$\Xi_c^{0'}$	$c\{sd\}$	6	$(1/2, -1/2)$	-1	1	2561
Ω_c^0	css	6	$(0, 0)$	-2	1	2667
Ξ_{cc}^{++}	ccu	3	$(1/2, 1/2)$	0	2	3616
Ξ_{cc}^+	ccd	3	$(1/2, -1/2)$	0	2	3616
Ω_{cc}^+	ccs	3	$(0, 0)$	-1	2	3706

Λ_c is well known. The mean from nine measurements (32–40) is $m(\Lambda_c) = 2285.0 \pm 0.6$ MeV. The mass of the Σ_c is obtained by measuring the mass difference $m(\Sigma_c) - m(\Lambda_c)$ in $\Sigma_c \rightarrow \Lambda_c \pi$ decays. The Σ_c decays appear as a narrow peak in the distribution of the difference of effective masses, $m(\Lambda_c \pi) - m(\Lambda_c)$, where the Λ_c candidates are selected by cuts in the observed mass spectra. Figure 1 shows the $m(\Lambda_c \pi^\pm) - m(\Lambda_c)$ spectra observed by the ARGUS group (41). Peaks corresponding to Σ_c^{++} and Σ_c^0 decay are visible at around 168 MeV. Corresponding results for Σ_c^{++} and Σ_c^0 have been reported by the CLEO group (42) and for Σ_c^0 by the E691 group (43). An earlier result by the E400 group (44) agrees with the recent experiments on the Σ_c^{++} mass, but their value for the Σ_c^0 mass is 10 MeV larger (3 standard deviations) and is omitted in the calculation of the averages, which are $m(\Sigma_c^{++}) - m(\Lambda_c) = 168.0 \pm 0.3$, $m(\Sigma_c^0) - m(\Lambda_c) = 167.5 \pm 0.3$, and $m(\Sigma_c^0) - m(\Sigma_c^{++}) = -0.4 \pm 0.5$ MeV.

So far, no experiment has observed the decay $\Sigma_c^+ \rightarrow \Lambda_c \pi^0$, with the exception of one candidate seen ten years ago in an experiment in the BEBC bubble chamber (40). The measured mass difference was $m(\Sigma_c^+) - m(\Lambda_c) = 168 \pm 3$ MeV, in agreement with the more precise Σ_c^{++} and Σ_c^0 results.

Table 3 Charmed $3/2^+$ baryon states

Notation	Quark content	SU(3)	(I, I_3)	S	C	Mass (MeV)
Σ_c^{*++}	cuu	6	(1, 1)	0	1	2545
Σ_c^{*+}	cud	6	(1, 0)	0	1	2545
Σ_c^{*0}	cdd	6	(1, -1)	0	1	2545
Ξ^{*+}	cus	6	(1/2, 1/2)	-1	1	2666
Ξ^{*0}	cds	6	(1/2, -1/2)	-1	1	2666
Ω_c^{*0}	css	6	(0, 0)	-2	1	2778
Ξ_{cc}^{*++}	ccu	3	(1/2, 1/2)	0	2	3744
Ξ_{cc}^{*+}	ccd	3	(1/2, -1/2)	0	2	3744
Ω_{cc}^{*+}	ccs	3	(0, 0)	-1	2	3838
Ω_{ccc}^{++}	ccc	1	(0, 0)	0	3	4797

The Ξ_c masses are known from three precise measurements by the CLEO, ARGUS, and NA32 groups (27–29, 45, 46): $m(\Xi_c^+) = 2466.2 \pm 2.2$ MeV; $m(\Xi_c^0) = 2472.8 \pm 1.7$ MeV; $m(\Xi_c^0) - m(\Xi_c^+) = 6.3 \pm 2.3$ MeV.

The Ω_c^0 has been seen only by experiment WA62, where a cluster of three events was observed at a mass of 2740 ± 20 MeV (30).

The remaining mass entries in Tables 1 and 2 are taken from theoretical calculations (23, 47) based on a central two-body potential supplemented by the spin-spin interaction

$$H_{ss} = \sum_{i<j} \frac{16\pi\alpha_s}{9m_i m_j} \mathbf{s}_i \mathbf{s}_j \delta^3(\mathbf{r}_{ij}), \qquad\qquad 1.$$

where one has just added two-body potentials between quarks i and j with masses m_i (m_j) and spins s_i (s_j) separated by the distance r_{ij}. The spin-spin interaction term results from the Breit-Fermi reduction of the one-gluon-exchange contribution. Starting with the seminal work of De Rujula et al (48), many authors have emphasized the fact that the hyperfine splitting resulting from Equation 1 is crucial to the understanding of the mass-breaking pattern of both charm and charmless hadrons (19, 49). As long as the spin-spin interaction term is taken into account, a variety of models with differing degrees of sophistication will basically reproduce the pattern of charm baryon masses in Tables 1 and 2. Of the observed charm baryons, the Λ_c and the Ξ_c states are weakly decaying. According to theoretical

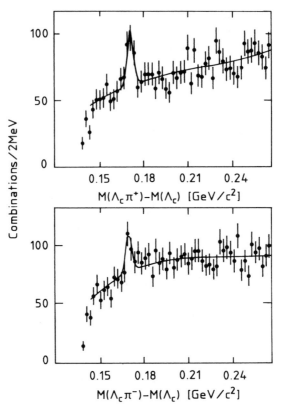

Figure 1   Distributions of the mass differences observed in the ARGUS experiment.

expectations, the unobserved Ω_c, Ξ_{cc}, Ω_{cc}, and Ω_{ccc} states are also anticipated to be weakly decaying.

Because of the spatial δ function in Equation 1, the matrix elements of
the spin-spin interaction term are proportional to the square of the baryon
wave function at the origin. The experimental hyperfine splittings thus
provide a reliable measure of the wave function at the origin of the ground-
state baryons, the value of which is needed in lifetime estimates (discussed
in Section 3).

3.   LIFETIMES

3.1   *Experimental Lifetime Measurements*

The Λ_c lifetime is determined from four recent experiments: E691 (34),
NA32 (32), NA14/2 (33), and E687 (51). The mean value is $\tau(\Lambda_c) =
(2.00 \pm 0.18) \times 10^{-13}$ s. The Ξ_c^+ lifetime is known from three experiments:

WA62 (52), E400 (53), and NA32 (46). The mean lifetime is $\tau(\Xi_c^+) = (2.6^{+0.9}_{-0.5}) \times 10^{-13}$ s. The NA32 group also obtained the first measurement of the Ξ_c^0 lifetime, $\tau(\Xi_c^0) = (0.82^{+0.59}_{-0.30}) \times 10^{-13}$ s, based on four observed events (28).

3.2   Theoretical Lifetime Estimates

In the large mass limit, one expects all heavy hadrons of the same flavor to have identical lifetimes. The spread in the experimental lifetime values of the Λ_c^+, $\Xi_c^{0,+}$, and Ω_c charm baryons signals that $1/m_c$ effects are still important in the weak inclusive decays of charm baryons (as they are for charm mesons). The preasymptotic effects enter in the form of W-exchange contributions (54), and additional contributions come from the interference of decay quarks and spectator quarks. These are sensitive to the probability that the charm and light quarks in the baryon wave function will come together at one point: to the square of the wave function at the origin $|\Psi(0)|^2$ with mass dimension $[m^3]$.

From dimensional arguments, one then finds

$$\Gamma_{\text{FQD}} \approx G_F^2 m_c^5$$

$$\Gamma_{\text{exch,int}} \approx G_F^2 m_c^2 |\Psi(0)|^2, \qquad\qquad 2.$$

where Γ_{FQD} denotes the "free quark decay" parton model decay rate, Γ_{exch} and Γ_{int} denote the W-exchange and interference rates, and m_c refers to the charm quark mass.

Explicit calculations show that $\Gamma_{\text{FQD}} \approx \Gamma_{\text{exch,int}}$ in the charm baryon sector (55, 56) and that $\Gamma_{\text{FQD}} \approx \Gamma_{\text{int}}$ in the charm meson sector (57). Using the fact that the wave function at the origin of the heavy-light bound state becomes independent of the heavy quark mass as the heavy quark mass becomes large (50), one can scale Equation 2 to the bottom quark sector. One then finds $\Gamma_{\text{exch,int}}/\Gamma_{\text{FQD}} \approx (m_c/m_b)^3 \approx O(5\%)$, which implies that the lifetime differences in the bottom sector are expected to be quite small.

The difficulty in obtaining reliable rate and lifetime estimates for the charm baryons is clearly evidenced by the fact that the preasymptotic effects, which are down by several powers of $1/m_c$, are so important. This makes an analysis in terms of a $1/m_c$ expansion difficult. Nevertheless one can attempt to obtain a qualitative picture of the lifetime differences of charm baryons in the form of a lifetime hierarchy (55, 56).

The starting point in the analysis is the usual effective nonleptonic Hamiltonian (58)

$$H_{\text{eff}} = \sqrt{2} G_F V_{cs} V_{ud}^* [c_- O_- + c_+ O_+], \qquad\qquad 3.$$

where $O_\pm$ are local 4-quark operators

$$O_\pm = (\bar{u}_L \gamma_\mu d_L)(\bar{s}_L \gamma^\mu c_L) \pm (\bar{s}_L \gamma_\mu d_L)(\bar{u}_L \gamma^\mu c_L) \qquad\qquad 4.$$

with $\bar{q}_L\gamma_\mu q_L = \frac{1}{2}\bar{q}\gamma_\mu(1-\gamma_5)q$, and $V_{q_\alpha q_\beta}$ are elements of the Kobayashi-Maskawa mixing matrix with $V_{cs} \simeq V_{ud} \simeq \cos\theta_c$ and θ_c the Cabibbo angle. The coefficients $c_\pm$ describe the leading logarithmic evolution of the non-leptonic Hamiltonian from the W mass scale down to the charm mass scale $\mu \simeq O(m_c)$ (58). We take $c_+ = 0.74$ and $c_- = 1.80$ as in the work by Guberina et al (56).

The effective nonleptonic Hamiltonian in Equation 3 induces the inclusive nonleptonic decay contributions drawn in Figure 2 for, say, the inclusive Λ_c^+ decays. Simple expressions can be obtained for these rates when one neglects u, d, and s quark masses and uses a nonrelativistic wave function for the charm baryons. For example, for the Λ_c^+ decay one then has a nonleptonic (n.l.) rate (56)

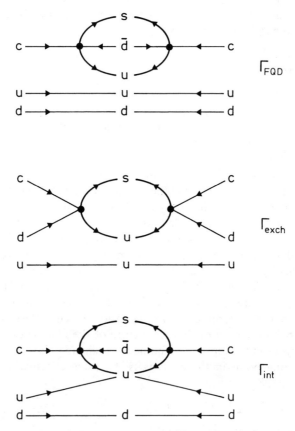

Figure 2   Free quark decay, W-exchange, and interference contributions to inclusive nonleptonic Λ_c^+ decays.

$$\Gamma_{n.l.}^{\Lambda_c^+} = \Gamma_{FQD}^{\Lambda_c^+} + \Gamma_{exch}^{\Lambda_c^+} + \Gamma_{int}^{\Lambda_c^+}$$

$$= (2c_+^2 + c_-^2)\frac{G_F^2 m_c^5}{192\pi^3} + c_-^2 \frac{G_F^2 m_c^2 |\Psi(0)|^2}{4\pi}$$

$$- c_+(2c_- - c_+)\frac{G_F^2 m_c^2 |\Psi(0)|^2}{4\pi}$$

$$= (1.58 + 3.01R - 0.99R) \times 10^{-12}\,\text{s}^{-1}, \qquad\qquad 5.$$

with $m_c = 1.6$ GeV, $c_\pm$ values as above, and $R = |\Psi(0)|^2/10^{-2}$ GeV^3.

As is evident from Equation 5, the resulting nonleptonic rate is quite sensitive to the value of the wave function at the origin $|\Psi(0)|^2$. From a fit to the hyperfine splitting, as discussed in Section 2, one has $|\Psi(0)|^2 \simeq 10^{-2}$ GeV^3. Adding a nominal semileptonic rate of $2 \times 15\%$ of the nonleptonic FQD rate, one finds $\tau_{\Lambda_c^+} = 2.46 \times 10^{-13}$ s, which is somewhat larger than the experimental value $\tau_{\Lambda_c^+}(\text{exp.}) = (2.00 \pm 0.2) \times 10^{-13}$ s. Smaller values of the wave function at the origin are obtained in a bag model (56) ($|\Psi(0)|^2 \simeq 0.4 \times 10^{-2}$ GeV^3) or if one equates the baryon and meson wave function at the origin (55) ($|\Psi(0)|^2 \simeq 0.4 \times 10^{-2}$ GeV^3 with $f_D = 165$ MeV). Values close to $|\Psi(0)|^2 \simeq 10^{-2}$ GeV^3 are also obtained in a non-relativistic model with a funnel-type potential (55) or by using electromagnetic mass differences in the hyperfine formula (59). It is clear that using the smaller values of $|\Psi(0)|^2$ makes the agreement with the experimental rate worse.

Applying the same calculation to the other weakly decaying charm baryons, Guberina, Rückl & Trampetic (56) found a lifetime hierarchy $\tau(\Omega_c) \approx \tau(\Xi_c^0) < \tau(\Lambda_c) < \tau(\Xi_c^+)$, whereas Voloshin & Shifman (55) obtained $\tau(\Omega_c) < \tau(\Xi_c^0) < \tau(\Lambda_c) \approx \tau(\Xi_c^+)$.

The main effects leading to the lifetime extremes in the inequalities are easily identified: the large Ω_c^0 decay rate is due to a large positive interference effect among the s quarks (the s quark from the weak decay vertex can interfere with either of the spectator s quarks) and the small Ξ_c^+ decay rate is due to the absence of a W-exchange contribution in this case.

One must stress that there are a number of theoretical uncertainties in the above lifetime calculations related to the size of the preasymptotic effects, which cannot be discussed in detail here. Voloshin & Shifman are rather optimistic and claim that their inequality chain can confidently be replaced by $\tau(\Omega_c) = (0.6\text{--}1.0)\tau(\Xi_c^0) = (0.6\text{--}1.0)\tau(\Lambda_c) \approx \tau(\Xi_c^+)$ (55).

The absence of large preasymptotic effects in bottom baryon nonleptonic decay rates is gratifying. As long as there is no experimental evidence to the contrary, one may confidently assume almost equal lifetimes for all weakly decaying bottom baryons and, for that matter, bottom mesons.

4. SEMILEPTONIC DECAYS

4.1 *Inclusive Semileptonic Rates*

The inclusive semileptonic branching ratio for $\Lambda_c \to e^+ + X$ has been measured by the Mark II collaboration (60) as $4.5 \pm 1.7\%$. This leads to the inclusive rate

$$\Gamma^{sl}_{\Lambda_c}(\text{inclusive}) = (22.5 \pm 8.7) \times 10^{10} \text{ s}^{-1}. \qquad\qquad 6.$$

The corresponding inclusive semileptonic charm meson rates are $\Gamma^{sl}_{D^\pm}(\text{inclusive}) = (18.1 \pm 1.54) \times 10^{10} \text{ s}^{-1}$ and $\Gamma^{sl}_{D^0}(\text{inclusive}) = (18.3 \pm 2.9) \times 10^{10} \text{ s}^{-1}$ (24). These numbers are in good accord with the free quark decay (FQD) rates calculated according to the parton model including first-order QCD corrections (61). Using $\Lambda = 200$ MeV, $m_c = 1.65$ GeV, and $m_s = 0.45$ GeV, one obtains $\Gamma^{sl}_{FQD} = 17.4 \times 10^{10} \text{ s}^{-1}$.

In order to indicate the sensitivity of the FQD rate to the assumed values of quark masses, we also give the results of the same calculation for the mass values ($m_c = 1.60$ GeV; $m_s = 0.45$ GeV) and ($m_c = m_{\Lambda_c} = 2.285$ GeV; $m_s = m_\Lambda = 1.116$ GeV), which are 14.5×10^{10} and $29.0 \times 10^{10} \text{ s}^{-1}$, respectively.

The Mark II collaboration has also measured semileptonic semi-inclusive branching ratios (60). They find $BR(\Lambda_c \to pe^+ + X) = 1.8 \pm 0.9\%$, including protons from Λ decay, and $BR(\Lambda_c \to \Lambda e^+ + X) = 1.1 \pm 0.8\%$.

4.2 *Exclusive Semileptonic Decays*

4.2.1 AMPLITUDES, RATES, AND ANGULAR DECAY DISTRIBUTIONS   Let us first describe the spin complexity of the problem. This entails an enumeration of the number of independent amplitudes and a discussion of how to measure them through polarization-type measurements. We remind the reader that polarization measurements in semileptonic $B \to D^*$ transitions (62, 63) have been crucial for the development of a plausible theory of heavy meson transition form factors (64–70).

Let us begin our discussion by defining a standard set of invariant form factors for the weak current-induced baryonic $1/2^+ \to 1/2^+$ and $1/2^+ \to 3/2^+$ transitions. One has

$$\langle \Lambda_s(P_2)|J^{V+A}_\mu|\Lambda_c(P_1)\rangle = \bar{u}(P_2)\,[\gamma_\mu(F^V_1 + F^A_1\gamma_5) + i\sigma_{\mu\nu}q^\nu(F^V_2 + F^A_2\gamma_5)$$

$$+ q_\mu(F^V_3 + F^A_3\gamma_5)]u(P_1)$$

$$\langle \Omega^-(P_2)|J^{V+A}_\mu|\Omega_c(P_1)\rangle = \bar{u}^\alpha(P_2)\,[g_{\alpha\mu}(G^V_1 + G^A_1\gamma_5) + P_{1\alpha}\gamma_\mu(G^V_2 + G^A_2\gamma_5)$$

$$+ P_{1\alpha}P_{2\mu}(G^V_3 + G^A_3\gamma_5) + P_{1\alpha}q_\mu(G^V_4 + G^A_4\gamma_5)]\gamma_5 u(P_1), \qquad 7.$$

where J_μ^V and J_μ^A are vector and axial-vector currents and $q_\mu = (P_1 - P_2)_\mu$ is the 4-momentum transfer. Here and throughout this review we find it convenient to use particle labels instead of generic names. The form factors $F_i^{V,A}$ and $G_i^{V,A}$ are functions of q^2, and the superscripts V and A stand for vector and axial vector.

The invariants F_3^V, F_3^A, G_4^V, and G_4^A multiplying q_μ contribute to semileptonic decays at $O(m_\ell^2/q^2)$ and are thus difficult to measure. Muon mass effects have been investigated in the corresponding semileptonic $D \to K(K^*)$ decays and have been found to be $\leq 5\%$ of the total rate (65). The biggest effect occurs for the partial rate into the longitudinal current component, where the contribution of the muon mass amounts to $O(10\%)$ and is largest at small q^2. This is different in semileptonic $b \to c$ decays, in which lepton mass effects can be conveniently probed in the τ channel (71, 72). Nevertheless, the invariants $F_3^{V,A}$ and $G_4^{V,A}$ multiplying q_μ are important in $\Delta C = 1$ charm-changing nonleptonic decays; there they contribute through the so-called factorizing contributions in the nonleptonic $B_c \to B + M(0^-)$ decays, where $M(0^-)$ refers to a pseudoscalar meson with spin-parity $J^P = 0^-$.

Rates and angular decay distributions are given in terms of bilinear forms of the form factors. First consider the decay of an unpolarized charm baryon $B_c \to B(\to B'M) + \ell^+ + \nu_\ell$, where the cascade decay $B(\to B'M)$ is used to analyze the polarization of the daughter baryon B. For semileptonic $1/2^+ \to 1/2^+$ transitions, the full four-fold decay distribution, differential in the momentum transfer squared (q^2) and the angles θ, χ, and θ_Λ shown in Figure 3, reads (73)

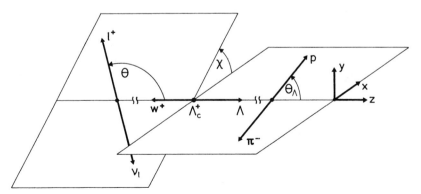

Figure 3 Definition of polar angles θ_Λ and θ and azimuthal angle χ in the decay $\Lambda_c^+ \to \Lambda(\to p\pi) + W^+(\to \ell^+ \nu_\ell)$.

$$\frac{d\Gamma[\Lambda_c^+ \to \Lambda(\to p\pi^-)+\ell^+ +\nu_\ell]}{dq^2\,d\cos\theta\,d\chi\,d\cos\theta_\Lambda} = BR(\Lambda \to p\pi^-)\frac{1}{4\pi}$$

$$\times \left[\frac{3}{8}(1+\cos^2\theta)\frac{d\Gamma_U}{dq^2}(1+\alpha_c^U\alpha_\Lambda\cos\theta_\Lambda)+\frac{3}{4}\sin^2\theta\frac{d\Gamma_L}{dq^2}(1+\alpha_c^L\alpha_\Lambda\cos\theta_\Lambda)\right.$$

$$-\frac{3}{2\sqrt{2}}\sin(2\theta)\cos\chi\sin\theta_\Lambda\alpha_\Lambda\frac{d\Gamma_I}{dq^2}-\frac{3}{4}\cos\theta\frac{d\Gamma_P}{dq^2}\left(1+\frac{d\Gamma_U}{d\Gamma_P}\alpha_\Lambda\cos\theta_\Lambda\right)$$

$$\left.+\frac{3}{\sqrt{2}}\sin\theta\cos\chi\sin\theta_\Lambda\alpha_\Lambda\frac{d\Gamma_A}{dq^2}\right], \qquad\qquad 8.$$

where

$$\frac{d\Gamma_i}{dq^2} = \frac{1}{2}\frac{G^2}{(2\pi)^3}|V_{cs}|^2\frac{pq^2}{12M_1^2}H_i. \qquad\qquad 9.$$

G is the Fermi coupling constant ($G = 1.026 \times 10^{-5}m_p^{-2}$) and p is the momentum of the daughter baryon in the Λ_c rest-frame.

The helicity rates H_i in Equation 9 are defined as follows

$$H_U = 2(|H^V_{1,1/2}|^2+|H^A_{1,1/2}|^2)$$

$$H_L = 2(|H^V_{0,1/2}|^2+|H^A_{0,1/2}|^2)$$

$$H_I = -\text{Re}(H^V_{1,1/2}H^{A*}_{0,1/2}-H^A_{1,1/2}H^{V*}_{0,1/2})$$

$$H_P = 4\text{Re}(H^V_{1,1/2}H^{A*}_{1,1/2})$$

$$H_A = \text{Re}(H^V_{1,1/2}H^{V*}_{0,1/2}-H^A_{1,1/2}H^{A*}_{0,1/2}), \qquad\qquad 10.$$

where the $H^{V,A}_{\lambda_W\lambda_\Lambda}$ are the helicity amplitudes of the current-induced transition, λ_W is the helicity of the current ($\lambda_W = 0$ longitudinal, $\lambda_W = \pm 1$ transverse) or, equivalently, of the off-shell W boson and λ_Λ is the helicity of the daughter baryon. The relation between the set of helicity and invariant form factors can be found elsewhere (e.g. 73, 74). For $\bar{c} \to \bar{s}$ transitions, the signs in the last two terms of Equation 8 have to be reversed. The labeling of the helicity rates H_i describes the polarization of the off-shell W boson in the decay: U (unpolarized transverse), L (longitudinal), I (transverse-longitudinal interference), P (parity odd), A (parity asymmetric). We have also introduced α_c^U and α_c^L, the q^2-dependent transverse and longitudinal asymmetry parameters

$$\alpha_c^U = \frac{2\text{Re}H^V_{1,1/2}H^{A*}_{1,1/2}}{|H^V_{1,1/2}|^2+|H^A_{1,1/2}|^2} \qquad\qquad 11.$$

$$\alpha_c^L = \frac{2\mathrm{Re}H_{0,1/2}^V H_{0,1/2}^{A*}}{|H_{0,1/2}^V|^2 + |H_{0,1/2}^A|^2}.$$ 12.

The term α_Λ is the known asymmetry parameter for the nonleptonic decay of the daughter baryon defined in analogy to Equation 12. Triple, double, and single decay distributions as well as the rate may be obtained from Equation 8 with the appropriate integrations.

We have assumed that the form factors and helicity amplitudes are real since the physical threshold is at $q^2 = (M_1 + M_2)^2 > q_{max}^2 = (M_1 - M_2)^2$. Thus so-called time reversal–odd (T-odd) contributions proportional to $\sin \theta \sin \chi \sin \theta_\Lambda$ and $\sin (2\theta) \sin \chi \sin \theta_\Lambda$ have been omitted (73). We note that the presence of such contributions could signal possible CP violations in the decay process (75).

The structure of the decay distribution in Equation 8 is quite similar to the corresponding four-fold decay distribution for the cascade decay $D \rightarrow K^*(\rightarrow K\pi) + \ell^+ + \nu_\ell$ (65, 68–70), which has been proven so useful in disentangling the form factor structure in the semileptonic $D \rightarrow K^*$ decays (74). Equation 8 defines a set of eight observables that are bilinear in the four independent q^2-dependent real form factors. A measurement of these eight observables would considerably overdetermine the form factors. Note though that the complexity of the problem is reduced close to the phase-space boundaries. At zero recoil, $q^2 \approx q_{max}^2$, only the s-wave contribution of the final state ($\Lambda + W_{off-shell}$) remains, and at maximum recoil $q^2 = 0$ only the longitudinal contribution H_L survives. The relevant dynamical information may be extracted by an analysis of the suitably defined asymmetries, as proposed in (75), or by multidimensional fits to the data, as was done by the E691 collaboration in their analysis of semileptonic $D \rightarrow K^*$ decays (76). The E691 collaboration had 204 data events, of which 21 were assumed to be background, and obtained already significant results on form factor ratios.

An additional set of polarization observables can be defined for the decay of polarized charm baryons. For example, hadronically produced Λ's have been observed to be polarized where the polarization necessarily has to be transverse to the production plane because of parity invariance in the production process. It may well be that hadronically produced Λ_c^+ show a similar polarization effect (77, 78).

Charm baryons from weak decays of bottom baryons are expected to be polarized. In this case one has to orient the Λ_c decay and the subsequent decay of the daughter baryon relative to the Λ_c polarization as drawn in Figure 4, where the orientation angles θ_p θ_Λ, and χ are defined. For the corresponding four-fold angular decay distribution one finds (73)

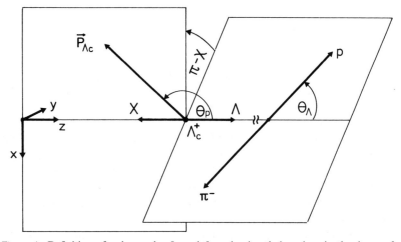

Figure 4 Definition of polar angles θ_Λ and θ_p and azimuthal angle χ in the decay of a polarized $\Lambda_c^\uparrow \to \Lambda(\to \pi^-)+X$. The left plane is determined by polarization vector $\mathbf{P}_{\Lambda_c}$ of the Λ_c.

$$
\frac{d\Gamma[\Lambda_c^\uparrow \to \Lambda(\to p\pi^-)+\ell^+ +\nu_\ell]}{dq^2\, d\cos\theta_p\, d\chi\, d\cos\theta_\Lambda} = \frac{1}{8\pi}\, BR(\Lambda \to p\pi^-)
$$

$$
\cdot\left[\frac{d\Gamma_U + d\Gamma_L}{dq^2} + \alpha_\Lambda \cos\theta_\Lambda\left(\alpha_c^U \frac{d\Gamma_U}{dq^2} + \alpha_c^L \frac{d\Gamma_L}{dq^2}\right)\right.
$$

$$
- P_c\cos\theta_p\left(\alpha_c^U \frac{d\Gamma_U}{dq^2} - \alpha_c^L \frac{d\Gamma_L}{dq^2}\right) + P_c\alpha_\Lambda \cos\theta_p \cos\theta_\Lambda\left(-\frac{d\Gamma_U}{dq^2} + \frac{d\Gamma_L}{dq^2}\right)
$$

$$
\left. - P_c\alpha_\Lambda \sin\theta_p \sin\theta_\Lambda \cos\chi \frac{d\Gamma_{LI}}{dq^2}\right], \qquad\qquad 13.
$$

where P_c is the polarization of the Λ_c^+. The longitudinal interference rate is $d\Gamma_{LI}/dq^2 = 2(|H_{0,1/2}^V|^2 - |H_{0,1/2}^A|^2)$. Measurements of the angular decay distribution relative to the initial spin polarization vector would allow for measurement of the size and sign of the polarization of the Λ_c if its decay structure is known, or vice versa, if P_c were known, to further constrain the decay amplitudes of the Λ_c. The information contained in the transition $W^+ \to \ell^+ \nu_\ell$ has not been used in Equation 13, that is the dependence on the angles (θ, χ') of the $W^+ \to \ell^+ \nu_\ell$ decay has been integrated out. If this angular dependence is kept explicitly, one would have a six-fold differential distribution. Corresponding decay distributions for semileptonic $1/2^+ \to$

$3/2^+$ transitions (polarized and unpolarized) can be found in (73). Let us mention that the decay distributions in Equations 8 and 13 can also be derived using a frame-independent representation in terms of spin and momentum correlations (79).

4.2.2 MODEL RESULTS   The accessible q^2 range in semileptonic charm baryon decays is not small $[m_\ell^2 \le q^2 \le (M_1 - M_2)^2]$. One thus has a large experimental leverage to study the q^2 dependence of the form factors. Also, the full spin-dependent form factor structure can be investigated because the center-of-mass momentum p becomes large enough when q^2 moves away from the zero recoil point $q^2 = (M_1 - M_2)^2$. One therefore anticipates large q^2 effects from the form factors' q^2 dependence in contrast to non-charm semileptonic hyperon decays. Therefore uncertainties of the form factors' q^2 dependence can be predicted to have an important influence on the analysis of form factor ratios in charm baryon decays. For large enough momenta p, all partial waves in the decay $B_1 \to B_2 + W_{\text{off-shell}}(q^2)$ come into play. This is different in ordinary hyperon decays, where the accessible q^2 range is small and only the lower partial waves contribute to any significant degree. Methods used to extract the various form factors through polarization measurements were discussed in Section 4.2.1.

There have been a number of theoretical attempts to model the form factors in the semileptonic $1/2 \to 1/2^+$ and $1/2^+ \to 3/2^+$ transitions employing flavor symmetry and/or quark models. In Table 4 we list the rate predictions of various models for the semileptonic (s.l.) decays $\Lambda_c^+ \to \Lambda + \ell^+ + \nu_\ell$, $\Xi_c^+ \to \Xi^0 + \ell^+ + \nu_\ell$ and the $1/2^+ \to 3/2^+$ decay $\Omega_c \to \Omega^- + \ell^+ + \nu_\ell$ in the limit of zero lepton mass.

The first column contains earlier predictions that exploited SU(4) flavor symmetry at $q^2 = 0$ to relate $\Delta C = 1$ to the known $\Delta C = 0$ amplitudes (80). The results were then extrapolated to $q^2 \ne 0$ by using suitable form factors. The predictions by Yamada are similar (81). The rates are too large because of the use of SU(4) at $q^2 = 0$. There are large mass corrections

Table 4   Exclusive semileptonic decay rates in units of $10^{10} s^{-1}$

	Buras (80)	Gavela (82)	PHGA (84) NRQM	PHGA (84) MBM	Singleton (85)	HK (86)
$\Lambda_c^+ \to \Lambda$	60	15	17	13	30	22
$\Xi_c^+ \to \Xi^0$	235	28	28	22	34	33
$\Omega_c^0 \to \Omega^-$	280	49	-	-	-	48

at $q^2 = 0$ that reduce the rates substantially (82, 83). Flavor symmetry should rather be applied at q^2_{max}. Nonrelativistic quark model calculations predict rates close to q^2_{max} and are then extended to $q^2 \neq q^2_{max}$ via form factors (84, 85). In order to compare predictions, we have taken the liberty of rescaling the results of (84, 85) by eliminating their assumed large QCD correction and SU(3)-breaking effects, respectively, which we considered to be unrealistically large. The resulting rates for $\Lambda_c^+ \to \Lambda$ scatter around 20×10^{10} s^{-1}, including the value quoted by Hussain & Körner (86) and discussed below. The calculated rates would imply a saturation of the inclusive rate in Equation 6 by the exclusive mode. The semi-inclusive rates quoted in Section 4.1 (60), however, preclude such a possibility. Clearly a more precise measurement of the semileptonic $\Lambda_c^+ \to \Lambda$ (exclusive and inclusive) rate and a reappraisal of the theoretical calculations are needed.

Even though the estimated semileptonic $\Lambda_c^+ \to \Lambda$ rates obtained by the model calculations (82–86) after the rescaling are comparable, the models differ considerably in the details of their predictions for polarization observables. It may be possible to test some aspects of the various model predictions in the future by using the angular measurements discussed above.

Large semileptonic rates for $\Omega_c^0 \to \Omega^- + \ell^+ + \nu_\ell$ arise in the calculations (82, 86). This is basically because there are several possibilities for the initial c quark to make a transition to the final s quark, regardless of the model.

In the past few years, there have been some interesting theoretical developments concerning the physics of hadrons containing heavy quarks and current-induced transitions among them (10–16). In the limit of infinitely massive quarks, symmetries above and beyond those usually associated with QCD arise. In the heavy quark effective theory (HQET) that results, these symmetries are termed heavy flavor and spin symmetry. The heavy flavor symmetry is not a spectrum symmetry; it is a flavor symmetry relating heavy quarks with different flavors but with the same velocity. The conserved charges associated with the heavy flavor symmetry normalize the s-wave vector transition form factors at zero velocity transfer [or at the zero recoil point $q^2_{max} = (M_1 - M_2)^2$]. The physics underlying this normalization condition was known and formulated earlier (87): at zero recoil there is complete overlap between the wave functions of the light degree of freedoms of the bound state of a heavy and light quark. The spin symmetry results from the fact that the hyperfine interaction between a pair of light and heavy quarks vanishes when the heavy quark mass becomes large. The assumption of no spin communication between heavy and light quarks is familiar from the spectator quark model and had been usefully applied to current-induced transitions between heavy-light meson and baryon states some time ago (88–90). In the last year, the previous observations concerning heavy-light hadron systems were combined for the first time and cast into the elegant theory of the HQET.

Heavy baryons can be viewed as bound states of a heavy quark and a light diquark system all moving with the same velocity. The light diquark can be either spin zero or spin one both of which combine with the heavy quark to form a spin-1/2 heavy baryon Λ_Q, and a degenerate pair of spin-1/2 and spin-3/2 heavy baryons (Σ_Q and Σ_Q^*), respectively. The consequences of the HQET for current-induced transitions between heavy-light baryons are best formulated in the velocity language. They read as follows (31, 91, 92):

(a) $1/2^+ \to 1/2^+$; spin-zero diquark

$$\langle \Lambda_{Q'}(P_2)|J_\mu^{V+A}|\Lambda_Q(P_1)\rangle = F_\Lambda \bar{u}_2 \gamma_\mu (1 - \gamma_5) u_1 \qquad 14.$$

(b) $1/2^+ \to 1/2^+$; spin-one diquark

$$\langle \Sigma_{Q'}(P_2)|J_\mu^{V+A}|\Sigma_Q(P_1)\rangle = -\frac{1}{3}\bar{u}_2\Big[F_L \gamma_\mu (1 - \gamma_5)$$

$$-\frac{2}{\omega+1}(F_L+F_T)(v_{1\mu}+v_{2\mu}) + \frac{2}{\omega-1}(F_L-F_T)(v_{1\mu}-v_{2\mu})\gamma_5 \Big] u_1 \qquad 15.$$

(c) $1/2^+ \to 3/2^+$; spin-one diquark

$$\langle \Sigma_Q^*(P_2)|J_\mu^{V+A}|\Sigma_Q(P_1)\rangle = \frac{1}{\sqrt{3}}\bar{u}_2^\alpha\Big[2F_T g_{\mu\alpha}(1+\gamma_5)$$

$$+\frac{1}{\omega+1}(F_L+F_T)v_{1\alpha}\gamma_\mu\gamma_5 - \frac{1}{\omega-1}(F_L-F_T)v_{1\alpha}\gamma_\mu$$

$$+\frac{2}{\omega^2-1}(F_L-\omega F_T)v_{1\alpha}v_{2\mu}(1+\gamma_5)\Big] u_1, \qquad 16.$$

where the form factors F_L and F_T describe the longitudinal and transverse spin-one diquark transitions (31). In the heavy quark mass limit there are thus a total of three universal independent form factors F_Λ, F_L, and F_T, compared to the many independent form factors in the general covariant expansion (7) to which they can be related (31). The universal form factors are functions of the velocity transfer variable $\omega = v_1 v_2 = (M_1^2 + M_2^2 - q^2)/2M_1 M_2$, where $v_\mu = P_\mu/M$. They are normalized to one at zero recoil $q^2 = q_{max}^2$ or $\omega = 1$, that is $F_\Lambda(\omega = 1) = F_L(\omega = 1) = F_T(\omega = 1) = 1$. Renormalization effects on the current matrix elements were discussed by Falk et al (13). They can be summed by renormalization group techniques and lead to scale- and velocity-dependent logarithmic corrections. The systematics of $1/m$ corrections to the limiting structure described by Equations 14–16 are now being explored (93, 94). An interesting result is that the zero recoil normalization condition of Voloshin & Shifman (87) discussed above remains intact at $O(1/m)$ (94).

In heavy to light transitions, in which the HQET can only be applied to the heavy parent baryon, there are more independent form factors. In the three cases discussed above (Equations 14–16) there are now two, four, and six independent form factors, respectively, with no normalization condition. For example, in $\Lambda_c^+ \to \Lambda$ transitions the predicted form factor structure now reads (31, 92)

$$\langle \Lambda(P_2)|J_\mu^{V+A}|\Lambda_c(P_1)\rangle = \bar{u}_2[F_1(q^2)+F_2(q^2)P_1]\gamma_\mu(1-\gamma_5)u_1. \qquad 17.$$

Because of the lack of normalization conditions for F_1 and F_2, there is no rate prediction. However, Equation 17 has very interesting consequences for the longitudinal asymmetry parameter α_c^L, which is predicted to be $\alpha_c^L = -1$ at $q^2 = 0$, independent of the values of F_1 and F_2 (95). Thus the Λ emerges with a 100% negative polarization at $q^2 = 0$. The same statement holds for the Ξ in the semileptonic transition $\Xi_c \to \Xi$. For the semileptonic decay $\Omega_c \to \Omega^-$ one finds close to zero polarization at $q^2 = 0$ (96). These predictions of the HQET could be easily checked through angular measurements discussed in Section 4.2.1 once enough data become available.

It is quite tempting to apply the results of the HQET also to the $c \to s$ transition, regarding the strange baryons as heavy baryons. There is some support for such an approach from the $c \to s$ meson decays, where the latest data on $D \to K(K^*)$ (97) transitions are in agreement with the naively applied HQET (65), whereas earlier data showed a structure distinctively different from predictions for heavy to heavy quark transitions (76). A basic plausible tenet of the HQET is the assumption that, on the scale of the heavy quark or particle mass, the light quark system has a small longitudinal and transverse momentum spread around its central equal velocity value. This is certainly not the case for a strange baryon or meson. It may very well be, however, that the longitudinal and transverse momentum smearing averages out effectively. In addition, explicit quark model wave functions show approximate unit overlap at zero recoil, regardless of the masses of the quarks involved in the transition.

With these qualifications in mind, the heavy quark to heavy quark transition HQET predictions have also been applied to semileptonic $\Lambda_c \to \Lambda$ transitions using a dipole-type q^2 dependence for the form factor $F_\Lambda(q^2)$ (86). The rate prediction found with this approach and given in Table 4 does not deviate much from the predictions of the other models, whereas the predicted polarization effects are different. The same model has also been applied to the transitions $\Xi_c^+ \to \Xi^0$ and $\Omega_c \to \Omega^-$ (86). The transverse and longitudinal diquark form factors $F_T(q^2)$ and $F_L(q^2)$ are related via $F_T(q^2) = F_L(q^2)$ with $F_T(q_{max}^2) = F_L(q_{max}^2) = 1$ according to the assumption of independent quark motion (31), where there is no spin communication among the light quarks. The resulting rate predictions are also shown in Table 4.

5. EXCLUSIVE NONLEPTONIC DECAYS

5.1 Experimental Results

For the Λ_c, branching ratios of various channels relative to the decay mode $\Lambda_c \to pK^-\pi^+$ have been measured by the Mark II (98), R415 (99), ARGUS (35, 106), NA32 (100), E691 (101), and CLEO (39) groups; the results are summarized in Table 5. In addition, the branching ratio $\Gamma(\Lambda_c \to p\bar{K}^0\pi^+\pi^-)/\Gamma(\Lambda_c \to \Lambda\pi^+\pi^+\pi^-)$ was measured in the BIS-2 experiment to be 4.3 ± 1.2 (102). The $pK^-\pi^+$ channel has resonance contributions from $\Delta^{++}K^-$ and $p\bar{K}^0(892)$. Other decay modes with comparable branching ratios are $\Lambda\pi^+\pi^+\pi^-$ and $p\bar{K}$ + pions.

A surprisingly large branching ratio is observed for the Cabibbo-suppressed $p\pi^+\pi^-\pi^0$ decay mode. In Figure 5 we show several mass spectra observed by the NA32 group (100). The clean $\Lambda_c \to pK^-\pi^+$ sample demonstrates the selective power of lifetime cuts used in the NA32 analysis.

Absolute branching ratios are poorly known. To measure them, one has not only to observe a sample of Λ_c decays, say in the $pK^-\pi^+$ channel, but also to know how many Λ_c in total were produced in that sample. Two different approaches have been used. The Mark II group measured the increase in baryon production above the $\Lambda_c\bar{\Lambda}_c$ threshold, attributing the increase to $\Lambda_c\bar{\Lambda}_c$ pair production, and obtained $BR(\Lambda_c \to pK^-\pi^+) = 2.2 \pm 1.0\%$ (103). More recently, another approach was used by the ARGUS and CLEO groups: They identified $\Lambda_c \to pK^-\pi^+$ (+charge conjugation) decays at the $\Upsilon(4S)$ resonance and thus originating from B decay. They estimated the inclusive branching ratio $\bar{B} \to \bar{\Lambda}_c + X$ from the number of baryons produced at the $\Upsilon(4S)$ resonance, assuming that all baryonic B decays produce a Λ_c. With this assumption, ARGUS obtained $BR(\Lambda_c \to pK^-\pi^+) = 4.1 \pm 2.4\%$ (104) and CLEO found $4.4 \pm 1.7\%$ (39), the mean value being $BR(\Lambda_c \to pK^-\pi^+) = 4.2 \pm 1.4\%$. It should be noted that the branching ratio will be larger if the above-mentioned assumption does not hold.

With a value of $\sim 4\%$ for the $\Lambda_c \to pK^-\pi^+$ branching ratio, so far up to 20% of all Λ_c decay modes have been observed. The remaining modes

Table 5 Measured Λ_c branching ratios relative to $\Lambda_c \to pK^-\pi^+$

	Mark II (98)	R 415 (99)	ARGUS (35, 106)	NA 32 (100)	E 691 (101)	CLEO (39)
$p\overline{K^0}$	$.50 \pm .25$		$.62 \pm .15$		$.55 \pm .22$	$.44 \pm .09$
$p\overline{K^0}(892)$ [a]	$.18 \pm .10$	$.42 \pm .24$				
$\Delta^{++}K^-$ [a]	$.17 \pm .07$	$.40 \pm .17$				
$\Lambda\pi^+\pi^+\pi^-$			$.61 \pm .16$	$.12^{+.35}_{-.12}$	$.82 \pm .40$	$.65 \pm .16$
$p\overline{K^0}\pi^+\pi^-$				$1.4 \pm .4$		$.43 \pm .13$
$p\overline{K^0}\pi^+\pi^-\pi^0$				$.57^{+.32}_{-.20}$		
$\Lambda\pi^+$			$.18 \pm .05$			$.17 \pm .04$
$\Xi^- K^+\pi^+$						$.15 \pm .05$
$pK^-\pi^+\pi^0$				$.72^{+.32}_{-.22}$		
$pK^-\pi^+\pi^0\pi^0$				$.23^{+.20}_{-.13}$		
$pK^-\pi^+\pi^+\pi^-$				$.027 \pm .015$		
$p\pi^+\pi^-$				$.065 \pm .050$		
$pf_0(975)$ [b]				$.05 \pm .05$		
$p\pi^+\pi^-\pi^0$				$.67^{+0.32}_{-0.25}$		
$p\pi^+\pi^-\pi^+\pi^-$				$.035 \pm .022$		
pK^+K^-				$.05 \pm .04$		
$p\phi$ [b]				$.04 \pm .03$		

[a] Decay modes included in $pK^-\pi^+$.

[b] Decay modes included in preceding decay mode.

are probably decays with higher particle multiplicities or with neutral pions or a neutron in the final state. It should be noted that the Mark II experimental inclusive semileptonic rates discussed in Section 4.1 and the $\Lambda_c \to pK^-\pi^+$ rates obtained by the same group in the same experiment have been normalized to the same baryon sample, so that any changes in that normalization would change all Mark II branching ratios in the same direction.

Not much is known about Ξ_c decays. The Ξ_c^+ was first observed in the

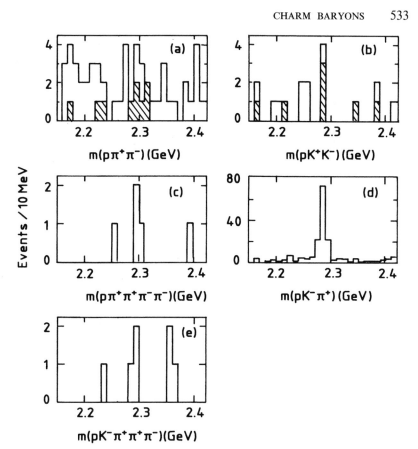

Figure 5 Invariant mass spectra observed in experiment NA32. The hatched histograms correspond to combinations compatible with (*a*) $pf^0(975)$ and (*b*) $p\phi$.

$\Lambda K^-\pi^+\pi^+$ decay channel [WA62 (26, 30), E400 (44)] and recently in Ξ_c^+ + pions channels [CLEO (27, 45), ARGUS (29)]. The NA32 group observed the Ξ_c^0 in the $pK^-\bar{K}^0(892)$ channel (28) and the Ξ_c^+ in the $\Xi^-\pi^+\pi^+$ and $\Sigma^+K^-\pi^+$ channels (46). They quote a relative branching ratio $\Gamma(\Xi_c^+ \to \Sigma^+K^-\pi^+)/\Gamma(\Xi^-\pi^+\pi^+) = 0.09^{+0.13}_{-0.06}$. The Ω_c was observed in the $\Xi^-\bar{K}^0(892)\pi^+$ decay [WA62(30)].

The first decay asymmetry measurements in charm baryon decays were recently reported. The decay asymmetry parameter α_c in $\Lambda_c \to \Lambda\pi^+$ decay was measured by the CLEO group (105) to be $\alpha_c = -1.0^{+0.4}_{-0.0}$ (<0 at 99% confidence level) and by the ARGUS group (106) to be $\alpha_c = -0.97 \pm 0.41$. Figure 6 shows the $\Lambda\pi^+$ invariant mass distributions observed by the

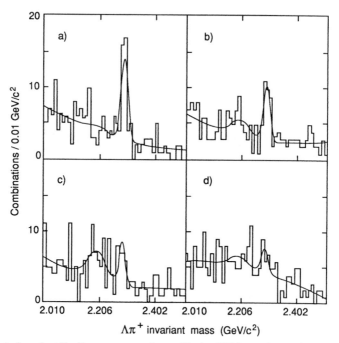

Figure 6 Invariant $(\Lambda\pi^+)$ mass spectra observed in the CLEO experiment, for different bins of $z = \cos(\Lambda, p)$. (*a*) $-1.0 < z < -0.5$. (*b*) $-0.5 < z < 0$. (*c*) $0 < z < 0.5$. (*d*) $0.5 < z < 1$.

CLEO group, in bins of $z = \cos\theta^{CMS}(\Lambda, p)$. A clear signal is seen in the bin $-1 < z < -0.5$, a smaller signal in the bin $-0.5 < z < 0$, but practically nothing for $z > 0$. The broader enhancements below the Λ_c mass peak would correspond to decays $\Lambda_c \to \Sigma^0\pi^+$, with the γ from $\Sigma^0 \to \Lambda\gamma$ decay not seen. The authors do not consider these enhancements as evidence for $\Lambda_c \to \Sigma^0\pi^+$ decay.

5.2   *Decay Rates and Decay Distributions*

Let us begin by counting the number of independent amplitudes in the four classes of nonleptonic ground-state to ground-state transitions:

$1/2^+ \to 1/2^+ + 0^-$: 2 complex amplitudes

$1/2^+ \to 3/2^+ + 0^-$: 2 complex amplitudes

$1/2^+ \to 1/2^+ + 1^-$: 4 complex amplitudes

$1/2^+ \to 3/2^+ + 1^-$: 6 complex amplitudes.

Using standard methods (e.g. 107) one can then derive angular decay distributions, which, upon integration, give the decay rates. Again we prefer an explicit frame-dependent representation of the decay distributions instead of the frame-independent representation discussed by Bjorken (79). We begin by considering the nonleptonic decay of unpolarized charm baryons. In the simplest case one has the decay $1/2^+ \to 1/2^+(\to 1/2^+ + 0^-) + 0^-$ as for example in $\Lambda_c^+ \to \Lambda(\to p\pi^-) + \pi^+$. Referring to Figure 3, one now sees a cascade only on one side as the pion's decay goes unobserved. Consequently one has only a single polar angle distribution. One obtains

$$\frac{d\Gamma[\Lambda_c^2 \to \Lambda(\to p\pi^-) + \pi^+]}{d\cos\theta_\Lambda} = \frac{1}{2} BR(\Lambda \to p\pi^-)\Gamma(\Lambda_c^+ \to \Lambda\pi^+)$$

$$\times \ (1 + \alpha_c\alpha_\Lambda \cos\theta_\Lambda), \quad 18.$$

where α_c and α_Λ are the asymmetry parameters in the decays $\Lambda_c^+ \to \Lambda\pi^+$ and $\Lambda \to p\pi^-$, respectively, defined in analogy to Equation 12. The definition of the polar angle θ_Λ is given in Figure 3 with the replacement $W^+ \to \pi^+$. The cascade decay $\Lambda \to p\pi^-$ acts as an analyzer of the longitudinal polarization of the daughter baryon Λ whose polarization is given by the asymmetry parameter α_c. This angular decay distribution was utilized experimentally to measure the asymmetry parameter α_c in the decay $\Lambda_c \to \Lambda(\to p\pi^-) + \pi^+$ (105, 106).

Somewhat more complicated is the decay distribution in the double cascade $1/2^+ \to 1/2^+(\to 1/2^+ + 0^-) + 1^-(\to 0^- + 0^-)$ as for example in $\Lambda_c^+ \to \Lambda(\to p\pi^-) + \rho^+(\to \pi^+\pi^0)$. One has (73)

$$\frac{d\Gamma[\Lambda_c^+ \to \Lambda(\to p\pi^-) + \rho^+(\to \pi^+\pi^0)]}{d\cos\theta\, d\chi\, d\cos\theta_\Lambda} = \frac{1}{2\pi} BR(\Lambda \to p\pi^-)BR(\rho^+ \to \pi^+\pi^0)$$

$$\times \frac{p}{32\pi M_1^2}\left[\frac{3}{4}\sin^2\theta H_U(1 + \alpha_c^U\alpha_\Lambda\cos\theta_B)\right.$$

$$+ \frac{3}{2}\cos^2\theta H_L(1 + \alpha_c^L\alpha_\Lambda\cos\theta_\Lambda) - \frac{3}{4\sqrt{2}}\sin(2\theta)\cos\chi\cos\theta_\Lambda\alpha_\Lambda H_I$$

$$\left. + \frac{3}{4\sqrt{2}}\sin(2\theta)\sin\chi\cos\theta_\Lambda\alpha_\Lambda H_{I'}\right], \quad 19.$$

where the helicity rates H_U, H_L, and H_I and the asymmetry parameters α_c^U and α_c^L are defined in analogy to Equation 10–12. Angles are defined as in Figure 3 with the replacement of $(W^+ \to \ell^+\nu_\ell)$ by $(\rho^+ \to \pi^+\pi^0)$. Clearly the six observables defined by the decay distribution do not suffice to determine the four complex decay amplitudes of the process.

The observable $H_{I'}$ is proportional to the *imaginary* part of the longitudinal-transverse interference term (see the third line of Equation 10) and thus a so-called T-odd observable. It obtains contributions from *CP*-violating interactions and/or from effects of final-state interaction. The Standard Model *CP*-violating contributions are expected to be quite small and thus $H_{I'}$ would be a good measure of the strength of final-state interaction effects. Alternatively, one may extract possible *CP*-violating effects by comparing Λ_c^+ and $\bar{\Lambda}_c^+$ cascade decays.

We now briefly turn to the decays of *polarized* charm baryons in which the decay of the daughter baryon is used as an analyzer and the meson decay goes unanalyzed. The orientation angles are defined in Figure 4. The angular decay distribution for the $1/2^+ \to 1/2^+ (\to 1/2^+ + 0^-) + 0^-$ transition is well known from the analysis of the nonleptonic decays of the cascade hyperon Ξ and reads (73, 108)

$$\frac{d\Gamma[\Lambda_c^{\uparrow} \to \Lambda(\to p\pi^-) + \pi^+]}{d\cos\theta_p \, d\cos\theta_\Lambda \, d\sin\chi} = \frac{1}{8\pi} \Gamma_{\Lambda_c \to \Lambda\pi^+} BR(\Lambda \to p\pi^-)$$

$$\times [1 + \alpha_c\alpha_\Lambda \cos\theta_\Lambda + P_c(\alpha_c\cos\theta_p + \alpha_\Lambda\cos\theta_p\cos\theta_\Lambda)$$

$$+ \alpha_\Lambda \sin\theta_p \sin\theta_\Lambda(\gamma_c \cos\chi + \beta_c \sin\chi)], \qquad\qquad 20.$$

where β_c and γ_c are the usual nonleptonic decay parameters (e.g. 79). In a noncascade charm baryon decay, for example $\Lambda_c \to p\bar{K}^0$, one would be left with a decay distribution $W(\theta_c) = (1 + P_c\alpha_c\cos\theta_p)$. This would allow for a determination the asymmetry parameter α_c only if P_c were known.

The remaining angular decay distributions (polarized and unpolarized decaying charm baryon) can be found in the report by Körner & Krämer (73).

5.3 Symmetry Considerations

In the nonleptonic Hamiltonian of Equation 4, we included only the dominant contribution proportional to $\simeq \cos^2\theta_c$. Once-suppressed transitions proportional to $\simeq \cos\theta_c \sin\theta_c$, not written in Equation 4, are the transitions $c \to d u \bar{d}$ and $c \to s u \bar{s}$, and the double-suppressed decays $c \to d u \bar{s}$ are proportional to $\simeq \sin^2\theta_c$. In the sum rule approach, one relates different nonleptonic decay amplitudes by using flavor symmetry relations based on the flavor symmetry group SU(4) and/or its SU(3) and SU(2) subgroups. The $\Delta C = 1$ SU(3) content of the antisymmetric SU(4) representation 20″ is 3 and 6*, and that of the symmetric SU(4) representation

84 is 3 and 15. The dominant pieces are the 6* and 15 SU(3) representa-
tions. The I-, U-, and V-spin content of the dominant piece is $\Delta I = 1$,
$\Delta U = 1$, and $\Delta V = 0, 1$. Sum rules relating different charm-changing nonleptonic decay amplitudes have been derived using various techniques (89,
109–112); the simplest technique appears to be an analysis using the three
$SU(2)_{I,U,V}$ subgroups (89). Nonet symmetry for the mesons can be incorporated in the usual way by excluding disconnected flavor flow diagrams
(see 89).

An interesting observation concerns the rates of the two members Λ_c^+
and Ξ_c^0 of the same U-spin doublet. In the case of 20″ dominance of H_{eff},
U-spin arguments reveal that for the dominating $c \rightarrow su\bar{d}$ transitions one
can derive equality of total rates and partial rates into any particular spin
channel (110). Considering the present nonequality of Λ_c^+ and Ξ_c^0 lifetimes,
20″ dominance of the effective nonleptonic Hamiltonian may not be a
good approximation.

Further sum rules may be obtained by relating Cabibbo-favored, suppressed and doubly suppressed decay amplitudes when the expected
Cabibbo suppression factors are removed. Similarly one may even attempt
to relate charm-changing $\Delta C = 1$ processes to ordinary $\Delta C = 0$ nonleptonic hyperon decays although the large mass difference between charm
and ordinary baryons makes such an approach problematic.

Still another class of sum rules may be obtained by considering parity-
conserving and parity-violating amplitudes separately and assuming full
SU(4) symmetry of the transition in conjunction with the charge conjugation symmetry of H_{eff}, which is $C = +1$ and $C = -1$ for the parity-
conserving and parity-violating parts (89, 109). One then obtains a vanish-
ing parity-violating decay amplitude for $\Lambda_c^+ \rightarrow \Lambda \pi^+$ in direct conflict
with the recent nonvanishing asymmetry measurement in this decay (105,
106). Thus SU(4) is not a useful flavor symmetry for charm-changing weak decays because of its large mass-breaking factor $(m_c - m_s)/m_c \simeq$
70%.

While SU(4) is not a useful symmetry, SU(3) flavor symmetry may
still be useful for charm-changing decays (89, 111). But even then one
encounters the problem of which mass dimension the SU(3) flavor symmetric amplitude should carry. Related to this is the extraction of flavor
symmetric amplitudes from rates even though one does not know which
mass scale $\bar{M}$ is appropriate for the $(p/\bar{M})^{2l+1}$ phase-space factor.
Until more is known about the appropriate mass scaling factors of an amplitude, the SU(3) flavor symmetry approach to nonleptonic and semileptonic decays involving $\Delta C = 1$ transitions provides a rule of thumb at
best.

5.4   *Quark Model and Current Algebra Results*

In the quark model, the effective current × current Hamiltonian (Equation 3) gives rise to the five types of flavor diagrams drawn in Figure 7. We have chosen to label the quark lines for the specific transition $\Lambda_c^+ \to \Lambda\pi^+$ for illustrative purposes. The wavy lines are included in order to indicate how the effective quark currents of the current × current Hamiltonian act (3). As a next step, one wants to interpret the diagrams as Feynman diagrams, with gluon exchanges. The general dynamical problem in all its complexity is far from being solved, so one has to resort to some approximation. The quark lines in Figure 7 transmit spin information from one hadron to the other. This is realized in the spectator quark model, which postulates that there is no spin communication between quark lines. Quark pairs are created from the vacuum with $^3P_0$ quantum numbers. Finally, these postulates can be cast into a covariant form if the quarks in a hadron are assumed to propagate with equal velocity, which is the hadron's velocity.

In terms of quark model spin wave functions, the decay amplitudes for the process $B_1 \to B_2 M$ corresponding to Figure 7 can then be written as (89)

$$
T_{B_1 \to B_2 + M} = H_1 \bar{B}_2^{ABC'} B_{1ABC} \bar{M}_D^{D'} \left(O_{C'D'}^{CD} - \frac{1}{N_C} O_{D'C'}^{CD} \right)
$$

$$
+ \frac{1}{N_C} H_2 \bar{B}_2^{AB'D} B_{1ABC} \bar{M}_D^{D'} O_{B'D'}^{BC}
$$

$$
+ \frac{1}{N_C} H_2' \bar{B}_2^{ABC'} B_{1ABC} \bar{M}_D^{B} O_{B'C'}^{CD}
$$

$$
+ \frac{1}{N_C} H_3 \bar{B}_2^{A'B'C'} B_{1ABC} \bar{M}_{C'}^{C} O_{A'B'}^{AB}, \qquad\qquad 21.
$$

where the first, second, third, and fourth terms correspond to the contributions of diagrams Ia, Ib, IIa, IIb, and III in Figure 7 in that order. B_{ABC} and M_A^B are quark model wave functions for the baryons and mesons. Each index "A" stands for a pair of indices (α, a), where α and a denote the spin and flavor degrees of freedom. Color degrees of freedom have already been summed resulting in the typical factors $1/N_C$. The matrix O_{AB}^{CD} describes the spin-flavor structure of the effective current × current Hamiltonian, Equation 3. H_1, H_2, H_2', and H_3 are wave function overlaps corresponding to diagrams Ia, Ib, IIa, IIb and III and may depend on the masses of the decay process. Equation 21 can be viewed as an algebraic realization of the diagrams shown in Figure 7: each line in Figure 7

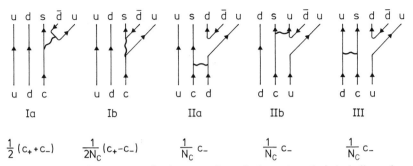

$$\frac{1}{2}(c_+ + c_-) \qquad \frac{1}{2N_c}(c_+ - c_-) \qquad \frac{1}{N_c}c_- \qquad \frac{1}{N_c}c_- \qquad \frac{1}{N_c}c_-$$

Figure 7 Quark diagrams contributing to nonleptonic decay $\Lambda_c \to \Lambda\pi^+$, including color-flavor weight factors.

corresponds to a contraction of doubly occurring spin-flavor indices in Equation 21, where one sums over the spin-flavor indices.

The first term in Equation 21 is not problematic. It corresponds to the so-called factorization contribution and can be calculated in terms of the current matrix elements of Section 4. We have brought the contributions of diagrams IIa, IIb, and III into tenable forms by making use of the above assumptions, but Equation 21 can also be derived from a more general point of view without some of those assumptions. One should note that in transitions between ground-state baryons, diagrams IIa, IIb, and III obtain contributions only from O^- [transforming as $20''$ in SU(4)], because of the symmetric nature of the ground-state baryons (112, 113). Both operators O^+ and O^- contribute to diagrams Ia and Ib. The sum of the contributions of Ia and Ib is proportional to $\chi_\pm = [c_+(1 + 1/N_C) \pm c_-(1 - 1/N_C)]/2$, where the sign depends on the charge of the final-state meson, charged $(+)$ or neutral $(-)$.

The results of calculating diagrams IIa, IIb, and III depend on the details of the quark model wave functions used as input. As a first approximation, one may use SU(2)$_W$ spin wave functions (89). They correspond to boosting static quark model wave functions to a collinear equal velocity frame as mentioned above (114). When explicit mass factors are scaled out of the baryon wave functions according to the HQET, one can set $H_2 = H_2'$ in Equation 21, because of CP invariance. After some straightforward algebraic manipulations involving the evaluation of the amplitude in Equation 21 with the SU(2)$_W$ wave functions, one can calculate the nonleptonic transition amplitudes for the decays $1/2^+ \to (1/2^+, 3/2^+) + (0^-, 1^-)$.

In order to be able to discuss some general features of the solutions, we treat the decay $1/2^+ \to 1/2^+ + 0^-$ in more detail. Writing the amplitude $T_{B_1 \to B_2 + M} = \bar{u}_2(A + B\gamma_5)u_1$ one obtains the following amplitude expressions:

$$A = A^{\text{fac}} - \frac{1}{8} \frac{H_2}{M_1 M_2 \sqrt{M_3}} \{ M_1 I_3 [M_1^2 - (M_2 + M_3)^2]$$

$$- M_2 \hat{I}_3 [M_2^2 - (M_1 + M_3)^2] \}$$

$$B = B^{\text{fac}} + \frac{1}{3} \frac{H_2}{4 M_1 M_2 \sqrt{M_3}} \{ Q_+ [M_1 (I_3 + 2I_4) + M_2 (\hat{I}_3 + 2\hat{I}_4)] \}$$

$$+ \frac{1}{3} \frac{H_3}{M_1 M_2 \sqrt{M_3}} 3 M_1 M_2 (M_1 + M_2 + M_3) I_5, \quad 22.$$

where the factorizing contributions A^{fac} and B^{fac} (corresponding to diagrams Ia and Ib) are obtained from the current-induced form factors discussed in Section 4.2.2. We have defined $Q_+ = (M_1 + M_2)^2 - M_3^2$. The invariant flavor wave function contractions (Clebsch-Gordan coefficients) denoted by I_i and $\hat{I}_i$ are defined by Körner et al (89). I_3 and I_4 are associated with diagram IIa, $\hat{I}_3$ and $\hat{I}_4$ with diagram IIb, and I_5 with diagram III. Diagram III can be seen to contribute only to the parity-conserving amplitude B, whereas diagrams I and II contribute to both parity-conserving and parity-violating amplitudes. The parity-conserving and -violating amplitudes can be seen to be even and odd with respect to the generalized charge conjugation operation $(M_1; I_{3,4}; I_5) \leftrightarrow (M_2; \hat{I}_{3,4}; I_5)$ as expected from the CP-conserving property of the nonleptonic Hamiltonian. For example, for the decay $\Lambda_c^+ \to \Lambda \pi^+$ one finds $I_3 = \hat{I}_3$ and thus the parity-violating amplitude A in the symmetry limit $M_1 = M_2$ vanishes, as remarked on earlier. With $M_1 \gg M_2$ this statement no longer holds true.

The contributions of diagrams IIb and III are nonleading on the scale of the mass M_1 of the parent baryon. As a helicity analysis shows, they are nonleading because the contributions IIb and III are suppressed by helicity as a result of the $(V - A) \times (V - A)$ nature of the underlying quark transition (115). This implies that only the factorizing contribution in Ia and Ib and the nonfactorizing contribution in IIa survive when $M_2/M_1 \to 0$. This conclusion holds in general for all ground-state decay channels. These leading contributions can be seen to lead to an exclusive decay mode power behavior $\Gamma \sim M_1$ when M_2 and M_3 are kept fixed (as is the case for exclusive nonleptonic meson decays). Compared to the inclusive non-leptonic rate $\Gamma_{\text{FQD}}^{\text{nl}} \sim M_1^5$ one infers that the exclusive branching ratio of a particular two-body channel decreases very rapidly as M_1 becomes large. No general statement can be made on the population of helicities for the daughter baryon from the leading contributions. This depends on the decay channel under consideration (115). The fact that the nonfactorizing contribution IIa survives in the heavy quark mass limit thwarts the hope that nonleptonic baryon decays can be described by the factorizing contribution alone (79, 116) unless one can invoke $1/N_C$ arguments. Non-leptonic meson decays are simpler in this regard (117).

The flavor structure in the parity-violating amplitude A has a remarkable property: there exists a one-to-one flavor correspondence with terms arising in the current algebra plus soft pion approach. This was first noticed empirically in the $\Delta C = 0$, $\Delta Y = 1$ (118) and in the $\Delta C = 0$, $\Delta Y = 0$ (119) transitions and was later proven in general (89). The correspondence between the quark model and current algebra approach works in the following way: the contributions proportional to I_3 and $\hat{I}_3$ have the flavor structure of the "equal time commutator" term when the symmetry limit $M_1 = M_2$ is taken. The factorizing contribution A_{fac} has the same interpretation in both schemes. In a similar vein, the nonfactorizing parity-conserving contributions can readily be interpreted as baryon pole contributions.

The details of the current algebra plus soft pion approach as applied to the charm baryon sector have been described in (89, 112, 120–122). In Table 6 we list some of the current algebra calculations and compare them to a quark model calculation with best fit values for the overlap parameters $H_2 = H_2'$ and H_3 in Equation 21. The branching ratios listed are relative to the $\Lambda_c^+ \to pK^-\pi^+$ mode, for which we take a nominal branching ratio of 4.2%. The spread in the predictions indicates the model dependence of the results. In the next few years the advent of new data will certainly constrain the current algebra calculations further. The quark model (123) fits the data quite well (at the cost of two parameters) except for the decay

Table 6 Current algebra and quark model predictions for non-leptonic Λ_c^+ decays. The numbers cited are branching rates relative to $\Lambda^+ \to pK^-\pi^+$ and asymmetry parameters α_c

	HK (120)	Cheng (121)	EK (122)	PTR (112)	Quark Model (123)
$\Lambda_c^+ \to \Lambda\pi^+$	0.8	0.5	0.15	0.5	0.17
(α_c)			(-1.0)	(-1.0)	(-0.70)
$\Lambda_c^+ \to p\bar{K}^0$	1.4	1.7	0.08	1.3	0.49
(α_c)			(-0.82)	(-0.61)	(-1.0)
$\Lambda_c^+ \to p\bar{K}^0(892)$				0.13	0.72
$\Lambda_c^+ \to p\phi$					0.05
$\Lambda_c^+ \to \Delta^{++}K^-$		0.14			0.63

$\Lambda_c^+ \to p\bar{K}^0(892)$, which is too large compared to the present experimental results (see Table 5). The indicated range of its rate in Table 6 reflects the uncertainty of the vector current coupling of the $K(892)$. All model calculations predict negative asymmetry parameter values close to their maximum values of -1 for the decays $\Lambda_c^+ \to \Lambda\pi^+$ and $\Lambda_c^+ \to p\bar{K}^0$. They are thus in agreement with the measured asymmetry in the decay $\Lambda_c^+ \to \Lambda\pi^+$ (105, 106). The Cabibbo-suppressed decay $\Lambda_c^+ \to p\phi$ is interesting: it obtains only contributions from the factorizing graphs Ia and Ib in Figure 7. Its measured rate can be accounted for quite well in the factorization approximation using the form factors described in Section 4.2.2 (86). The decay $\Lambda_c^+ \to \Delta^{++}K^-$ receives contributions only from the nonfactorizing diagram IIa in Figure 7 and thus gives a measure of the nonfactorizing contributions to the nonleptonic decays. Its rate indicates that the factorizing approximation to nonleptonic charm baryon decays advocated by Mannel et al and Acker et al (116) may not always be a good approximation. In fact, the quark model results imply that factorizing and nonfactorizing contributions enter with approximately equal weight, depending of course on the decay mode under consideration (123).

6. SUMMARY AND OUTLOOK

In this review we have concentrated on the decay properties of the weakly decaying ground-state baryons, focusing on recent theoretical advances in the understanding of current-induced transitions among heavy to heavy and heavy to light hadrons. Future experiments will reveal how relevant these new theoretical ideas are for the charm baryons.

The CLEO and ARGUS experiments will continue to run. Their results on Λ_c, Σ_{cl}, and Ξ_c demonstrate the potential of e^+e^- experiments in the Υ region for charmed baryon physics. Photoproduction experiments have contributed and will continue to contribute to our knowledge on Λ_c and Σ_c. Experiment E687 at Fermilab has taken a large sample of data and published a first result on the Λ_c lifetime. More results with much higher statistics are expected.

In hadroproduction, we expect results from experiment E769 at Fermilab. A successor experiment, E791, is just beginning. A promising way to study charmed strange baryons is the use of hyperon beams. At CERN, experiment WA89 had its first run in 1990 in a Σ^- beam. A similar experiment, E781, is being prepared at Fermilab. There is a chance that some of these experiments will see the missing Σ_c^*, Ξ_c', Ξ_c^*, and Ω_c^* states. The masses of these states will provide important clues to the behavior of the charm quark in baryons.

A measurement of the lifetimes of all four weakly decaying charm baryons will yield very interesting information on the wave-function at the origin, $|\Psi(0)|^2$, and on interference effects, as discussed in Section 3.2.

New decay modes will be found through increased statistics and in particular through the introduction of photon detectors and hadron calorimeters to identify decays with π^0, n, Σ^+, Σ^0, or Σ^- in the final state. Absolute branching ratios will be further studied in $B \to \Lambda_c + X$ decays. It would also be very interesting to repeat at high luminosities the studies of Λ_c production in e^+e^- above the $\Lambda_c\bar{\Lambda}_c$ threshold. The present plans for a tau charm factory do not, however, aim at energies high enough for that purpose.

The expected increase in statistics should make it possible to analyze angular distributions in nonleptonic and maybe even semileptonic decays. We remind the reader that more than twenty years ago, the first measurements of g_A/g_V in semileptonic hyperon decays with accuracies of about 0.2 were made with samples of less than 200 events. Recently, important results on form factors in semileptonic D decay were obtained by the E691 collaboration on a sample of 200 events (76).

In all, we can expect a great deal of interesting new data on charmed baryons in the next few years.

ACKNOWLEDGMENTS

J.G.K. would like to acknowledge the help of his students S. Balk and M. Krämer without whose untiring efforts this review could not have been finished on time. We have benefited from discussions with D. Gromes, F. Hussain, T. Mannel, A. Martin, O. Nachtmann, J. M. Richard, Z. Ryzak, K. Schilcher, H. Schröder, B. Stech, H. J. Stiewe, G. Thompson, and Y. L. Wu. We thank our colleagues from the BIS-2, NA32, ARGUS, and CLEO groups for enlightening discussions of their results. This work was supported in part by the Bundesministerium für Forschung und Technologie.

Literature Cited

1. Aubert, J. J., et al., *Phys. Rev. Lett.* 33: 1404 (1974)
2. Augustin, J. E., et al., *Phys. Rev. Lett.* 33: 1406 (1974)
3. Cazzoli, E. G., et al., *Phys. Rev. Lett.* 34: 1125 (1975)
4. Goldhaber, G., et al., *Phys. Rev. Lett.* 37: 255 (1976)
5. Peruzzi, I., et al., *Phys. Rev. Lett.* 37: 569 (1976)
6. Niu, K., Mikumo, E., Maeda, Y., *Prog. Theor. Phys.* 46: 1644 (1971)
7. Herb, S. W., et al., *Phys. Rev. Lett.* 39: 252 (1977); Innes, W. R., et al., *Phys. Rev. Lett.* 39: 1240, 1640(E) (1977)
8. Behrends, S., et al., *Phys. Rev. Lett.* 50: 881 (1983); Giles, R., et al., *Phys. Rev.* D30: 2279 (1984)
9. Albrecht, H., et al., *Phys. Lett.* 185B: 218 (1987)
10. Isgur, N., Wise, M. B., *Phys. Lett.* 237B: 527 (1990)
11. Bjorken, J. D., in *Proc. Les Recontres de Physique de la Vallee de Aosta*, La Thuile, Italy. Gif-sur-Yvette: Ed. Frontieres (1990), p. 583

12. Georgi, H., *Phys. Lett.* B240: 447 (1990)
13. Falk, A. F., et al., *Nucl. Phys.* B343: 1 (1990)
14. Grinstein, B., *Nucl. Phys.* B339: 253 (1990)
15. Eichten, E., Hill, B., *Phys. Lett.* B240: 511 (1990)
16. Hussain, F., et al., *Phys. Lett.* B249: 295 (1990)
17. Bloch, F., Nordsieck, A., *Phys. Rev.* 52: 54 (1937)
18. Cahn, R. N., ed., e^+e^- *Annihilation: New Quarks and Leptons* (A volume in the Annual Reviews Special Collections Program). Menlo Park: Benjamin-Cummings (1985)
19. Kwong, W., Rosner, J. L., Quigg, C., *Annu. Rev. Nucl. Part. Sci.* 37: 325 (1987)
20. Morrison, R. J., Witherell, M. S., *Annu. Rev. Nucl. Part. Sci.* 39: 183 (1989)
21. Shifman, M. A., *Nucl. Phys. B (Proc. Suppl.)* 3: 289 (1988); *Sov. Phys. Usp.* 30: 91 (1987)
22. Klein, S. R., *Int. J. Mod. Phys.* A5: 1457 (1990)
23. Fleck, S., Richard, J. M., *Particle World* 1: 67 (1990)
24. Particle Data Group, *Phys. Lett.* B239: (1990)
25. Knapp, B., et al., *Phys. Rev. Lett.* 37: 882 (1976)
26. Bourquin, M., *Phys. Lett.* B122: 455 (1983)
27. Avery, P., et al., *Phys. Rev. Lett.* 62: 863 (1989)
28. Barlag, S., et al., *Phys. Lett.* B236: 495 (1990)
29. Albrecht, H., et al., *Phys. Lett.* B247: 121 (1990)
30. Biagi, S., et al., *Z. Phys.* C28: 175 (1985)
31. Hussain, F., et al., *Z. Phys.* C51: 321 (1991)
32. Barlag, S., et al., *Phys. Lett.* B218: 374 (1989)
33. Alvarez, M. P., et al., *Phys. Lett.* B246: 256 (1990)
34. Anjos, J. C., et al., *Phys. Rev. Lett.* 60: 1379 (1988)
35. Albrecht, H., et al., *Phys. Lett.* B207: 109 (1988)
36. Aguilar-Benitez, M., et al., *Z. Phys.* C40: 321 (1988)
37. Jones, G. T., et al., *Z. Phys.* C36: 593 (1987)
38. Bosetti, P. C., et al., *Phys. Lett.* B109: 234 (1982)
39. Avery, P., et al., preprint CLNS 90/992-CLEO 90-5 (1990)
40. Calicchio, M., et al., *Phys. Lett.* B93: 521 (1980)
41. Albrecht, H., et al., *Phys. Lett.* B211: 489 (1988)
42. Bowcock, T., et al., *Phys. Rev. Lett.* 62: 1240 (1989)
43. Anjos, J. C., et al., *Phys. Rev. Lett.* 62: 1721 (1989)
44. Diesburg, M., et al., *Phys. Rev. Lett.* 59: 2711 (1987)
45. Alam, M. S., et al., *Phys. Lett.* B226: 401 (1989)
46. Barlag, S., et al., *Phys. Lett.* B233: 522 (1989)
47. Richard, J. M., Taxil, P., *Phys. Lett.* B128: 453 (1983)
48. De Rujula, A., Georgi, H., Glashow, S. L., *Phys. Rev.* D12: 147 (1975)
49. Capstick, S., Isgur, N., *Phys. Rev.* D34: 2809 (1986)
50. Politzer, H. D., Wise, M. B., *Phys. Lett.* B206: 68 (1988)
51. Frabetti, P. L., et al., *Phys. Lett.* B251: 639 (1990)
52. Biagi, S. F., et al., *Phys. Lett.* B150: 230 (1985)
53. Coteus, P., et al., *Phys. Rev. Lett.* 59: 1530 (1987)
54. Barger, V., Leveille, J. P., Stevenson, P. M., *Phys. Rev. Lett.* 44: 226 (1980)
55. Voloshin, M. B., Shifman, M. A., *Sov. Phys. JETP* 64: 698 (1986); *Sov. J. Nucl. Phys.* 41: 120 (1985)
56. Guberina, B., Rückl, R., Trampetić, J., *Z. Phys.* C33: 297 (1986)
57. Kobayashi, T., Yamazaki, N., *Prog. Theor. Phys. Lett.* 65: 775 (1981)
58. Gaillard, M. K., Lee, B. W., *Phys. Rev. Lett.* 33: 108 (1974); Altarelli, G., Maiani, L., *Phys. Lett.* B52: 35 (1974)
59. Cortes, J. L., Sanchez-Guillen, J., *Phys. Rev.* D24: 2982 (1981)
60. Vella, E., et al., *Phys. Rev. Lett.* 48: 1515 (1982)
61. Suzuki, M., *Nucl. Phys.* B145: 420 (1978); Cabibbo, N., Maiani, L., *Phys. Lett.* B79: 109 (1978)
62. Albrecht, H., et al., *Phys. Lett.* B219: 121 (1989)
63. Bertoletto, D., et al., *Phys. Rev. Lett.* 63: 1667 (1989)
64. Körner, J. G., Schuler, G. A., *Z. Phys.* C38: 511 (1988); erratum *Z. Phys.* C41: 690 (1989)
65. Körner, J. G., Schuler, G. A., *Z. Phys.* C46: 93 (1990)
66. Körner, J. G., Schuler, G. A., *Phys. Lett.* B226: 185 (1989)
67. Grinstein, B., et al., *Phys. Rev.* D39: 799 (1989)
68. Hagiwara, K., Martin, A. D., Wade, M. F., *Phys. Lett.* B228: 144 (1989); *Nucl. Phys.* B327: 569 (1989)
69. Gilman, F. J., Singleton, R. L., *Phys. Rev.* D41: 142 (1990)
70. Köpp, G., et al., *Z. Phys.* C48: 327 (1990)

71. Körner, J. G., Schuler, G. A., *Phys. Lett.* B231: 306 (1989); Dominguez, C. A., Körner, J. G., Schilcher, K., *Phys. Lett.* B248: 399 (1990)
72. Hagiwara, K., Martin, A. D., Wade, M. F., *Z. Phys.* C46: 299 (1990)
73. Körner, J. G., Krämer, M., preprint MZ-TH/91-06 (1991)
74. Hussain, F., Körner, J. G., Migneron, R., *Phys. Lett.* B248: 406 (1990); erratum B252: 723 (1990)
75. Körner, J. G., Schilcher, K., Wu, Y. L., *Phys. Lett.* B242: 119 (1990)
76. Anjos, J. C., et al., *Phys. Rev. Lett.* 65: 2630 (1990)
77. Aleev, A. N., et al., *Sov. J. Nucl. Phys.* 43: 395 (1986)
78. Chauvat, P., et al., *Phys. Lett.* B199: 304 (1987)
79. Bjorken, J. D., *Phys. Rev.* D40: 1513 (1989)
80. Buras, A. J., *Nucl. Phys.* B109: 373 (1976)
81. Yamada, K., *Phys. Rev.* D22: 1676 (1980)
82. Gavela, M. B., *Phys. Lett.* B83: 367 (1979)
83. Avilez, C., Kobayashi, T., *Phys. Rev.* D19: 3448 (1979)
84. Pérez-Marcial, R., et al., *Phys. Rev.* D40: 2955 (1989)
85. Singleton, R., *Phys. Rev.* D43: 2939 (1991)
86. Hussain, F., Körner, J. G., *Z. Phys. C.* In press (1991)
87. Voloshin, M. B., Shifman, M. A., *Sov. J. Nucl. Phys.* 47: 511 (1988)
88. Gudehus, T., *Phys. Rev.* 184: 1788 (1969)
89. Körner, J. G., Kramer, G., Willrodt, J., *Z. Phys.* C2: 117 (1979)
90. Ali, A., et al., *Z. Phys.* C1: 269 (1979)
91. Isgur, N., Wise, M. B., *Nucl. Phys.* B348: 276 (1991); Georgi, H., *Nucl. Phys.* B348: 293 (1991)
92. Mannel, T., Roberts, W., Ryzak, Z., *Nucl. Phys.* B355: 38 (1991)
93. Georgi, H., Grinstein, B., Wise, M. B., *Phys. Lett.* B252: 456 (1990); Boyd, C. G., Brahm, D. E., *Phys. Lett.* B254: 468 (1991)
94. Boyd, C. G., Brahm, D. E., *Phys. Lett.* B257: 393 (1991); Körner, J. G., Thompson, G., *Phys. Lett. B.* In press (1991)
95. Körner, J. G., Krämer, M., preprint MZ-TH/91-05 (1991)
96. Hussain, F., et al., preprint MZ-TH/91-25 (1991)
97. Bai, Z., et al., *Phys. Rev. Lett.* 66: 1011 (1991)
98. Weiss, J. M., in *Proc. Baryon 1980*

99. Basile, M., et al., *Nuovo Cimento* 62A: 14 (1981)
100. Barlag, S., et al., *Z. Phys.* C48: 29 (1990)
101. Anjos, J. C., et al., *Phys. Rev.* D41: 801 (1990)
102. Aleev, A. N., et al., *Z. Phys.* C23: 333 (1984)
103. Abrams, G. S., et al., *Phys. Rev. Lett.* 44: 10 (1980)
104. Albrecht, H., et al., *Phys. Lett.* B210: 263 (1988)
105. Avery, P., et al., *Phys. Rev. Lett.* 65: 2842 (1990)
106. Stiewe, J. (ARGUS), presented at PANIC XII Conf., MIT, June 25–29, 1990
107. Jackson, J. D., in *High Energy Physics*, ed. C. de Witt, R. Gatto. New York: Gordon & Breach (1965), p. 325; Martin, A. D., Spearman, D., *Elementary Particle Theory*. Amsterdam: North-Holland (1970); Pilkuhn, H., *The Interactions of Hadrons*. Amsterdam: North-Holland (1967)
108. Lednicky, R., *Sov. J. Nucl. Phys.* 43: 817 (1986)
109. Iwasaki, Y., *Phys. Rev. Lett.* 34: 1407 (1975); erratum 35: 246 (1975)
110. Altarelli, G., Cabibbo, N., Maiani, L., *Phys. Lett.* B57: 277 (1978)
111. Savage, M. J., Springer, R. P., *Phys. Rev.* D42: 1527 (1990)
112. Pakvasa, S., Tuan, S. F., Rosen, S. P., *Phys. Rev.* D42: 3746 (1990)
113. Körner, J. G., *Nucl. Phys.* B25: 282 (1970); Pati, J. C., Woo, C. H., *Phys. Rev.* D3: 2920 (1971)
114. Hussain, F., Körner, J. G., Thompson, G., *Ann. Phys.* 206: 334 (1991)
115. Körner, J. G., Goldstein, G. R., *Phys. Lett.* B89: 105 (1979)
116. Mannel, T., Roberts, W., Ryzak, Z., *Phys. Lett.* B255: 593 (1991); Acker, A., et al., *Phys. Rev.* D43: 3083 (1991)
117. Dugan, M. J., Grinstein, B., *Phys. Lett.* B255: 583 (1991)
118. Körner, J. G., Gudehus, T., *Nuovo Cimento* 11A: 597 (1972)
119. Körner, J. G., Kramer, G., Willrodt, J., *Phys. Lett.* B81: 365 (1979)
120. Hussain, F., Khan, K., *Nuovo Cimento* 88A: 213 (1985); Hussain, F., Scadron, M., *Nuovo Cimento* 79A: 248 (1984)
121. Cheng, H. Y., *Z. Phys.* C29: 453 (1985)
122. Ebert, D., Kallies, W., *Z. Phys.* C29: 1571 (1985)
123. Körner, J. G., Krämer, M., MZ-TH/91-07 (1991)

Conf., Toronto, July 14–16, 1980, p. 319 (1980)

CUMULATIVE INDEXES

CONTRIBUTING AUTHORS, VOLUMES 32–41

CHAPTER TITLES, VOLUMES 32–41

ANNUAL REVIEWS INC.

A NONPROFIT SCIENTIFIC PUBLISHER

4139 El Camino Way
P.O. Box 10139
Palo Alto, CA 94303-0897 • USA

Annual Reviews Inc. publications may be ordered directly from our office; through booksellers and subscription agents, worldwide; and through participating professional societies. Prices subject to change without notice.   ARI Federal I.D. #94-1156476

- **Individuals:** Prepayment required on new accounts by check or money order (in U.S. dollars, check drawn on U.S. bank) or charge to credit card — American Express, VISA, MasterCard.
- **Institutional buyers:** Please include purchase order.
- **Students:** $10.00 discount from retail price, per volume. Prepayment required. Proof of student status must be provided (photocopy of student I.D. or signature of department secretary is acceptable). Students must send orders direct to Annual Reviews. Orders received through bookstores and institutions requesting student rates will be returned. You may order at the Student Rate for a maximum of 3 years.
- **Professional Society Members:** Members of professional societies that have a contractual arrangement with Annual Reviews may order books through their society at a reduced rate. Check with your society for information.
- **Toll Free Telephone orders:** Call 1-800-523-8635 (except from California) for orders paid by credit card or purchase order and customer service calls only. California customers and all other business calls use 415-493-4400 (not toll free). Hours: 8:00 AM to 4:00 PM, Monday-Friday, Pacific Time. **Written confirmation** is required on purchase orders from universities before shipment.
- **FAX: 415-855-9815   Telex: 910-290-0275**
- **We do not ship on approval.**

Regular orders: Please list below the volumes you wish to order by volume number.
Standing orders: New volume in the series will be sent to you automatically each year upon publication. Cancellation may be made at any time. Please indicate volume number to begin standing order.
Prepublication orders: Volumes not yet published will be shipped in month and year indicated.
California orders: Add applicable sales tax. **Canada:** Add GST tax.
Postage paid (4th class bookrate/surface mail) **by Annual Reviews Inc.** UPS domestic ground service available (except Alaska and Hawaii) at $2.00 extra per book. Airmail postage or UPS air service also available at prevailing costs. UPS must have street address. P.O. Box, APO or FPO not acceptable.

ANNUAL REVIEWS SERIES		Prices postpaid, per volume USA & Canada / elsewhere		Regular Order Please Send	Standing Order Begin With
		Until 12-31-90	After 1-1-91	Vol. Number:	Vol. Number:
Annual Review of ANTHROPOLOGY					
Vols. 1-16	(1972-1987)	$31.00/$35.00	$33.00/$38.00		
Vols. 17-18	(1988-1989)	$35.00/$39.00	$37.00/$42.00		
Vol. 19	(1990)	$39.00/$43.00	$41.00/$46.00		
Vol. 20	(avail. Oct. 1991)	$41.00/$46.00	$41.00/$46.00	Vol(s). _____	Vol. _____
Annual Review of ASTRONOMY AND ASTROPHYSICS					
Vols. 1, 5-14	(1963, 1967-1976)				
16-20	(1978-1982)	$31.00/$35.00	$33.00/$38.00		
Vols. 21-27	(1983-1989)	$47.00/$51.00	$49.00/$54.00		
Vol. 28	(1990)	$51.00/$55.00	$53.00/$58.00		
Vol. 29	(avail. Sept. 1991)	$53.00/$58.00	$53.00/$58.00	Vol(s). _____	Vol. _____
Annual Review of BIOCHEMISTRY					
Vols. 30-34, 36-56	(1961-1965, 1967-1987) ..	$33.00/$37.00	$35.00/$40.00		
Vols. 57-58	(1988-1989)	$35.00/$39.00	$37.00/$42.00		
Vol. 59	(1990)	$39.00/$44.00	$41.00/$47.00		
Vol. 60	(avail. July 1991)	$41.00/$47.00	$41.00/$47.00	Vol(s). _____	Vol. _____
Annual Review of BIOPHYSICS AND BIOPHYSICAL CHEMISTRY					
Vols. 1-11	(1972-1982)	$31.00/$35.00	$33.00/$38.00		
Vols. 12-18	(1983-1989)	$49.00/$53.00	$51.00/$56.00		
Vol. 19	(1990)	$53.00/$57.00	$55.00/$60.00		
Vol. 20	(avail. June 1991)	$55.00/$60.00	$55.00/$60.00	Vol(s). _____	Vol. _____

Annual Review of **CELL BIOLOGY**

Vols. 1-3	(1985-1987)	$31.00/$35.00	$33.00/$38.00
Vols. 4-5	(1988-1989)	$35.00/$39.00	$37.00/$42.00
Vol. 6	(1990)	$39.00/$43.00	$41.00/$46.00
Vol. 7	(avail. Nov. 1991)	$41.00/$46.00	$41.00/$46.00 Vol(s). _____ Vol. _____

Annual Review of **COMPUTER SCIENCE**

Vols. 1-2	(1986-1987)	$39.00/$43.00	$41.00/$46.00
Vols. 3-4	(1988, 1989-1990)	$45.00/$49.00	$47.00/$52.00 Vol(s). _____ Vol. _____

Series suspended until further notice. SPECIAL OFFER: Volumes 1-4 are available at the special promotional price of $100.00 USA & Canada / $115.00 elsewhere, when all 4 volumes are purchased at one time. Orders at the special price must be prepaid.

Annual Review of **EARTH AND PLANETARY SCIENCES**

Vols. 1-10	(1973-1982)	$31.00/$35.00	$33.00/$38.00
Vols. 11-17	(1983-1989)	$49.00/$53.00	$51.00/$56.00
Vol. 18	(1990)	$53.00/$57.00	$55.00/$60.00
Vol. 19	(avail. May 1991)	$55.00/$60.00	$55.00/$60.00 Vol(s). _____ Vol. _____

Annual Review of **ECOLOGY AND SYSTEMATICS**

Vols. 2-18	(1971-1987)	$31.00/$35.00	$33.00/$38.00
Vols. 19-20	(1988-1989)	$34.00/$38.00	$36.00/$41.00
Vol. 21	(1990)	$38.00/$42.00	$40.00/$45.00
Vol. 22	(avail. Nov. 1991)	$40.00/$45.00	$40.00/$45.00 Vol(s). _____ Vol. _____

Annual Review of **ENERGY**

Vols. 1-7	(1976-1982)	$31.00/$35.00	$33.00/$38.00
Vols. 8-14	(1983-1989)	$58.00/$62.00	$60.00/$65.00
Vol. 15	(1990)	$62.00/$66.00	$64.00/$69.00
Vol. 16	(avail. Oct. 1991)	$64.00/$69.00	$64.00/$69.00 Vol(s). _____ Vol. _____

Annual Review of **ENTOMOLOGY**

Vols. 10-16, 18	(1965-1971, 1973)		
20-32	(1975-1987)	$31.00/$35.00	$33.00/$38.00
Vols. 33-34	(1988-1989)	$34.00/$38.00	$36.00/$41.00
Vol. 35	(1990)	$38.00/$42.00	$40.00/$45.00
Vol. 36	(avail. Jan. 1991)	$40.00/$45.00	$40.00/$45.00 Vol(s). _____ Vol. _____

Annual Review of **FLUID MECHANICS**

Vols. 2-4, 7	(1970-1972, 1975)		
9-19	(1977-1987)	$32.00/$36.00	$34.00/$39.00
Vols. 20-21	(1988-1989)	$34.00/$38.00	$36.00/$41.00
Vol. 22	(1990)	$38.00/$42.00	$40.00/$45.00
Vol. 23	(avail. Jan. 1991)	$40.00/$45.00	$40.00/$45.00 Vol(s). _____ Vol. _____

Annual Review of **GENETICS**

Vols. 1-21	(1967-1987)	$31.00/$35.00	$33.00/$38.00
Vols. 22-23	(1988-1989)	$34.00/$38.00	$36.00/$41.00
Vol. 24	(1990)	$38.00/$42.00	$40.00/$45.00
Vol. 25	(avail. Dec. 1991)	$40.00/$45.00	$40.00/$45.00 Vol(s). _____ Vol. _____

Annual Review of **IMMUNOLOGY**

Vols. 1-5	(1983-1987)	$31.00/$35.00	$33.00/$38.00
Vols. 6-7	(1988-1989)	$34.00/$38.00	$36.00/$41.00
Vol. 8	(1990)	$38.00/$42.00	$40.00/$45.00
Vol. 9	(avail. April 1991)	$41.00/$46.00	$41.00/$46.00 Vol(s). _____ Vol. _____

Annual Review of **MATERIALS SCIENCE**

Vols. 1, 3-12	(1971, 1973-1982)	$31.00/$35.00	$33.00/$38.00
Vols. 13-19	(1983-1989)	$66.00/$70.00	$68.00/$73.00
Vol. 20	(1990)	$70.00/$74.00	$72.00/$77.00
Vol. 21	(avail. Aug. 1991)	$72.00/$77.00	$72.00/$77.00 Vol(s). _____ Vol. _____